Fundamentals of III-V Semiconductor MOSFETs

Serge Oktyabrsky • Peide D. Ye
Editors

Fundamentals of III-V Semiconductor MOSFETs

Springer

Editors
Serge Oktyabrsky
College of Nanoscale Science & Engineering
University at Albany - SUNY, Albany
255 Fuller Road
Albany, NY 12203
USA
soktyabrsky@uamail.albany.edu

Peide D. Ye
Birck Nanotechnology Center
Purdue University
1205 W. State Street
West Lafayette IN 47907-2057
BRK 2050
USA
yep@purdue.edu

ISBN 978-1-4419-1546-7 e-ISBN 978-1-4419-1547-4
DOI 10.1007/978-1-4419-1547-4
Springer New York Dordrecht Heidelberg London

Library of Congress Control Number: 2010920631

Printed on acid-free paper

Springer is part of Springer Science+Business Media (www.springer.com)

Preface

Is it true that III-V semiconductor materials are back in play for mainstream digital ICs? Or it is just another round of interest to other-than-silicon materials when Si CMOS technology is approaching just another "fundamental limit". There is no simple answer. Moreover, the answer depends not only on physics, materials and technologies, but on economics, demand from other industries, etc. Anyway, we would like the reader to answer these questions on his/her own. The book will help by presenting the fundamentals and current status of research on III-V compound semiconductor metal-oxide-semiconductor field-effect transistors (MOSFETs). We believe it is just the right time to summarize results and provide guidelines for the future efforts because of the following recent developments in digital electronics:

- After almost 50 years of research, it is finally clear that there are technologies to make better-than-silicon MOSFETs. Although the efforts were not sustained during this long time period with ups and downs, this area is now very active with yet growing interest of researches and engineers in electronic industry and academia. The number of papers published recently on III-V MOSFETs are way higher than at any given time in the past.
- Silicon oxide is out of play in mainstream Si CMOS. That means that the key materials advantage of silicon for CMOS circuits is gone. Introduction of high-k oxides into Si ICs makes the perspectives of III-V integration significantly more feasible.
- Further scaling of transistors relaxes some of the requirements to the gate stack, such as interface trap density, D_{it}. In fact, due to increased oxide capacitance, the circuits can handle much higher levels of D_{it} which previously considered as detrimental.
- Si IC companies, mainly INTEL and IBM and their consortia, have shown interest beyond just research.

Apparently, there are still a lot of challenges to be overcome before manufacturing becomes viable. According to Robert Chau, Director of transistor research and nanotechnology at INTEL Corporation, there are following four major challenges (CSIC 2005 Tech. Digest): (1) compatible high-quality gate dielectric; (2) scaling with acceptable I_{ON}/I_{OFF} ratio; (3) p-channel with a reasonable transport; and

(4) integration onto Si substrate. This book addresses research covering the first three of these challenges. We believe the integration with Si involves significantly different materials problems, than those covered in this book. In addition, there are a few good books and reviews on this topic (E. Towe (Ed.) Heterogeneous optoelectronics integration, SPIE Press, 2000; E. Fitzgerald, ECS Trans. 19, 345, 2009; F. Letertre, AIP Conf. Proc. 1068, 185, 2008).

The book begins with a concise historic review (Chap. 1) of challenges and breakthroughs that led to the evolution of today's III-V MOSFETs. Two chapters on device simulations (Chaps. 2, 3) present performance analysis of MOSFETs with different III-V channels with the focus on benefits, potential showstoppers, and novel promising device structures (Chap. 2); and device physics and technology issues for InGaAs HEMTs based on close comparison with recent experimental results (Chap. 3). The chapters on ab initio density function theory simulations include concise introduction into DFT (Chaps. 4, 5) and simulation results on oxide/III-V interfaces with a particular focus on amorphous oxides (Chap. 5), and on bulk and surface properties of HfO_2 and ZrO_2 high-k oxides and metal-oxide interfaces (Chap. 4). Chapter 6 reviews the interfacial chemistry of III-V's, with particular attention to native and deposited oxide gate dielectrics, and correlation with electrical properties of these interfaces. Chapter 7 proposes an empirical model to correlate the experimental work on III-V MOSFETs with the existing oxide/III-V interface models. It follows by six chapters (Chaps. 8–13) on high-k/III-V integration and device work on III-V MOSFETs. Chapter 8 begins with comparison of HEMT for logic applications to MOSFET technology with emphasis on current transport and interface passivation. Chapter 9 presents the new progress on InGaAs, Ge and GaN MOSFETs with MBE $Ga_2O_3(Gd_2O_3)$ or ALD Al_2O_3 as gate dielectrics. Chapter 9 discusses the critical process issues for self-aligned III-V MOSFET and presents the device work on self-aligned GaAs enhancement-mode MOSFETs using regrown source and drain regions. Detailed work on HfO_2 with silicon interface passivation on various III-V substrates are summarized in Chap. 11. The new progress on III-V p-channel MOSFETs, one of the grand challenges in III-V CMOS technology, is reviewed in Chap. 12. Chapter 13 describes materials growth, deposition and fabrication technology, device characteristics, reliability, and applications of insulated gate group III-nitride field effect transistors. The book is ended by the Chap. 14 as a III-V circuit chapter, where the complete technology-circuit assessment of III-V FETs and the co-design approach from the device/SPICE models, logic/memory circuit analysis and technology requirements are presented.

After over 40 years of success of Si/SiO_2 material system in digital circuits, high-k oxides became attractive for further CMOS scaling at the end of 1990s and instigated explosive research growth that resulted in its successful commercialization. We hope that the III-V research is currently at a similar stage as high-k's were in 1990s, and that the combined efforts in academia and industry will make the long-standing GaAs MOSFET dream a commercial technology.

We have benefited greatly from suggestions and discussions with Dmitri A. Antoniadis, Robert Chau, Jesus del Alamo, Eugene Fitzgerald, Max Fischetti. This book has become possible due to support of Focus Center Research Program and

Intel Corporation. We also appreciate the permission granted to us from the respective journals and authors to reproduce their original figures cited in this book. We are further indebted to Mr. Steven Elliot of Springer for sustained efforts to publish the book.

Albany and West Lafayette
September, 2009

Serge Oktyabrsky and Peide D. Ye

Contents

Contributors

Jesús A. del Alamo Department of Electrical Engineering and Computer Science, Massachusetts Institute of Technology, Cambridge, Massachusetts, USA.

Dimitri A. Antoniadis Department of Electrical Engineering and Computer Science, Massachusetts Institute of Technology, Cambridge, Massachusetts, USA.

Evgueni A. Chagarov Department of Chemistry and Biochemistry, University of California, San Diego, La Jolla, California, USA.

P. Chang Department of Materials Science and Engineering, National Tsing-Hua University, Hsinchu, Taiwan.

Donald Cheng Department of Electrical and Computer Engineering, University of Illinois at Urbana-Champaign, Urbana, Illinois, USA.

K. Y. (Norman) Cheng Department of Electrical and Computer Engineering, University of Illinois at Urbana-Champaign, Urbana, Illinois, USA.

Alexander A. Demkov Department of Physics, The University of Texas at Austin, Austin, Texas, USA.

Milton Feng Department of Electrical and Computer Engineering, University of Illinois at Urbana-Champaign, Urbana, Illinois, USA.

Remis Gaska Sensor Electronic Technology, Inc., Columbia, SC, USA.

Niti Goel International SEMATECH, Austin, Texas, USA.

Christopher J. Hinkle Department of Materials Science and Engineering, University of Texas at Dallas, Richardson, Texas, USA.

Minghwei Hong Department of Materials Science and Engineering, National Tsing-Hua University, Hsinchu, Taiwan.

Mao-Lin Huang Department of Materials Science and Engineering, National Tsing-Hua University, Hsinchu, Taiwan.

R. Jain Sensor Electronic Technology, Inc., Columbia, SC, USA.

Dae-Hyun Kim Microsystems Technology Laboratories, Massachusetts Institute of Technology, Cambridge, Massachusetts, USA.

Donghyun Kim Department of Electrical Engineering, Stanford University, Stanford, CA, USA.

Sergei Koveshnikov College of Nanoscale Science and Engineering, University at Albany—SUNY, Albany, New York, USA.

Tejas Krishnamohan Department of Electrical Engineering, Stanford University, Stanford, CA, USA.

Jaydeep P. Kulkarni Intel Corporation, Hillsboro, OR, USA.

Andrew C. Kummel Department of Chemistry and Biochemistry, University of California, San Diego, La Jolla, California, USA.

J. Raynien Kwo Department of Physics, National Tsing-Hua University, Hsinchu, Taiwan.

Jack C. Lee Cockrell School of Engineering, The University of Texas at Austin, Ausin, Texas, USA.

W. C. Lee Department of Materials Science and Engineering, National Tsing-Hua University, Hsinchu, Taiwan.

Chichih Liao Department of Electrical and Computer Engineering, University of Illinois at Urbana-Champaign, Urbana, Illinois, USA.

Tsung-da Lin Department of Materials Science and Engineering, National Tsing-Hua University, Hsinchu, Taiwan.

Yang Liu Department of Electrical and Computer Engineering, Purdue University, West Lafayette, Indiana, USA.

Mark S. Lundstrom Department of Electrical and Computer Engineering, Purdue University, West Lafayette, Indiana, USA.

Xuhui Luo Department of Physics, The University of Texas at Austin, Austin, Texas, USA.

Marko Milojevic Department of Materials Science and Engineering, University of Texas at Dallas, Richardson, Texas, USA.

Yoshio Nishi Department of Electrical Engineering, Stanford University, Stanford, CA, USA.

Injo Ok Cockrell School of Engineering, The University of Texas at Austin, Austin, Texas, USA.

Serge Oktyabrsky College of Nanoscale Science and Engineering, University at Albany—SUNY, Albany, New York, USA.

Himadri S. Pal Department of Electrical and Computer Engineering, Purdue University, West Lafayette, Indiana, USA.

Kaushik Roy School of Electrical and Computer Engineering, Purdue University, West Lafayette, IN, USA.

Sergei Rumyantsev ECSE Department and Broadband Center, Rensselaer Polytechnic Institute, Troy, New York, USA.

Krishna C. Saraswat Department of Electrical Engineering, Stanford University, Stanford, CA, USA.

Onise Sharia Department of Physics, The University of Texas at Austin, Austin, Texas, USA.

Michael Shur ECSE Department and Broadband Center, Rensselaer Polytechnic Institute, Troy, New York, USA.

Grigory Simin Department of Electrical Engineering, University of South Carolina, Columbia, SC, USA.

Wilman Tsai International SEMATECH, Austin, Texas, USA.

Eric M. Vogel Department of Materials Science and Engineering, University of Texas at Dallas, Richardson, Texas, USA.

Robert M. Wallace Department of Materials Science and Engineering, University of Texas at Dallas, Richardson, Texas, USA.

Wei-E Wang IMEC, Leuven, Belgium.

Jerry M. Woodall School of Electrical and Computer Engineering, Purdue University, West Lafayette, Indiana, USA.

Yanqing Wu School of Electrical and Computer Engineering, Purdue University, West Lafayette, Indiana, USA.

Min Xu School of Electrical and Computer Engineering, Purdue University, West Lafayette, Indiana, USA.

Yi Xuan School of Electrical and Computer Engineering, Purdue University, West Lafayette, Indiana, USA.

Peide D. Ye School of Electrical and Computer Engineering, Purdue University, West Lafayette, Indiana, USA.

Chapter 1
Non-Silicon MOSFET Technology: A Long Time Coming

Jerry M. Woodall

Abstract A summary of the important materials science issues associated with the realization of viable III-V MOSFET technologies is presented. The key science breakthrough was the unambiguous identification of which components of the non-stoichiometric native oxides were responsible for surface Fermi level pinning (FLP). The components that cause FLP are the anion oxides and the elemental anion associated with a particular III-V compound semiconductor. For GaAs, these are As_2O_3 and elemental As respectively. The physics of FLP is also applicable to Schottky barriers. Although many attempts were made to explain FLP, the most comprehensive theory is that the elemental anion acts to cause FLP via its Schottky barrier workfunction. During the past decade several technologies have succeeded in mitigating these chemical barriers and III-V MOSFET technology is now a component of the MOSFET menu.

1.1 Introduction

The purpose of this chapter is to provide the reader with a brief, non-comprehensive overview of the history, scientific and technological barriers, pre-device solutions, and seminal research results that led to the evolution of today's non-silicon MOSFET (NSMOSFET) technology.

If you are living on planet Earth, you must know that silicon MOSFET (SMOSFET) technology, with a 2008 worldwide chip sales of more that \$250 billion, is essentially the only semiconductor technology used by all electronics based industries. By contrast, the total 2008 sales of *all* compound semiconductor devices and chips are about \$20 billion, and most of this revenue was generated by heterojunction based photonic devices and chips. In contrast, none of the revenue was produced by NSMOSFETs sales (save some devices sold for testing and military applications).

J. M. Woodall (✉)
School of Electrical and Computer Engineering, Purdue University, 465 Northwestern Ave., West Lafayette, IN 47907, USA
e-mail: woodall@purdue.edu

S. Oktyabrsky, P. D. Ye (eds.), *Fundamentals of III-V Semiconductor MOSFETs*,
DOI 10.1007/978-1-4419-1547-4_1, © Springer Science+Business Media LLC 2010

Why is Si technology the dominant semiconductor technology? To answer this at the highest level, let me tell the reader the answer I give my undergraduate students during the first lecture of Purdue's core semiconductor course. Even though Si is an inferior electronic and photonic material compared to, for example, GaAs, Si is still king because Si is cheap and the rust on Si is electronically exquisite, whereas the rust on nearly all compound semiconductors of interest is either inferior or non-functional electronically. In other words, both price-performance advantages of SMOSFET technology and the lack of a viable NSMOSFET technology during the early R&D efforts have hampered its development. This chapter will discuss why the electronic and chemical properties of compound semiconductor rust were not suitable for the "O" material for NSMOSFETs.

Another question, possibly rhetorical, comes to mind at this point. Even if the application specific performance of NSMOSFET technology could be far superior to planned performance for SMOSFET technology, will it supplant Si technology? Is it too little or too late? This situation can be likened to the current global energy crisis. Let us equate Si technology with fossil fuel technology (ignoring carbon footprint issues). Both are the overwhelming dominant incumbents. Now let us equate alternative solar energy technology with NSMOSFET technology. The central question for both fossil fuel and Si technology is whether there is a business model for solar energy technology and NSMOSFET technology that will result in supplanting the current fossil fuel and Si incumbents. Or, will Si, for example, stay on a revised "Moore's Law" path? This question will be addressed in a later chapter in this book.

1.2 Brief and Non-Comprehensive History of the NSMOSFET

Any discussion of the history of any MOSFET technology, or of any other transistor technology, must begin with the Heil and Lilienfeld Patents [1, 2]. Examination of Heil's patent clearly indicates that a MOSFET was being described. Amusingly, Heil's 1934 patent date is 13 years before the commonly accepted birth date of the transistor. It is less amusing to note that neither Heil nor Lilienfeld were included in the Nobel Prize for the invention of the transistor.

The next seminal event, also not recognized by the Nobel Committee, was the first public report of the SMOSFET by Kang and Atalla [3]. This report was seminal in that the SMOSFET became the dominant material for commercial MOSFETs. Very rarely does the early work ever evolve into the dominant technology.

The 1960s and 1970s were a period of intense development and rapid progress of SMOSFET technology. Hundreds of worldwide workers in corporate and university laboratories contributed to this progress. However, except for isolated successes limited to laboratory scale III-V and II-VI compound thin film transistor (TFT) devices, attempts during this period to develop a NSMOSFET technology were essentially unsuccessful. An exception to this notable lack of progress in NSMOSFET development was the work of Brody and Kunig, who, in 1966, realized both enhancement and depletion mode MOSFETs made of InAs [4]. Let us now explore why the promise of NSMOSFET technology has been slow in coming.

1.3 Surface Fermi Level Pinning: The Bane of NSMOSFET Technology Development

With respect to NSMOSFET technology, the realization of a flat band compound semiconductor sample, i.e., a uniformly doped sample whose band edge was invariant with position, in air and at equilibrium, had been elusive up to around 10 years ago. The reason is that in an air environment "surface states" form with energies that lie at the band edges or within the band gap and have a characteristic energy determined by the material. For example, oxidized GaAs has a near mid-gap surface state energy, while oxidized InAs has states at or just above the conduction band edge.

Understanding the origin of compound semiconductor Fermi level pinning has been a holy grail for hundreds of surface physicists for the past 40 years. In fact this phenomenon was considered to be so scientifically and technologically important that the Office of Navel Research (ONR) has sponsored an annual conference, "Physics of Compound Semiconductor Interfaces" (PCSI), for over 35 years to discuss it. The complicating factor in this quest is the fact that unlike for SMOSFET technology, where the Si-SiO_2 interface states or "traps" can be passivated by hydrogen, compound semiconductor surface states are impervious to such an approach. This has led many research scientists, including this author, to the conclusion that the etiology of compound semiconductor surface states is fundamentally different from that of Si or SMOSFETs. The quest for understanding Fermi level pinning has led to at least five unique pinning models, including a surface defect model, a metal induced gap state model, the standard Schottky barrier model [5], a modified Schottky model [6], and a couple of bulk defect models. Except for the modified Schottky barrier model, I will not discuss these other models at length, nor reference them. The reason is that since the other models posit that surface Fermi level pinning is an intrinsic property, they are not capable of addressing unpinned MOS-C interfaces which form the basis for the recent successful experimental results for inversion mode NSMOFET technology, the central focus of this book.

The technological impact of surface states is both manifold and detrimental to device performance. First, surface Fermi level pinning can result in reverse band bending producing electric fields that can drift optically or electrically generated minority carriers into surface states where they are lost. This will either limit or disable performance of, for example, LEDs, lasers, photo-detectors, bipolar transistors, and solar cells. Many of these limitations have been mitigated through the use of heterojunctions. However, current heterojunction FETs are application limited with respect to ideal NSMOSFETs. Next, because of surface Fermi level pinning, when non-reactive metals (and even some reactive ones) are applied, the Schottky barrier height is essentially invariant with metal work function. For example, the n-type Schottky barrier height for nearly all metals deposited on air exposed GaAs surfaces has a value of 0.8 ± 0.1 eV regardless of their work functions [6]. Thus, the original Schottky barrier theory is not applicable to this situation. However, it has been shown that if the native oxides on, for example, GaAs, InGaAs, and InAs are removed in an UHV environment followed by deposition of non-reactive metals of differing workfunction at 77 K, the interface Fermi level (Schottky barrier height) is the value expected by the standard Schottky barrier model [7]. This result rules out the general applicability of the other intrinsic pinning models referred to above and supports the modified Schottky model of Ref. [6].

Technologically, Fermi level pinning leading to an invariant Schottky barrier height versus metal workfunction is not all bad. For example even though the GaAs transistor menu has not included NSMOSFET until recently, Fermi level pinning enabled a robust niche market for GaAs MESFETs. This has allowed a broad choice of electronically acceptable metal gate materials that can be optimized on basis of other important technological considerations such as reliability, reproducibility, etc. So, while NSMOSFET research scientists have been struggling to realize a viable technology for GaAs, many niche market companies have enjoyed good revenue from GaAs MESFETs. Or, as the saying goes, they turned lemons into lemonade.

The seminal breakthrough that lead to an unpinned (100) GaAs-oxide interface, hence, to a GaAs based NSMOSFET technology, came in 1986 by a group led by this author. Their investigation revealed that by removing both the metallic arsenic and arsenic oxide components in the native GaAs oxide, the resulting GaAs-arsenic free interface was flat band in air [8]. The technique they used to achieve this result was a simple photo-washing procedure in which either an n-type or p-type GaAs wafer was mounted on a laboratory photoresist spinner. The spinner was turned on and deionized water was applied to the GaAs surface while being illuminated by a low intensity He-Cd 440 nm laser. This breakthrough did not lead to an NSMOSFET technology immediately, though. When left exposed to air, oxygen diffusion back into the GaAs-oxide interface led to both elemental As and arsenic oxide chemistry regeneration, and thus back to interface Fermi level pinning. However, it did point the way to the critical process parameters and environment needed to realize an ideal NSMOSFET.

Since Fermi level pinning at GaAs-oxide interfaces was unambiguously correlated with the presence of equilibrium arsenic species, particularly excess elemental arsenic, the unpinning experiments provided strong evidence in support of the "Effective Work Function (EWF)" model developed and published by Freeouf and this author [6]. This model is simply the Schottky model modified to account for the actual metal in intimate contact at the M/S interface. For nearly all III-V oxide interfaces (except GaP) at equilibrium, the dominant chemistry leads to excess elemental anions at the interface. Therefore, in order to correctly apply the Schottky model when, for example, gold is deposited on a GaAs sample covered with a few monolayers of its native oxide, the work function of Au or any other metal must be replaced by the workfunction of As to get the right Schottky barrier height. This also leads to the experimentally observed invariance of the Schottky barrier height with work function for Schottky diodes fabricated by deposition of metals on top of the native oxide. The barrier height will be fixed by the work function of arsenic. However, the real success of the EWF model compared to other major models is that it validates the original Schottky barrier model when the oxide is removed and the metals are deposited at low temperatures [6].

Finally, let us examine Fig. 1.1 as a way to understand how the presence of a metal layer, as modeled by dispersed metal particles of arsenic (As), at a GaAs-oxide interface, would affect the observed band diagram and C-V behavior of, for example, an n-GaAs MOS-C device. We assume a thin native GaAs oxide with As particles at the interface of the GaAs with an ohmic metal contact on the backside. We then continue to deposit a good quality dielectric on top of the native oxide, i.e., the "O" layer. Finally, we deposit a metal gate "M" layer on top of the oxide to complete the MOS-C structure. We now connect the leads from a capacitance meter to the top gate metal and the backside ohmic metal contact. Note that the arsenic metal "plate" forms an M/S contact to the GaAs and that this plate is not electrostatically connected to any part of the DC supply voltage circuit. Therefore, under DC bias, the Fermi level is invariant with respect to position in both the top metal

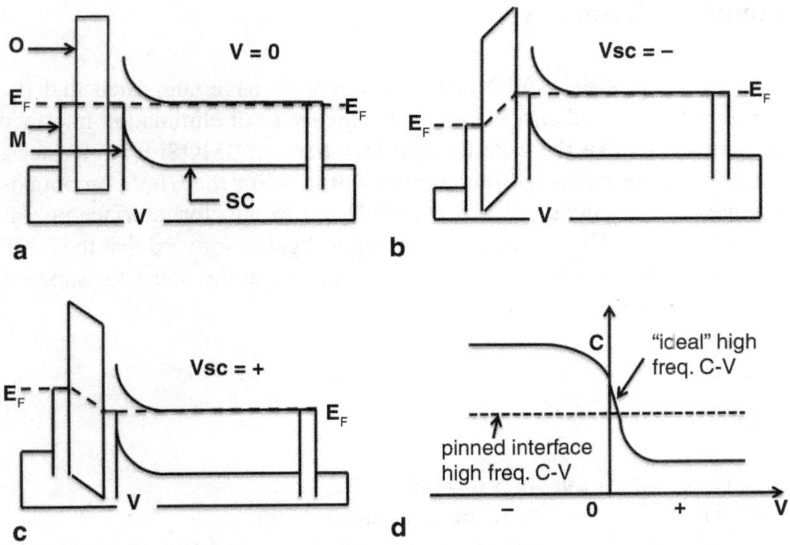

Fig. 1.1 N-type GaAs MOS-C model with Fermi level pinning due to arsenic particles at the GaAs-oxide interface. The As particles form an M/S Schottky diode and pin the Fermi level near mid-gap at the M/S interface with a Schottky barrier height that corresponds to the workfunction of As. The As particles float electrostatically under all bias voltages. The overall MOS-C is modeled as a MOM capacitor in series with an M/S Schottky diode. **a** band diagram of the model under no bias. **b** same as **a** except the GaAs is biased negatively with respect to the top metal gate contact. **c** same as **b** except the GaAs is biased positively with respect to the top metal gate contact. **d** C-V (*dashed line*) for the interface pinned model compared schematically with an ideal unpinned MOS-C. Note the invariance of the C vs. V for the pinned interface. This is a result of a position invariant Fermi level in the GaAs and pinning at the interface, hence, an invariant GaAs depletion width for all DC biases

contact and throughout the M/S GaAs Schottky diode, and varies with bias only in the "O" layer. Since the Fermi level is pinned at the GaAs-oxide interface, the series capacitance of the oxide plus the GaAs depletion region will remain invariant with respect to the DC applied voltage, see Fig. 1.1a–c. Thus, the invariant dashed line in Fig. 1.1d will be the observed C vs V. This behavior was commonly observed in many early GaAs MOS-C reports [8]. From a device point of view, the pinned interface situation can be modeled as metal-oxide-metal (MOM) capacitor series with an M/S Schottky in which the M contact is floating electrostatically. The solid line in Fig. 1.1d represents a schematic C-V curve for an ideal MOS-C.

The reader should note that owing to MOS-C issues other than interface Fermi level pinning, such as a bulk charge in the oxide layer, bulk semiconductor defects near the oxide interface, etc., the actual observed C-V characteristics of improved MOS-C structures will be more complicated. However, the purpose of this chapter was served by showing how interface Fermi level pinning hampered NSMOSFET development until its origin and a technique to eliminate the problem was discovered.

1.4 Concluding Remarks

Some authors of recent successful NSMOSFET reports have suggested that the achievement of inversion in their technologies is the result of eliminating interface Fermi level pinning. Unlike the case for interface traps in SMOSFET interfaces, this author has shown in published literature that at least for the III-V compounds and alloys removal of arsenic species at the oxide-semiconductor interface unpins the interface Fermi level. This provides strong support going forward that the EWF model is the proper framework to make further progress in the quest for superior NSMOSFET technologies.

References

1. J.E. Lilienfeld, U.S. Patent 1,900,018, 7 Mar 1933
2. O. Heil, British Patent 439, Dec 1935 (German Patent date, 1934)
3. D. Kang, M.M. Atalla, Silicon-silicon dioxide field induced surface device. IRE-AIEE solid state device research conference, Carnegie Institute of Technology, Pittsburgh, 1960
4. T.P. Brody, H.E. Kunig, Appl. Phys. Lett. **9**, 259 (1966)
5. W. Schottky, Z. Phys. **118**, 539 (1942)
6. J.L. Freeouf, J.M. Woodall, Appl. Phys. Lett. **39**, 727 (1981)
7. L.J. Brillson, M.L. Slade, R.E. Viturro, M.K. Kelly, N. Tache, G. Margaritondo, J.M. Woodall, P.D. Kirchner, G.D. Pettit, S.L. Wright, Appl. Phys. Lett. **48**, 1458 (1986)
8. S.D. Offsey, J.M. Woodall, A.C. Warren, P.D. Kirchner, T.I. Chappell, G.D. Pettit, Appl. Phys. Lett. **48**, 475 (1986)

Chapter 2
Properties and Trade-Offs of Compound Semiconductor MOSFETs

Tejas Krishnamohan, Donghyun Kim and Krishna C. Saraswat

Abstract In order to continue the scaling of silicon-based CMOS and maintain the historic progress in information processing and transmission, innovative device structures and new materials are required. A channel material with high mobility and therefore high injection velocity can increase ON current and reduce delay. Currently, strained-Si is the dominant technology for high performance MOSFETs. However, looking into future high mobility III-V materials can offer several advantages over even very highly strained Si.

The experimental research space looking at various III-V materials as candidates for high mobility channels in nanoscale MOSFETs is extremely vast and complex. In this work, we will use various simulation techniques to narrow down and identify the top III-V material candidates for both n- and p-MOSFETs, in terms of increased drive current and lower leakage power dissipation. We will explore binary and ternary III-V channel materials, and effectively exploit strain engineering along different orientations to enhance their performance. We will also address the problem of parasitics, and discuss innovative quantum-well device structures, which together with the application of strain engineering can enable energy-efficient nanoscale III-V n- and p-MOSFETs.

2.1 Introduction

As we continue to aggressively scale transistors in accordance with Moore's Law to sub-20-nm dimensions, it becomes increasingly difficult to maintain the required device performance. Currently, the increase in drive currents for faster switching speeds at lower supply voltages is largely at the expense of an exponentially growing leakage current, which leads to a large standby power dissipation. There is an important need to explore novel channel materials and device structures that would

T. Krishnamohan (✉)
Center for Integrated Systems, Department of Electrical Engineering,
Stanford University, Stanford, CA 94305, USA
e-mail: tejask@stanford.edu

S. Oktyabrsky, P. D. Ye (eds.), *Fundamentals of III-V Semiconductor MOSFETs*,
DOI 10.1007/978-1-4419-1547-4_2, © Springer Science+Business Media LLC 2010

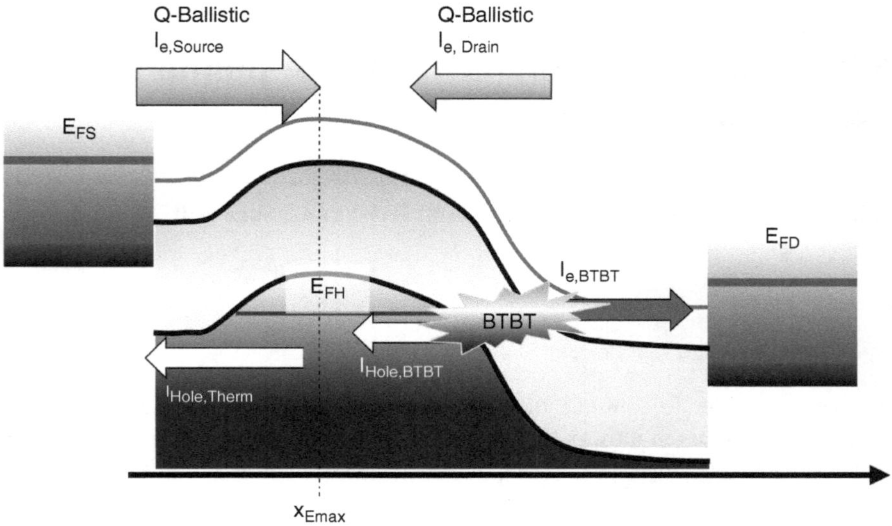

Fig. 2.1 Dominant leakage paths in nanoscale high mobility CMOS devices. High mobility (low-bandgap) materials suffer from excessive BTBT leakage

provide us with energy-efficient nanoscale MOSFETs. Due to their significant transport advantage, high mobility materials are very actively being researched as channel materials for future highly scaled CMOS [1–10]. However, most high mobility materials also have a significantly smaller bandgap compared to Si, leading to very high band-to-band tunneling (BTBT) leakage currents, which may ultimately limit their scalability.

The different significant components of channel leakage are shown in Fig. 2.1 (Assuming an ideal high-k gate dielectric technology, gate leakage has not been addressed in the present study.) Conventionally, the exponentially rising diffusion (or subthreshold) current due to lowered threshold voltages has been the dominant leakage mechanism in Si MOSFETs. However, with continued device scaling into the nanometer regime and the increasing E-fields in the channel, there is also an exponential growth in the tunneling components (BTBT and direct source/drain (S–D) tunneling) of leakage. Figure 2.2a, b shows the material parameters of low effective mass (high mobility) semiconducting channel materials. A universal trend is observed with respect to the bandgap and the dielectric constant of low effective mass semiconductors. As we push toward lower effective mass (high mobility) channel materials, the bandgap rapidly drops and the dielectric constant increases.

As seen in Table 2.1, III-V materials have significantly smaller effective mass and higher electron mobility compared to Si and Ge. However, most high mobility materials like Ge, InAs, and InSb also have a significantly lower bandgap compared to Si. Due to the increasing E-fields in the channel and the smaller bandgap in these high mobility materials, the BTBT leakage current can become

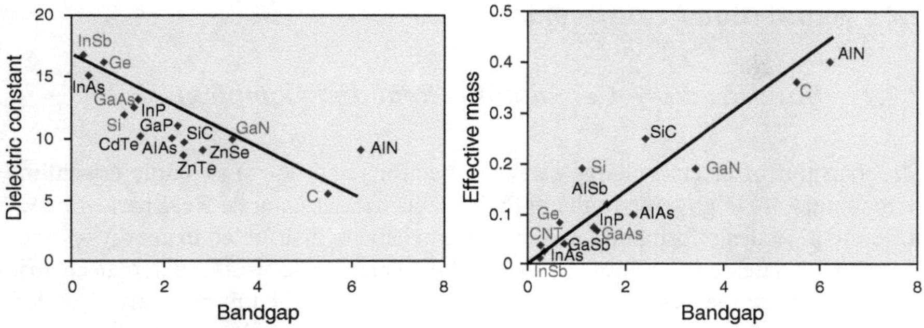

Fig. 2.2 Tradeoffs between effective mass (m*), bandgap (E_G), and dielectric constant (κ_s) in semiconducting materials

excessive and can ultimately limit the scalability of high mobility channel materials. Another point to note is that since most high mobility materials also typically have a higher permittivity (κ_s), they also suffer from worse short channel effects (SCEs).

At an initial glance, due to their extremely small transport mass leading to high injection velocity (v_{inj}), III-V materials appear to be very attractive candidates as channel materials for highly scaled n-MOSFETs [1, 11, 12]. However, III-V materials have many significant and fundamental issues, which may prove to be severe bottlenecks to their implementation. These tradeoffs need to be systematically and thoroughly investigated from a theoretical standpoint in order to help in providing useful guidelines, which can efficiently streamline the resources invested by several research groups throughout the world in their experimental development.

In this chapter, we will look at novel III-V channel materials for n- and p-MOSFETs, and the application of various strain along different surface/channel orientations to improve their performance, in terms of increased drive current and lower leakage. We will also discuss novel device structures, and approaches to address the problem of parasitics, which can enable energy-efficient nanoscale III-V n- and p-MOSFETs.

Table 2.1 Carrier mobility and bandgap of some commonly used semiconductors

Material/Property	Si	Ge	GaAs	InAs	InSb
m_{eff}^*	0.19	0.08	0.067	***0.023***	***0.014***
μ_n (cm^2/Vs)	1600	3900	9200	***40,000***	***77,000***
E_G (eV)	1.12	0.66	1.42	***0.36***	***0.17***
ε_r	11.8	16	12.4	14.8	***17.7***

2.2 Simulation Framework

2.2.1 Bandstructure Calculation (Real and Complex)

The transport of electric charge carriers (electrons and holes) in single crystalline semiconductors is governed inherently by their bandstructures. Real part of bandstructure provides information on how the carriers are distributed in energy-k space (density of states and valleys) and how the carriers react to the external electric fields (effective masses). The imaginary part of bandstructure helps us to understand the tunneling process of carriers between different bands.

The plane-wave based empirical pseudopotential method (EPM) [13] is one of the useful ways to accurately estimate the details of the bandstructure. While the tight-binding method or k·p theory requires many fitting parameters, EPM only requires a few fitting parameters and provides more accurate bandstructures, thus EPM is more suitable when it is needed to study the entire first Brillouin zone and the effect of the strain.

The matrix elements of the pseudo-potential including the spin–orbit interactions are:

$$\mathrm{H}_{(\mathbf{G}',s'),(\mathbf{G},s)} = \frac{\hbar^2}{2m_0}\beta|\mathbf{G}+\mathbf{k}|^2\delta_{\mathbf{G}',G}\delta_{s',s} + \frac{\Omega_c}{\Omega}\sum_{\mathbf{r}_a} e^{-i\mathbf{q}\cdot\mathbf{r}_a}V_{f,a}(\mathbf{q})\delta_{s',s}$$
$$+ \frac{\Omega_c}{2\Omega}\sum_{\mathbf{r}_a} e^{-i\mathbf{q}\cdot\mathbf{r}_a}[-i\mu_{\mathbf{r}_a}[(\mathbf{G}'+\mathbf{k})\times(\mathbf{G}+\mathbf{k})]\cdot\sigma'_{s',s}].$$

The last term in the elements represents the spin–orbit coupling when inner core states and d-core states are neglected and only outermost p-core states are treated [14–16].

The parameters, $V_{f,a}(\mathbf{q})$, represent the pseudo-potential form factors and they need to be defined for any continuous wave-vectors **q**, in order to model the strained semiconductor where the strains modify the reciprocal-lattice vectors **G**. For the form factors, we take a form suggested in [17]

$$V_{f,\alpha}(\mathbf{q}) = \frac{1}{\Omega_\mathrm{c}}\frac{a_0(q^2-a_1)}{a_2 \mathrm{e}^{a_3 q^2}-1}$$

Under applied stress, a solid is strained, resulting in a change of volume and shape. According to the small deformation theory, when the second-order terms are neglected, the position of an atom in strained semiconductor can be expressed using a strain tensor, ϵ and an internal ionic displacement vector, **u** [18].

$$\mathbf{r}'_a = (\delta+\epsilon)\mathbf{r}_a + \zeta\mathbf{u},$$

where $\mathbf{r}_\mathrm{a}$ is a vector indicating the location of atom, a, in unstrained lattice while $\mathbf{r}'_\mathrm{a}$ is the corresponding vector for the strained system.

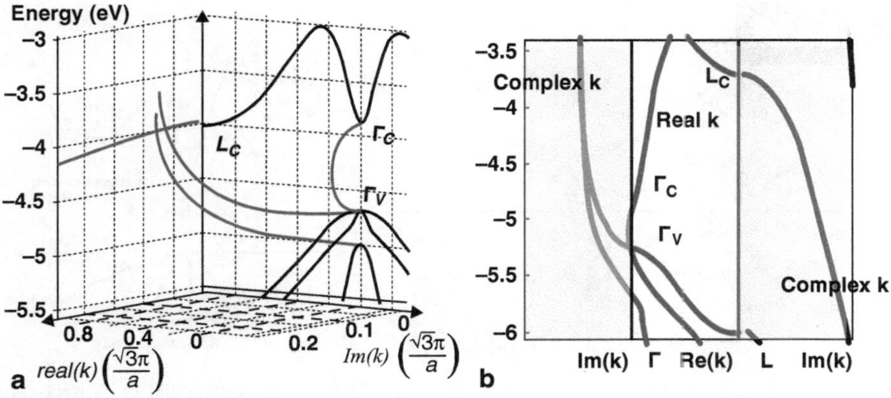

Fig. 2.3 **a** 3D plot of complex bandstructure of germanium along the [111] direction. **b** 2D plot of complex bandstructure of InAs

Figure 2.3 shows the result of the bandstructure calculation done by EPM for Ge and InAs. Using this method, both complex part and real part of bandstructure can be obtained.

2.2.2 *Band-to-Band Tunneling (Off-State Leakage)*

In semiconductors, most of states in valence bands are filled with electrons. Under a strong applied external electric field, electrons in valence bands can tunnel through the forbidden bandgap and when they reach the conduction bands, electron-hole pairs are created (Fig. 2.4) [19]. In in-direct band-gap materials such as Si, since the minimum of the conduction band is located at Δ point away from the top of valence band which is located at Γ_v, the momentum conservation law prohibits the tunneling between two bands (Fig. 2.4). Phonons created by thermal lattice vibration can provide the momentum difference and allow two bands to be coupled. This tunneling process incorporating phonons is called in-direct BTBT or phonon assisted BTBT, which is the dominant BTBT process in Si [20].

We evaluated the interband matrix elements between quantized states for both direct and indirect tunneling, following Kane's approach [19, 20]. Fermi's Golden rule is used to determine the band to band transition rate between states. The final BTBT carrier generation rate was calculated by adding up the transition rates for all the possible transitions [20, 21]. We included all the possible transition from valence band to conduction bands such as Γ_V-Γ_C, Γ_V-L and Γ_V-X (Fig. 2.4b). While counting, for direct tunneling the momentum conservation selection rule was applied, while for the indirect tunneling the selection rule was relieved.

The efficiency of BTBT process is altered by large amounts in few nanometer thick slabs compared to bulk semiconductors [20]. An ultra-thin body has a larger

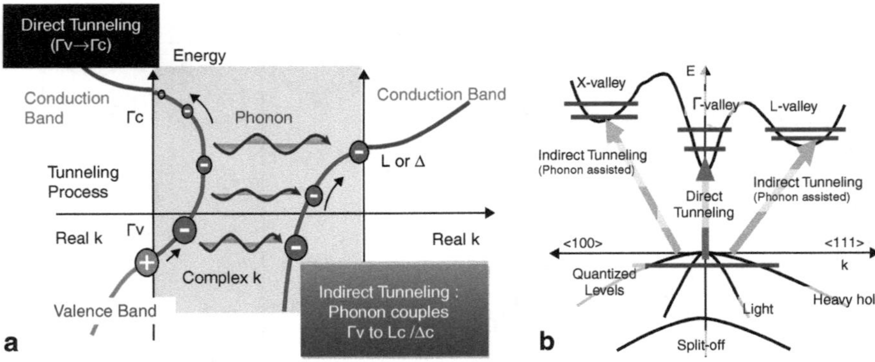

Fig. 2.4 **a** Illustration of band-to-band tunneling paths. The electron can tunnel to indirect-gap conduction band (L or Δ valleys) through phonon interactions. **b** All Possible (Direct and Indirect) BTBT Tunneling Path, Γ-Γ, Γ-X and Γ-L, are captured by the model

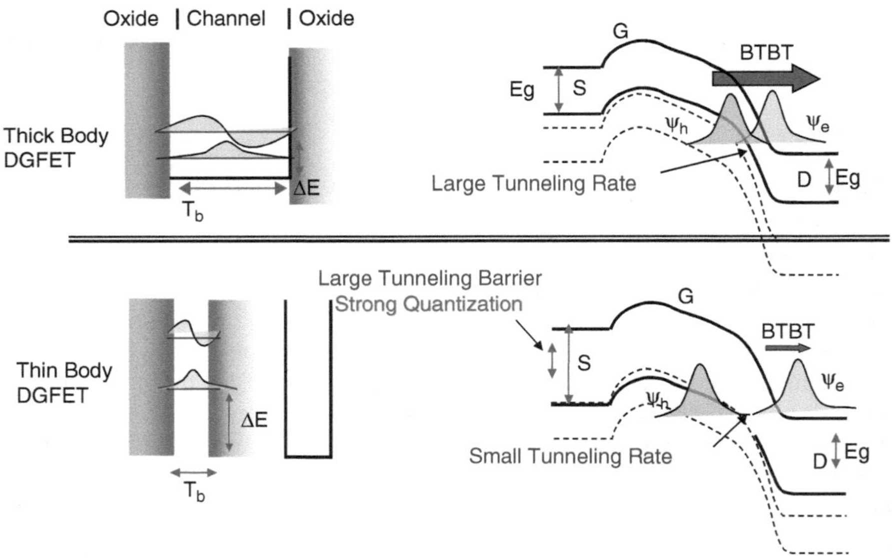

Fig. 2.5 Ultra-thin body and larger quantization increases the effective bandgap and lowers the tunneling rate

effective band gap and smaller DOS than bulk, due to strong quantization. With larger effective band gap, the wave function decays faster inside the forbidden gap and lowers BTBT rate (Fig. 2.5). To take into account the quantization effect caused by ultra-thin body, the wavefunctions and the energy levels of quantized subband states were obtained by solving 1D Schrödinger equation along z-direction for both electrons and holes [22].

2.2.3 *Quantum Ballistic Current (On-State Drive Current)*

If quantum confinement is strong only in one direction (z direction) the states are written as a sum of 2D subbands, the edges of which are obtained by solving the Schrodinger equation in the z-direction for each mesh point along the x axis. For materials with degenerate minima in the conduction band, the states can be computed with the directionally decoupled effective mass approximation [22] for each minimum, taking into account mass anisotropy and non-parabolicity. The same procedure is applied to the valence band maxima.

Carrier wavefunctions are decoupled using subband decomposition and take the form:

$$\Psi_{\alpha n}(x,y,z) = \sqrt{t_b}\phi_{\alpha n}(z;x)\,\psi_{\alpha n\varepsilon}(x)\exp(ik_{y0}y)u_{\alpha n}(\vec{\mathbf{k}}_{\square},\vec{\mathbf{r}})$$

where α, n and t_b represent degenerate valleys, quantum number for subbands, thickness of the channel, respectively. The wave-functions are normalized in z direction and unit-cell functions $u_{\alpha n}(\vec{\mathbf{k}}_{\square},\vec{\mathbf{r}})$ are normalized in the space. 1D Schrodinger equation in the z confined direction provides eigen-functions $\phi_{\alpha n}(z;x)$ and eigen-energies $E_{\alpha,n}(x)$.

The electron levels in 1D quantized Double Gate FETs with small effective mass like in III-V materials can be several tenths of an electron volt above the bulk-conduction band edge. The E-k dispersion relationship for the two-band Kane model [23] can be expressed in the simple form:

$$E(1+\alpha E) = \frac{\hbar^2 k^2}{2m^*}$$

where m* is the effective mass and a coefficient α contains nonparabolicity corrections. This expression can be generalized for the multi-dimensions. For example, the dispersion relation along y direction for the system depicted in Fig. 2.6 can be modeled as:

$$E_y = \frac{\hbar^2 k_y^2}{m_y\left(1+\alpha_x E_x+\alpha_y E_y+\alpha_z \Delta E_{\alpha n}\right)}$$

We assume that source and drain contacts are assumed to be ideal reservoirs in thermal equilibrium with the Fermi energies E_{FS} and E_{FD}. They supply carriers to the channel and absorb carriers from the channel without reflections. The charge density injected from one-side of the contact is given as:

$$|Q| = \frac{q}{2\pi^2\hbar^2}\sum_{\text{valley}(\alpha)}\sum_{n_z}\sqrt{m_x m_y}\int_{E_x=-\infty}^{\infty}\frac{|\varphi_x(x;E_x)\,\phi_z(z;x)|^2}{\sqrt{|E_x|}}$$
$$\times\int_{E_y=0}^{\infty}\frac{\left(\Delta_{E_{xz}}+\alpha_x E_x+\alpha_y E_y\right)\left(\Delta_{E_{xz}}+2\alpha_y E_y\right)}{\sqrt{E_y}\left(\Delta_{E_{xz}}+\alpha_y E_y\right)\left[1+\exp\left(\frac{E_{n_z}+E_x+E_y-E_F}{kT}\right)\right]}dE_y dE_x,$$

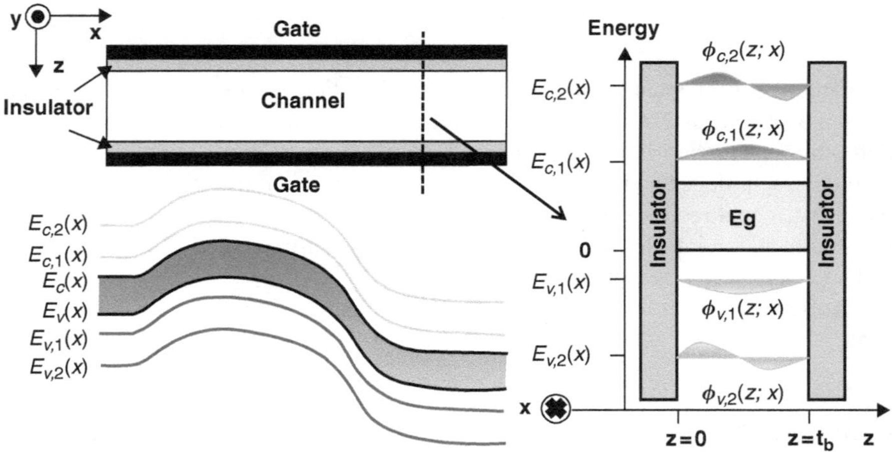

Fig. 2.6 Schematic view of the thin channel double-gate MOSFET (DGFET).Thickness of the channel is t_b. Current flows along x-direction and the channel is quantized along the z-direction

where $|\varphi(x;E_x)|^2 = |P_{\alpha n\varepsilon}(x)|^2\left(2 - |P_{\alpha n\varepsilon}(x_{contact})|^2\right)$. Since the device has two contacts, the total charge density is obtained by adding contributions from both sides.

The quantum-ballistic current is given by the Landauer's formula [24] where the current is the sum of the current component flowing in each sub-channel [21]. Each current component is expressed as the product of the unit charge, the number of carriers flowing through the sub-channel per unit time and the transmission coefficient. The total net current is obtained by subtracting the drain side contribution from the source side contribution.

$$I_{S\to D} = 2_{(\text{spin})}$$
$$\times\, qW \sum_{\text{valley}(\alpha)} \sum_{n_z} \int_{z=0}^{t_b} |\phi_z(z;x)|^2 \int_{-\infty}^{\infty} \begin{pmatrix} \vec{v}_{S\to D}\vec{N}(E) f(E_{FS},E) \\ -\vec{v}_{S\leftarrow D}\vec{N}(E) f(E_{FD},E) \end{pmatrix} T(E)dEdz,$$

where

$$\int_{E_x=-\infty}^{\infty} \vec{v}_{S\to D}\vec{N}(E) f(E_{FS},E) T(E)dE$$
$$= \int_{-\infty}^{\infty}\int_{0}^{\infty} \frac{1}{2^{\frac{3}{2}}(\pi\hbar)^2} |P^{+}_{\alpha n\varepsilon}(x_D)|^2 \frac{\sqrt{m_y}(\Delta_{E_{xz}} + 2\alpha_y E_y)}{\sqrt{E_y(\Delta_{E_{xz}} + \alpha_y E_y)}}$$
$$\times\, \frac{1}{1 + \exp\,\exp\left(\frac{E_y - E'_{FS}}{kT}\right)} dE_y dE_x,$$

$$E'_{FS} = E_{FS} - E_{\alpha n} - E_x \quad \text{and} \quad \left|\psi^{source}_{\alpha n\varepsilon}(x)\right|^2$$
$$\cong \frac{v(x_{source})}{v(x)} |P_{\alpha n\varepsilon+}(x)|^2 \left(2 - |P_{\alpha n\varepsilon+}(x_{drain})|^2\right).$$

2.3 Power-Performance Tradeoffs in Binary III-V Materials (GaAs, InAs, InP and InSb) vs. Si and Ge

Although their small transport mass leads to high injection velocity (v_{inj}), III-V materials have a low density of states (DOS) in the Γ-valley, tending to reduce the inversion charge (Q_{inv}) and hence reduce drive current [25–29]. Furthermore, the small direct band gaps of Ge and III-V materials inherently give rise to very large band to band tunneling (BTBT) leakage current compared to Si [20, 30]. They also have a high permittivity and hence are more prone to short channel effects (SCE). Quantum confinement in these ultra-thin nanoscale DGFETs plays a very important role. In III-V materials, large quantum confinement makes electrons populate and conduct in heavier L- or X-valleys. Furthermore, there is a significant increase in the conductivity mass in quantized sub-bands due to large non-parabolicity of Γ-valley [29, 31]. Quantum confinement effects also reduce BTBT leakage in ultra thin channel [20, 21, 30]. In this section, we will discuss Double Gate (DG) n-MOSFETs with Ge and III-V materials (GaAs, InP, $In_xGa_{(1-x)}As$, InAs and InSb) and compared to Si [21, 31–33].

2.3.1 Inversion Charge and Injection Velocity

Generally, low DOS in III-V materials reduces the effective gate capacitance (C_{Geff}), resulting in lower inversion charge at the given gate voltage (Fig. 2.7). For example, InAs can store only one third of electrons that Si can store. Figure 2.8 shows injection velocities (v_{inj}) at V_{Gate} of 0.7 V. Due to their small transport effective mass, III-V materials and Ge have several times larger injection velocity than Si. Among all, transport effective masses of InAs and InSb are only 0.023 m_0 and 0.014 m_0, respectively and they exhibit six times larger v_{inj} than Si. As we scale channel thickness (T_S), the Γ-valley energies in III-V materials rise up rapidly due to very low m_z resulting in most of the Q_{inv} in thin films moving to the heavier L-valley. Once the L-valleys are resided by electrons, III-V materials start to behave like Ge—large C_{Geff} and large Q_{inv}. The L-valleys start to contribute for GaAs at T_S = 7 nm and for

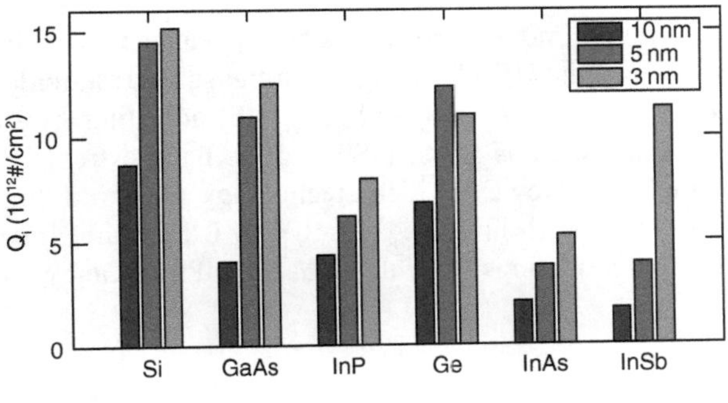

Fig. 2.7 Inversion charge (Q_i) at V_{Gate} = 0.7 V and T_{ox} = 0.7 nm. Si channel stores four times more electrons than InAs. As T_S scales, Q_i becomes bigger due to reduced V_{th} and non-parabolicity of band

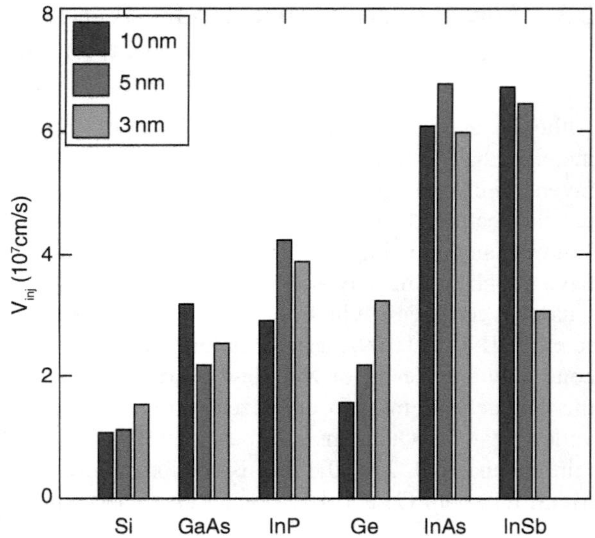

Fig. 2.8 Injection velocities (v_{inj}) at $V_{Gate} = 0.7\,V$ and $T_{ox} = 0.7\,nm$. InAs and InSb have six times larger v_{inj} than Si due to their small transport mass. As T_S scales, lowered threshold voltage (v_{th}) increases v_{inj}, while nonparabolicity reduces v_{inj}

InSb at 3 nm. Scaling T_S improves subthreshold slopes and DIBL effects, permitting larger Q_{inv} or higher Fermi level at the given gate voltages; thus carriers are injected at a higher v_{inj} from source to drain in thin body (Fig. 2.8). However, in small bandgap materials such as InAs and InSb, due to strong non-parabolicity in Γ-valley, high Fermi levels and lifted-up quantized energy states in thin body make electrons heavier and injection velocity is reduced.

2.3.2 I_{ON}, $I_{OFF,\,BTBT}$ and Delay

Device simulation results for $L_G = 15\,nm$, $T_{OX} = 0.7\,nm$, $T_S = 3$–$5\,nm$ and $V_{DD} = 0.3$–$0.7\,V$ are shown in Fig. 2.9 (I_{ON}), Fig. 2.10 ($I_{OFF,\,BTBT}$). There have been concerns that III-V materials would underperform than Si because of their low Q_{inv} [25, 28, 31–34]. However our study shows that despite of low Q_{inv}, thanks to their large v_{inj}, III-V materials like InAs, InSb and InP can flow up to 80% larger drive current than Si.

The minimum achievable standby leakage by BTBT ($I_{OFF,\,BTBT}$) is at the intersection of the BTBT leakage with the sub-threshold leakage. DGFET cannot have lower leakage current than $I_{OFF,\,BTBT}$ by adjusting threshold voltage. Small bandgap materials such as InAs, InSb and Ge have extremely large $I_{OFF,\,BTBT}$ higher than 0.1 μA/μm. However, if the technology allows to thin channel thickness to 3 nm, it is possible to limit $I_{OFF,\,BTBT}$ below 0.1 μA/μm. Based on our study, larger than 1000× reduction is possible in GaAs, InP and Ge by scaling T_S from 10 to 3 nm.

2.3.3 *Effect of Scaling Film Thickness and V_{DD}*

Scaling T_S enhances the device performance by eliminating the SCE effect and reducing BTBT leakage. $I_{OFF,BTBT}$ in Ge, InAs, GaAs and InSb can be reduced by over 1000× by scaling Ts to 3 nm (Fig. 2.10), due to enlarged bandgap by quantization of sub-bands. As T_S scales, I_{ON} is increased by 2.5 times (Fig. 2.9), but concurrently electrons become heavier, resulting in losing the advantage of III-Vs over Si. Scaling V_{DD} is desirable since it reduces BTBT leakage (Fig. 2.10) by relieving maximum electric field applied in the device and it decreases switching energy. III-Vs and Ge with V_{DD} of 0.5 V provides with higher drive current than Si does with V_{DD} of 0.7 V (Fig. 2.9b), which means a transistor made with III-Vs and Ge can consume only 1/2 of active energy consumed by a transistor with Si. However large threshold voltages of III-Vs prevent V_{DD} from scaling below 0.5 V where III-Vs

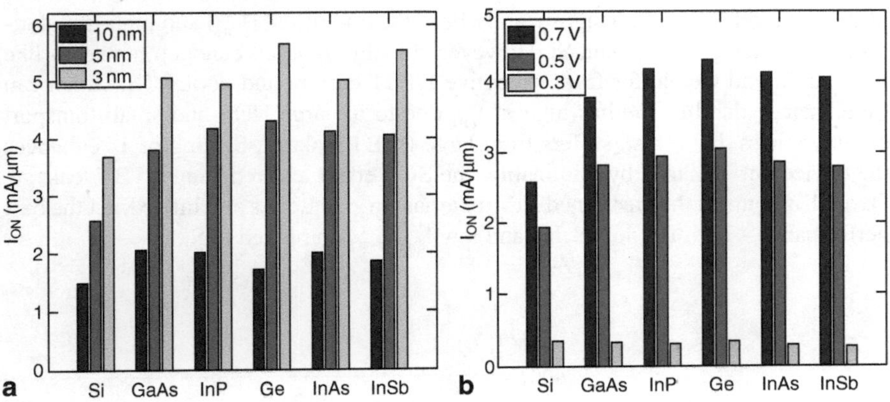

Fig. 2.9 Drive currents (I_{ON}) for various **a** Ts and **b** V_{DD}. $L_G = 15$ nm, $T_{OX} = 0.7$ nm

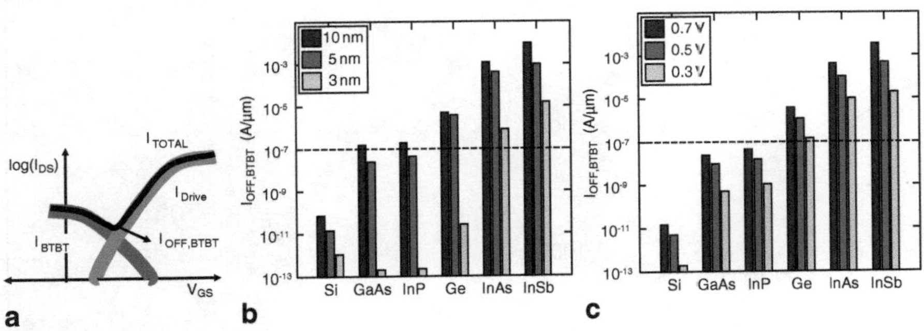

Fig. 2.10 **a** $I_{OFF,BTBT}$ is minimum achievable leakage current defined as a intersecting point between BTBT current and Drive current. **b** $I_{OFF,\,BTBT}$ for various T_S. $V_{DD} = 0.5$ V. **c** $I_{OFF,\,BTBT}$ for various V_{DD}. $T_S = 5$ nm

start to drive smaller currents than Si. InP exhibits good performance (high I_{ON}, short delay and low $I_{OFF,BTBT}$) in any scaling scenarios.

2.3.4 *Power-Performance Tradeoff of Binary III-V Materials vs. Si and Ge*

Relationships between delay and $I_{OFF,BTBT}$ for different materials are depicted in Fig. 2.11 Subthreshold leakage is fixed to be 0.1 μA/μm and V_{DD} is varied from 0.3 to 0.7 V. The best material should show short delay and low $I_{OFF,BTBT}$ at the same time. Si exhibits slowest switching and lowest $I_{OFF,BTBT}$. Although delays for InAs and InSb are small, they suffer extremely large $I_{OFF,BTBT}$. InP exhibit fast switching time, large I_{ON} at the reduced $I_{OFF,BTBT}$.

Our results show that with oxide thickness of 0.7 nm, small density of states (DOS) of these materials does not significantly limit the on-current (I_{ON}) and high-μ materials still show higher I_{ON} than Si. However, the high-μ small bandgap materials like InAs, InSb and Ge, suffer from excessive BTBT current and poor SCE, which can limit their scalability. Ge has highest I_{ON} due to its large DOS and small transport effective mass, but it also suffers from large BTBT leakage. Scaling of T_S enhances the device performance by eliminating the SCE effect and reducing BTBT leakage. Overall, among all the unstrained IIIV materials $In_{0.25}Ga_{0.75}As$ and InP exhibit the best performance—high I_{ON}, low delay and low $I_{OFF,BTBT}$ compared to Si.

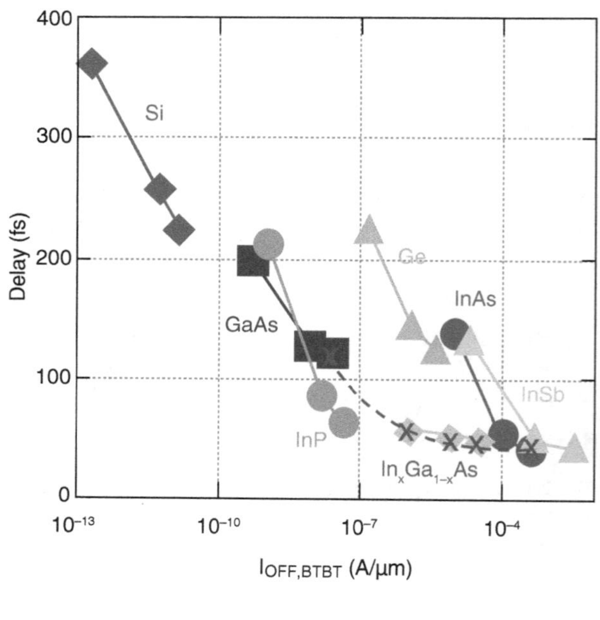

Fig. 2.11 Intrinsic delay vs. $I_{OFF,BTBT}$ trade-off in various materials. $L_g = 15$ nm, $T_{ox} = 0.7$ nm, $T_S = 5$ nm. V_{DD} is varied from 0.3 to 0.7 V

2.4 Power-Performance of Strained Ternary III-V Material ($In_xGa_{1-x}As$)

$In_xGa_{1-x}As$ is a very promising candidate for future NFETs because it allows for a very good tradeoff between the excellent transport properties of InAs and the low leakage of GaAs [27, 29, 31]. Just like in the case of Si or Ge, strain engineering can be used to further enhance the performance of III-V materials in terms of both, increasing the drive current and reducing the off-state leakage. By varying strain conditions and orientations for the different compositions, the best performing strained $In_xGa_{1-x}As$ was identified. Unlike p-MOS [35], in the case of n-MOS uniaxial strain does not significantly improve performance [36].

In this section, we will discuss power-performance tradeoffs in nanoscale DG NFETs with channel materials composed of $In_xGa_{1-x}As$ under various biaxial strain conditions with different orientations.

2.4.1 Strained InGaAs Band Structures

LEPM [8] is used to obtain the bandstructure of strained InGaAs. Figure 2.12 shows the band shifts of GaAs (001) and $In_{0.75}Ga_{0.25}As$ (111) by applying a biaxial strain.

Application of strain can strongly modify the bandgap (Eg) of material. Figure 2.13a shows Eg of the different 5 nm thin $In_xGa_{1-x}As$ as a function of biaxial (compressive/tensile) strain and orientation. On quantization, due the small mass of the electrons in the Γ-valley, the carriers start occupying the high L- and X-valleys. Hence, in III-V materials, apart from the Eg, the separation between the Γ- and L-valley ($\Delta E_{\Gamma L}$) is a

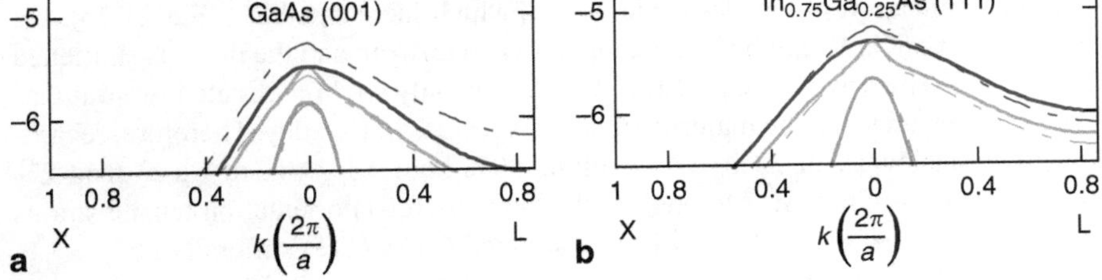

Fig. 2.12 Band shifts in biaxially strained **a** GaAs (001) and **b** $In_{0.75}Ga_{0.25}As$ (111) calculated by local empirical pseudopotential method. For GaAs (001), 4% biaxial compressive (BiC) strain is applied and for $In_{0.75}Ga_{0.25}As$ (111), 4% biaxial tensile (BiT) strain is applied. *Solid lines* represent relaxed material and *dotted lines* represent strained material

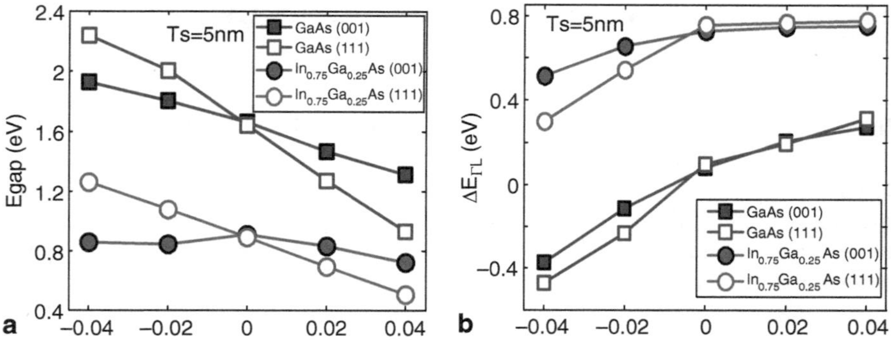

Fig. 2.13 **a** Quantized bandgap as a function of strain ($e_{||}$) in 5 nm film. Compressive strain increases bandgap. Especially (111) orientation is strongly modified **b** Γ-L Separations as a function of strain ($e_{||}$) in 5 nm film. Tensile strain widens the separation. With tensile strain, $\Delta E_{\Gamma L}$ in GaAs can be over 0.3 (eV)

very important parameter in determining the transport. Figure 2.13b shows $\Delta E_{\Gamma L}$ under different strain/orientation conditions. Due to strong non-parabolicity in the Γ-valley, transport effective mass is bigger in thin layers than in bulk material. Tunnel mass (m_{Tunnel}) determines how well electron and hole penetrate into bandgap and cause BTBT leakage. Tunnel mass is also strongly modified with application of strain [36].

2.4.2 I_{ON} and $I_{OFF,BTBT}$ with Strain Engineering and Channel Orientation

Device simulation results for $L_G = 15\,nm$, $T_{OX} = 0.7\,nm$, $T_S = 5\,nm$, $V_{DD} = 0.7\,V$ ($I_{OFF} = 10^{-7}$ A/um) are shown in Fig. 2.14a (I_{ON}) and Fig. 2.14b (I_{OFF}). In relaxed GaAs, $\Delta E_{\Gamma L}$ is not large enough to prevent the population of electrons in L-valleys. In spite of the occupation in L-valleys providing large channel charge, relaxed GaAs suffers from low injection velocity (Fig. 2.8), which results in overall low drive current. This implies that with 0.7 nm-thick-gate oxide, having small mass and thus, fast injection velocity is more effective for larger I_{ON} than having large mass and large DOS. Indeed, with the application of tensile-biaxial strain, which increases $\Delta E_{\Gamma L}$ (Fig. 2.13b), the drive current of GaAs can be boosted up to over 4 mA/μm and the delay is shortened to 55fs [36]. The drive current of $In_{0.75}Ga_{0.25}As$ is only slightly affected by strain, as large $\Delta E_{\Gamma L}$ ensures that the majority of carriers resides in Γ-valley. Therefore, regardless of strain levels, $In_{0.75}Ga_{0.25}As$ exhibits 30% higher I_{ON} (4.3 mA/μm) than 2% biaxial tensile strained Si. Although both (111) and (001) orientation tensile strains enhance the I_{ON} of GaAs, biaxial-tensile strained GaAs (111) suffers large $I_{OFF,BTBT}$ due to reduction in bandgap and decrease in tunnel mass. Relaxed $In_{0.75}Ga_{0.25}As$ have small bandgap, leading to too high $I_{OFF,BTBT}$ (Fig. 2.14b). (111) compressive strain effectively reduces leakage current in $In_{0.75}Ga_{0.25}As$ by concurrently increasing bandgap and tunneling mass. Due to the enlarged bandgap and tunneling mass, the leakage current for 4% biaxially compressive-strained $In_{0.75}Ga_{0.25}As$ (111) is below

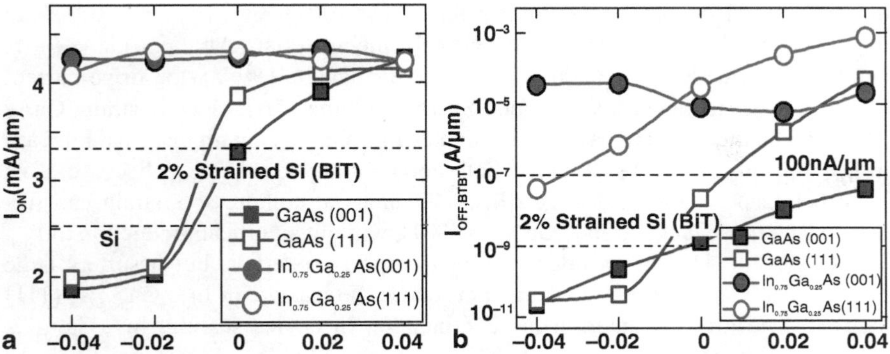

Fig. 2.14 **a** I_{ON} as a function of strain ($e_{||}$) in strained GaAs and $In_{0.75}Ga_{0.25}As$ NMOS DGFET. For comparison, I_{ON}s for 2% biaxially tensile-strained (BiT) Si and unstrained Si are plotted as *dotted lines*. **b** $I_{OFF, BTBT}$ as a function of strain in strained GaAs and $In_{0.75}Ga_{0.25}As$. For comparison, $I_{OFF, BTBT}$ for 2% Biaxially tensile-strained Si is drawn as *dotted line*. 4% (111) compressive strain reduces $I_{OFF, BTBT}$ by over 1000×

the leakage spec. of 0.1 μA/μm. However, (001) compressive strain does not notably reduce the BTBT leakage current, since despite the increased bandgap, it simultaneously reduces the tunneling mass, leading to no significant reduction in BTBT.

2.4.3 *Power-Performance of Strained Ternary III-V Material (InGaAs)*

$In_xGa_{1-x}As$ is a very promising candidate for future NFETs. Figure 2.15 shows the best channel materials and how these materials can be improved by strain engi-

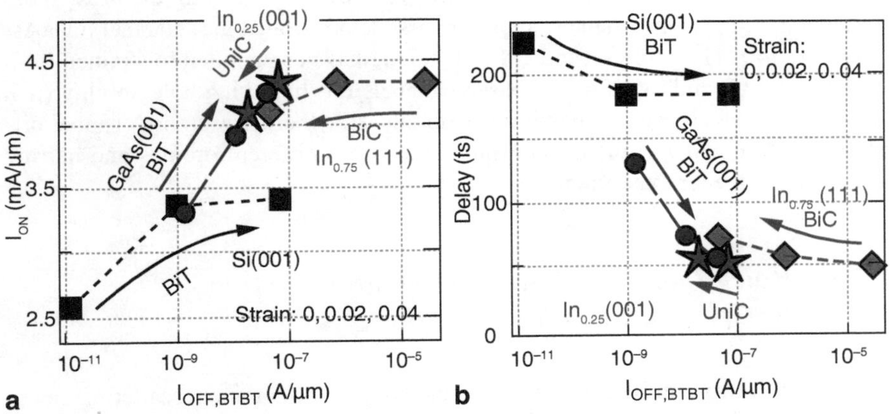

Fig. 2.15 **a** I_{ON} vs. $I_{OFF,BTBT}$ and **b** Delay vs. $I_{OFF,BTBT}$ of the best performing n-DGFETs with materials, biaxially tensile-strained GaAs (001), uniaxially compressive-strained $In_{0.25}Ga_{0.75}As$ (001) and biaxially compressive-strained $In_{0.75}Ga_{0.25}As$ (111). Strain levels are 0, 0.02 and 0.04. Values for biaxially tensile-strained Si are given for comparison

neering. Based on our simulation results, GaAs (001), $In_{0.25}Ga_{0.75}As$ (001) and $In_{0.75}Ga_{0.25}As$ (111) are selected as the best channel materials. Although GaAs (001) exhibits the lowest $I_{OFF,BTBT}$ due to its large bandgap (>1.4 eV), the drive current of GaAs (001) is higher than strained Si. By applying biaxial tensile-strain, GaAs (001) can be improved to have I_{ON} as high as $In_{0.25}Ga_{0.75}As$, with marginal increase in $I_{OFF,BTBT}$. $In_{0.25}Ga_{0.75}As$ (001) has both good I_{ON} and low $I_{OFF,BTBT}$ because of its large bandgap (>1 eV) and large $\Delta E_{\Gamma\text{-}L}$. 4% uniaxial compressive strain can further reduce the leakage in $In_{0.25}Ga_{0.75}As$ (001), without significant reduction in I_{ON}. $In_{0.75}Ga_{0.25}As$ (111) has excellent carrier transport properties, but it suffers large $I_{OFF,BTBT}$. Biaxial compressive strain can reduce the leakage in $In_{0.75}Ga_{0.25}As$ (111) below 0.1 μA/μm. Considering the continuation in device scaling, $In_{0.75}Ga_{0.25}As$ (111) would be the better material choice over other $In_xGa_{1-x}As$ compositions. As the gate length scales down close to 10 nm, further scaling in channel thickness would be necessary to control the short channel effects. Larger quantization effect in thinner body will further increase the bandgap of $In_{0.75}Ga_{0.25}As$ (111), leading to even smaller leakage current [36]. By contrast, the scalability of devices with the channel materials of GaAs (001) and $In_{0.25}Ga_{0.75}As$ (111) would be limited, since the strong quantization effect in Γ-valley will diminish I_{ON} due to reduced $\Delta E_{\Gamma\text{-}L}$.

2.5 Strained III-V for p-MOSFETs

As seen in the previous sections, III-V compounds are one of the most promising materials for future high-speed, low-power n-MOSFETs due to their high electron mobility. Recently, high performance III-V n-FETs have been demonstrated. However, for CMOS logic, there is a significant challenge of identifying high mobility III-V p-FET candidates [37–39]. Strain in silicon has been successful in significantly enhancing the p-MOS performance and is now employed ubiquitously in the industry. Use of strain to reduce hole effective masses by splitting the heavy-hole (hh) and light-hole (lh) valence bands was first demonstrated in p-channel InGaAs/(Al)GaAs [40, 41]. More recently, the technique has been applied to strained InSb [42], GaSb [43] and InGaSb [44] based channels for improving hole nobility (μ_h). Given the many choices available for materials, stoichiometry, strain, channel orientation, a modeling effort is necessary to evaluate different options and narrow down the choice for experimentation.

2.5.1 Hole Mobility in Ternary III-V Materials (InGaAs vs. InGaSb)

Unlike for electrons in III-V where polar scattering is the dominant scattering mechanism (non polar optical phonons do not interact due to s-like spherical symmetry of conduction band in III-V's [12]) both deformation potential and polar scattering mechanisms are important for hole mobility. Varying group III element while keep-

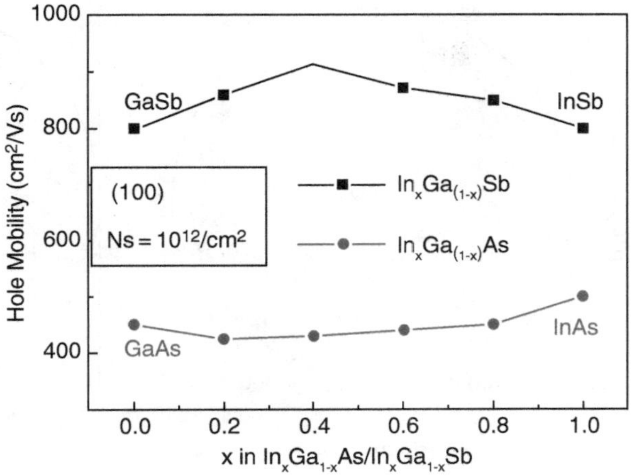

Fig. 2.16 Mobility for $In_xGa_{1-x}Sb$ and $In_x\ Ga_{1-x}As$. Sb's have higher (~2X) hole mobilities than As's [37]

ing the group V element same (i.e., InAs → GaAs) μ_h does not change appreciably while a large change is observed when the group V element is varied (i.e., InP → InAs → InSb). This is a consequence of the fact that the valence bands of III-V mostly derive from the p-orbitals of the anion [45]. Figure 2.16 plots the low field μ_h of $In_xGa_{1-x}As$ and $In_xGa_{1-x}Sb$ (Sheet Charge (N_S) = 10^{12} cm^{-2}). The antimonides have significantly higher mobilities than the arsenides.

2.5.2 *Hole Mobility Enhancement in III-Vs with Strain*

Next we look at the effect of biaxial and uniaxial strain on μ_h enhancement in these III-V materials. Figure 2.17 shows plots of the μ_h enhancement with respect to the unstrained case (for uniaxial case the strain is always applied along the channel direction). In all cases compression is better than tension for μ_h enhancement. In the case of biaxial strain, the enhancement in (100) is isotropic and maximum enhancement with 2% biaxial strain is ~2X. Much higher μ_h enhancement is achieved with same% of uniaxial strain (~4X with 2% uniaxial compression—best case Fig. 2.17). Also, in the case of uniaxial strain, the enhancement is highly anisotropic. Also note that ~2 times lower stress is required to produce the same amount of strain in InSb/GaSb as compared to Si, due to their lower elasticity constants. Figure 2.18 shows that a hole mobility greater than 1600 cm²/V s can be obtained by application of biaxial strain in InGaSb (achievable through lattice mismatch between channel and barrier layers during MBE growth).

In summary, our comprehensive analysis of the hole mobility enhancement in ternary III-V materials shows that Sb's are significantly better than As's. Further, application of strain significantly enhances the mobility in (In/Ga/In_XGa_{1-X})Sb's. A hole mobility greater than 1600 cm²/V s can be obtained by application of biaxial strain in InGaSb. Further enhancement should be possible using uniaxial strain. Strained Sb's are extremely promising candidates for nanoscale III-V p-MOSFETs.

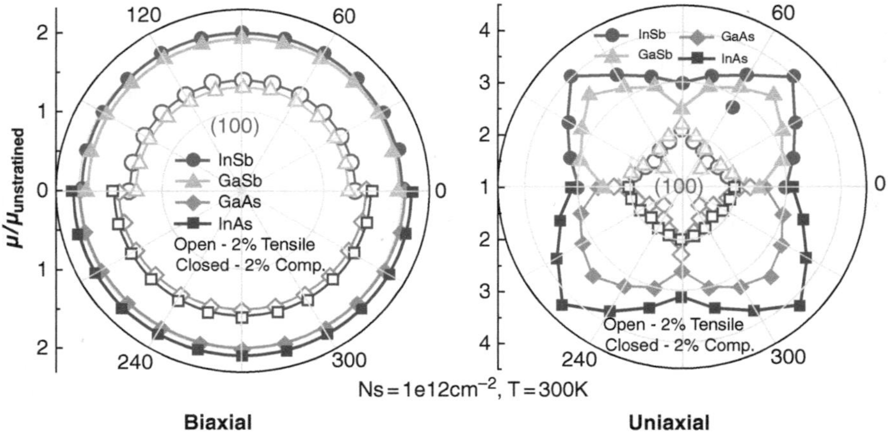

Fig. 2.17 Enhancement of mobility with respect to the unstrained case with biaxial (*left*) and uniaixal (*right*) strain. Enhancement with compression is always better than tension. Upto ~4X/~2X enhancement possible with 2% uniaxial/biaxial compression. Enhancement with strain (especially for uniaxial) depends highly on the choice of channel direction [37]

Fig. 2.18 μ_h enhancement with biaxial strain; achievable by lattice mismatch between channel and barrier layers during MBE growth

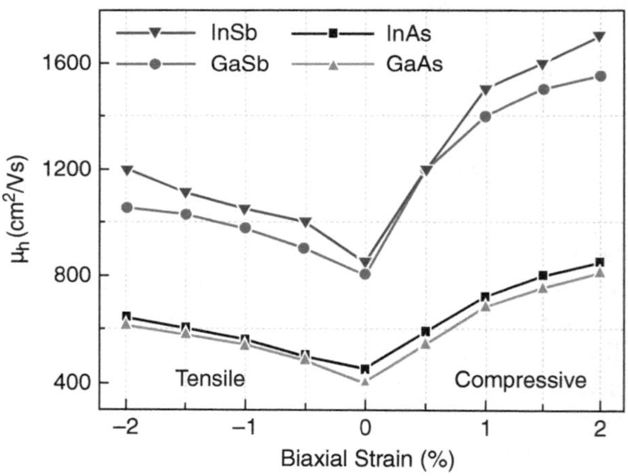

2.6 Novel Device Structure and Parasitics

2.6.1 Quantum Well (QW) Strained Heterostructure III-V FETs

Strain in general results in reduction in the E_G and hence enhanced off-state leakage (I_{BTBT}). Confinement on the other hand results in increased E_G as shown in Fig. 2.5, and hence reduces I_{BTBT}. The heterostructure QW-FET of Fig. 2.19 proposes a unique

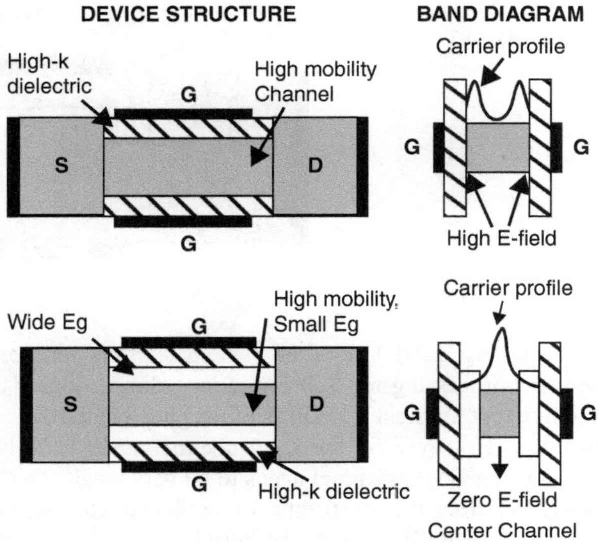

Fig. 2.19 Conventional and QW double gate FET structures

and novel device structure to combine strain and quantum mechanical confinement to obtain desired transport properties with reduced off-state leakage [8, 10, 46, 47]. In these structures the transport can be confined to the center of the channel in a high mobility material flanked by a high E_G material. The capping layer helps in providing a very good high-k gate dielectric interface and reduced mobility degradation due to interface states. The mobility is further enhanced due to strain, reduced electric field in the center of the double gate structure due to symmetry and the channel being away from the dielectric interface. The bandgap of the center channel can be increased due confinement by keeping it very thin. However, QW FETs are not completely free from their tradeoffs. QW FETs are found to have dramatic mobility degradation at QW thickness of less than ~4 nm due to strong quantum confinement effects [10, 48, 49]. To take complete advantage of high mobility III-V materials, heterostructure quantum-well FETs, which can simultaneously achieve high drive currents and low off-state leakage should be investigated, similar to the case of Ge [10, 46].

2.6.2 *Parasitic Resistance*

Parasitic resistance in the S/D regions of the conventional MOSFET has been identified as one of the primary problems of the non-scaling of drive currents in transistor scaling. This problem can be significantly worse in III-V n-MOSFETs due to their low n-type dopant concentration. Replacing the S/D regions of transistors with metals has been suggested as one of the techniques, which might help address this

Fig. 2.20 Metal source-drain III-V n-MOSFET

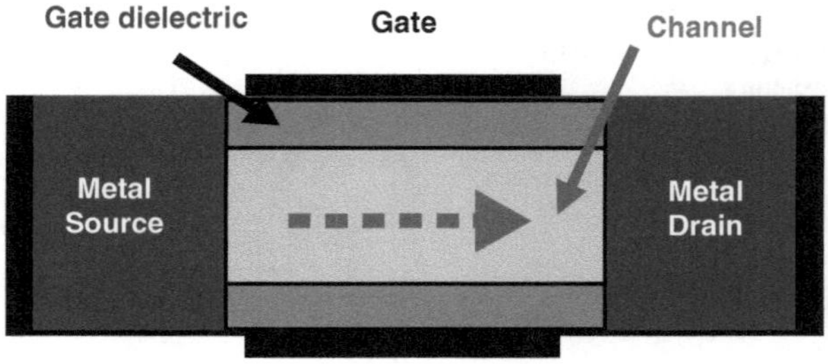

problem (Fig. 2.20). Use of the metal S/D offers additional advantages of low-temperature processing for S/D formation, elimination of the parasitic bipolar action and inherent physical scalability of the gate lengths due to the abrupt silicide-silicon interface. However, for the I_{ON} of the metal S/D to be better than diffused S/D the Schottky barrier to channel needs to be very small. The free-electron wavefunction penetrates from the metal into the semiconductor, which generates metal-induced gap-states (MIGS) and pins the Fermi level near the charge neutrality level (E_{CNL}) [50]. In the case of Ge we have found that the Fermi level at metal-Ge Schottky barriers is pinned near the valence band of Ge for a variety of metals, including, Ni, Co and Ti [51, 52]. In the case of III-Vs such as InAs, the Fermi level pins very close to conduction band [53]. This could provide a very small barrier to the electrons in the channel in a III-V n-MOSFETs. Due to the small barrier height to electrons coupled with the high inversion electron mobility, Schottky S/D IIIV n-MOS transistors are very attractive future device candidates.

2.6.3 Parasitic Capacitance

It has been shown that scaling MOSFET below 32 nm may increase parasitic gate capacitance and it may significantly limit the performance [54, 55]. Parasitic gate capacitance can undermine the advantage in delay of III-V materials over silicon. Figure 2.21 shows how much the delays worsen by adding parasitic gate capacitance. In this analysis, transit time is time required for electron to pass through the channel and defined as, Lg/v_{inj}. Gate delay is $\Delta Q/I_{eff}$. To take into account parasitic effect, we assumed parasitic gate capacitance to be half of gate capacitance of Si and added the parasitic gate capacitance to gate capacitance of all the materials as:

$$\text{Delay} = (\Delta Q_{\text{Channel}} + \Delta Q_{\text{Parasitic}})/I_{\text{eff}},$$

where $\Delta Q_{Parasitic} = 1/2\ \Delta Q_{Si}$ and $I_{eff} = 1/2 \times [I_D(V_{DD}, V_{DD}/2) + I_D(V_{DD}/2, V_{DD})]$; $I_D(V_G, V_D)$ [56].

Although parasitic gate capacitance reduces the gap between high mobility materials and Si, III-V materials and Ge still outperform Si in delay.

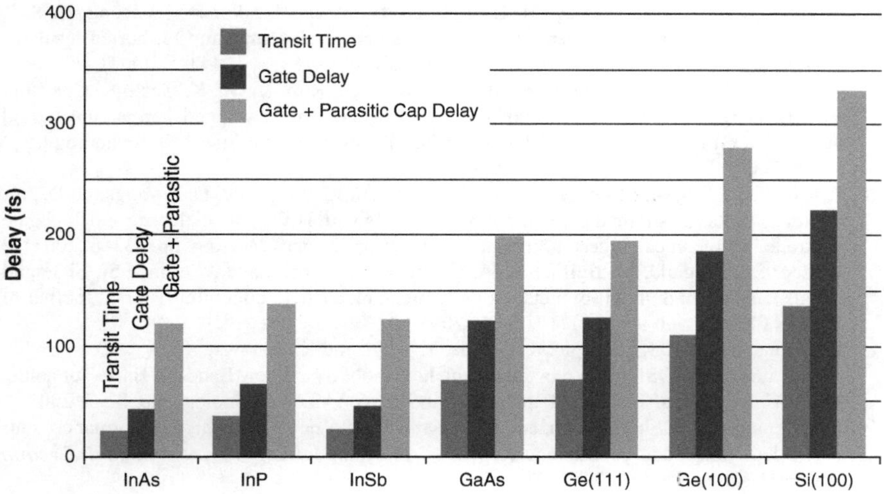

Fig. 2.21 Delay in consideration of parasitic gate capacitance. $L_g = 15\,nm$, $T_{ox} = 0.7\,nm$, $T_S = 5\,nm$. $V_{DD} = 0.7\,V$

2.7 Conclusion

MOSFETs utilizing high mobility III-V channel materials can take us to the sub-20 nm regime. Ternary III-V materials, such as strained $In_xGa_{1-x}As$ could be suitable for n-MOSFETs and strained InGaSb for p-MOSFETs. However, to take full advantage of high mobility/small bandgap channel materials, novel device structures, such as heterostructure quantum well (QW) FETs along with strain engineering will be needed, in order to achieve high drive currents while maintaining low off-state leakage. For III-V materials to become mainstream, important challenges, such as good surface passivation, low parasitic resistance/capacitance, and heterogeneous integration on Si platform must be overcome.

References

1. R. Chau, "Benchmarking nanotechnology for high performance and low-power logic transistor applications," *IEEE Transactions on Nanotechnology*, vol. 4, pp. 153–8, 2005.
2. S. Datta, G. Dewey, M. Doczy, B. Doyle, B. Jin, J. Kavalieros, R. Kotlyar, M. Metz, N. Zelick, and R. Chau, "High mobility Si/SiGe strained channel MOS transistors with HfO2/TiN gate stack" in *Technical Digest—International Electron Devices Meeting*, pp. 653–6, 2003.
3. T. Tezuka, N. Sugiyama, T. Mizuno, and S. Takagi, "Ultrathin body SiGe-on-insulator pMOSFETs with high-mobility SiGe surface channels," *IEEE Transactions on Electron Devices*, vol. 50, pp. 1328–33, 2003.

4. H. Shang, K.-L. Lee, P. Kozlowski, C. D. Emic, I. Babich, E. Sikorski, M. Ieong, H.-S. P. Wong, K. Guarini, and W. Haensch, "Epitaxial silicon and germanium on buried insulator heterostructures and devices," Applied Physics Letters, vol. 83, pp. 5443–5, 2003.
5. H. Shang, J. O. Cho, X. Wang, P. M. Mooney, K. Lee, J. Rim, K. Ott, K. Chan, K. Guarinin, and M. Ieong, "Channel design and mobility enhancement in strained germanium buried channel MOSFETs" in Digest of Technical Papers—Symposium on VLSI Technology, pp. 204–5, 2004.
6. A. Ritenour, S. Yu, M. L. Lee, Z. Lu, W. Bai, A. Pitera, E. Fitzgerald, D. Kwong, and D. Antoniadis, "Epitaxial strained germanium p-MOSFETs with HfO2 gate dielectric and TaN gate electrode" in Technical Digest—International Electron Devices Meeting, pp. 433–6, 2003.
7. M. Lee, E. Fitzgerald, M. Bulsara, C. A. Currie, and A. Lochtefeld, "Strained Si, SiGe and Ge channels for high mobility metal oxide semiconductor field effect transistors," Journal of Applied Physics, vol. 97, pp. 11101-1-27, 2005.
8. T. Krishnamohan, Z. Krivokapic, K. Uchida, Y. Nishi, and K. Saraswat, "Low defect ultrathin fully strained Ge MOSFET on relaxed Si with high mobility and low Band-To-Band-Tunneling (BTBT)" in Digest of Technical Papers—Symposium on VLSI Technology, pp. 82–3, 2005.
9. T. Krishnamohan, C. Jungemann, and K. C. Saraswat, "A novel, very high performance, sub-20 nm depletion-mode double-gate (DMDG)" in *Technical Digest—International Electron Devices Meeting*, pp. 687–90, 2003.
10. T. Krishnamohan, Z. Krivokapic, K. Uchida, Y. Nishi, and K. C. Saraswat, "High-mobility ultrathin strained Ge MOSFETs on bulk and SOI with low band-to-band tunneling leakage: Experiments," *IEEE Transactions on Electron Devices*, vol. 53, pp. 990–9, 2006.
11. S. Datta, T. Ashley, J. Brask, L. Buckle, M. Doczy, M. Emeny, D. Hayes, K. Hilton, R. Jefferies, T. Martin, T. J. Phillips, D. Wallis, P. Wilding, and R. Chau, "85 nm gate length enhancement and depletion mode InSb quantum well transistors for ultra high speed and very low power digital logic applications" in *Technical Digest—International Electron Devices Meeting*, pp. 763–6, 2005.
12. D.-H. Kim, J. D. Alamo, J.-H. Lee, and K.-S. Seo, "Logic suitability of 50-nm In0.7Ga0.3AsHEMTs for Beyond-CMOS applications," *IEEE Transactions on Electron Devices*, vol. 54, pp. 2606–13, 2007.
13. M. L. Cohen and T. K. Bergstresser, "Band structures and pseudopotential form factors for fourteen semiconductors of the diamond and zinc-blende structures," *Physical Review*, vol. 141, pp. 789, 1966.
14. J. R. Chelikowsky and M. L. Cohen, "Nonlocal pseudopotential calculations for the electronic structure of eleven diamond and zinc-blende semiconductors," *Physical Review B*, vol. 14, pp. 556, 1976.
15. J. P. Walter and M. L. Cohen, "Calculated and measured reflectivity of ZnTe and ZnSe," *Physical Review B*, vol. 1, pp. 2661, 1970.
16. G. Weisz, "Band structure and fermi surface of white tin," *Physical Review*, vol. 149, pp. 504, 1966.
17. L.-W. Wang, J. Kim, and A. Zunger, "Electronic structures of [110]-faceted self-assembled pyramidal InAs/GaAs quantum dots," *Physical Review B*, vol. 59, pp. 5679, 1999.
18. O. H. Nielsen and R. M. Martin, "Stresses in semiconductors: Ab initio calculations on Si, Ge, and GaAs," *Physical Review B*, vol. 32, pp. 3792, 1985.
19. E. O. Kane, "Zener tunneling in semiconductors," *Journal of Physics and Chemistry of Solids*, vol. 12, pp. 181–8, 1959.
20. D. Kim, T. Krishnamohan, Y. Nishi, and K. C. Saraswat, "Band to band tunneling limited off state current in ultra-thin body double gate FETs with high mobility materials: III-V, Ge and strained Si/Ge," presented at *IEEE SISPAD*, Monterey, CA, 2006.
21. D. Kim, "Theoretical performance evaluations of NMOS double gate FETs with high mobility materials: Strained III-V, Ge and Si," in *Electrical Engineering*, Ph. D. Dissertation (to be published). Stanford, Stanford University, 2009.
22. A. Rahman, M. S. Lundstrom, and A. W. Ghosh, "Generalized effective-mass approach for n-type metal-oxide-semiconductor field-effect transistors on arbitrarily oriented wafers," *Journal of Applied Physics*, vol. 97, pp. 1–12, 2005.

23. G. Bastard, "Theoretical investigations of superlattice band structure in the envelope-function approximation," *Physical Review B*, vol. 25, pp. 7584, 1982.
24. R. Landuer, "Electrical resistance of disordered one-dimensional Lattices," *Philosophical Magazine*, vol. 21, pp. 863, 1970.
25. M. V. Fischetti and S. E. Laux, "Monte Carlo simulation of transport in technologically significant semiconductors of the diamond and zinc-blende structures–II: Submicrometer MOSFET's," *IEEE Transactions on Electron Devices*, vol. 38, pp. 650–60, 1991.
26. A. Asenov, K. Kalna, I. Thayne, and R. J. W. Hill, "Simulation of implant free III-V MOSFETs for high performance low power Nano-CMOS applications," *Microelectronic Engineering*, vol. 84, pp. 2398–403, 2007.
27. M. D. Michielis, D. Esseni, and F. Driussi, "Analytical models for the insight into the use of alternative channel materials in ballistic nano-MOSFETs," *IEEE Transactions on Electron Devices*, vol. 54, pp. 115–23, 2007.
28. S. E. Laux, "A simulation study of the switching times of 22- and 17-nm gate-length SOI nFETs on high mobility substrates and Si," *IEEE Transactions on Electron Devices*, vol. 54, pp. 2304–20, 2007.
29. A. Pethe, T. Krishnamohan, D. Kim, S. Oh, and H.-S. P. Wong, "Investigation of the performance limits of III-V double-gate n-MOSFETs" in *Technical Digest—International Electron Devices Meeting*, pp. 605–8, 2005.
30. D. Kim, T. Krishnamohan, L. Smith, H.-S. P. Wong, and K. C. Saraswat, "Band to band tunneling study in high mobility materials: III-V, Si, Ge and strained SiGe," in *2007 65th DRC Device Research Conference.* South Bend, IN, pp. 57–8, 2007.
31. D. Kim, T. Krishnamohan, and K. C. Saraswat, "Performance evaluation of III-V double-gate n-MOSFETs" presented at *2008 Annual Device Research Conference (DRC)*, Santa Barbara, CA, 2008.
32. D. Kim, T. Krishnamohan, and K. C. Saraswat, "Performance evaluation of 15 nm gate length double-gate n-MOSFETs with high mobility channels: IIIV, Ge and Si," *ECS Transactions*, vol. 16, pp. 47–55, 2008.
33. T. Krishnamohan and K. C. Saraswat, "High mobility Ge and III-V materials and novel device structures for high performance nanoscale MOSFETS" in *ESSDERC 2008*, Edinburgh, UK, pp. 38–46, 2008.
34. M. V. Fischetti, T. P. O'Regan, S. Narayanan, C. Sachs, S. Jin, J. Kim, and Y. Zhang, "Theoretical study of some physical aspects of electronic transport in nMOSFETs at the 10-nm gate-length," *IEEE Transactions on Electron Devices*, vol. 54, pp. 2116–36, 2007.
35. T. Krishnamohan, D. Kim, T. V. Dinh, A.-T. Pham, B. Meinerzhagen, C. Jungemann, and K. C. Saraswat, "Comparison of (001), (110) and (111) uniaxial- and biaxial-strained-Ge and strained-Si PMOS DGFETs for all channel orientations: Mobility enhancement, drive current, delay and off-state leakage" presented at *International Electron Devices Meeting (IEDM)*, San Francisco, CA, 2008.
36. D. Kim, T. Krishnamohan, and K. C. Saraswat, "Performance evaluation of uniaxial- and biaxial-strained In(x)Ga(1−x)As NMOS DGFETs" in *International Conference on Simulation of Semiconductor Processes and Devices (SISPAD)*, pp. 101–4, 2008.
37. A. Nainani, D. Kim, T. Krishnamohan, and K. C. Saraswat, "Hole mobility and its enhancement with strain for technologically relevant III-V semiconductors" presented at *2009 International Conference on Simulation of Semiconductor Processes and Devices (SISPAD'09)*, San Diego, CA, 2009.
38. P. P. Ruden, M. Shur, D. K. Arch, R. R. Daniels, D. E. Grider, and T. E. Nohava, "Quantum-well p-channel AlGaAs/InGaAs/GaAs heterostructure insulated-gate field-effect transistors," *IEEE Transactions on Electron Devices*, vol. 36, pp. 2371–9, 1989.
39. A. Nainani, S. Raghunathan, D. Witte, M. Kobayashi, T. Irisawa, T. Krishnamohan, K. C. Saraswat, B. R. Bennett, M. Ancona, and J. B. Boos, "Engineering of strained III-V heterostructures for high hole mobility" presented at *2009 IEEE International Electron Devices Meeting*, Baltimore, MD, 2009.
40. G. C. Osbourn, "Electron and hole effective masses for two-dimensional transport in strained-layer superlattices," *Superlattices and Microstructures*, vol. 1, pp. 223–6, 1985.

41. M. Jaffe, J. E. Oh, J. Pamulapati, J. Singh, and P. Bhattacharya, "In-plane hole effective masses in InxGa1-xAs/Al0.15Ga0.85As modulation-doped heterostructures," *Applied Physics Letters*, vol. 54, pp. 2345–6, 1989.
42. M. Radosavljevic, T. Ashley, A. Andreev, S. D. Coomber, G. Dewey, M. T. Emeny, M. Fearn, D. G. Hayes, K. P. Hilton, M. K. Hudait, R. Jefferies, T. Martin, R. Pillarisetty, W. Rachmady, T. Rakshit, S. J. Smith, M. J. Uren, D. J. Wallis, P. J. Wilding, and R. Chau, "High-performance 40 nm gate length InSb p-channel compressively strained quantum well field effect transistors for low-power (VCC = 0.5 V) logic applications" in *Technical Digest—2008 IEEE International Electron Devices (IEDM)*, pp. 727–30, 2008.
43. B. R. Bennett, M. G. Ancona, J. B. Boos, C. B. Canedy, and S. A. Khan, "Strained GaSb/AlAsSb quantum wells for p-channel field-effect transistors," *Journal of Crystal Growth*, vol. 311, pp. 47–53, 2008.
44. B. R. Bennett, M. G. Ancona, J. B. Boos, and B. V. Shanabrook, "Mobility enhancement in strained p-InGaSb quantum wells," *Applied Physics Letters*, vol. 91, pp. 042104, 2007.
45. G. Bastard, *Wave Mechanics applied to semiconductor heterostructures*. New York, Halsted Press, Wiley, 1988.
46. T. Krishnamohan, D. Kim, C. Nguyen, C. Jungemann, Y. Nishi, and K. C. Saraswat, "High-mobility low band-to-band-tunneling strained-germanium double-gate heterostructure FETs: Simulations," *IEEE Transactions on Electron Devices*, vol. 53, pp. 1000–9, 2006.
47. T. Krishnamohan, D. Kim, C. Jungemann, Y. Nishi, and K. C. Saraswat, "Strained-Si, relaxed-Ge or strained-(Si)Ge for future nanoscale p-MOSFETs?" in *Digest of Technical Papers—2006 Symposium on VLSI Technology*, pp. 144–5, 2006.
48. F. A. T. M and A. V. K., "Quantum engineering of nanoelectronic devices: The role of quantum confinement on mobility degradation," *Microelectronics Journal*, vol. 32, pp. 679–86, 2001.
49. K. Uchida and S.-I. Takagi, "Carrier scattering induced by thickness fluctuation of silicon-on-insulator film in ultrathin-body metal–oxide–semiconductor field-effect transistors," *Applied Physics Letters*, vol. 82, pp. 2916, 2003.
50. J. Tersoff, "Schottky barriers and the continuum of gap states," *Physical Review Letters*, vol. 52, pp. 465–8, 1984.
51. A. Pethe and K. C. Saraswat, "High-mobility, low parasitic resistance Si/Ge/Si heterostructure channel Schottky source/drain PMOSFETs" in *65th Device Research Conference*, 2007, pp. 55–6.
52. A. Dimoulas, P. Tsipas, A. Sotiropoulos, and E. K. Evangelou, "Fermi-level pinning and charge neutrality level in germanium," *Applied Physics Letters*, vol. 89, pp. 252110-1-3, 2006.
53. H. H. Wieder, "Surface and interface barriers of InxGa1-xAs binary and ternary alloys," *Journal of Vacuum Science and Technology B*, vol. 21, pp. 1915–9, 2003.
54. A. Khakifirooz and D. A. Antoniadis, "MOSFET performance scaling. Part I. Historical trends," *IEEE Transactions on Electron Devices*, vol. 55, pp. 1391–400, 2008.
55. A. Khakifirooz and D. A. Antoniadis, "MOSFET performance scaling-Part II: Future directions," *IEEE Transactions on Electron Devices*, vol. 55, pp. 1401–8, 2008.
56. D. A. Antoniadis, I. Aberg, C. Ni Chleirigh, O. M. Nayfeh, A. Khakifirooz, and J. L. Hoyt, "Continuous MOSFET performance increase with device scaling: The role of strain and channel material innovations," *IBM Journal of Research and Development*, vol. 50, pp. 363–76, 2006.

Chapter 3
Device Physics and Performance Potential of III-V Field-Effect Transistors

Yang Liu, Himadri S. Pal, Mark S. Lundstrom, Dae-Hyun Kim, Jesús A. del Alamo and Dimitri A. Antoniadis

Abstract The device physics and technology issues for III-V transistors are examined from a simulation perspective. To examine device physics, an InGaAs HEMT structure similar to those being explored experimentally is analyzed. The physics of this device is explored using detailed, quantum mechanical simulations based on the non-equilibrium Green's function formalism. In this chapter, we: (1) elucidate the essential physics of III-V HEMTs, (2) identify key technology challenges that need to be addressed, and (3) estimate the expected performance advantage for III-V transistors.

3.1 Introduction

Driven by tremendous advances in lithography, the semiconductor industry has followed Moore's law by shrinking transistor dimensions continuously for the last 40 years. The big challenge going forward is that continued scaling of planar, silicon, CMOS transistors will be more and more difficult because of both fundamental limitations and practical considerations as the transistor dimensions approach ten nanometers. The issues at small gate lengths are many fold. First, transistor scaling increases the number of gates on a chip and the operating frequency. To prevent the chip from overheating, the power dissipation should be limited, which requires lowering the power supply voltage while maintaining the ability to deliver high on-currents for each new generation of technology. Secondly, the drain bias decreases the energy barrier height between the source and channel in a transistor due to 2D electrostatics. Degraded short channel effects become more significant as the gate length gets shorter, and the increased off-state leakage has pushed the stand-by power to its practical limit. Thirdly, the accompanying scaled oxide thickness provides better gate control of the channel potential, but this inevitably increases

Y. Liu (✉)
Department of Electrical and Computer Engineering, Purdue University,
West Lafayette, Indiana, USA
e-mail: liuy@purdue.edu

S. Oktyabrsky, P. D. Ye (eds.), *Fundamentals of III-V Semiconductor MOSFETs*,
DOI 10.1007/978-1-4419-1547-4_3,

the gate leakage and makes it very difficult to obtain both high on-currents and low off-currents at lowered supply voltage. Lastly, the parasitic resistance and capacitance have become comparable to, or even larger than the continuously decreasing intrinsic channel capacitance and resistance, which may provide a practical limit to scaling [1]. A 45 nm process based on high-k, metal gate, and strained silicon was introduced in 2007 [2]. With such technologies, scaling will continue to the 32 nm node and beyond [3]. Conventional silicon-based CMOS scaling will, however, become very difficult at the 15 nm node and beyond. Further improvements in transistor speed and performance may have to come from new channel materials.

To address the scaling challenge, both industry and academia have been investigating alternative device architectures and materials, among which III-V compound semiconductor transistors stand out as promising candidates for future logic applications because their light effective masses lead to high electron mobilities and high on-currents, which should translate into high device performance at low supply voltage. Recent innovations on III-V transistors include sub-100 nm gate-length, high performance InGaAs buried channel [4, 5] and surface channel MOSFETs [6], sub-80 nm E-mode InGaAs/InAs HEMTs [7–10], and InSb p-channel HEMTs [11] with outstanding logic performance at short channel lengths and low supply voltages. At the same time, theoretical work has predicted the performance of III-V transistors with respect to Si at near future technology nodes with the focus on device design, bandstructure effects, source engineering, etc [12–21].

In this chapter, we will examine device physics issues of III-V transistors from a simulation perspective by addressing a very specific question: how would an In-rich, InGaAs MOSFET operate if the technological challenges identified in other chapters of this volume are solved. To examine device physics, we will use an InGaAs HEMT structure similar to that being used by the MIT and Intel groups [10, 22]. We will begin by examining the device physics using detailed quantum mechanical simulations based on the non-equilibrium Green's function formalism [23]. Our objectives are threefold: (1) to elucidate the physics of III-V HEMTs, (2) to identify key technology challenges that need to be addressed, and (3) to determine what the performance advantage (if any) for III-V transistors would be.

3.2 InGaAs HEMTs

3.2.1 Device Structure

The HEMT structure for logic applications studied in this chapter is shown in Fig. 3.1a [22]. The high-mobility channel consists of $In_{0.53}Ga_{0.47}As/In_{0.7}Ga_{0.3}As/In_{0.53}Ga_{0.47}As$ quantum well of 2 nm/8 nm/3 nm in thickness and is sandwiched between two $In_{0.52}Al_{0.48}As$ barrier layers on the top and bottom with thickness of 4 nm and 500 nm respectively. The gate length of these devices ranges from 40 to 130 nm.

A silicon δ-doped layer of $5 \times 10^{12}\ cm^{-2}$ is placed 3 nm away from the channel in the upper barrier layer to provide carriers for the source and drain. In real devices,

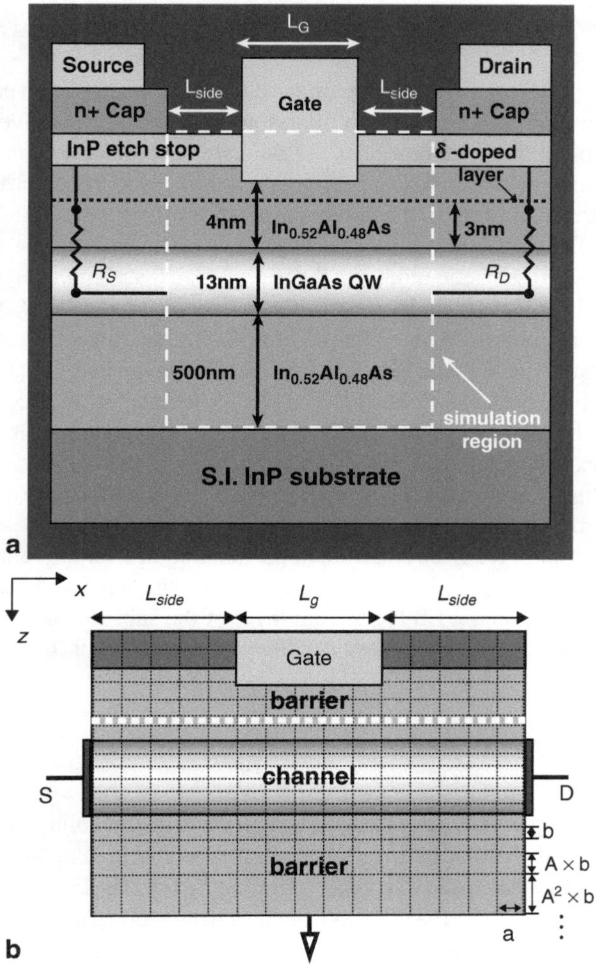

Fig. 3.1 **a** The structure of the MIT InGaAs HEMTs designed for logic application (after [22]). The simulation domain is indicated by the *dashed square* which is modeled with quantum ballistic transport. The heterostructure stack is simply replaced with two series resistances. **b** The simplified device structure with proper boundary conditions. The delta-doped layer is indicated by the *white dashed line*

the current flow is 2D from the raised source to the drain through the doped heterostructure stack and then laterally to the channel. Rather than attempting to simulate the contacts and the associated metal semiconductor contact resistance, we place ideal contacts at the two ends of the channel to simplify the structure. The simulated "intrinsic" device structure, where quantum ballistic transport is expected to dominate, is indicated by the dashed square in Fig. 3.1a. The effects of the heterostructure contact stack on the intrinsic device are approximated by simply adding two series resistance R_S and R_D to both ends. This approach neglects some source design issues such as source access [24] and so-called source starvation [14] that may be important in practice. Nevertheless, it is a reasonable starting point and should provide us with upper limit projections. The comparisons with experiment to be discussed later show that neglecting source design issues is acceptable at this stage of technology development.

3.2.2 Simulation Approach

The 2D simulation program used for these studies evolved from the nanoMOS simulation program [25]. The Poisson equation is first solved with the charge from an initial flat band profile as the estimated potential. To compute the carrier density, the uncoupled mode space approach is then used to solve the quantum transport problem assuming ballistic transport. The resulting spatial charge distribution from the 2D charge density weighed by the wavefunction in the quantum well is inserted in the Poisson equation, and the new charge distribution leads to a new potential profile. This process continues until the desired convergence is achieved (typically when the maximum difference in potential for the last two iterations is under 0.1 meV). The quantum ballistic current is then readily calculated for each subband within the mode space NEGF formalism. Final post-processing steps utilize the two fitting parameters (the metal work function and the series resistance) to fit the simulation results to the experimental data. The simulation procedure was discussed in detail in [26]. Note that different from [26], we adopt an embedded gate structure to capture the fringing effects which become important in short gate length devices. Figure 3.1b shows the simplified device structure in the simulation with the following boundary conditions: (1) the potential of the embedded gate region is fixed according to the gate bias and workfunction, (2) the bottom layer of the substrate is grounded, and (3) zero normal electrical field boundary conditions are applied for the rest of the boundary.

3.2.3 Materials Parameters

Non-parabolicity effects are important in the conduction band of III-V materials, and the use of bulk effective masses would lead to significant errors for ultra-thin-body structures [27]. Before simulating the HEMTs, we extract the channel effective masses of the III-V HEMT devices from atomistic $sp^3d^5s^*$ tight-binding simulations using the NEMO-3D program [28]. The bandstructure calculated with the atomistic tight-binding model incorporates the non-parabolicity effects as well as the strain effects due to the lattice mismatch between the $In_{0.53}Ga_{0.47}As/In_{0.7}Ga_{0.3}As$ layers. The tight-binding bandstructure calculation also shows that higher valleys are well above the Γ valley subbands and therefore make a negligible contribution to the carrier and current density. The effective mass is extracted from the first subband by fitting a parabola from the band bottom to up to 0.1 eV higher. The extracted equivalent effective masses for the quantum well channel are $0.053m_0$ for transport and transverse directions, and $0.067m_0$ for the confinement direction, in contrast with the value of $0.041m_0$ obtained from a linear interpolation of the bulk effective masses of InAs and GaAs [29]. The confinement effective mass of $In_{0.52}Al_{0.48}As$ barrier is $0.075m_0$. The dielectric constant of the InGaAs channel used in the simulation is assumed to be $\varepsilon = 14.3$ and that of the $In_{0.52}Al_{0.48}As$ barrier is 12.7 [29]. The conduction band discontinuity between $In_{0.53}Ga_{0.47}As/In_{0.52}Al_{0.48}As$ layers is assumed to be $\Delta E_C = 0.50$ eV [30, 31].

3.2.4 Results

We compare the simulation results with experimental data by examining the *I–V* characteristics. Two fitting parameters are used: (1) the workfunction of the gate metal and (2) the series resistance. The gate workfunction is first determined by tuning its value so that the subthreshold regime of the intrinsic $\log I_d$–V_{GS} overlaps that of the experimental data. Below subthreshold, the current is so small that the *I–V* characteristics are not affected by the series resistance. Once we fit the subthreshold regime by determining the gate workfunction, we choose an appropriate series resistance to include in the ballistic intrinsic *I–V* and adjust to best match the linear region within the above-threshold I_d regime (low V_{DS} and high V_{GS}).

Figure 3.2 compares the simulation with the experimental data [32] in $\log I_d$–V_{GS} and linear I_d–V_{DS} plots for nominal gate lengths of $L_g = 40$ nm and $L_g = 130$ nm InGaAs HEMTs after tuning the workfunction and including the series resistance. Good quantitative agreement is achieved with adjustment of the nominal gate length and insulator thickness by about 10%. For the device with a nominal gate length of $L_g = 40$ nm and $t_{ins} = 4$ nm, the best fit was obtained with $L_g = 45$ nm and $t_{ins} = 3.6$ nm. For the device with a nominal gate length of 130 nm a $t_{ins} = 4$ nm, the best fit was obtained with $L_g = 125$ nm and $t_{ins} = 4.6$ nm. These values are within the measurement error and reasonable [33]. The fitted series resistance $R_S = R_D = 220\,\Omega\,\mu$m is identical to the value quoted from the measurement [34]. We also observe that the difference in current between experimental data and simulation

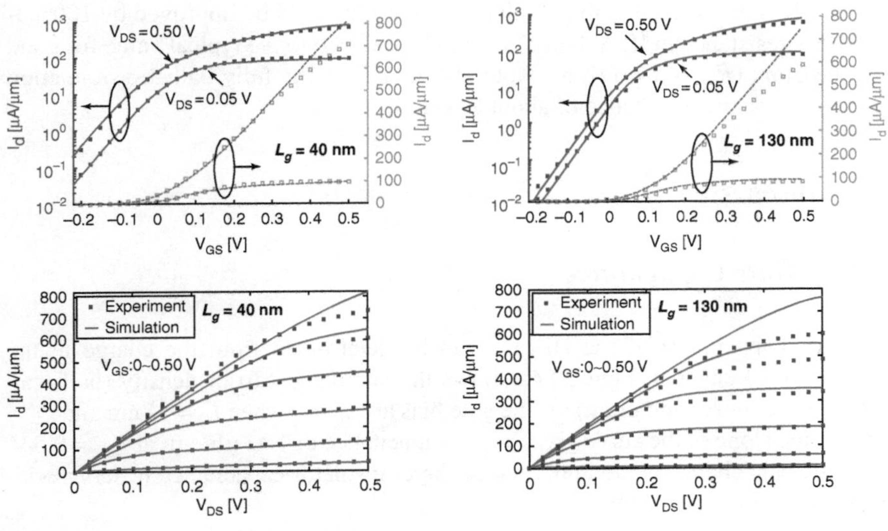

Fig. 3.2 Comparison of the *I–V* characteristics between experimental (*square symbols*) and simulation results (*solid and dash lines*) for 40 nm (*left*) and 130 nm (*right*) HEMTs shows quantitatively good agreement by adjusting the gate length and insulator thickness in a reasonable range

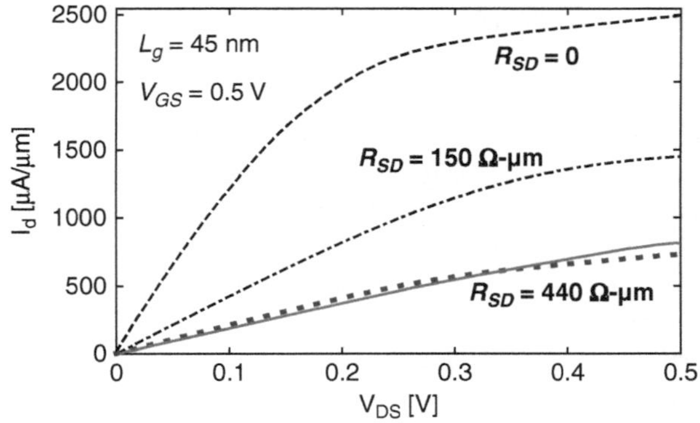

Fig. 3.3 I_d–V_{DS} of 45 nm HEMTs at $V_{GS} = 0.5$ V with different series resistance; the *square symbols* are experimental data. The on-current is expected to be improved by about 100% with reduced series resistance matched to Si transistors

results at high V_{DS} increases as both V_{GS} and L_g increases, which might be due to three reasons. First, the assumed constant series resistance used in the simulation might not hold when the current becomes large. Second, as the gate length increases from 40 to 130 nm the device may become less ballistic. Finally the source may not be able to supply the desired on-current. Nevertheless, the simulation demonstrates that the ballistic model with attached series resistance is a good first order description of the HEMTs' I–V characteristics.

Note that the series resistance in these transistors is quite large, and it presents a significant limit on device performance. This is shown in Fig. 3.3, where the simulated I_d–V_{DS} for L_g = 45 nm at V_{GS} = 0.5 V is compared for different assumed values of R_{SD} (the square symbols are experimental data for the nominal L_g = 40 nm device). It is observed that the predicted on-current could be improved by 100% if the series resistance in III-V HEMTs could be reduced to the typical value for good Si transistors (R_{SD} ~ 150 Ω µm). Note also that even the fully ballistic simulation displays a channel resistance of about 80 Ω µm.

3.3 Discussion

3.3.1 Gate Capacitance

The gate capacitance of the HEMTs can be determined from the charge in the quantum well channel. Figure 3.4a shows the simulated carrier density (half-way between the source and drain) vs. the gate bias at $V_{DS} = 0$ when $L_g = 45$ nm and t_{ins} = 3.6 nm; the slope of the curve gives a gate capacitance of 1.41 µF/cm^2 at $V_{GS} = 0.5$ V. The gate capacitance is the upper barrier layer insulator capacitance in series with the semiconductor capacitance:

$$\frac{1}{C_G} = \frac{1}{C_{ins}} + \frac{1}{C_S}, \tag{3.1}$$

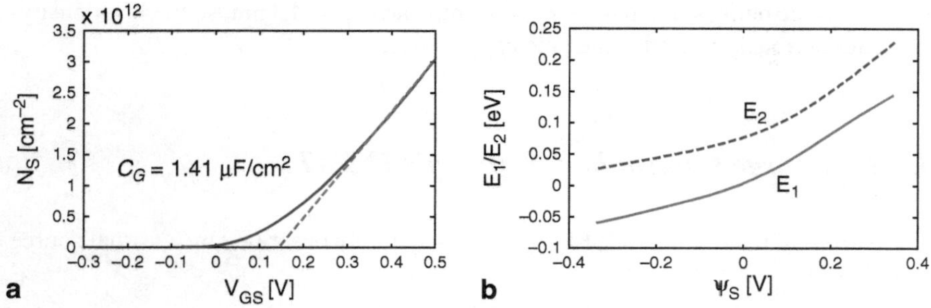

Fig. 3.4 **a** The charge density in the quantum well gives $C_G = 1.41\,\mu F/cm^2$ at $V_{GS} = 0.5\,V$. **b** The 1st and 2nd subband energy as a function of the surface potential of the quantum well

where $C_{ins} = \dfrac{\varepsilon_{ins}}{t_{ins}} = 3.12\ \mu F/cm^2$. The semiconductor capacitance C_S can be expressed as [35]:

$$C_S = \frac{q^2 m^*}{\pi \hbar^2} \cdot \sum_i f(E_i; E_F) \cdot (1 - \partial E_i / \partial \psi_S), \tag{3.2}$$

where $C_Q = q^2 m^*/\pi\hbar^2$ is the so called quantum capacitance at 0 K in a 2D system, $f(E_i; E_F)$ is the Fermi function for subband i, and $\partial E_i/\partial \psi_S$ is the change of the ith subband energy E_i with respect to the surface potential ψ_S. Equation (3.2) may be viewed as the product of C_Q and a factor that depends on how the shape of the quantum well changes with gate bias. The second factor is often interpreted as describing how the centroid of the charge changes with gate voltage. At $V_{GS} = 0.5\,V$ only two subbands are occupied in the quantum well, and Fig. 3.4b shows the two subband energies as a function of ψ_S. The semiconductor capacitance is then readily calculated from Eq. (3.2) as $C_S = 3.05\,\mu F/cm^2$, which is, close to the quantum capacitance $C_Q = 3.53\,\mu F/cm^2$. Using Eq. (3.1) we find $C_G = 1.54\,\mu F/cm^2$, close to $C_G = 1.41\,\mu F/cm^2$ as calculated directly from the charge in the quantum well. An independent determination of the gate capacitance obtained by de-embedding the capacitance from S-parameter measurements yields a somewhat lower value of $C_G = 1.08\,\mu F/cm^2$. The reason for this discrepancy is still not understood.

Our calculations show that C_S is comparable to C_{ins} for this transistor. We also find that $C_G < C_{ins}$. This occurs because of the small density of states effective mass in III-V materials [14, 16]. The result is a significant degradation of the total gate capacitance. One can describe this effect as an effective increase of the insulator thickness according to

$$C_G = \frac{\varepsilon_{ins}}{t_{ins} + \Delta t_{ins}}. \tag{3.3}$$

For the device being studied here, $t_{ins} = 3.6\,\text{nm}$ and $\Delta t_{ins} = 4.4\,\text{nm}$, so the low density-of- states seriously degrades the gate capacitance.

3.3.2 Charge Control in a Nanoscale HEMT

The current of nanoscale MOSFETs can be accurately described by a virtual source model as [1]

$$I_D/W = Q_i(x_0) \cdot \langle \upsilon\,(x_0) \rangle \tag{3.4a}$$

where I_D is the drain current, W is the device width, and $Q_i(x_0)$ is the charge per unit area at the virtual source. In the ballistic limit, the virtual source model becomes the top-of-the-barrier ballistic model [36, 37] and for on-current conditions,

$$I_{ON}/W = Q_i(x_0) \cdot \upsilon_{inj} \tag{3.4b}$$

where $\langle \upsilon\,(x_0) \rangle = \upsilon_{inj}$ is the so-called ballistic injection velocity and is a key figure of merit for nanoscale MOSFETs [1].

When analyzing experimental results, the charge at the virtual source (top of the barrier) and injection velocity at the same location are estimated. Consider first the charge at the virtual source. It can be estimated from experiment C–V (long channel) from [38, 39]

$$Q_i(x_0) = \int_0^{V_{gs}^*} C'_{gsd}\Big|_{V_{ds}=(\text{long-chan.})} dV_{gs}, \tag{3.5}$$

where $V_{gs}^* = V_{gs} + \Delta V_T - I_{on}R_S$ accounts for the correction of V_T roll-off, *DIBL*, and series resistance.

For the 45 nm HEMTs studied here, the simulated on current at $V_{GS} = V_{DS} = 0.5\,\text{V}$ is $I_{ON}/W = 813\,\mu\text{A}/\mu\text{m}$, the intrinsic biases are $V_{GS,in} = 0.32\,\text{V}$, $V_{DS,in} = 0.14\,\text{V}$ with $R_S = R_D = 220\,\Omega\,\mu\text{m}$. Figure 3.5 plots the first subband profile vs. position along with the electron density vs. position as a function of position along the

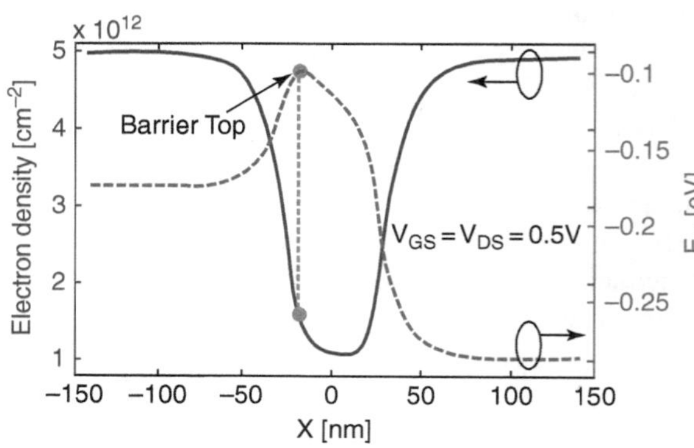

Fig. 3.5 The sheet charge density at the virtual source is determined from the spatial sheet charge density at the top of the potential barrier

device channel. From this plot, we find the charge at the virtual source (top of the potential barrier) to be $1.60 \times 10^{12}\,\text{cm}^{-2}$. From Eq. (3.5) and the intrinsic simulated C–V of a long channel device ($L_g = 125$ nm), the electron density extracted at $V_{GS} = V_{DS} = 0.5$ V is about $1.90 \times 10^{12}\,\text{cm}^{-2}$, which is reasonably close to the charge density obtained directly from the top of the potential barrier. As will be discussed in Sect. 3.3.7, the charge at the top of the barrier under high drain bias is less than the equilibrium charge because the semiconductor capacitance is reduced under high drain bias.

3.3.3 Velocity at the Virtual Source

The ballistic injection velocity υ_{inj} is of particular interest in MOSFETs, and can be evaluated at the top of the source-channel potential barrier ($x = x_0$):

$$I_D/W = Q_i(x_0) \cdot \upsilon_{inj} \tag{3.6}$$

where I_D, W, and $Q_i(x_0)$ have the same meaning as in Eq. (3.4a). For the 40 nm HEMTs studied, the on current at $V_{GS} = V_{DS} = 0.5$ V is $I_{ON}/W = 813\,\mu$A/μm, the intrinsic biases are $V_{GS,in} = 0.32$ V, $V_{DS,in} = 0.14$ V with $R_S = R_D = 220\,\Omega\,\mu$m. Figure 3.6 plots the first subband profile and average velocity as a function of position along the device channel. The ballistic injection velocity is readily read from the average velocity at the top of the barrier, which gives $\upsilon_{inj} = 3.17 \times 10^7$ cm/s, and is close to the experimental reported value [40]. In comparison, the injection velocity extracted with the simulated on-current (813 μA/μm) and the charge from the integration of the long channel device C–V as in last section ($1.90 \times 10^{12}\,\text{cm}^{-2}$) is $\upsilon_{inj} = 2.67 \times 10^7$ cm/s. Note that the III-V HEMTs have much larger ballistic injection velocity than that of the strained Si MOSFETs in spite of the small intrinsic gate and drain biases in III-V HEMTs that result from the large series resistance.

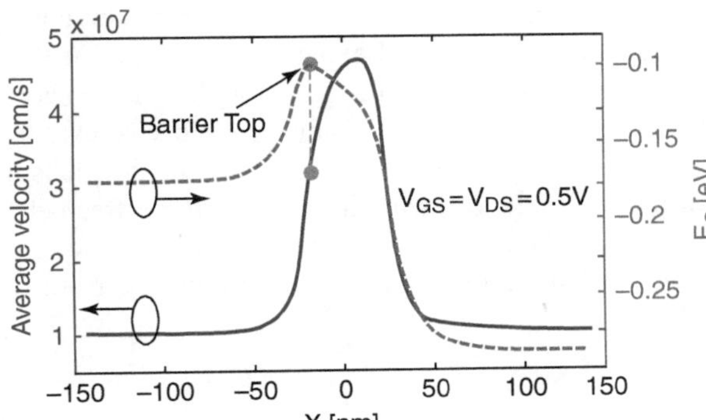

Fig. 3.6 The ballistic injection velocity is the average velocity at the top of the potential barrier

3.3.4 Ballistic Mobility

The "ballistic mobility" is a reflection of the ballistic quantum contact conductance, and imposes severe limitation on the apparent channel mobility in III-V transistors when the gate length scales down to ballistic regime. The apparent channel mobility μ_{app} is defined from the linear region of the I_d–V_{DS} as

$$I_d \equiv \frac{W}{L_g}\mu_{app}C_G\left(V_{GS} - V_T\right)V_{DS}. \tag{3.7}$$

Since the current in a ballistic FET is independent of channel length, it is clear that the apparent mobility must be a channel length dependent quantity. The apparent channel mobility μ_{app} is the combination of the "ballistic mobility" μ_B and the bulk mobility μ_0 through the Mathiessen's rule [41]:

$$\frac{1}{\mu_{app}} = \frac{1}{\mu_B} + \frac{1}{\mu_0}. \tag{3.8}$$

From ballistic theory, μ_B can be calculated from

$$\mu_B = \frac{\upsilon_T \cdot L_g}{V_{DS}} \frac{\sum\limits_i \left(\Im_{1/2}(\eta_{FS,i}) - \Im_{1/2}(\eta_{FD,i})\right)}{\sum\limits_i \left(\Im_0(\eta_{FS,i}) + \Im_0(\eta_{FD,i})\right)}, \tag{3.9}$$

where $\upsilon_T = \sqrt{2k_BT/\pi m_c}$, m_c is the transport effective mass; L_g is the gate length; V_{DS} is the intrinsic drain bias; and $\Im_{1/2}(x), \Im_0(x)$, are the Fermi-Dirac integrals of order 1/2 and 0, with $\eta_{FS,i} = (E_{FS} - E_i)/k_BT$, $\eta_{FD,i} = \eta_{FS,i} - q_0V_{DS}/k_BT$. Assuming $V_{DS} \ll k_BT/q_0$, and only one subband occupied, Eq. (3.9) reduces to [42, 43]:

$$\mu_B = \frac{\upsilon_T \cdot L_g}{2k_BT/q_0} \cdot \frac{\Im_{-1/2}(\eta_{FS,1})}{\Im_0(\eta_{FS,1})}. \tag{3.10}$$

In the nondegenerate limit, the ratio of Fermi-Dirac integrals reduces to unity.

Equation (3.8) can well explain the mobility of the transistors in both ballistic and diffusive limit; when the gate length L_g is so short that the transistor is in the ballistic limit, the apparent channel mobility is just the ballistic mobility. When L_g is much longer than the mean-free-path, the device is in the diffusive limit, and L_g in Eq. (3.9) is replaced with the carrier's mean free path, λ_0, so the apparent channel mobility will be largely determined by the bulk mobility.

The ballistic mobility can also be obtained from the simulated linear ballistic I_d–V_{DS}:

$$I_d = Q_i \cdot \mu_B \cdot V_{DS}/L_g, \tag{3.11}$$

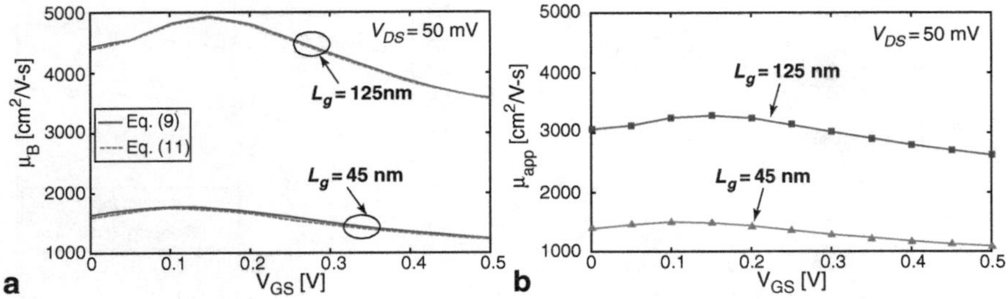

Fig. 3.7 **a** The ballistic mobilities μ_B determined from Eqs. (3.9) and (3.11) are almost identical for 45 nm and 125 nm HEMTs respectively. **b** The apparent channel mobility μ_{app} is gate length dependent and significantly degraded from the bulk mobility μ_0 (= 10,000 cm^2/V s) due to the small μ_B. The series resistance is $R_S = R_D = 220\,\Omega\,\mu$m

where Q_i is the charge density at the top of the barrier. Figure 3.7a compares μ_B vs. V_{GS} as obtained from Eqs. (3.9) and (3.11) respectively for the 45 and 125 nm HEMTs; the two methods give very close results. Note that the gate-length dependent μ_B is much smaller than the bulk mobility (~10,000 cm^2/V s), and therefore it degrades the apparent mobility significantly. This effect is shown in Fig. 3.7b, where μ_{app} is calculated from Eq. (3.8) with $\mu_0 = 10{,}000$ cm^2/V s, and plotted as a function of V_{GS} for 45 and 125 nm HEMTs. The point is that the large bulk mobilities of III-V materials will not be reflected in the apparent mobility that describes the linear region of a FET.

3.3.5 Source Design Issues

Fischetti and Laux have pointed out the importance of source design considerations such as access geometry and source starvation for III-V transistors [14, 18, 19]. The first issue refers to the fact that the source access geometry may restrict the flow of carriers into the channel. Source starvation refers to the condition when the source is unable to inject electrons into longitudinal momentum states in the channel—these states become depleted, or "starved". In addition, a third effect may also occur.

Transistors operate by modulating potential energy barriers [44, 45]. As the gate voltage increases, the potential energy barrier decreases, and the charge in the channel increases. When the gate voltage increases to the point where the barrier is removed and the channel charge is equal to the charge in the source, the transistor drops. In other words, there can't be more charge in the channel than in the source. This effect has been called "source exhaustion" [46, 47]. Its effect on the transistor's IV characteristics is similar to that produced by the "source starvation" effect discussed by Fischetti [14, 19], but it is simply consequence of electrostatics.

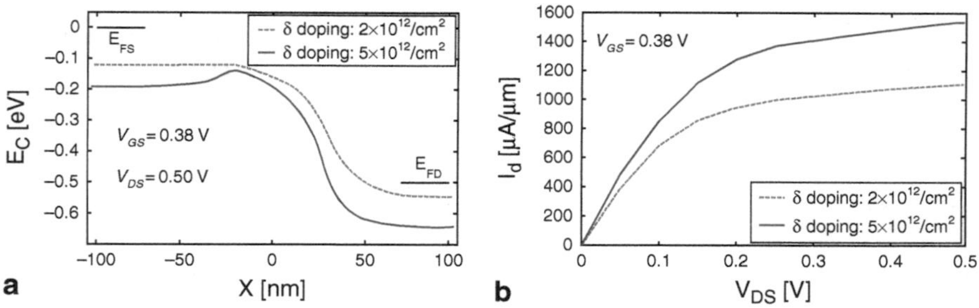

Fig. 3.8 **a** The first subband profile of the HEMT transistor indicates that the potential barrier of the low doping HEMT vanishes at smaller gate bias than the high doping HEMT. **b** The I_d–V_{DS} plot shows low delta-doping HEMT has smaller current than the one with higher delta doping at the same gate bias

Our simulations do not capture the source access and source starvation effects, but they do include the possibility of source exhaustion. Source exhaustion is illustrated in Fig. 3.8a, where the first subband profiles along the channel for delta-doping equal to $2 \times 10^{12}\,\mathrm{cm}^{-2}$ and $5 \times 10^{12}\,\mathrm{cm}^{-2}$ at $V_{GS} = 0.38\,\mathrm{V}$, $V_{DS} = 0.50\,\mathrm{V}$ (intrinsic) are compared for the $L_g = 45\,\mathrm{nm}$ HEMT. For lower delta-doping, the barrier in the channel is smaller and reaches the same level as the "source" region beyond the gate at $V_{GS} = 0.38\,\mathrm{V}$, while for the higher doping the barrier still exists. The electron sheet density at the almost flat potential barrier is $1.7 \times 10^{12}\,\mathrm{cm}^{-2}$, which is very close to the delta doping density, and the transistor begins to lose transconductance as the channel barrier vanishes. In ballistic simulations, this effect in simulation results in non-convergent results if V_{GS} continues to increase (the effect is, however, simply a matter of electrostatics and is observed in drift/diffusion simulations as well.). In comparison, with a higher delta-doping of $5 \times 10^{12}\,\mathrm{cm}^{-2}$, the carrier sheet density at the top of the barrier is $2.3 \times 10^{12}\,\mathrm{cm}^{-2}$ at the same $V_{GS} = 0.38\,\mathrm{V}$, and the larger barrier in the channel ensures the proper function of the transistor at increased gate bias. The doping effect on the HEMT's I_d–V_{DS} characteristics is further shown in Fig. 3.8b, where it is noted that with other conditions remaining the same, higher delta-doping HEMT has larger current than that of the lower doping one. These simulations illustrate the importance of achieving high carrier densities in the source of III-V FETs.

3.3.6 Role of S/D Tunneling

Source-drain tunneling degrades transistor performance by increasing the off-current and sub-threshold swing. It is an important limiting factor in devices with low transport effective mass and short gate lengths. The S/D tunneling effect in III-V HEMTs is explored by examining the energy-resolved current density for a

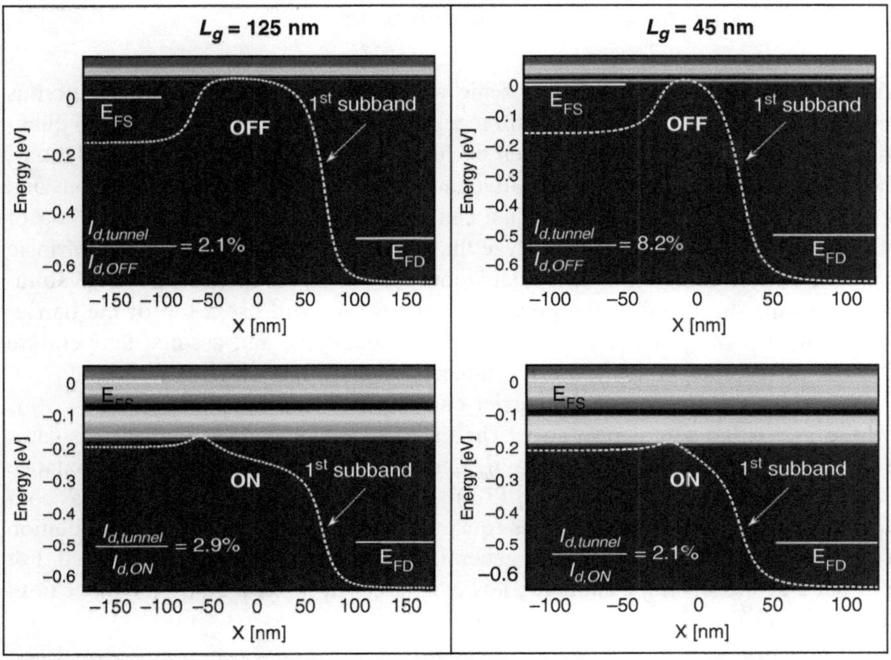

Fig. 3.9 The S/D tunneling is illustrated by the energy-resolved current density in the short and long gate length HEMTs at off and on states. The S/D tunneling effect is larger in shorter HEMTs at off state due to short gate length and larger DIBL

long (L_g = 125 nm) and a short (L_g = 45 nm) gate length HEMT under both off and on states. Figure 3.9 plots the energy-resolved current density for the HEMTs at off (V_{GS} = 0 V, V_{DS} = 0.5 V, intrinsic) and on (V_{GS} = V_{DS} = 0.5 V, intrinsic) states. The first subband along the device is also shown. The S/D tunneling current component is the fraction of the current contributed by carriers with energy below the top of the barrier for each subband. It is observed that under on state conditions, S/D tunneling is an insignificant fraction of the total current for both HEMTs—the current is mainly contributed by carriers with energy larger than the small barrier under high gate bias. Under off state conditions, the S/D tunneling current accounts a larger fraction of the total current than in the on state for L_g = 45 nm device, and the fractional contribution is more in the L_g = 45 nm device than in the L_g = 125 nm device. The tunneling distance in shorter gate length HEMT is further reduced under off state conditions due to the larger drain-induced barrier lowering (DIBL) in the shorter devices, which is indicated by the Fermi level, top of the potential barrier, and the barrier thickness in this energy range in Fig. 3.9. For the III-V HEMTs being examined here, S/D tunneling is not significant; but it can be foreseen that as a target device with superior performance over Si at the 15 nm gate length regime and beyond, the III-V compound semiconductor channel transistors will have a larger S/D tunneling besides the gate leakage at off-state.

3.3.7 Back of the Envelope Calculations

As device dimensions continue to scale well into the nanoscale regime, rigorous treatment of transport using quantum mechanical simulations is necessary to quantitatively predict and benchmark their performance. Simple theoretical calculations, however, often provide a more intuitive understanding of the device operations and estimates of key figures of merit like charge, mobility, and velocity at the top of the barrier. In this section, we analyze the performance of the $L_g = 45$ nm intrinsic ballistic HEMT using analytical calculations, and compare them with NEGF simulation results discussed in the previous sections. We will use a top of the barrier model with a single parabolic band in these equations, and assume temperature $T = 0$ K to keep the mathematics as simple as possible.

The charge at the top of the barrier can be expressed as $Q_i = C_G(V_{GS} - V_T)$, where C_G is the gate capacitance. The gate capacitance consists of the insulator (dielectric) capacitance ($C_{ins} = \varepsilon_{ins}/t_{ins}$) and semiconductor (channel) capacitance ($C_S = -dQ_i/d\psi_s$) in series (see Eq. (3.1)). The semiconductor capacitance has contributions from the density of states (quantum capacitance C_Q) and the modulation of subband energies (Eq. (3.2)). In general, a numerical simulation is required. For our back of the envelope estimate, however, we will replace C_S by its upper limit C_Q which at $T = 0$ K, can be expressed as

$$C_Q = \frac{q^2 m^*}{\pi \hbar^2} = \frac{\varepsilon_{ins}}{\Delta t_{ins}} \tag{3.12}$$

Using the material parameters provided in Sect. 3.2.3, Δt_{ins} and C_G are calculated to be 3.2 nm and 1.66 μF/cm², reasonably close to the $\Delta t_{ins} = 4.4$ nm $C_G = 1.41$ μF/cm² obtained from the simulations. The result is actually closer than expected. From simulation, we know that two subbands are occupied. But we also know that the subband-modulation term in Eq. (3.2) is approximately 0.5 at high V_{GS} for both subbands from Fig. 3.4b. The two factor-of-two errors cancel, which is why the final result is rather close to the simulations.

With this C_G and an estimated threshold voltage $V_T = 0.057$ V (at $V_{DS} = 0.05$ V), the carrier density at the top of the barrier is $N_S = 4.5 \times 10^{12}$ cm^{-2} at $V_{GS} = 0.5$ V, compared with simulation results of $N_S = 3.1 \times 10^{12}$ cm^{-2} mainly due to the difference in the capacitance. Under on-current conditions ($V_{DS} = 0.5$ V), the density of states at the top of the barrier is only filled by carriers with positive momentum (going from source to drain), thus reducing C_Q by half, and the corresponding gate capacitance decreases to $C_G = 1.1$ μF/cm². The threshold voltage is also reduced to $V_T = -0.008$ V due to effects of 2D electrostatics (*DIBL* = 145 mV/V). The charge under on-state conditions ($V_{GS} = V_{DS} = 0.5$ V) is 3.5×10^{12} cm^{-2}, compared to 3.2×10^{12} cm^{-2} from the simulations. The conclusion is that estimating the inversion charge in the on state by integrating the equilibrium *C–V* curve as in Eq. (3.5) will over-estimate the charge because the semiconductor capacitance decreased under high drain bias.

Although there is no scattering in a ballistic conductor, it has finite conductance, and hence a ballistic mobility can be extracted. This ballistic mobility μ_B is related to the ballistic channel resistance as (see Eq. (3.7)):

$$\mu_B = \frac{1}{R_{CH}W}\frac{L_g}{q_0 N_S} \tag{3.13}$$

where R_{CH} is inversely related to the conductance ($G_{CH} = 1/R_{CH}$). The conductance of a ballistic conductor is proportional to the number of transverse propagating modes M, the proportionality constant being the quantum of conductance $2q^2/h$ (with spin). As the transverse modes are separated by $(2\pi/W)$ in the momentum (k) space, the number of modes for a maximum transverse wave vector k_F is given by $M = 2k_F/(2\pi/W) = Wk_F/\pi$ [48]. Here, k_F is the maximum wave vector filled by carriers at $T = 0$ K, and is determined by the Fermi energy with respect to the top of the barrier. For a 2D electron gas, the density of states in momentum space is given by $A/4\pi^2$, hence the carrier density is related to k_F as

$$N_S = 2 \times \frac{1}{A}\frac{A}{4\pi^2}\pi k_F^2 = \frac{k_F^2}{2\pi}. \tag{3.14}$$

For $N_S = 4.5 \times 10^{12}\,\text{cm}^{-2}$ (back of the envelope calculation at $V_{DS} = 0.05$ V, $V_{GS} =$ 0.50 V), the number of transverse propagating modes per unit length is 169 μm^{-1}, which corresponds to a channel resistance of $R_{CH}{\cdot}W = 76\,\Omega$ μm. For $N_S = 3.1 \times 10^{12}\,\text{cm}^{-2}$ (simulation results at $V_{DS} = 0.05$ V, $V_{GS} = 0.50$ V), the channel resistance is $R_{CH}{\cdot}W = 92\,\Omega$ μm. For comparison, the channel resistance extracted directly from the slope of linear I_d–V_{DS} at the same gate and drain biases at 300 K is $R_{CH}{\cdot}W = 77\,\Omega$-μm, which shows a close matching for analytical calculations. The ballistic mobility is then computed from Eq. (3.13) to be $\mu_B = 822\ \text{cm}^2/\text{V s}$, and the corresponding $\mu_{app} = 760\ \text{cm}^2/\text{V s}$ (using $\mu_0 = 10{,}000\ \text{cm}^2/\text{V s}$, as in Sect. 2.3.4), compared with the simulation results $\mu_B = 1178\ \text{cm}^2/\text{V s}$ and $\mu_{app} = 1054\ \text{cm}^2/\text{V s}$.

The injection velocity at the top of the barrier (average carrier velocity) is given by $\upsilon_{inj} = 4\upsilon_F/3\pi$ [48], where υ_F is the maximum carrier velocity corresponding to the maximum occupied wave-vector k_F ($\upsilon_F = \hbar k_F/m^*$). Note that at high V_{DS}, N_S at the top of the barrier is dominantly from source-injected carriers, therefore $N_s = k_F^2/4\pi$. Using the equations above, υ_{inj} can be readily obtained for a given carrier density. Under on-current conditions, the υ_{inj} at 0 K corresponding to the back of the envelope calculation of $N_S = 3.5 \times 10^{12}\,\text{cm}^{-2}$ is $\upsilon_{inj} = 6.1 \times 10^7\,\text{cm/s}$, compared to 4.9×10^7 cm/s obtained with the simulations, which agree fairly well (note again that the abnormally high current, charge density, and injection velocity here are due to our intrinsic discussion without series resistance; in real device all these quantities are substantially lowered by the large series resistance as shown in previous sections).

3.4 Conclusions

In this chapter we have investigated the performance as well as the device physics of recently reported InGaAs HEMTs by using a self-consistent quantum ballistic NEGF model based on effective masses in mode space. Good quantitative agreement between simulation and experimental data indicates that the III-V HEMTs with gate length ~40 nm operate rather close to the ballistic limit. Compared to the simulation results, the smaller drive current reported from experiments at either higher gate bias or longer gate length devices is probably due to phonon scattering degradation. Note that the large series resistance severely limits the III-V HEMTs performance; optimizing the source/drain contacts structure to minimize the series resistance will be critical for future III-V transistors designing, and may amplify the difference between theory and experiment under high gate bias.

The small effective mass in the III-V compound semiconductors has both positive and negative effects on the device performance. The direct positive effect is that the III-V HEMTs ballistic injection velocity is as high as $\sim 3 \times 10^7$ cm/s, as determined from both simulation and experiments. The DOS bottle-neck is a negative effect that degrades the gate capacitance by effectively increasing the upper barrier layer thickness by almost 100%. The resulting electron density at the virtual source is comparably small at on-state, which limits the drive current.

We also found that the apparent channel mobility in III-V HEMT devices is significantly degraded from its very large bulk mobility due to the comparably very small "ballistic mobility", which becomes important as the device channel length scales down to the ballistic limit regime. The δ-doping effects on the source designing were also investigated. Lower δ-doping will improve S and $DIBL$, but the current is smaller due to the smaller energy range between the source Fermi level and the top of the barrier. Besides, the top of the barrier tends to disappear at relatively smaller intrinsic gate bias, after which the device will become dysfunctional. Next the S/D tunneling in III-V HEMTs was found insignificant at gate length of 40 nm; it is however, foreseen to become severe when the gate length approaches 15 nm and beyond. Finally, the intrinsic simulation results can be well explained by theoretical calculations with simple device physics based on a top-of-the-barrier model, which helps the understanding of the device operational mechanism as a reference to the 2D simulation.

The next challenge to address is the lowering of the series resistance, because Fig. 3.3 shows that if the series resistance can be decreased to values typical of silicon MOSFETs, then III-V FETs would offer high drive current at power supply of one-half of silicon.

Acknowledgement This work was supported by the Focus Center Research Program (FCRP) through the center for Materials, Structures, and Devices (MSD). Computational support was provided by the Network for Computational Nanotechnology which is supported by the National Science Foundation under Grant No. EEC—0634750. One of the authors (MSL) acknowledges illuminating discussions with M.V. Fischetti at the University of Massachusetts and T. Rakshit at Intel.

References

1. D. A. Antoniadis and A. Khakifirooz, "MOSFET performance scaling: Limitations and future options," *IEEE International Electron Devices Meeting 2008, Technical Digest*, pp. 253–256, 2008.
2. K. Mistry, C. Allen, C. Auth, B. Beattie, D. Bergstrom, M. Bost, M. Brazier, M. Buehler, A. Cappellani, R. Chau, C. H. Choi, G. Ding, K. Fischer, T. Ghani, R. Grover, W. Han, D. Hanken, M. Hatttendorf, J. He, J. Hicks, R. Huessner, D. Ingerly, P. Jain, R. James, L. Jong, S. Joshi, C. Kenyon, K. Kuhn, K. Lee, H. Liu, J. Maiz, B. McIntyre, P. Moon, J. Neirynck, S. Pei, C. Parker, D. Parsons, C. Prasad, L. Pipes, M. Prince, P. Ranade, T. Reynolds, J. Sandford, L. Schifren, J. Sebastian, J. Seiple, D. Simon, S. Sivakumar, P. Smith, C. Thomas, T. Troeger, P. Vandervoorn, S. Williams, and K. Zawadzki, "A 45 nm logic technology with high-k plus metal gate transistors, strained silicon, 9 Cu interconnect layers, 193 nm dry patterning, and 100% Pb-free packaging," *2007 IEEE International Electron Devices Meeting*, vol. 1 and 2, pp. 247–250, 2007.
3. S. Natarajan, M. Armstrong, M. Bost, R. Brain, M. Brazier, C.-H. Chang, V. Chikarmane, and M. Childs, "A 32 nm logic technology featuring 2nd-generation High-k+Metal-Gate transistors, enhanced channel strain and 0.171 um^2 SRAM cell size in a 291 Mb array," *2008 IEEE International Electron Devices Meeting*, pp. 941–943, 2008.
4. M. Passlack, P. Zurcher, K. Rajagopalan, R. Droopad, J. Abrokwah, M. Tutt, Y. B. Park, E. Johnson, O. Hartin, A. Zlotnicka, P. Fejes, R. J. W. Hill, D. A. J. Moran, X. Li, H. Zhou, D. Macintyre, S. Thoms, A. Asenov, K. Kalna, and I. G. Thayne, "High mobility III-V MOSFETs for RF and digital applications," *2007 IEEE International Electron Devices Meeting*, vol. 1 and 2, pp. 621–624, 2007.
5. Y. Sun, E. W. Kiewra, J. P. de Souza, J. J. Bucchignano, and K. E. Fogel, "Scaling of In_{0.7}Ga_{0.3}As buried-channel MOSFETs," *2008 IEEE International Electron Devices Meeting*, pp. 367–370, 2008.
6. Y. Xuan, T. Shen, M. Xu, Y. Q. Wu, and P. D. Ye, "High-performance surface channel in-rich In_{0.75}Ga_{0.25}As MOSFETs with ALD High-k as gate dielectric," *2008 IEEE International Electron Devices Meeting*, pp. 371–374, 2008.
7. D. H. Kim and J. A. del Alamo, "Scaling behavior of $In_{0.7}Ga_{0.3}$AsHEMTs for logic," *2006 IEEE International Electron Devices Meeting*, vol. 1 and 2, pp. 587–590, 2006.
8. D. H. Kim and J. A. del Alamo, "Logic performance of 40 nm InAsHEMTs," *2007 IEEE International Electron Devices Meeting*, vol. 1 and 2, pp. 629–632, 2007.
9. D. H. Kim and J. A. del Alamo, "30 nm E-mode InAs PHEMTs for THz and future logic applications," *2008 IEEE International Electron Devices Meeting*, pp. 719–722, 2008.
10. G. Dewey, M. K. Hudait, K. Lee, R. Pillarisetty, W. Rachmady, M. Radosavljevic, T. Rakshit, and R. Chau, "Carrier transport in high-mobility III-V quantum-well transistors and performance impact for high-speed low-power logic applications," *IEEE Electron Device Letters*, vol. 29, pp. 1094–1097, Oct 2008.
11. M. Radosavljevic, T. Ashley, A. Andreev, S. D. Coomber, G. Dewey, and M. T. Emeny, "High-performance 40 nm gate length InSb P-Channel compressively strained quantum well field effect transistors for low-power (Vcc = 0.5 V) logic applications," *2008 IEEE International Electron Devices Meeting*, pp. 727–730, 2008.
12. A. Pethe, T. Krishnamohan, D. Kim, S. Oh, H.-S. P. Wong, Y. Nishi, and K. C. Saraswat, "Investigation of the performance limits of III-V double-gate n-MOSFETs," *IEEE IEDM Technical Digest*, pp. 605–608, 2005.
13. K. D. Cantley, Y. Liu, H. S. Pal, T. Low, S. S. Ahmed, and M. S. Lundstrom, "Performance analysis of III-V materials in a double-gate nano-MOSFET," *2007 IEEE International Electron Devices Meeting*, vol. 1 and 2, pp. 113–116, 2007.
14. M. V. Fischetti, L. Wang, B. Yu, C. Sachs, P. M. Asbeck, Y. Taur, and M. Rodwell, "Simulation of electron transport in high-mobility MOSFETs: Density of states bottleneck and source

starvation," *2007 IEEE International Electron Devices Meeting*, vol. 1 and 2, pp. 109–112, 2007.
15. M. Luisier, N. Neophytou, N. Kharche, and G. Klimeck, "Full-band and atomistic simulation of realistic 40 nm InAs HEMT," *2008 IEEE International Electron Devices Meeting*, pp. 887–890, 2008.
16. M. V. Fischetti and S. E. Laux, "Monte-Carlo simulation of transport in technologically significant semiconductors of the diamond and zincblende structures. II. Submicrometer MOSFETs," *IEEE Transactions on Electron Devices*, vol. 38, pp. 650–660, Mar 1991.
17. P. M. Solomon and S. E. Laux, "The ballistic FET: Design, capacitance and speed limit," *2001 IEEE International Electron Devices Meeting,* pp. 5.1.1–5.1.4, 2001.
18. S. E. Laux, "A simulation study of the switching times of 22-and 17-nm gate-length SOI nFETs on high mobility substrates and Si," *IEEE Transactions on Electron Devices*, vol. 54, pp. 2304–2320, Sep 2007.
19. M. V. Fischetti, T. P. O'Regan, S. Narayanan, C. Sachs, S. Jin, J. Kim, and Y. Zhang, "Theoretical study of some physical aspects of electronic transport in nMOSFETs at the 10-nm gate-length," *IEEE Transactions on Electron Devices*, vol. 54, pp. 2116–2136, Sep 2007.
20. N. Neophytou, T. Rakshit, and M. Lundstrom, "Performance analysis of 60-nm gate-length III-V InGaAs HEMTs: Simulations versus experiments," *IEEE Transactions on Electron Devices*, vol. 56, pp. 1377–1387, 2009.
21. H. S. Pal, T. Low, and M. S. Lundstrom, "NEGF analysis of InGaAs Schottky barrier double gate MOSFETs," *IEEE International Electron Devices Meeting 2008, Technical Digest*, pp. 891–894, 2008.
22. D. H. Kim and J. A. del Alamo, "Lateral and vertical scaling of $In_{0.7}Ga_{0.3}As$ HEMTs for Post-Si-CMOS logic applications," *IEEE Transactions on Electron Devices*, vol. 55, pp. 2546–2553, Oct 2008.
23. S. Datta, "Nanoscale device modeling: The Green's function method," *Superlattices and Microstructures*, vol. 28, pp. 253–278, Oct 2000.
24. R. Venugopal, S. Goasguen, S. Datta, and M. S. Lundstrom, "Quantum mechanical analysis of channel access geometry and series resistance in nanoscale transistors," *Journal of Applied Physics*, vol. 95, pp. 292–305, Jan 2004.
25. Z. B. Ren, R. Venugopal, S. Goasguen, S. Datta, and M. S. Lundstrom, "nanoMOS 2.5: A two-dimensional simulator for quantum transport in double-gate MOSFETs," *IEEE Transactions on Electron Devices*, vol. 50, pp. 1914–1925, Sep 2003.
26. Y. Liu and M. Lundstrom, "Simulation of III-V HEMTs for high-speed low-power logic applications," *ECS Transactions*, vol. 19, pp. 331–342, 2009.
27. Y. Liu, N. Neophytou, G. Klimeck, and M. S. Lundstrom, "Band-structure effects on the performance of III-V ultrathin-body SOI MOSFETs," *IEEE Transactions on Electron Devices*, vol. 55, pp. 1116–1122, May 2008.
28. G. Klimeck, S. S. Ahmed, H. Bae, N. Kharche, R. Rahman, S. Clark, B. Haley, S. H. Lee, M. Naumov, H. Ryu, F. Saied, M. Prada, M. Korkusinski, and T. B. Boykin, "Atomistic simulation of realistically sized nanodevices using NEMO 3-D—Part I: Models and benchmarks," *IEEE Transactions on Electron Devices*, vol. 54, pp. 2079–2089, Sep 2007.
29. I. Vurgaftman, J. R. Meyer, and L. R. Ram-Mohan, "Band parameters for III-V compound semiconductors and their alloys," *Journal of Applied Physics*, vol. 89, pp. 5815–5875, Jun 2001.
30. D. F. Welch, G. W. Wicks, and L. F. Eastman, "Calculation of the conduction-band discontinuity for $Ga_{0.47}In_{0.53}As$-$A_{l0.48}In_{0.52}As$ heterojunction," *Journal of Applied Physics*, vol. 55, pp. 3176–3179, 1984.
31. C. K. Peng, A. Ketterson, H. Morkoc, and P. M. Solomon, "Determination of the conduction-band discontinuity between $In_{0.53}Ga_{0.47}As$," *Journal of Applied Physics*, vol. 60, pp. 1709–1712, Sep 1986.
32. D. H. Kim and J. A. del Alamo, "Scalability of sub-100 nm thin-channel InAs PHEMTs," *IEEE International Conference on Indium Phosphide & Related Materials*, pp. 132–135, 2009.
33. D. A. Antoniadis and D. H. Kim, *Private Communication*, 2009.

34. D. H. Kim and J. A. del Alamo, "30 nm InAs pseudomorphic HEMTs on an InP substrate with a current-gain cutoff frequency of 628 GHz," *IEEE Electron Device Letters*, vol. 29, pp. 830–833, Aug 2008.
35. W. Y. Quan, D. M. Kim, and H. D. Lee, "Quantum *C–V* modeling in depletion and inversion: Accurate extraction of electrical thickness of gate oxide in deep submicron MOSFETs," *IEEE Transactions on Electron Devices*, vol. 49, pp. 889–894, May 2002.
36. A. Rahman, J. Guo, S. Datta, and M. S. Lundstrom, "Theory of ballistic nanotransistors," *IEEE Transactions on Electron Devices*, vol. 50, pp. 1853–1864, Sep 2003.
37. K. Natori, "Ballistic metal-oxide-semiconductor field-effect transistor," *Journal of Applied Physics*, vol. 76, pp. 4879–4890, Oct 1994.
38. A. Lochtefeld and D. A. Antoniadis, "On experimental determination of carrier velocity in deeply scaled NMOS: How close to the thermal limit?," *IEEE Electron Device Letters*, vol. 22, pp. 95–97, Feb 2001.
39. A. Lochtefeld, I. J. Djomehri, G. Samudra, and D. A. Antoniadis, "New insights into carrier transport in n-MOSFETs," *IBM Journal of Research and Development*, vol. 46, pp. 347–357, Mar–May 2002.
40. J. A. del Alamo, *FCRP e-Workshop*, Apr 2009.
41. M. S. Shur, "Low ballistic mobility in submicron HEMTs," *IEEE Electron Device Letters*, vol. 23, pp. 511–513, Sep 2002.
42. J. Wang and M. Lundstrom, "Ballistic transport in high electron mobility transistors," *IEEE Transactions on Electron Devices*, vol. 50, pp. 2185, Oct 2003.
43. M. Zilli, D. Esseni, P. Palestri, and L. Selmi, "On the apparent mobility in nanometric n-MOSFETs," *IEEE Electron Device Letters*, vol. 28, pp. 1036–1039, Nov 2007.
44. E. O. Johnson, "Insulated-gate field-effect transistor—Bipolar transistor in disguise," *Rca Review*, vol. 34, pp. 80–94, 1973.
45. M. Lundstrom and Z. B. Ren, "Essential physics of carrier transport in nanoscale MOSFETs," *IEEE Transactions on Electron Devices*, vol. 49, pp. 133–141, Jan 2002.
46. T. J. Walls, V. A. Sverdlov, and K. K. Likharev, "MOSFETs below 10 nm: Quantum theory," *Physica E-Low-Dimensional Systems & Nanostructures*, vol. 19, pp. 23–27, Jul 2003.
47. Y. Naveh and K. K. Likharev, "Modeling of 10-nm-scale ballistic MOSFET's," *IEEE Electron Device Letters*, vol. 21, pp. 242–244, May 2000.
48. M. Lundstrom and J. Guo, "Nanoscale transistors: Device physics, modeling and simulation," Springer, New York, 2005.

Chapter 4
Theory of HfO_2-Based High-k Dielectric Gate Stacks

Alexander A. Demkov, Xuhui Luo and Onise Sharia

Abstract Continuous scaling in semiconductor technology, associatsed with Moore's law, brought new materials in every functional element of Si-based metal-oxide-semiconductor (MOS) field effect transistors (FET). In particular, for a gate dielectric instead of traditional SiO_2 new HfO_2-based oxides with higher dielectric constants are now used. The introduction of these so-called high-k dielectrics opened the possibility of using high mobility semiconductors instead of Si in MOS technology. In this chapter we discuss density functional calculations of high-k oxides hafnia and zironia. After briefly describing theoretical methods used in our calculations we discuss the bulk properties, surfaces and interfaces of hafnia and zirconia relevant to advanced gate stack engineering.

4.1 Introduction

Scaling of the complementary metal oxide semiconductor (CMOS) technology is at the heart of Moore's law. However, after reaching the oxide thickness of 12 Å the scaling in Si CMOS has more or less stopped due to the prohibitively large gate leakage current caused by direct tunneling across the gate oxide. To circumvent this problem, a new dielectric with a higher dielectric constant had to be introduced. Hafnium dioxide or hafnia (HfO_2) is currently used instead of silica. The dielectric constant of hafnia is around 20 and it will serve as a gate dielectric for the next several technology nodes. However, if technology is to follow the Moore's law, the next step is the introduction of high mobility semiconductors as a replacement for silicon itself [1]. Ironically, this step has been made possible by the introduction of high-k dielectrics such as HfO_2, since it has effectively ended the dominance of the native oxide in the gate stack of the inversion-mode device. The physics and chemistry of transition metal oxides such as hafnia is much more complicated than that

A. A. Demkov (✉)
Department of Physics, The University of Texas at Austin, Austin, Texas 78712, USA
e-mail: demkov@physics.utexas.edu

S. Oktyabrsky, P. D. Ye (eds.), *Fundamentals of III-V Semiconductor MOSFETs*,
DOI 10.1007/978-1-4419-1547-4_4,

of Si_3N_4 or SiO_2, and theoretical calculations of their properties have proven to be extremely useful in gate stack development. The work horse of the modern computational materials science is density functional theory (DFT) within the local density approximation (LDA) and pseudopotential (PP) approximation. In this chapter we shall review our recent theoretical results in the area of high-k dielectrics. The rest of the chapter is organized as follows. In Sect. 4.2 the DFT-LDA-PP scheme is briefly outlined. In Sect. 4.3 we discuss our recent theoretical studies of bulk properties of hafnia and zirconia. In Sect. 4.4 we summarize our investigation of hafnia surface. We conclude with the review of the band alignment issues endemic to HfO_2-based gate stacks.

4.2 Theoretical Background

4.2.1 Density Functional Theory

Density functional theory introduced by Walter Kohn and co-workers formulates the many-body problem of interacting electrons and ions in terms of a single variable, namely the electron density [2, 3]. The Hohenberg-Kohn theorem states that: (1) the electron density alone is necessary to find the ground state energy of a system of N electrons, and (2) that the energy is a unique functional of the density [2]. Unfortunately, the precise form of that functional is not presently known. However, we do have reasonably good approximations. Although the Hohenber-Kohn theorem doesn't offer a specific method to compute the electron density, the solution for a slow varying density is given by the Kohn-Sham formalism [3]. Here an auxiliary system of non-interacting electrons in the effective potential is introduced, and potential is chosen in such a way that the non-interacting system has exactly the same density as the system of interacting electrons in the ground state.

The Kohn-Sham (KS) equations below need to be solved iteratively until the self-consistent charge density is found:

$$\left[-\frac{1}{2}\nabla^2 + v_{\mathit{eff}}(r)\right]\varphi_i(r) = \varepsilon_i\varphi_i(r) \tag{4.1}$$

with the effective potential given by:

$$v_{\mathit{eff}}(r) = v(r) + \int \frac{n(r')}{|r-r'|}dr' + \frac{\delta E_{xc}[n]}{\delta n(r)} \tag{4.2}$$

where $v(r)$ is the external potential (e.g., due to ions) and $E_{xc}[n]$ is the exchange correlation energy functional. The exact form of this functional is not known and has to be approximated. The density is given by:

$$n(r) = \sum_{occ} |\varphi_i(r)|^2 \tag{4.3}$$

where the sum is over the N lowest occupied eigenstates. For the slowly varying density Kohn and Sham introduced the local density approximation (LDA):

$$E_{xc}[n] = \int \varepsilon_{xc}(n(r))n(r)dr \tag{4.4}$$

where $\varepsilon_{xc}[n]$ is the exchange and correlation energy per particle in a uniform electron gas of density n. It is important to keep in mind that it is the electron density that is the "output" of the KS equations. Strictly speaking, the eigenvalues of the KS equations $\{\varepsilon_i\}$ have no direct physical meaning; nevertheless they are often very useful when the single particle electronic spectra (band structures) are discussed. The reasons behind the tremendous success of the Kohn-Sham theory are easy to identify. By solving essentially a single electron equation not much different from that of Hartree, but including the effects of exchange and correlation, one gets an upper estimate of the ground state energy of a many-body system. The theory is variational, and thus forces acting on the atoms can be calculated. The equation however, is non-linear and an iterative solution is needed.

Typically, the KS equations are projected onto a particular functional basis set, and the resulting matrix problem is solved. In terms of the basis, when solving KS equations one has two options. It is possible to discritize the equations in real space (this amounts to using δ-functions as a basis set) and solve them directly; these are so-called real space techniques [4]. Alternatively, one can choose a complete set of conventional functions. There are two major functional basis set types presently employed. For periodic systems plane waves offer an excellent expansion set which along with the fast Fourier transformations affords an easy to program computational scheme, the accuracy of which can be systematically improved by increasing the number of plane waves [5]. For systems with strong, localized potentials such as those of the first row elements, a large number of plane waves is necessary in the expansion, and calculations require the use of ultra-soft pseudopotentials (see below) to be feasible. The second choice is to use local orbitals such as e.g., atomic orbitals or any other spatially localized functions. Among advantages of a localized basis set are a smaller number of basis functions, and sparsity of the resulting matrix due to the orbital's short range. The disadvantages are the complexity of multi-center integrals one needs, and absence of the systematic succession of approximations, since the set is typically either under-complete or over-complete. In both cases calculations are computer intensive.

Since only the valence electrons are involved in bonding, and these electrons see a weaker potential due to screening by the core electrons), one can substitute the full Coulomb potential due to ions $v(r)$ with a smooth pseudopotential. This effectively reduces the number of electrons one needs to consider to the valence electrons only. The practical importance of this approximation should not be overlooked, a typical diagonalization algorithm scales as N^3 with the size of the matrix, thus for silicon we get a factor of 42 for the speed-up! The most straightforward way to introduce pseudopotential is due to Philips and Kleinman [6]. Today pseudopotentials used in electronic structure calculations may be broadly divided in three classes: the hard norm-conserving pseudopotentials [7], soft pseudopotentials [8], and Vanderbilt-type ultra-soft pseudopotentials [9]. The "softness" refers to how rapidly the potential changes in real space. The analogy comes from expanding a step function in a Fourier series; it takes a large number of plane waves to eliminate spurious oscillations at the step edge. On the other hand a "softer" function such as e.g., hyperbolic tangent can be expanded with greater ease. In general, hard pseudopotentials are more transferable. The choice of pseudopotential is in part dictated by the choice of a basis set used in a calculation. The use of local orbitals allows for a much harder pseudopotential. We will return to this point when discussing supercells.

Once the solution of KS equations is found, the total energy in the LDA is given by:

$$E_{total} \approx \sum_i \varepsilon_i - \frac{1}{2} \iint \frac{n(r)n(r')}{|r-r'|} drdr' + \int n(r) \{\varepsilon_{xc}(n(r)) - \mu_{xc}(n(r))\} dr \tag{4.5}$$

where the exchange-correlation potential is given by $\mu_{xc} \equiv \frac{d}{dn}\{\varepsilon_{xc}(n(r))n(r)\}$. Now all ground state properties of the system can in principle be calculated. In particular, since we are using the Born-Oppenheimer approximation, the total energy of the electronic system which is a function of the ionic positions $\{\vec{R}_1, \ldots \vec{R}_i \ldots \vec{R}_N\}$, can be used as a an inter-atomic potential. Note that unlike potential functions used in classical molecular dynamics or molecular mechanics methods, the energy function $E_{total}(\vec{R}_1, \ldots \vec{R}_N)$ is not a sum of pair-wise interactions $\frac{1}{2}\Sigma_{i,j}V_{i,j}$ but a true many-body interaction energy computed quantum mechanically! One can easily calculate a force acting on any atom i in the direction α using the so-called Hellman-Feynman theorem ($\frac{\partial E}{\partial \lambda} = \langle \varphi(\lambda)|\frac{\partial H}{\partial \lambda}|\varphi(\lambda)\rangle$) which is a rediscovery of the Ehrenfest result:

$$F_i^\alpha = \frac{\partial E_{total}}{\partial R_i^\alpha}, \quad \alpha = x, y, z \tag{4.6}$$

At this point one can find the lowest energy atomic configuration by employing an energy minimization technique such as damped molecular dynamics or a conjugate gradient method. Alternatively, a real molecular dynamics (MD) simulation can be launched. One has to keep in mind, however, that electronic frequencies $\frac{E_i - E_j}{\hbar}$ are much higher than a typical phonon frequency ω and for a stable simulation the time step needs to be a small fraction of the characteristic atomic period. The calculation then proceeds as follows. The KS energy is first calculated in a self-consistent manner for the initial atomic configuration, the Hellman-Feynman forces are evaluated, and atoms are moved to the next time step via some MD algorithm (Verlet, Gear, etc. [10]). At the new atomic configuration the KS equations are solved again, and the procedure is repeated. Needless to say, these are very expensive calculations.

They offer certain advantages if a temperature dependence of a particular quantity is sought, since MD can be performed at finite temperature. For example, the Fourier transform of the velocity auto-correlation function gives the vibration spectrum, thus calculations performed at different temperature would give the temperature dependence of the phonon frequency.

Other properties such as for example, the dielectric constant, can now be evaluated. In this case the use of the periodic boundary conditions does cause some complications; or rather it is the absence of the surface in an infinite periodic solid that is the problem. Vanderbilt has shown that the change in electronic polarization can be calculated using the geometric or Berry phase of electrons [11]:

$$P_\alpha^{el} = \frac{i}{\Omega} \sum_{ki} \left\langle u_{ki} \left| \frac{\partial}{\partial k_\alpha} \right| u_{ki} \right\rangle, \tag{4.7}$$

where Ω is the unit cell volume, k is the Bloch vector, and u_{ki} is the cell periodic part of the Bloch wave function. Once the change in polarization with respect to a reference state of the system is determined, Born effective charges $Z_{i\alpha}^{*M}$ can be evaluated, and the dielectric constant is given by:

$$\varepsilon_{\alpha\beta} = \varepsilon_{\alpha\beta}^{\infty} + \frac{4}{\pi} \sum_i \frac{Z_{i\alpha}^{*M} Z_{i\beta}^{*M}}{\omega_i^2 - \omega^2} \tag{4.8}$$

The electronic contribution $\varepsilon_{\alpha\beta}^{\infty}$ can be computed using the linear response theory. The values thus computed typically overestimate experiment by about 20%, mainly due to the error in the band gap. A semi-empirical "scissor" correction is then used in which the conduction bands are moved up in energy by hand to match the experimental spectrum.

All calculations discussed in this chapter are performed using density functional theory with the plane-wave based Vienna ab initio package (VASP) [12, 13].

4.2.2 Modeling Interfaces and Surfaces

The plane wave method is particularly well suited for studying periodic systems. Clearly a surface is a system in which the periodicity in the direction perpendicular to the surface is broken. To perform surface calculations with a plane wave basis set a large simulation cell or a supercell is introduced in order to maintain artificial periodicity. This supercell contains a slab of bulk material (with several unit cells of the corresponding crystal) and vacuum slab in the direction perpendicular to the surface. The periodic boundary condition in the direction normal to the surface is applied for the supercell dimension, rather than the physical crystal cell side. The lateral periodicity depends on the area of the surface one intends to simulate. Thus the "universe" is filled with infinite parallel slabs of chosen thickness, separated by infinite parallel slabs of vacuum. It is crucial that the length of a supercell in the direction normal to the surface is large enough to eliminate any spurious interactions between the cells across the vacuum region. The thickness of a slab should be sufficient for bulk properties to be restored in the middle of it. The supercell obviously creates two surfaces, and it is advisable to use a symmetric termination of the slab if possible.

In principle, the larger the supercell is chosen the better it approximates true surface (or rather a set of two identical surfaces). However, the calculation also becomes more demanding, as we shall now demonstrate. In the case of a periodic system we write the eigenfunctions $\Psi_{n,\mathbf{k}}(\mathbf{r})$ of the KS equations as Bloch functions:

$$\Psi_{n,\mathbf{k}}(\mathbf{r}) = u_{n,\mathbf{k}}(\mathbf{r})e^{i\mathbf{kr}} \tag{4.9}$$

where $u_{n,\mathbf{k}}(\mathbf{r})$ is a lattice periodic function, n is the band index, and a wave vector $\mathbf{k}$ belongs to the first Brillouin zone (BZ). Since $u_{n,\mathbf{k}}(\mathbf{r})$ is periodic, it can be expanded over the reciprocal lattice:

$$u_{n,k}(r) = \sum_{G'} \varphi_{n,k}(G')e^{i\vec{G}'\vec{r}} \tag{4.10}$$

Where, **G'** are the reciprocal lattice vectors. This expansion goes to infinity! Note that we actually deal with two types of infinities here. One is due to the infinite periodic nature of the crystal and is captured by the wave vector **k**; the other comes from this expansion. For practical purposes the sum over **G'** is restricted to plane waves with kinetic energy below a given cutoff energy E_{cut}. Thus, defining the set $\Omega(\mathbf{G})$:

$$\Omega(G) := \left\{ G \,\middle|\, \frac{\hbar^2}{2m} |\vec{k} + \vec{G}|^2 \le E_{cut} \right\} \tag{4.11}$$

we obtain the following expansion of the Kohn-Sham wave functions:

$$\psi_{n,\vec{k}}(r) = \sum_{G \in \Omega(G)} \varphi_{n,k}(G)e^{i(\vec{G}+\vec{k})\vec{r}} \tag{4.12}$$

The cutoff energy E_{cut} controls the numerical convergence and depends strongly on the elements which are present in the system under investigation. For example, first row elements with strong potentials require higher cutoff energy. Here we immediately see the weakness of the supercell method. In the direction normal to the surface, the reciprocal cell vectors $|\vec{G}_\perp|$ are very short due to a large length of the direct space cell (often many multiples of the physical cell's lattice constant). Thus a very large number of plane waves is needed to reach the convergence. This is the price one has to pay for the artificial periodicity. The introduction of ultra-soft pseudopotentials made these calculations practical. The localized basis set does have advantage of being less sensitive to the simulation cell size; however, the range of the orbitals should be sufficient to describe the vacuum decay.

In what follows the DFT-LDA scheme is used to calculate the band discontinuity at the interface between two dissimilar materials. The discontinuity can be estimated using the reference potential method originally introduced by Kleinman [14]. Van de Walle and Martin proposed using the macroscopically averaged electrostatic potential as reference energy [15]. The method requires calculating a heterojunction AB in either slab (in this case you would have free surfaces) or supercell geometry to compute the average reference potential across the interface, and two additional bulk calculations to locate the valence band top (VBT) in materials A and B with respect to the average potential. For a supercell (or a slab) containing the interface one calculates the average potential using the formula:

$$\bar{V}(z) = \frac{1}{d_1 d_2} \int_{z-d_1/2}^{z+d_1/2} dz' \int_{z'-d_2/2}^{z'+d_2/2} dz'' V(z'') \tag{4.13}$$

Where $V(z)$ is obtained by the *xy*-plane averaging (a simple $\frac{1}{(a_x \cdot a_y)} \iint_{cell} dxdy$ integration) of the electrostatic potential:

$$V(r) = -\sum_i \frac{Z_i e^2}{|r - R_i|} + e^2 \int \frac{n(r')}{|r - r'|} dr' \tag{4.14}$$

The parameters d_1 and d_2 are the inter-planar distances along the z direction (normal to the interface) in materials A and B, respectively. This produces a smooth reference potential. Assuming that far away from the interface the potential reaches its bulk value, one can place the corresponding VBTs with respect to the average potential on both sides of the interface using the bulk reference, and thus determine the VBO. The conduction band offset has to be inferred using the experimental values of the band gaps, since those are seriously underestimated in the DFT-LDA calculations.

4.3 Properties of Bulk Hafnia and Zirconia

Hafnia is remarkably similar to zirconia both structurally and chemically. For instance, the difference between the lattice constants of these two materials is so small that the conventional powder X-ray diffraction can hardly detect their individual presence [16]. Since 1968 the fact that HfO_2 and ZrO_2 can form continuous solid solution systems has been confirmed [17]. Bulk crystalline hafnia and zirconia both undergo a succession of phase transitions, from the high temperature high symmetry cubic phase (space group *Fm3m*, see Fig. 4.1a) to slightly distorted structures with tetragonal (space group $P4_2/nmc$, see Fig. 4.1b) and monoclinic (space group $P2_1/c$, Fig. 4.1c) symmetry. The cubic phase has a fluorite-type structure in which each Hf^{4+} or Zr^{4+} ion is coordinated by eight equidistant O^{2-} ions, and each anion is tetrahedrally coordinated by four Hf cations (Fig. 4.1a). As shown in the Fig. 4.1a,

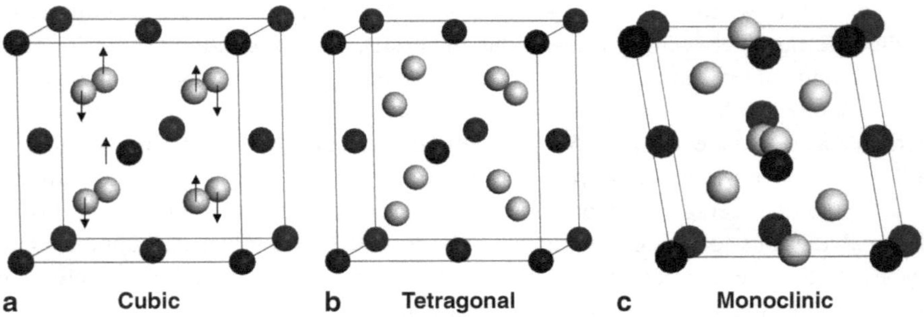

Fig. 4.1 Structure of three HfO_2 (ZrO_2) polymorphs: **a** cubic, **b** tetragonal and **c** monoclinic. Black balls and white balls denote Hf and O atoms, respectively. The *arrows* in Fig. 4.1a show the movement pattern of oxygen atoms corresponding to the imaginary frequency mode at the X k-point in the first BZ of the cubic phase [18] (Reprinted with permission. Copyright 2009 American Physical Society)

there are eight oxygen atoms inside the cubic cell. If four diagonal oxygen atoms are moved up and the other four are moved down, and the c/a ratio is not one any more, the cubic phase changes into tetragonal (Fig. 4.1b). The tetragonal phase is characterized by two lattice parameters and oxygen displacement along the *c* axis Δu. Once the tetragonal phase is distorted further with the *c* axis tilted and atoms shuffled, the monoclinic phase appears as the ground state of hafnia and zirconia (Fig. 4.1c).

Table 4.1 presents optimized theoretical structural parameters for HfO_2 and ZrO_2 along with the experimental data [18]. Because the first principle calculations give us the ground state of a system at zero temperature, the theoretical volume is smaller than the experimental value measured at room temperature for both ZrO_2 and HfO_2 [16, 19]. As seen in Table 4.1 monoclinic HfO_2 has a smaller unit cell than monoclinic ZrO_2 both theoretically and experimentally. However, the experimental volume of tetragonal HfO_2 is larger than that of tetragonal ZrO_2. We note that while calculations are done at 0 K, the tetragonal phase of HfO_2 only exists above around 2000 K and the tetragonal ZrO_2 can appear when temperature is over 1450 K (Table 4.2). Therefore, theory and measurements of the tetragonal phase for HfO_2 and ZrO_2 are all conducted at different temperature. We also compare the calculated metal-oxygen bond length with corresponding experimental values [20] in Table 4.3. The bonds are referred to as in Fig. 4.2 showing MO_7 coordination polyhedron of the monoclinic phase. Overall, the agreement is fair with the worst deviation of 3% found for the I_b bond in hafnia.

The fact that hafnia's unit cell is smaller than that of zirconia is related to the electron configuration of hafnium and zirconium metals. The outer electrons of hafnium and zirconium have the configurations of $4f^{14}5d^{2}6s^{2}$ and $4d^{2}5s^{2}$, respectively. The rare earth elements immediately preceding hafnium add electrons to the inner $4f$ shell from cerium through to lutetium. Since no outer electrons are added to compensate for the increased nuclear charge, there is a contraction of the atomic size known as a lanthanide contraction. The atomic radii of hafnium and zirconium

Table 4.1 Theoretical structural parameters for the HfO_2 and ZrO_2 polymorphs in comparison with the experimental data

	a (Å)	b (Å)	c (Å)	β (deg)	Data type and reference
Mono. HfO_2	5.1156	5.1722	5.2948	99.18	Experiment from [19]
	5.119	5.169	5.29	99.25	Experiment from [17]
	5.117	5.172	5.284	99.37	Experiment from [21]
	5.029	5.132	5.183	99.48	Our theoretical work
	5.106	5.165	5.281	99.35	Theory from [22]
Mono. ZrO_2	5.156	5.191	5.304	98.9	Experiment from [23]
	5.145	5.208	5.311	99.23	Experiment from [24]
	5.115	5.23	5.26	99.61	Our theoretical work
	5.1065	5.1678	5.27	99.21	Theory from [25]
Tetr. HfO_2	5.14		5.25		Experiment from [26]
	5.15		5.295(1760)		Experiment from [27]
	5.175		5.325(2000)		Experiment from [28]
	4.98		5.07		Our theoretical work
	5.056		5.127		Theory from [22]
Tetr. ZrO_2	5.0282		5.0987		Theory from [25]
	5.07		5.14		Our theoretical work
	5.094		5.177		Experiment from [19]

are 1.442 Å and 1.454 Å, respectively [29]. The electronegativity values are 1.23 for hafnium and 1.22 for zirconium [30].

The total energy (enthalpy) differences between the different phases are summarized in Table 4.2. Our calculations correctly reproduce the energetic ordering of the phases, increasing in energy from monoclinic to tetragonal to cubic. Because the

Table 4.2 Phase transition data for ZrO_2 and HfO_2

	Transition	Ref.	T (K)	ΔH (kJ/mol)	ΔS (J/mol/K)	Method
ZrO_2	M–T	28	1475	5.272 ± 0.544	3.56	Calorimetry
		Our work	1560	7.5		Theoretical calculation
	T–C	29	2650	5.564	2.09	Assessed
	M–C	Our work		14.26		Theoretical calculation
HfO_2	M–T	16	2052	8.208		Optimization
		Our work	2078	7.7 ± 0.6		DSC and TMA
		Our work	1920	10.21		Theoretical calculation
	T–C	16	3073	11.212		Optimization
		Our work	2930			Theoretical calculation
	M–C	Our work		18.11		Theoretical calculation

M monoclinic, *T* tetragonal, *C* Cubic, *DSC* differential scanning calorimetry, *TMA* thermomechanical analysis

Table 4.3 Experimental and theoretical bond lengths for the HfO_2 and ZrO_2 polymorphs. Ia–Ic and IIa–Iid refer to oxygen atoms shown in the Fig. 4.2. The unit is Å

	Experiment for Hf–O distance	Experiment for Zr–O distance	Theory for Hf–O distance	Theory for Zr–O distance
Ia	2.031	2.057	2.024	2.063
Ib	2.174	2.163	2.111	2.163
Ic	2.052	2.051	2.025	2.058
IIa	2.17	2.189	2.143	2.181
IIb	2.162	2.22	2.114	2.223
IIe	2.202	2.151	2.187	2.146
IId	2.254	2.285	2.189	2.230

Reprinted with permission from Ref. [18]. Copyright 2009 American Physical Society

enthalpy of a phase transition is equal to the difference in total energy between two phases (the PΔV term relating energy and entropy being negligible at atmospheric pressure), we can directly compare the experimental measurements of enthalpy and theoretical total energy as shown in Table 4.2. The agreement is rather good for both HfO_2 and ZrO_2. The energy difference between the polymorphs in hafnia is larger than that in zirconia.

Phonons of hafnia and zirconia are important to their thermodynamic properties and have been recently studied with Raman and infra-red (IR) spectroscopy [31, 32]. Experiments show that in the high frequency region the phonon frequencies of hafnia are higher than that of zirconia and in the low frequency region the frequencies of hafnia are lower than those of the corresponding modes of zirconia. Here we perform a systematic study of phonons in hafnia and zirconia, and compare theoretical values of the Raman and IR active modes to available experiment [18]. We use the Hellmann-Feynman forces to calculate the dynamical matrix. The short-range force constant matrix is computed in the 96-atom $2 \times 2 \times 2$ super-cell (one

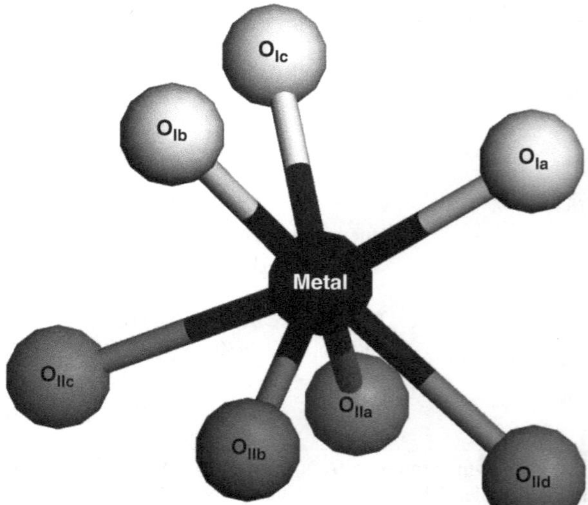

Fig. 4.2 The structure of oxygen atoms in the ZrO_7 and HfO_7 polyhedra in ZrO_2 and HfO_2. Three of seven oxygen atoms are threefold coordinated and four of them are fourfold coordinated [18] (Reprinted with permission. Copyright 2009 American Physical Society)

can chose a 12-atom unit cells for all polymorphs, see the next section). To calculate the force constant matrix each atom is displaced in turn along each Cartesian axis by ± 0.04 Å, and the numerical derivative of the force is calculated and averaged to eliminate the odd power unharmonicity. The dynamical matrix is then computed by the usual lattice Fourier transform:

$$\mathbf{D}^0_{\alpha\beta}(\mathbf{k};\mu\nu) = \frac{1}{\sqrt{M_\mu M_\nu}} \sum_{\mathbf{m}} \mathbf{B}(0,\mu;\mathbf{m},\nu) \times \exp\{-2\pi i\mathbf{k}\cdot[\mathbf{R}(0,\mu) - \mathbf{R}(\mathbf{m},\nu)]\} \tag{4.15}$$

Because hafnia is an ionic compound, one needs to consider the long range Coulomb contributions (particularly at the Γ point). The long range part of the dynamical matrix is given by [33]:

$$\mathbf{D}^{long}_{\alpha,\beta}(\mathbf{k};\mu\nu) = \frac{4\pi e^2}{V\varepsilon_\infty\sqrt{M_\mu M_\nu}} \frac{[\mathbf{k}\cdot\mathrm{Z}^*(\mu)]_\alpha[\mathbf{k}\cdot\mathrm{Z}^*(\upsilon)]_\beta}{|\mathbf{k}|^2} \exp\{-2\pi i\mathbf{g}\cdot[\mathbf{r}(\mu)-\mathbf{r}(\upsilon)]\}\exp(-\mathbf{k}^2/\rho^2) \tag{4.16}$$

Here, ρ is a parameter to control the range of the long range term. We choose $\rho = 0.06\,\text{Å}^{-1}$ and $\varepsilon_\infty = 5$ [22]. We use the Born effective charge tensors recently calculated by Zhao and Vanderbilt [22]:

$$Z^*_{\mathrm{Hf}} = \begin{pmatrix} 5.56 & -0.47 & 0.96 \\ -0.13 & 5.55 & 0.14 \\ 0.21 & 0.41 & 4.74 \end{pmatrix}, \quad Z^*_{\mathrm{O1}} = \begin{pmatrix} -3.09 & 0.97 & -0.58 \\ 1.37 & -2.73 & -0.71 \\ -0.18 & -0.61 & -2.24 \end{pmatrix},$$

$$Z^*_{\mathrm{O2}} = \begin{pmatrix} -2.48 & 0.20 & -0.39 \\ 0.21 & -2.82 & 0.35 \\ -0.07 & 0.42 & -2.58 \end{pmatrix} \tag{4.17}$$

In Fig. 4.3a–c we show the phonon spectra of cubic, tetragonal and monoclinic hafnia calculated including only the short-range contributions to the dynamical matrix. The phonon dispersion is plotted along the high symmetry directions in the first BZ. The path of the calculation for the cubic phase starts and ends at the Γ (0, 0, 0) point, going through $X(0,0,\frac{2\pi}{a})$, $L(\frac{\pi}{a},\frac{\pi}{a},\frac{\pi}{a})$ and $W(0,\frac{\pi}{a},\frac{2\pi}{a})$ as shown in Fig. 4.3a. For the tetragonal phase we start at the Γ point passing through $M(\frac{\pi}{a},\frac{\pi}{a},0)$, $X(0,\frac{\pi}{a},0)$, Γ, $Z(0,0,\frac{\pi}{a})$ and end at $A(\frac{\pi}{a},\frac{\pi}{a},\frac{\pi}{a})$ (see Fig. 4.3b). The high symmetry points along the path for monoclinic phase are denoted as Γ, $B(0,0,\frac{\pi}{a})$, $A(\frac{\pi}{a},0,\frac{\pi}{a})$, Γ, $E(\frac{\pi}{a},\frac{\pi}{a},\frac{\pi}{a})$, $Y(\frac{\pi}{a},0,0)$ and Γ. The inclusion of the long range correction to cubic phonons mostly influences the modes close to the Γ point as can be seen in Fig. 4.3d. So in the total density of states shown in Fig. 4.4 for tetragonal and monoclinic hafnia the long range correction is omitted. The phonon DOS is obtained by diagonalizing the dynamical matrix $D(\vec{k})$ on a dense 24 × 24 × 24 special k-point grid over the entire first BZ. In Fig. 4.4 the solid and dashed lines show the density of states for tetragonal and monoclinic phases, respectively. The DOS

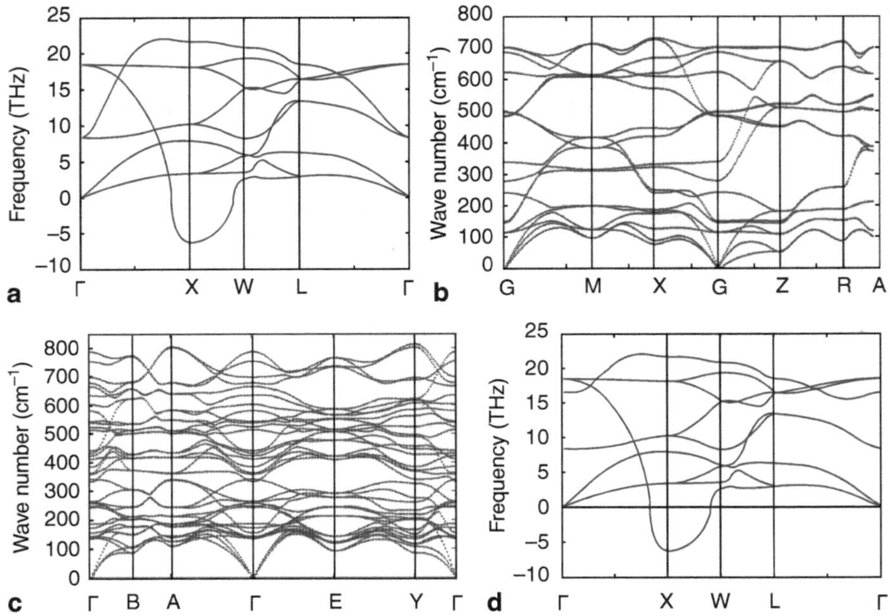

Fig. 4.3 **a** Calculated phonon dispersion of cubic hafnia without the long range interaction. **b** Phonon dispersion of tetragonal hafnia without the long range interaction. **c** Phonon dispersion of monoclinic hafnia without the long range interaction. **d** Calculated phonon dispersion of cubic hafnia with the long range interaction [18] (Reprinted with permission. Copyright 2009 American Physical Society)

of the monoclinic phase is blue-shifted with respect to that of the tetragonal phase. Both spectra may be roughly divided into two regions, the low frequency part (below 350 cm^{-1}) and the high frequency part. In the case of monoclinic hafnia two parts are separated by a quasi gap. The origin of this separation will be discussed later.

The heat capacity of HfO_2 and ZrO_2 can now be calculated using the phonon density of states g(ω) as follows [34]:

$$C_v = \frac{1}{4k_B T^2} \int_0^\infty d\omega \cdot g(\omega) \frac{\hbar^2 \omega^2}{\sinh^2\left(\dfrac{\hbar\omega}{2k_B T}\right)} \tag{4.18}$$

We compare the heat capacity of monoclinic HfO_2 and ZrO_2 in Fig. 4.5. The constant volume specific heat C_v of HfO_2 and ZrO_2 calculated using Eq. (4.18) is plotted in Fig. 4.6a, b for monoclinic phase and tetragonal phase, respectively. In the insert of Fig. 4.6a we compare our calculations with the experimental data [35–37]. In the case of zirconia the agreement with experiment is very good, at least in the low temperature regime when the harmonic approximation is expected to work well. The agreement is somewhat less impressive in the case of hafnia; however, the

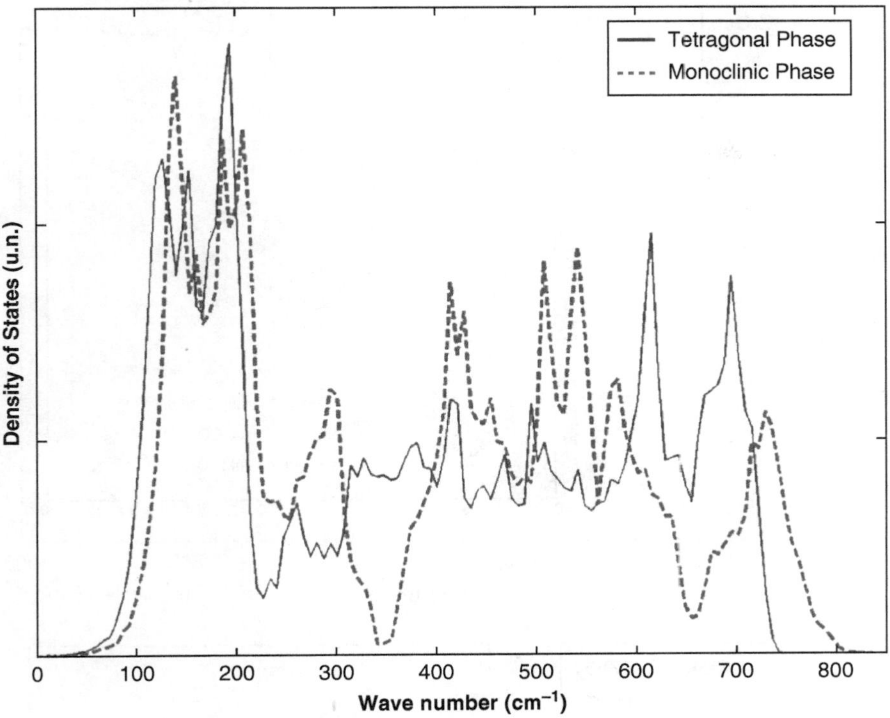

Fig. 4.4 The phonon density of states of monoclinic and tetragonal hafnia. The *dashed* and *solid lines* denote the phonon density of states of monoclinic and tetragonal hafnia,respectively [18] (Reprinted with permission. Copyright 2009 American Physical Society)

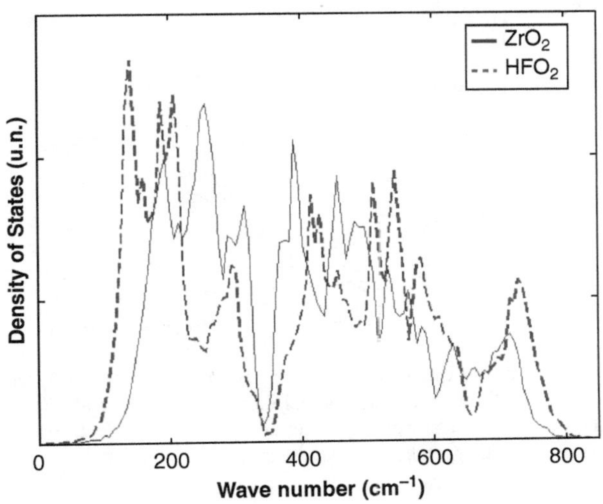

Fig. 4.5 The density of states of phonons of monoclinic zirconia and hafnia. The *dashed* and *solid lines* denote the phonon density of states in monoclinic hafnia and zirconia, respectively [18] (Reprinted with permission. Copyright 2009 American Physical Society)

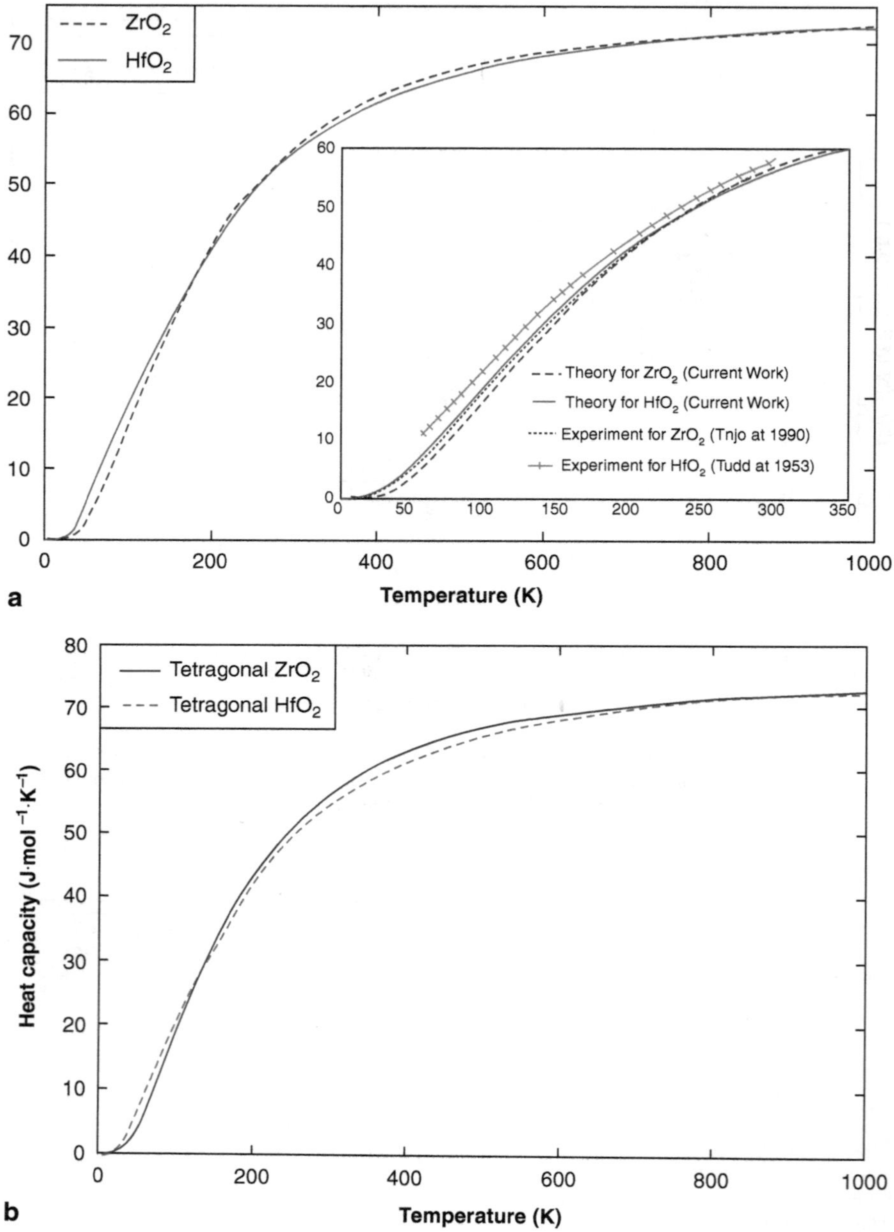

Fig. 4.6 **a** The calculated heat capacity of monoclinic HfO_2 and ZrO_2 from 0 to 1000 K. In the embedded figure the experimental data is compared with theoretical results from 0 to 350 K. **b** The calculated heat capacity of tetragonal HfO_2 and ZrO_2 from 0 to 1000 K [18] (Reprinted with permission. Copyright 2009 American Physical Society)

experimental data [37] are rather old and do not extend down to liquid helium temperature. An interesting feature of our calculation is a crossover at 190 K when C_v of zirconia becomes larger than that of hafnia. As we discuss later, the crossover can be attributed to the features in the phonon density of states and traced down to the interplay between the force constant and atomic mass.

Vibrational spectra of HfO_2 and ZrO_2 have been recently examined by Raman spectroscopy by Quintard et al. [31]. We list their results in Table 4.4; the Raman mode frequencies in ZrO_2 are slightly lower than those of HfO_2 in the high frequency region (above 350 cm^{-1}). However, in the low frequency region, the frequencies of ZrO_2 are about 10–25% higher than those of HfO_2. The results of our calculations of Raman modes of monoclinic HfO_2 and ZrO_2 are summarized in Table 4.5. We find a similar trend in mode frequencies when comparing the two oxides. Naively one would expect the vibrational spectrum of hafnia to be red-shifted with respect to that of zirconia because of a larger atomic mass of hafnium (for a simple harmonic oscillator we have $\omega = \sqrt{k/m}$ where k is the spring constant and m is the mass). However, the picture is more complicated. Our ab initio calculations reveal that the magnitude of the interatomic force constants in zirconia is approximately 10% smaller than that in hafnia. In Fig. 4.7 we compare the effective force constants for both materials by plotting the absolute values of the dynamical matrix at the Γ point without including the mass factor (we plot $|\sum_m B(0, \mu; m, \nu)|$, where B is the force constant coupling an atom μ in the central cell with an atom ν in the m-th cell). The highest value in hafnia is almost 200 eV/Å^2, while in zirconia it is only 160 eV/Å^2. This would suggest a blue-shift of the hafnia spectrum. However, for the actual

Table 4.4 Experimental Raman spectra [31] of the monoclinic phases of ZrO_2 and HfO_2

Number	Experiments		
	HfO_2 exp. (cm^{-1})	ZrO_2 exp. (cm^{-1})	HfO_2/ZrO_2
1	774 (s)	757	1.02
2	642 (s), B_g	616	1.04
3	672 (s), A_g	637	1.05
4	580 (m), A_g	556	1.04
5	552 (m), B_g	536	1.03
6	522 (m), B_g	505	1.03
7	500 (vs), A_g	476	1.05
8	398 (s)	385	1.03
9	384 (s), B_g	381	1.01
10	336 (s), A_g	348	0.97
11	256 (s), B_g	334	0.78
12	270 (m), A_g	305	0.88
13	242 (m)	270	0.9
14	168 (m), B_g	224	0.75
15	150 (m), A_g	190	0.79
16		179	
17	135 (s), A_g	179	0.75
18	108/83 (s), A_g	102	1.06/0.82

Reprinted with permission from Ref. [18]. Copyright 2009 American Physical Society
Band Intensity: *s* strong, *m* medium, *w* weak

Table 4.5 Calculated Γ-point phonon frequencies classified according to irreducible representations of symmetry group C_{2h} of monoclinic hafnia. A_g and B_g modes are Raman active, while A_u and B_u modes are IR active

	Wave number (cm^{-1}) of A_g (Raman Active modes)			Wave number (cm^{-1}) of B_g (Raman Active modes)			Wave number (cm^{-1}) of A_u (IR)		Wave number (cm^{-1}) of B_u (IR)	
	HfO_2	ZrO_2	HfO_2/ZrO_2	HfO_2	ZrO_2	HfO_2/ZrO_2	HfO_2	ZrO_2	HfO_2	ZrO_2
1	695.9	629.71	1.105	785.47	747.8	1.050	675.89	643.32	752.02	712.15
2	600.74	552.22	1.088	663.23	612.9	1.082	632.1	574.12	540.87	493.73
3	511.5	453.88	1.127	577.91	538.11	1.074	521.11	475.86	429.97	429.79
4	409.83	388.41	1.055	536.67	481.38	1.115	439.17	404.31	356.76	370.26
5	361	359.19	1.005	421.23	389.1	1.083	382.72	364.68	331.56	326.87
6	257.01	334.09	0.769	338.86	331.95	1.021	260.91	264.1	263.3	321.73
7	153.34	194.39	0.789	245.84	319.09	0.770	185.57	234.53	241.66	232.14
8	140.59	184.32	0.763	170.86	224.11	0.762	139.76	182.73	0	0
9	133.29	133.46	0.999	135.85	176.3	0.771	0	0	0	0

frequency calculation (diagonalization of the dynamical matrix) the force constant matrix is "re-normalized" by the mass factor as follows:

$$\mathbf{D}^0(\mathbf{k};\mu\nu) = \frac{1}{\sqrt{M_\mu M_\nu}} \sum_{\mathbf{m}} \mathbf{B}(0,\mu;\mathbf{m},\nu) \times \exp\{-2\pi i \cdot [\mathbf{R}(0,\mu) - \mathbf{R}(\mathbf{m},\nu)]\} \tag{4.19}$$

In Fig. 4.8 we show relative amplitudes of the atomic movement in all vibration modes of monoclinic hafnia at the Γ point. It is clearly seen that the low frequency modes correspond to the movement involving predominantly metal atoms. The mass ratio between Hf and Zr is approximately two. In other words, even though the force constant is smaller in zirconia, the dynamical matrix is enhanced by the mass factor in the low frequency range. In the high frequency region, the vibrational modes are associated with the movement of oxygen atoms (Fig. 4.8). Thus in both hafnia and zirconia the mass factors are the same, and the frequency is controlled by the force constant. Due to weaker force constants of zirconia, its high frequency modes have frequencies lower than those of hafnia. This is true beyond the Γ point as can be seen in the DOS plots in Fig. 4.5. Also, we can attribute the spectral gap at around 350 cm^{-1} in the phonon density of states for both hafnia and zirconia to the transition from the metal-dominated modes to oxygen dominated ones.

We have shown that in the low frequency region the phonon modes of zirconia have higher frequencies than those of hafnia. This result has peculiar implications for thermodynamic properties of two oxides. The heat capacity of zirconia should be smaller than that of hafnia at low temperature because only low frequency modes can be excited. On the other hand, at high temperature the heat capacity of zirconia should be larger than that of hafnia because in the high frequency region zirconia has higher frequency modes. As shown in Fig. 4.6 we find that the heat capacity of monoclinic HfO_2 crosses that of monoclinic ZrO_2 at 190 K. The result should not

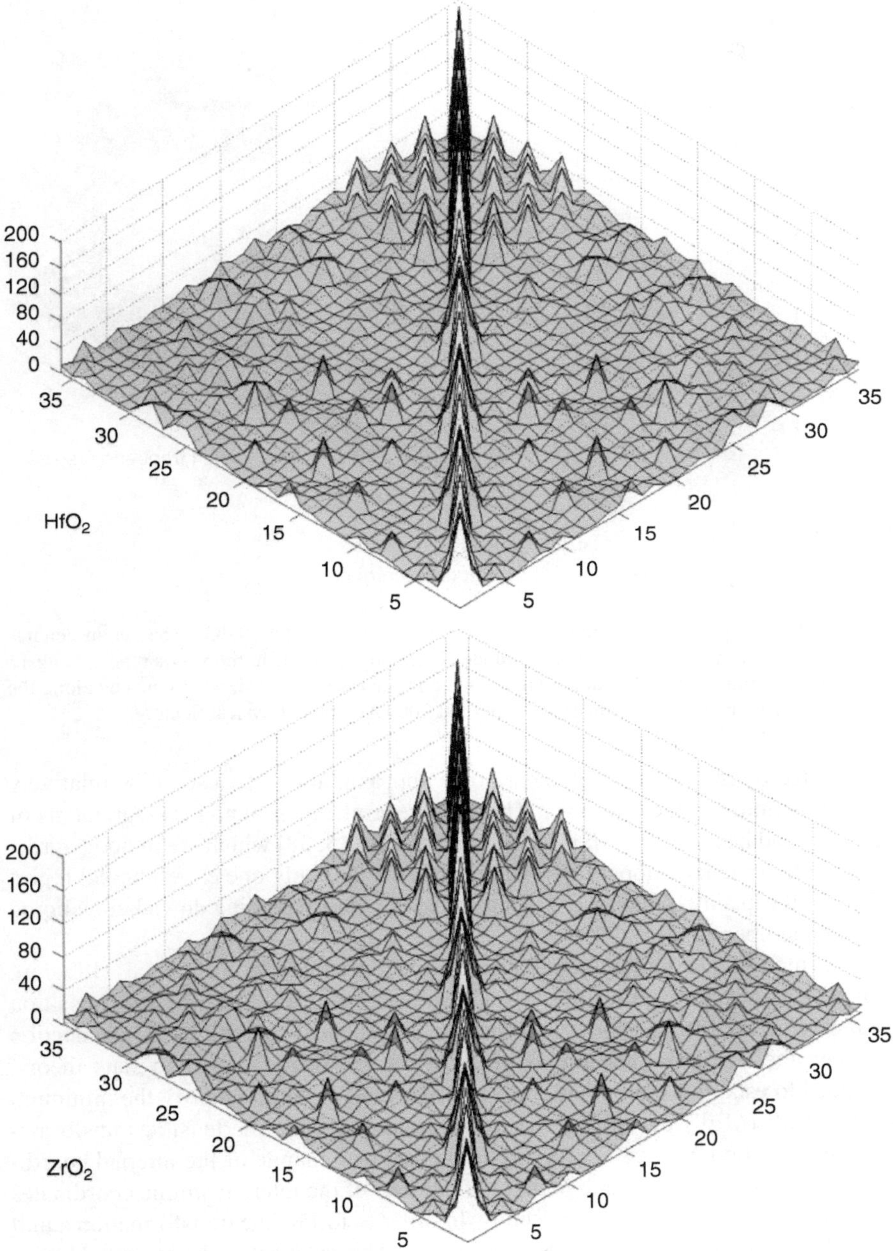

Fig. 4.7 The 36 × 36 force constant matrices of HfO_2 and ZrO_2 in eV/Å^2. The first 24 × 24 block corresponds to oxygen atoms, and the 12 × 12 diagonal block starting with row 25 corresponding to the metal atoms (Hf or Zr) [18] (Reprinted with permission. Copyright 2009 American Physical Society)

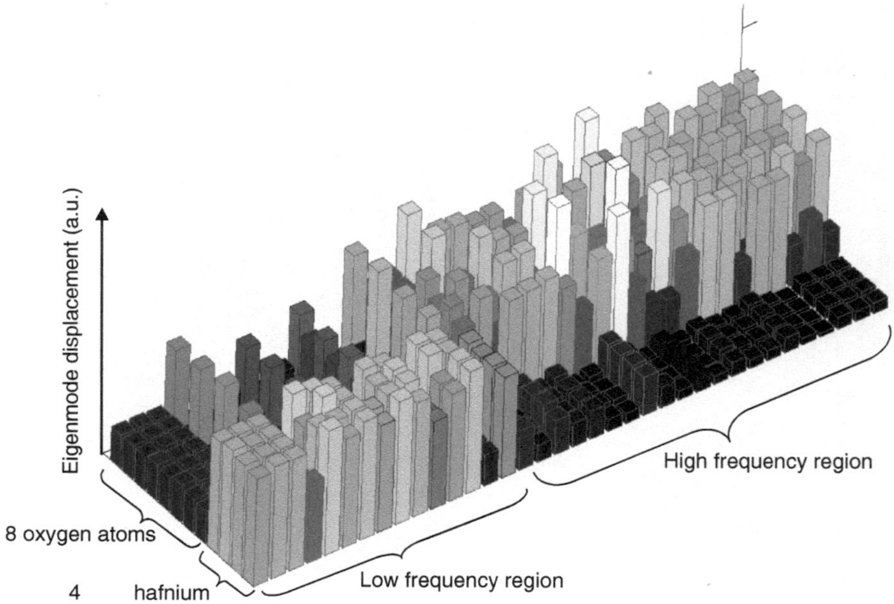

Fig. 4.8 The bar plot of 36 eigenmodes at the Γ point of monoclinic HfO_2. The height denotes the displacement of one atom in the eigenmode. There are 12 points in the x-axis representing 12 atoms in the primitive unit. 36 modes are arranged in the ascending order of frequency along the y-axis [18] (Reprinted with permission. Copyright 2009 American Physical Society)

be affected by the validity of the harmonic approximation because of the relatively low temperature of the crossover. There are several experimental measurements of enthalpy and heat capacity of monoclinic zirconia [35, 38] which are in good agreement with our calculations. However, we have found only one experimental report of the heat capacity of hafnia dating back to 1953, and extending down below liquid nitrogen temperature [37].

The monoclinic to tetragonal phase transformations in ZrO_2 and HfO_2 are believed to be martensitic [39–41]. However, the details such as the transition path or whether the transition is proper are not known. To explore the connection between the tetragonal and monoclinic phases we turn to transition state theory. We need to establish the potential energy surface (PES) and identify the minimum energy path (MEP) describing the transformation. Since martensitic transformations include both unit cell deformation (strain) and change of the internal coordinates ("shuffle"), the PES and MEP are functions of the internal atomic coordinates as well as of the unit cell lattice vectors. In order to follow the transformation a unit cell common to both phases needs to be chosen. This establishes the so-called lattice correspondence between the two phases. The set of lattice vectors (in Angström) for a 12 atom primitive unit cell of the monoclinic phase is:

$$\eta_m = \begin{pmatrix} 5.0258 & 0.0000 & 0.0000 \\ 0.0000 & 5.1323 & 0.0000 \\ -0.8616 & 0.0000 & 5.1109 \end{pmatrix} \begin{matrix} \cdots\cdots a_m \\ \cdots\cdots b_m, \\ \cdots\cdots c_m \end{matrix} \tag{4.20}$$

where a_m, b_m and c_m represent the three axes of the primitive cell of the monoclinic phase. Similarly, the set of the primitive unit cell vectors of the tetragonal phase is:

$$\eta_t = \begin{pmatrix} 3.5214 & 0.0000 & 0.0000 \\ 0.0000 & 3.5214 & 0.0000 \\ 0.0000 & 0.0000 & 5.0773 \end{pmatrix}, \tag{4.21}$$

However, there are only 6 atoms in this cell. To establish the one-to-one lattice correspondence between the monoclinic and tetragonal cells, instead of a primitive cell we use a $\sqrt{2}$-doubled cell for the tetragonal phase. The new cell vector set is:

$$\eta_t' = \begin{pmatrix} 4.9800 & 0.0000 & 0.0000 \\ 0.0000 & 4.9800 & 0.0000 \\ 0.0000 & 0.0000 & 5.0773 \end{pmatrix} \begin{matrix} \cdots\cdots a_t \\ \cdots\cdots b_t, \\ \cdots\cdots c_t \end{matrix} \tag{4.22}$$

There are 12 atoms in this cell, same as in the monoclinic one. During the monoclinic to tetragonal transformation there are three possible lattice orientation schemes (LOS) A, B and C depending on which monoclinic axis a_m, b_m or c_m is parallel to c_t as shown in Table 4.6 [39]. Bailey studied the LOS by transmission electron microscopy and found direct evidence for an orientation relationship consistent with LOS C [39], which we adopt in this chapter.

Experiments [40] suggest that during the monoclinic to tetragonal transformation in zirconia the atoms retain their neighbors in both phases (making it a proper martensitic transformation). Therefore, we postulate the atom correspondence leading to the minimal atomic movement in both HfO_2 and ZrO_2, and find one-to-one correspondence between the atoms of the tetragonal and monoclinic phases. In order to satisfy the condition of minimal atomic movement we shift the cell of the monoclinic phase to have a metal atom at the origin. We write the fractional atomic coordinates in monoclinic and tetragonal phases as $\{v_m^{i,\alpha}\}$ and $\{v_m^{i,\alpha}\}$, where i corresponds to the atom number, and $\alpha = x, y, z$.

In principle, the PES is function of nine parameters describing the unit cell and thirty six additional parameters describing the atomic positions. To simplify the picture we adopt the following approximation. The lattice distortion accompanying the phase transition is rather small and we assume a uniform transformation of the lattice vectors. Furthermore in the spirit of the minimal atomic movement we also assume a uniform transformation of the atomic coordinates. We now can write the energy as a function of two parameters:

$$E(x,y) = E(\{\eta_x\}, \{v_y\}), \tag{4.23}$$

Table 4.6 Possible lattice correspondence schemes between monoclinic and tetragonal cells [18] (Reprinted with permission. Copyright 2009 American Physical Society)

	Lattice correspondence		
LC A	$a_t \rightarrow b_m$	$b_t \rightarrow c_m$	$c_t \rightarrow a_m$
LC B	$a_t \rightarrow a_m$	$b_t \rightarrow c_m$	$c_t \rightarrow b_m$
LC C	$a_t \rightarrow a_m$	$b_t \rightarrow b_m$	$c_t \rightarrow c_m$

Here x and y are defined from the interpolation relations where $\eta_x = (1-x)\eta_m + x\eta_t$ is the lattice vector set, and $v_y^{i\alpha} = (1-y)v_m^{i\alpha} + yv_t^{i\alpha}$ are the fractional atomic coordinates describing the state of a system between two end phases; in this picture x and y change from 0 to 1, zero and one being the monoclinic and tetragonal phase sets, respectively. Using these parameters as two independent variables we map the PES of hafnia and zirconia in two dimensions as shown in Fig. 4.9. We identify the MEP on the energy surface thus generated, shown as a thick black curve in Fig. 4.9. One can clearly see the saddle point (the transition state) at the top of the MEP. The symmetry of the transition state is $P2_1/c$ (same as in the monoclinic phase). The transition barrier is equal to the energy difference between the monoclinic phase and the saddle point. We find the barrier in hafnia to be 0.21 eV per HfO_2 unit, and that in zirconia to be 0.17 eV per ZrO_2. The ratio of the barrier heights is close to the ratio of transition temperatures of hafnia and zirconia (see Table 4.2).

The transition temperature of monoclinic to tetragonal transition can be calculated using simple thermodynamic analysis. Free energies of two phases are equal at the transition temperature. In the harmonic approximation (ignoring other forms of disorder), free energy of a system is given by:

$$F_{Gibbs} = E + pV - TS = E_g + pV + E_{phonon} - TS_{phonon}, \tag{4.24}$$

The first term E_g is the internal energy of the ground state of HfO_2 obtained from ab initio calculations. The second term is negligible because the solid transformation discussed here occurs under ambient pressure and the volume change during the transformation is small. The last two terms together constitute the phonon contribution to free energy, and can be calculated from the phonon density of states as follows [34]:

$$F_{harm} = rk_BT \int_0^{\infty} g(\omega) \ln\left[2\sinh\left(\frac{\hbar\omega}{2k_BT}\right)\right] d\omega, \tag{4.25}$$

where r is the number of atoms in the unit cell, ω is the phonon frequency and $g(\omega)$ is the phonon DOS. In Fig. 4.10 we plot the difference in free energy per HfO_2 formula unit between the monoclinic and tetragonal phases. At the transition

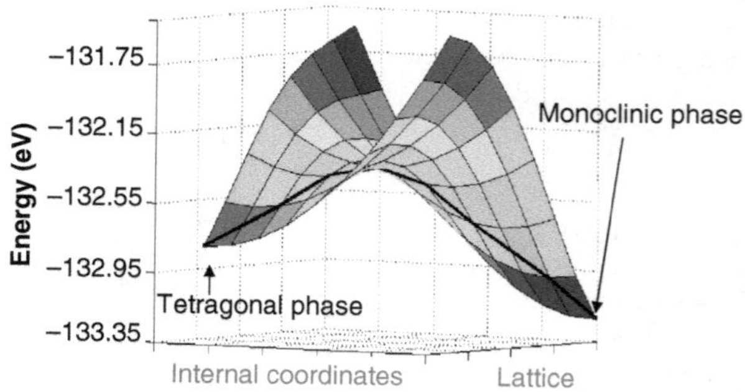

Fig. 4.9 The potential energy surface during the martensitic transformation. The minimum energy path between the monoclinic and tetragonal phases in hafnia is shown as a *black line* on the energy surface. The energy barrier during the martensitic transformation is calculated to be a 0.22 eV/mol [18] (Reprinted with permission. Copyright 2009 American Physical Society)

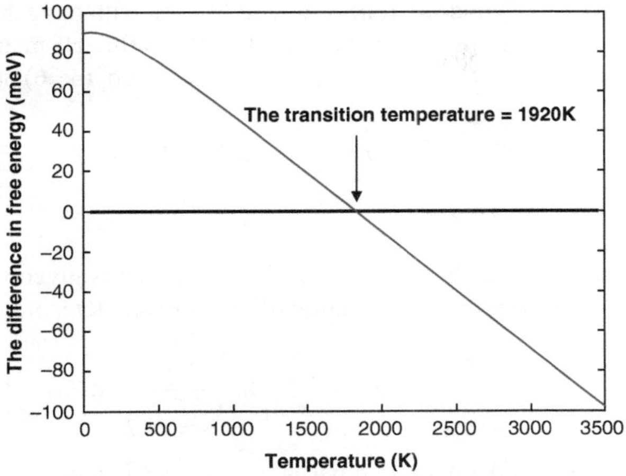

Fig. 4.10 The excess of free energy of the tetragonal HfO_2 with respect to the monoclinic phase as function of temperature, $\Delta F = F_t - F_m$ changes sign at the transition temperature $T_c = 1920\,K$ [18] (Reprinted with permission. Copyright 2009 American Physical Society)

temperature the difference in entropy between two phases cancels the difference in the total energy, and the free energy difference is zero. From Fig. 4.10 it follows that the monoclinic phase is stable in the low temperature region when the E_g term dominates. Above 1920 K the tetragonal phase becomes more stable as its free energy is lower. Recently, Sternik and Parlinski [42], using a similar approach, calculated the m–t transition temperature for ZrO_2 to be about 1560 K. Considering the high temperature of these transitions and the use of the harmonic approximation the agreement with experiment is rather good as can be seen from Table 4.2.

4.4 Surfaces

To simulate the surface we employ slab geometry and symmetric slabs. In the direction normal to the surface, a 15 Å thick vacuum layer is added to eliminate spurious slab-slab interactions. We use the conjugate-gradient algorithm to optimize the atomic structure for various surface terminations of hafnia polymorphs. The surface energy of a vacuum-cleaved surface is then estimated using the zero temperature "grand canonical" thermodynamic potential [43]:

$$E_s = \frac{1}{2A}\{E_{slab} - N_{O_2}\mu_{O_2} - N_{O_2}E_{O_2} - N_{Hf}\mu_{Hf} - N_{Hf}E_{Hf}\} \tag{4.26}$$

Here the energy is given per surface unit cell, and the factor of 1/2 is due to having two interfaces in the supercell. The chemical potentials of Hf and O_2 are related by the equilibrium condition $\mu_{Hf} + \mu_{O_2} = -E_{HfO_2}$. Here E_{HfO_2} is the formation energy of hafnia; the chemical potential of HfO_2 is set to zero since we are in equilibrium with the bulk. Therefore the surface free energy is function of one chemical potential only. We choose the chemical potential of oxygen, and it is set to change in the energy window from zero (equilibrium with oxygen supply) to negative energy of formation of hafnia E_{HfO_2} (below this value Hf metal will start forming on the surface). For hydrogen-passivated surfaces Eq. (4.26) needs to be modified [44]:

$$E_s = \frac{1}{2A}\{E_{slab} - N_{O_2}\mu_{O_2} - N_{O_2}E_{O_2} - N_{Hf}\mu_{Hf} - N_{Hf}E_{Hf} - N_{H_2}E_{H_2} - N_{H_2}\mu_{H_2}\} \tag{4.27}$$

Note that the chemical potential of hydrogen is given with respect to H_2 molecule and can be written as function of pressure and temperature as follows:

$$\mu_{H_2} = kT\left[\ln\left(\frac{pV_Q}{kT}\right) - \ln Z_{rot} - \ln Z_{vib}\right] \tag{4.28}$$

where p is the hydrogen partial pressure, T is temperature and Z_{rot} and Z_{vib} are the rotational and vibrational partition functions, respectively. We now discuss several surfaces of monoclinic and tetragonal phases in more details with the emphasis on the former since the dominant phase in the ALD-grown hafnia thin films is monoclinic.

4.4.1 Monoclinic Hafnia

Because of low symmetry of the monoclinic phase, cutting a surface from a monoclinic crystal theoretically is not a trivial task. The slabs thus generated are not unique. We construct several structures with different orientation and stoichiometry. Then we relax them using VASP, and use the total energies of optimized structures to calculate the surface energy of each termination as function of chemical potential according to Eq. (4.26). In Fig. 4.11a we plot the surface energy for different orientations for the monoclinic phase as function of the oxygen chemical potential, the zero value of which indicates equilibrium with the oxygen supply and thus describes the oxygen rich environment. As seen in Fig. 4.11a all stoichiometric surfaces have surface energies independent of the chemical potential. For the stoichiometric surfaces our calculations agree well with those presented in references [47, 48]. We compare our results to those in Ref. [48] in Table 4.7. The order and values of surface energy for the relaxed stoichiometric surfaces are in good agreement. In addition, we consider several non-stoichiometric and H-passivated surfaces as shown in Fig. 4.11a. Two surfaces with the lowest surface energy among the non H-passivated surfaces are the stoichiometric $(11\bar{1})$ and oxygen rich $(11\bar{2})$. Under extreme oxygen rich conditions the oxygen-terminated (001) surface becomes comparable to $(11\bar{1})$ in energy. The stoichiometric $(11\bar{1})$ surface has the lowest surface energy in most of the allowed range of the chemical potential (with the exception of the extreme oxygen rich regime). This might explain why thin hafnia films in our experiments (as well as in reports by others [45]) favor the texture axis normal to $(11\bar{1})$ planes. However, and under certain

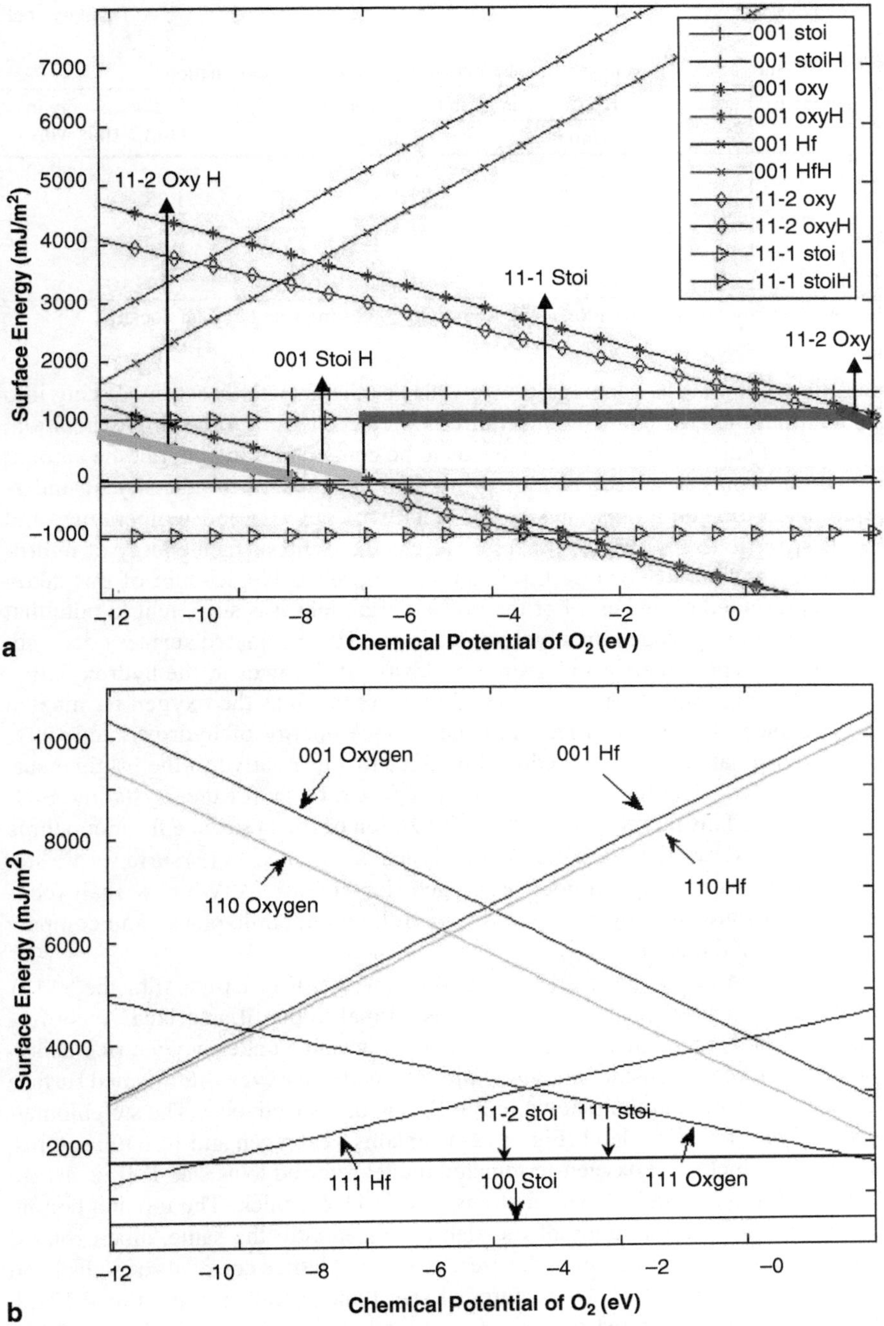

Fig. 4.11 **a** Surface energy of several surfaces of monoclinic HfO_2 as function of the oxygen chemical potential. Labels ending in "Oxy", "Hf" and "Stoi" refer to oxygen, hafnia, and stoichiometric terminations, respectively. Labels ending in 'H' refer to hydrogen passivation. *Thick lines* indicate the stable surfaces under certain ranger of chemical environment. **b** Surface energy of several surfaces of tetragonal HfO_2 with different termination and orientation. Labels indicate the orientation and termination of a particular surface [48] (Reprinted with permission. Copyright 2008 American Physical Society)

Table 4.7 Surface energy of monoclinic hafnia with stoichiometric termination

Stoichiometric surface	Surface energy in (J/m²) from Ref. [47]		Surface energy in (J/m²), (this work)
	Unrelaxed	Relaxed	
$(11\bar{1})$	1.46	0.993	1.04
(111)	1.562	1.199	1.25
(001)	2.169	1.416	1.45
$(11\bar{2})$			1.71
(100)	2.165	1.667	1.79

conditions [46] hafnia films may favor (001) as the growth direction. Taking into account that the $HfCl_4$ precursor requires water cycle during the film deposition, the effect of surface hydroxylation needs to be considered. In general, the amount of hydrogen on the surface of a growing film depends on the density of under-coordinated oxygen atoms, and on the details of purging and temperature, and thus is specific to the growth process. We calculate the surface energy of hydrogen-passivated surfaces using Eq. (4.2). A comprehensive account of this study will be presented elsewhere. For the present argument it is sufficient to note that the presence of hydrogen on stoichiomtetric and Hf-terminated surfaces does not change the surface energy and geometry drastically. However, the hydroxylation has a rather big impact on O-terminated surfaces, such as the oxygen-terminated (001) surface. As seen in Fig. 4.11a the surface energy of hydroxylated (001) oxygen-terminated surface is reduced by 3200 mJ/m² relative to the original surface. A similar effect has been previously reported by us for the crystalline SiO_2 surfaces [44]. This might explain the stabilization of (001) surface for monoclinic hafnia reported by other workers. Also, oxygen-terminated $(11\bar{2})$ surfaces are stabilized under either oxygen poor or oxygen rich conditions. We now shall focus on the most stable hydrogen free surfaces of the monoclinic phase, and compare their atomic and electronic structures.

As we have discussed in our paper [48], for the 40 Å hafnia film the $(11\bar{2})$ direction is consistent with the texture axes normal to the film surface. According to our thermodynamic analysis this termination is stable under oxygen rich conditions. Therefore we consider one stoichoimetric and one oxygen-terminated surface with the $(11\bar{2})$ orientation (shown in Fig. 4.12a, b, respectively). The stoichiometric model referred to as slab I (Fig. 4.12a) contains 32 oxygen and 16 hafnia atoms, and is 9.6 Å thick. The oxygen-terminated model referred to as slab II (Fig. 4.12b) contains 40 oxygen and 16 hafnia atoms, and is 11.2 Å thick. The top and bottom surface atomic configurations of the slab are essentially the same, albeit rotated with respect to one another, in both cases. The lateral lattice constants of slabs I and II are listed in Table 4.8. The structures of slab I and II shown in the Fig. 4.12a, b are fully relaxed. The most obvious difference between the two is that the latter structure has several peroxy bonds between pairs of oxygen atoms on the surface and the former has no such bonds. Peroxy bonds are rare in metal oxides [49]; they appear in this structure because there are not enough hafnium atoms to be bonded to oxygen on the surface, this is the energy lowering mechanism of the oxygen rich

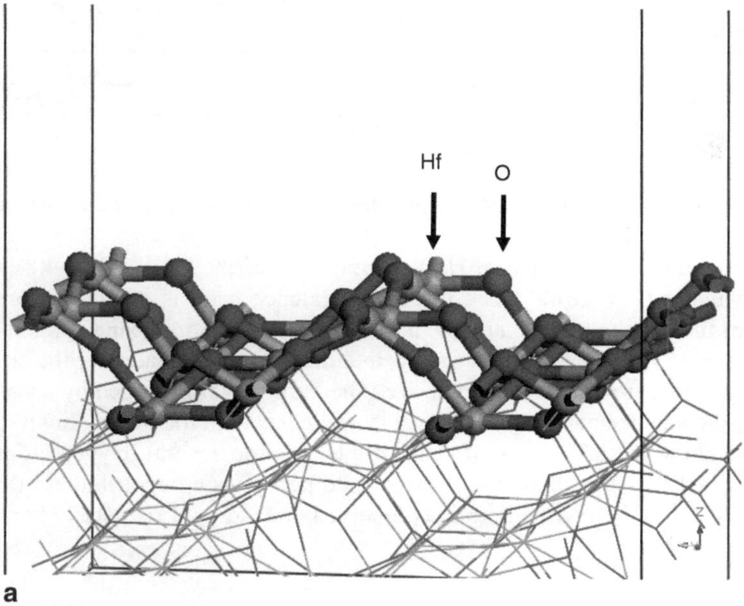

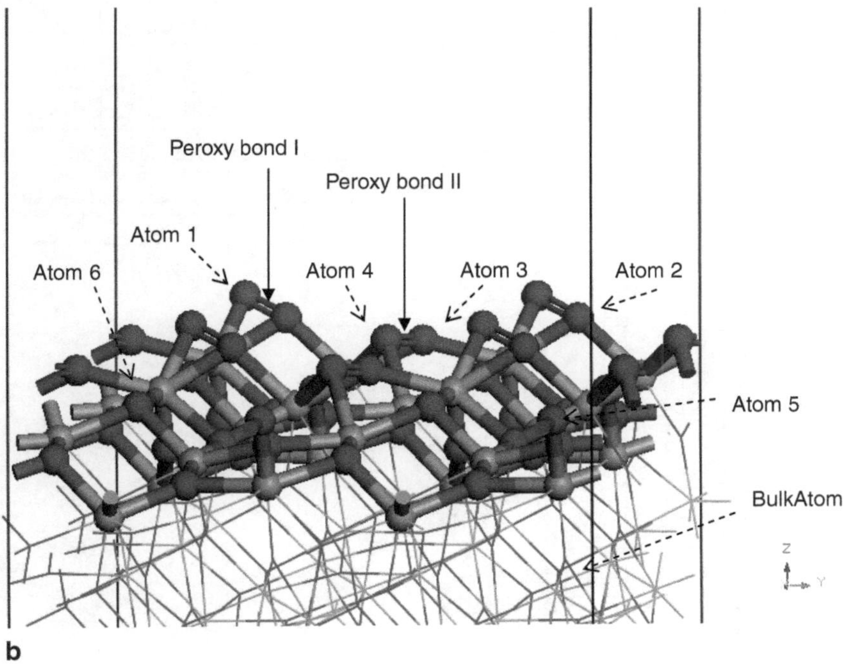

Fig. 4.12 **a** The relaxed structure of the stoichiometric $(11\bar{2})$ slab of monoclinic hafnia with 32 oxygen atoms and 16 hafnia atoms. **b** The relaxed structure of the oxygen rich oxygen-terminated $(11\bar{2})$ monoclinic slab with 36 oxygen atoms and 16 hafnia atoms [48] (Reprinted with permission. Copyright 2008 American Physical Society)

Table 4.8 The simulation cell parameters used to calculate the slab of the surface $(11\bar{2})$ of monoclinic hafnia

A	b	α	β	γ
8.49 Å	7.30 Å	90°	90°	94.78°

surface. The presence of peroxy bonds affects the electronic structure of the surface, as we shall discuss in detail below.

The theoretical GGA band gap of HfO_2 is only 4.1 eV (the LDA gap is 3.8 eV, and experimentally it is 5.8 eV). The top of the valence band is comprised predominantly of the oxygen p-state, and the bottom of the conduction band is mainly hafnium d-states [53]. In Fig. 4.13a we show the total density of states for the slab with the oxygen rich surface discussed above. The Fermi level is pinned by a state 3.8 eV above the valence band top. The level is a filled state of the peroxy bond. In Fig. 4.13b we show partial density of states projected onto several oxygen atoms labeled in Fig. 4.12b. Atoms 3 and 4 of the second peroxy bond contribute to this peak. The state 2.7 eV below the Fermi level corresponds to the first peroxy bond.

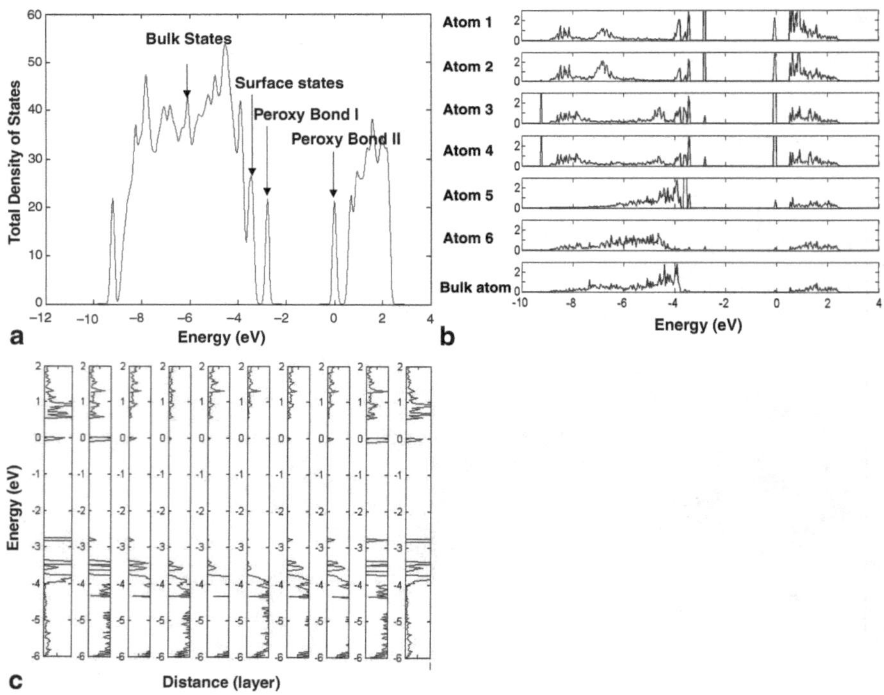

Fig. 4.13 **a** The total density of states of $(11\bar{2})$ oxygen rich monoclinic hafnia plotted in the energy window from −12.0 to 4.0 eV. **b** The partial density of states projected on the oxygen atoms labeled in Fig. 4.6b. **c** The oxygen-projected density of states plotted layer-by-layer along the z-axis of the oxygen-terminated slab [48] (Reprinted with permission. Copyright 2008 American Physical Society)

We can also see the "proper" surface states (atom 5) 0.2 eV above the valence band top. The valence band top is 3.8 eV below the Fermi level as can be seen from the lowest panel in Fig. 4.13b. It is useful to project the partial density of states (PDOS) onto oxygen atoms across the slab, and plot the $l = 1$ component (p-state) layer by layer as in Fig. 4.13c. We can see that the surface states coming from the peroxy bonds are rapidly decaying as we move deeper into the slab. Yet the "proper" surface state right at the band edge decays with a slower rate. Clearly the behavior of these surface states is quite different, and can be understood with the complex band structure. The complex band structure of the monoclinic phase suggests that the decay length of both evanescent states corresponding to the peroxy bonds should be about 2 Å, while the gap states close to the band edge decay much slower [51]. Figure 4.14 shows the electron density distribution corresponding to the peroxy states. In Fig. 4.14a we show the contour plot (in the surface plane) of the electron density corresponding to the lower peroxy state. It looks like an anti-bonding *ppπ* orbital of the peroxy dimer. In Fig. 4.14b we show the contour plot in the sub-surface plane of the electron density corresponding to the surface state at the Fermi level. It is clearly an anti-bonding *ppσ* orbital of the other peroxy bond II. Note that the valence band top doesn't show any bending across the film. This is because the peroxy state pins the Fermi level approximately mid-gap. Peroxy bonds on oxide surfaces are well documented [52–54]. They play important role in surface catalysis. However we are unaware of studies focused specifically on hafnia. Currently we are investigating surface vibrational properties of hafnia, the results may be used to identify peroxy bonds by Raman spectroscopy. We hope our work will inspire further experimental effort.

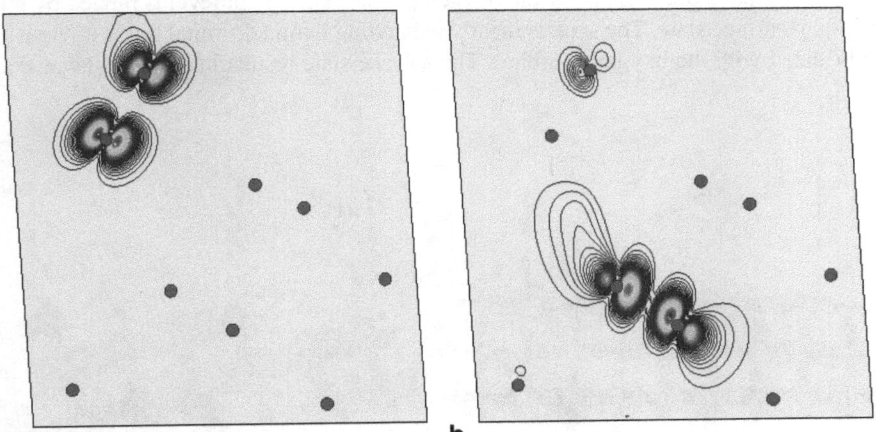

a **b**

Fig. 4.14 **a** The partial electron density distribution coming from the localized states 2.7 eV below the Fermi level for the oxygen rich $(11\bar{2})$ monoclinic surface (see Fig. 4.6b). Dots represent oxygen atoms at the surface. **b** The partial electron density distribution coming from the states right at the Fermi level of the oxygen rich $(11\bar{2})$ monoclinic surface [48] (Reprinted with permission. Copyright 2008 American Physical Society)

4.4.2 *Tetragonal Hafnia*

When compared with the monoclinic phase, the structure of tetragonal hafnia has higher symmetry. Thus the number of possible surface terminations of the tetragonal phase is limited. Most surfaces of tetragonal hafnia are either oxygen-terminated or hafnia-terminated, with the exception of the (100) surface that can only be cut stoichoimetric.

In Fig. 4.11b we plot the surface energy for different orientations of the tetragonal phase as function of the oxygen chemical potential. The plots with a positive slope correspond to hafnium-terminated surfaces and those with a negative slope correspond to oxygen-terminated surfaces. The surface energy of stoichiometric surfaces does not depend on the chemical potential. Among the non-stoichiometric surfaces the (111) family has lowest surface energy. However, the stoichiometric (100) surface is the most stable termination of tetragonal hafnia. This is in contrast with tetragonal zirconia where the most stable surface is (111) [55]. We should note that the surface described as (111) in Ref. [55] is actually stoichiometric. The real (111) surface of both t-hafnia and t-zirconia is built of facets (see Fig. 4.15a, b). We have considered both oxygen and hafnium terminated (111) as well as a stoichiometric termination. Surprisingly, in terms of surface energy the stoichiometric $(11\bar{2})$ and (111) terminations are indistinguishable (see Fig. 4.11b).

In Fig. 4.16 we analyze the average potential plot for the slab with the (111) surface for the oxygen and hafnium terminations. Figure 4.16a shows a combined plot of the plane-averaged electrostatic potential and its macroscopic average for hafnia-terminated (111) surface across the thickness of the slab along with the plane by plane projected oxygen density of states. Oxygen is chosen since the top of the valence band in hafnia is predominantly the oxygen p-state. The Fermi level is pinned by the mid-gap surface state. The state is mostly a dangling hafnia d-orbital but is obviously hybridized with the oxygen p-orbital. The surface state results in a double layer that

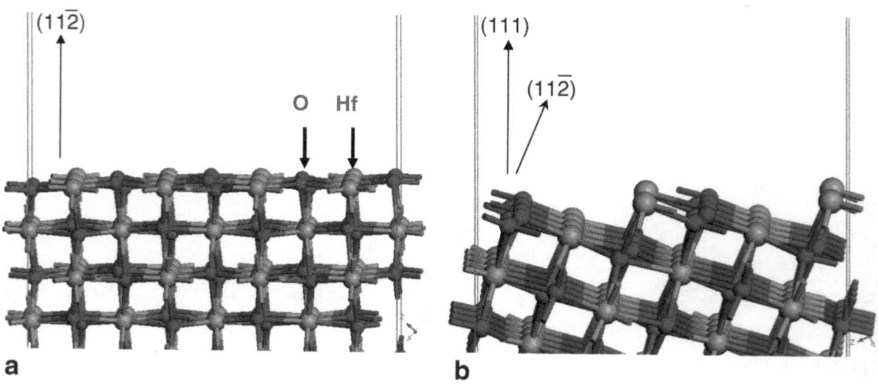

Fig. 4.15 **a** The relaxed $(11\bar{2})$ structure of the tetragonal hafnia stoichoimetric slab with 32 oxygen atoms and 16 hafnia atoms. **b** The relaxed (111) structure of the tetragonal stoichiometric hafnia slab with 32 oxygen atoms and 16 hafnia atoms [48] (Reprinted with permission. Copyright 2008 American Physical Society)

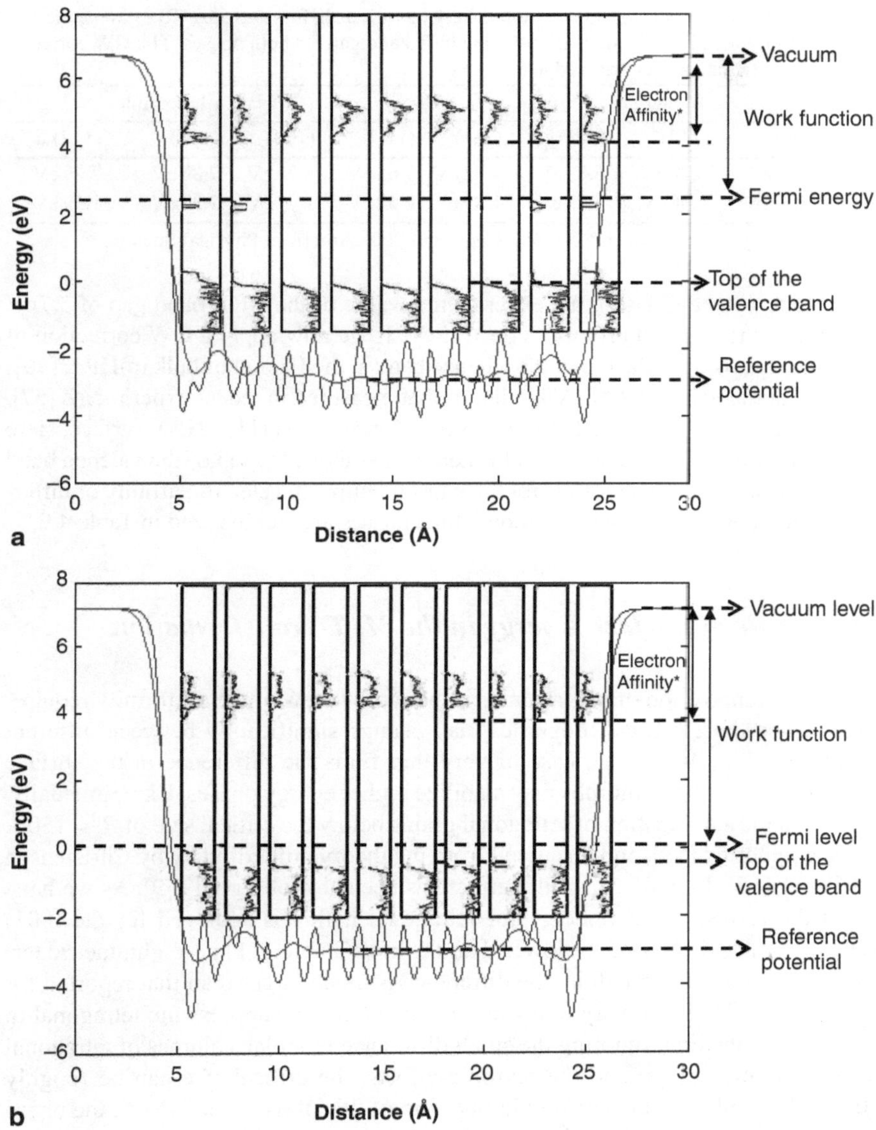

Fig. 4.16 **a** A composite graph of the plane averaged electrostatic potential used as the energy reference, its macroscopic average, and the plane by plane projected oxygen density of states for the Hf-terminated (111) surface of t-HfO_2. The surface state which is actually a dangling Hf d-orbital is clearly seen pinning the Fermi level. **b** Same as **a** for the O-terminated (111) surface of t-HfO_2 [48] (Reprinted with permission. Copyright 2008 American Physical Society)

lowers the bulk potential by about 0.2 eV. The work function can be estimated as a difference between the Fermi level and the value of the potential in the vacuum region. We also estimate the electron affinity, but for that we use the bulk value of the reference potential. In the bulk the valence band top is 2.4 eV higher than the average

Table 4.9 The work function and electron affinity of various hafnia surfaces. To estimate the electron affinity the 5.7 eV value is used for the bulk band gap for both phases. The GW correction is applied to the valence band edge (see text)

	Tetragonal					Monoclinic	
	$(11\bar{2})_{stoi}$	$(111)_{stoi}$	$(111)_{hf}$	$(111)_{oxy}$	$(100)_{stoi}$	$(11\bar{1})_{stoi}$	$(11\bar{2})_{oxy}$
Work function	6.78 eV	6.83 eV	4.82 eV	7.65 eV	6.71 eV	7.30 eV	7.29 eV
Electron affinity	1.08 eV	1.13 eV	1.99 eV	2.52 eV	1.71 eV	1.82 eV	1.59 eV

electrostatic potential. Using the experimental value of the HfO_2 band gap of 5.7 eV we estimate the electron affinity to be 1.95 eV. If we now apply a GW correction of 0.57 eV (calculated for the top of the valence band at the Γ point of bulk m-HfO_2 [56]) we arrive at the value of 2.52 eV similar to that measured in recent experiments [57]. Figure 4.16b shows similar data for the oxygen-terminated (111) t-HfO_2 surface. Here the highest occupied surface state is located 0.4 eV above the top of the valence band and is oxygen related. The results for the work function and electron affinity of different terminations of tetragonal and monoclinic phases are summarized in Table 4.9.

4.4.3 Role of Surface Energy in the M–T Transformation

The phase composition of a thin film is important in view of the uniformity requirements, since the electrical properties may change significantly between different polymorphs [22, 44]. In the case of very thin films the difference in the surface energy of different polymorphs may stabilize higher energy phases. Experimentally, in zirconia the stabilization of tetragonal grains below the critical size of $R_c = 150$ Å has been reported [58], and discussed from the theory point of view by Christensen and Carter [55]. In hafnia a similar effect has been also observed [59]. As we have shown the lowest surface energy for tetragonal hafnia is achieved for the (001) orientation. This is 500 mJ/m^2 lower than the most stable $(11\bar{1})$ stoichiometric termination of monoclinic hafnia; the difference is much larger than that reported for zirconia [51]. Thus in a very thin film one may hope to suppress the tetragonal to monoclinic transition. Ignoring the small difference in molar volumes of tetragonal and monoclinic hafnia, and the entropic effects, the critical size can be roughly estimated as follows. In a uniform hafnia film of thickness R and area A, the phase equilibrium equation defines the critical thickness:

$$A\gamma_t + AR_cE_t = A\gamma_m + AR_cE_m \tag{4.29}$$

where γ_t and γ_m are the lowest surface energies for the tetragonal and monoclinic phases, respectively. E_t and E_m are the bulk energy densities for the two phases, and R_c is the critical thickness. Thus the critical thickness R_c is given by:

$$R_c = \frac{\gamma_m - \gamma_t}{E_t - E_m} \tag{4.30}$$

Using our calculated bulk energies, and surface energies calculated for the stoichiometric $(11\bar{1})$ surface of monoclinic hafnia and (100) stoichiometric surface of tetragonal hafnia we estimate that for films thinner than 15 Å the presence of tetragonal phase may be expected. If a spherical grain is considered the critical radius is 45 Å. Experimentally, however, we observe the presence of the tetragonal phase only in 99 Å thick films. This might suggest the presence of small size crystallites in these films, and not in the thicker or thinner ones. Clearly, other mechanisms such as the presence of hydrogen and point defects (i.e., oxygen vacancies) could be responsible for the stabilization of a high symmetry phase. In addition, we have not considered the energy of grain boundaries which might be more appropriate than surface energies for this case. Therefore we view our findings as consistent with the surface energy driven suppression of the phase transition rather than as proving it.

4.5 Band Alignment at Hafnia Interfaces

When designing a gate stack of the MOS capacitor, one of the most important questions is the overall line-up of the electronic bands in various materials of which the stack is made. One of the most serious problems of GaAs FETs is the Fermi level pinning at the interface of GaAs and gate dielectric. Recently, significant improvements have been reached using a Si passivating interlayer [60–62]. In what follows, we will limit our discussion to GaAs/Si/high-k gate stacks. Hafnia can be deposited on Si by several techniques: atomic layer deposition (ALD), metalorganic chemical vapor deposition (MOCVD), or physical vapor deposition (PVD), using various precursors [63]. However, in all cases, a thin SiO_2 layer is present at the interface between the high-k film and semiconductor. The band offset between SiO_2 and HfO_2 clearly determines the overall alignment of the gate stack. From the research done on Si we know that the failure to correctly include the dipole layer at this oxide–oxide interface contributes to our inability to explain many experimental results in these gate stacks [64].

We introduce the problem of band alignment with a brief discussion of the classical problem of the Si–SiO_2 interface. A simple estimate of the conduction band offset using the metal induced gap state (MIGS) model is given by [65]:

$$\phi = (\chi_a - \Phi_a) - (\chi_b - \Phi_b) + S(\Phi_a - \Phi_b) \tag{4.31}$$

Here χ is the electron affinity, Φ_i is the charge neutrality level of material i measured from the vacuum level, S is an empirical dielectric pinning parameter describing the screening by the interfacial states, and subscripts a and b refer to Si and dielectric, respectively. If $S = 1$ the offset is given by a difference in electron affinities as was originally proposed by Schottky [66]. Alternatively, for $S = 0$ we get the strong pinning or the Bardeen limit [67]. The pinning parameter can be estimated by the empirical formula [65]:

$$S = \frac{1}{1 + 0.1(\varepsilon_\infty - 1)^2} \tag{4.32}$$

where ε_∞ is the high frequency component of the dielectric constant. Electron affinities are typically known experimentally. Tersoff proposed a simple way to estimate the charge neutrality level position associating it with the branch point of the complex band structure of the dielectric [68]. For β-crystobalite the complex band structure gives charge neutrality level 5.1 eV (the value is actually rescaled using the ratio of the experimental and calculated band gap) above the valence band maximum [23], thus placing it 4.8 eV below the vacuum if we assume an electron affinity χ of 0.9 eV. The imaginary wave vector along the *c*-axis of the tetragonal cell has a length of 1.3 $Å^{-1}$ at the branch point. The electron affinity and charge neutrality level of Si with respect to vacuum are 4.0 and 4.9 eV, respectively. The pinning parameter *S* of SiO_2 is 0.9, thus the conduction band offset comes out as 3.1 eV in rather good agreement with experiment.

4.5.1 SiO_2/HfO_2 Interface

To study band alignment from first principles we construct several atomistic models of the SiO_2/HfO_2 interface [69]. We build interface models using the β-crystobalite structure of SiO_2 (C9 symmetry) and cubic and monoclinic HfO_2. When constructing an insulator/insulator model, one needs to address two issues. First, there is lattice mismatch. The lattice constant of SiO_2 is 7.34 Å, while that of cubic HfO_2 (fluorite structure) is 4.98 Å. To resolve this problem we choose a reduced structure of SiO_2 rather than conventional unit cell, which is obtained by 45° rotation around $\bar{c}$. The new lattice constants are $a = b = 5.19$ Å and $c = 7.34$ Å. We use silica as "substrate" and match the lattice constants of hafnia to silica. This creates 4% tensile strain in hafnia which is compensated by vertical relaxation of hafnia. The second issue is stoichiometry; we need to avoid the interface states in the band gap. In our simulation cell this is achieved by choosing three oxygen atoms as the interfacial layer. In Fig. 4.17a we show the initial interface structure. As one can see, there are twice as many oxygen atoms in a hafnia layer when compared to that in silica. To maintain right stoichiometry we remove either atom *c* or atom *d*.

We construct a supercell composed of 6 layers of cubic hafnia and 12 layers of silica. Note that since there are two interfaces in the supercell, we can choose a "symmetric" removal scheme when d is removed (Fig. 4.17a) on both sides and "asymmetric" one when different oxygen atoms are removed. After the relaxation hafnia loses its perfect cubic structure and becomes monoclinic like (Fig. 4.17b). Most importantly, depending whether we choose "symmetric" or "asymmetric" scheme we obtain different oxygen coordination at the interface. In Fig. 4.1b, which corresponds to the "symmetric" removal scheme, there are two threefold oxygen atoms at the interfacial layer and one twofold. We refer to this structure as c-332. For "asymmetric" removals scheme (Fig. 4.17c) there are one threefold and two twofold oxy-

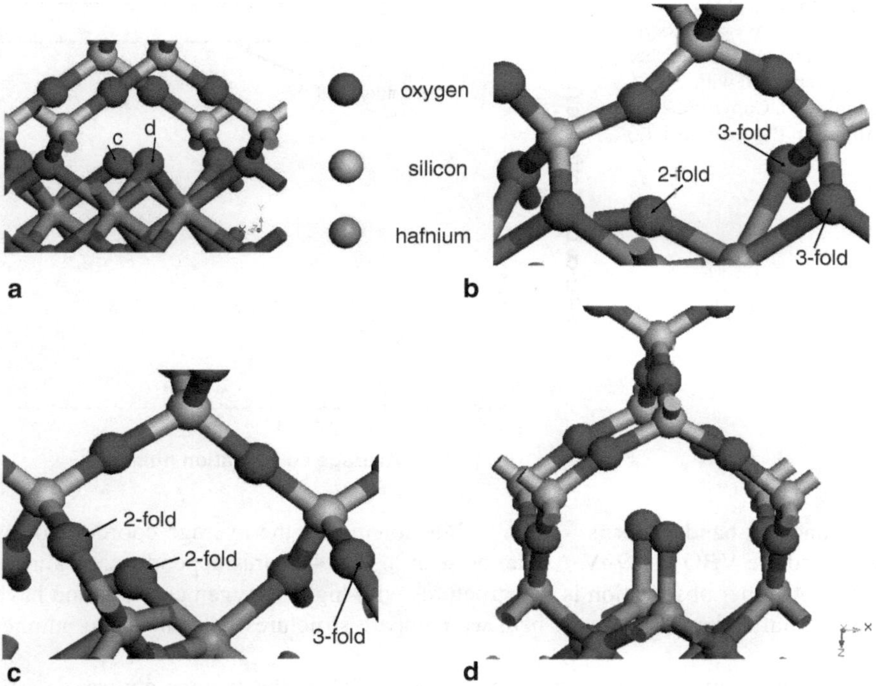

Fig. 4.17 **a** Interface structure before removing the "extra" oxygen atom. **b** m-332 interface corresponding to "symmetric" removal of oxygen atom. **c** m-322 structure, and **d** m-221 structure containing terminal oxygen atoms [69] (Reprinted with permission. Copyright 2007 American Physical Society)

gen atoms at the interfacial layer, which we refer as c-322 structure. In addition, we construct an interface with terminal oxygen atoms (Fig. 4.17d) c-221; there are two twofold bridging oxygen atoms and one terminal atom in c-221 structure.

Low symmetry of the monoclinic phase doesn't allow constructing a supercell with identical interfaces. Therefore, for the interface model with monoclinc hafnia we use slab geometry instead of a supercell. We match 8 layers hafnia in (001) direction to 10 layers of silica in (001) direction. On top we add 11.9 Å of vacuum. The interface structure is similar to that in cubic. We construct three types of interfaces: m-332, m-322, and m-222 (as before, numbers stand for coordination of three interfacial oxygen atoms).

We determine the band alignment using both the site projected density of states method [70] and average potential method [14, 15]. The valence band offset (VBO), i.e., the difference between the valence band maximum of hafnia and that of silica, ranges from –2.0 eV (for the negative value the valence band maximum of silica is above that of hafnia) to 1.0 eV from interface to interface. We note that the VBO is governed by the average oxygen coordination at the interface and is independent of whether hafnia is cubic or monoclinic. The higher oxygen coordination, the bigger the band offset (See Fig. 4.18). For example, for m-222 the average coordination

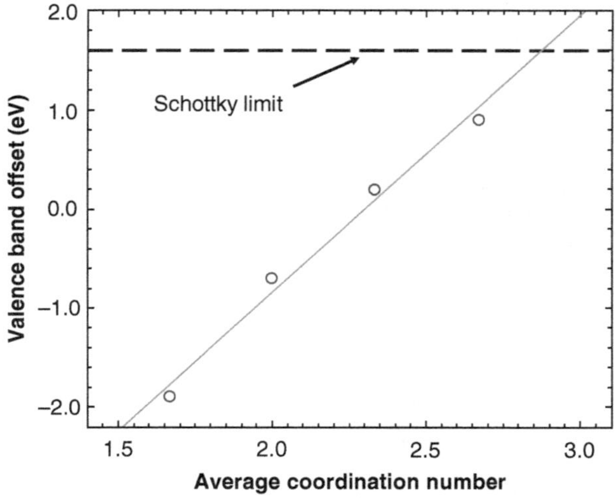

Fig. 4.18 Dependence of the VBO on the average oxygen coordination at the interface [69] (Reprinted with permission. Copyright 2007 American Physical Society)

is 2 and the band offset is −1.0 eV, while for m-332 the average coordination is 2.67 and the VBO is 0.9 eV. As can be seen in Fig. 4.18 this dependence is almost linear. Another observation is the structures with higher oxygen coordination have lower total energy (obviously, here we compare structures with the same number of atoms).

In order to explain this dependence of the VBO on the average oxygen coordination, we propose the following model. Interfacial layer can be considered as a plane capacitor which has a dipole. When silica and hafnia are far apart the valence band offset should be 1.6 eV corresponding to the Schottky limit. When the oxides are brought in contact the charge transfer creates a dipole layer, which results in a potential shift away from the Schottky limit. Silicon has higher electronegativity than hafnium (1.3 and 1.9 for Hf and Si, respectively) and electrons will tend to flow from Hf to Si across the oxygen bridge. Thus, the silica side is negatively charged and hafnia side is positively charged. One then can think of the interfacial oxygen layer as of a dielectric medium placed between the plates of a capacitor. The dielectric response of that layer depends on the oxygen coordination, the higher the coordination, the better it screens. Therefore, higher coordination corresponds to a smaller interface dipole.

4.5.2 *Effects of Al Doping at the SiO_2/HfO_2 Interface*

Among many challenges associated with integrating high-k materials is the requirement to identify the n- and p-type metal electrodes (metals which exhibit the band edge work functions). As we shall discuss below, this is not an easy task. One method to manipulate the overall gate stack alignment is to incorporate metals, such as Al and La, into the interface between SiO_2 and HfO_2 and try to adjust the interface dipole. In this section we discuss how incorporation of Al atoms into the oxide gate stack effects the overall band alignment [71].

To model aluminum doping of the silica/hafnia heterostructure we start with m-322 structure of the SiO_2/HfO_2 interface discussed above. We double m-332 in a lateral direction and substitute silicon or hafnium atoms with aluminum atoms. To ensure that the resulting structures are insulating we consider only stoichiometric aluminum complexes. This is achieved by accommodating the substitutions either by creating oxygen vacancy (V_{Si}, V_{Hf}, and V_{SiHf} in Fig. 4.19) or by incorporating oxygen and aluminum atoms (I_{Si} in the Figure). We consider doping not only at the interface but also in the bulk silica and hafnia (Fig. 4.20).

To analyze the relative stabilities of the doping complexes we consider the following reaction: we bring one Al_2O_3 molecule from the bulk alumina into SiO_2, substitute two Si atoms with Al atoms and remove one oxygen atom, then with the remaining two Si and four O atoms (three from Al_2O_3 and one from removal) we from two SiO_2 molecule which are in equilibrium with the bulk SiO_2. We do the same for doping in hafnia. The resulting formation energies are 0.38 eV, 0.84 eV, and 2.90 for the V_{Si}, V_{Hf}, and I_{Si}. However, for V_{SiO2} and V_{HfO2} they are 4.67 eV and 3.24 eV. This suggests that doping of the silica/hafnia stack occurs via the Al substitution (for Si) in silica accompanied by generation of oxygen vacancies at the interface with hafnia.

The band alignment analysis (see Table 4.10) shows that interfacial Al doping complexes significantly change the VBO with respect to the un-doped structure (m-332).

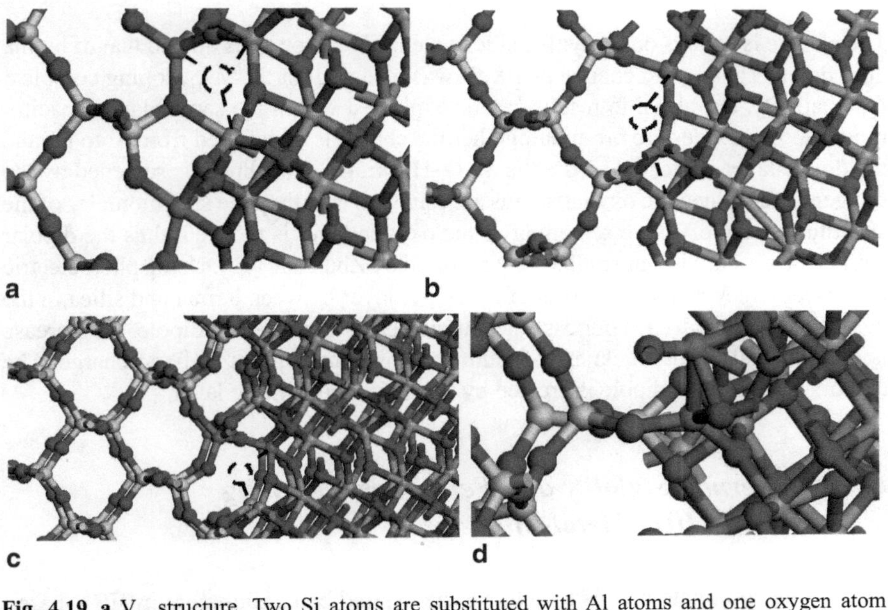

Fig. 4.19 **a** V_{Si} structure. Two Si atoms are substituted with Al atoms and one oxygen atom (*indicated with a dashed circle*) is removed; **b** V_{Hf} structure, **c** V_{SiHf} structure, and **d** I_{Si} structure (see text) [71] (Reprinted with permission. Copyright 2009 American Physical Society)

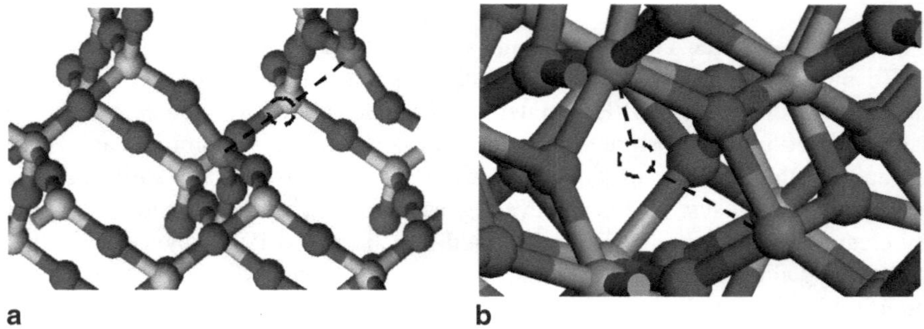

Fig. 4.20 **a** V_{SiO_2} complex in bulk SiO_2, two Si atoms are substituted with Al and one oxygen atom between them is removed. Calculations showed that this is the lowest energy configuration in bulk SiO_2. **b** V_{HfO_2} complex in bulk HfO_2 [71] (Reprinted with permission. Copyright 2009 American Physical Society)

Table 4.10 The valence band offset for different Al-containing SiO_2/HfO_2 interface structures [71] (Reprinted with permission. Copyright 2009 American Physical Society)

Interface	Average potential method (eV)	Projected density of states method (eV)
m-332	0.9	0.7
V_{Si}	–0.2	–0.4
I_{Si}	0.2	0.2
V_{Hf}	0.3	0.2
V_{SiHf}	–0.9	–1.0

The change is always one way: the silica valence band top goes up and that of hafnia goes down. The biggest change of 1.8 eV was obtained for the V_{SiHf} doping complex. Qualitatively, our ab initio results can be explained within the same plane capacitor model. When oxides are brought together the charge is transferred from Si to Hf and the dipole layer is created across the Si–O–Hf bridge. This dipole is screened by the oxygen layer. When the oxygen atoms are removed to satisfy the stoichiometry of the substitution, the effective screening of the oxygen layer is reduced, thus the dipolar shift is increased. Similar results were reported by Zhu et al. [72]. Using photoelectric microscopy they analyzed effects of the Al_2O_3 layer between hafnia and silica in the gate stack. Although as in our case the alumina layer increases the dipole, the increase is proportional to the thickness of alumina. This would point to fixed charge. The authors state that the dipole is created by the aluminum silicate layer.

4.5.3 Thermal Stability and Fermi Level Pinning at the HfO_2/Metal Interface

Identifying right gate metals to achieve the desired band alignment in HfO_2 based stacks has proven to be far from straightforward. For a p-type contact one needs to find a metal with a high work-function to align its Fermi level with the valence band maximum of the semiconductor. A similar problem exists in Si technology. There is only a handful of pure metals with work functions that are large i.e., W, Mo, Pd, Pt, Os, Re, Ru, Rh, Au, Co and Ni [73]. Co and Ni are fast diffusers and would not be the first choice, Au is difficult to etch, but W, Mo, Pd, Os, Re, Ru, Rh, and Pt are all potential contenders. However, as we have found another problem can arise due to the internal oxidation of the metal [74]. At the interface, oxygen atoms from HfO_2 can diffuse into the metal, thus leaving a vacancy in HfO_2 and creating an interstitial in metal. We refer to this defect structure as extended Frenkel pair or EFP. Obviously, the number of EFPs depends on the oxide formation energy; metals that are prone to oxidation such as Mo would stabilize FEPs. On the other hand, the vacancy left behind could ionize with electrons leaving for the closest low energy empty state (for example the Fermi level of the metal). Thus the large number of EFPs can significantly change the band alignment and shift the Fermi level. Note, that unlike the oxygen vacancy in silica, the one in hafnia can't reconstruct by forming a metal-metal bond, and thus is always electrically active. We investigate this issue by considering interfaces between HfO_2 and two pure metals: Mo with the high oxidation energy (2.57 eV) and Rh with the low oxidation energy (1.22 eV) [74–75].

First, we consider the HfO_2/Mo interface. We model it by depositing Mo atoms one by one on top of the (111) slab of tetragonal HfO_2. After each deposition we completely relax all atomic positions. After building three layers of Mo in such a way, we match to it a (110) oriented slab of Mo crystal. The overall thickness of the metallic layer is 9.9 Å. On top we add another 9 Å of vacuum to avoid the spurious slab-slab interaction. The projected density of states analysis shows that the VBO (the difference between the Fermi level of the metal and valence band maximum of HfO_2) is 3.0 eV, which translates into a 2.8 eV Schottky barrier measured from the Fermi level of the metal to the conduction band of the oxide. This would suggest that Mo is a good choice for p-type metal. However, when we calculate the EFP formation energy at this interface, it is only 1.47 eV. At 1200° this results $6.31\times10^{17}\,cm^{-3}$ defect concentration, or the planar density of defects of $7.19 \times 10^{11}\,cm^{-2}$. We also find that the Schottky barrier changes by one third of a volt by forming the EFP. In other words the effective work function of Mo on HfO_2 shows significant instability.

We next model the HfO_2/Rh interface using 15.47 Å of monoclinic HfO_2 oriented in (100) direction and 21.89 Å of Rh (FCC structure). Again, to avoid the gap states the interfacial layer is composed of oxygen atoms. Depending on the number of oxygen atoms, we construct different interface models: O4 with four oxygen atoms at the interface per cell (Fig. 4.21), O3 with three oxygen atoms, and O2 with two oxygen atoms. We calculate the valence band offset and the interface energies for all three structures. As one would expect, the 1.5 eV VBO for the oxygen rich interface O4 is the lowest, it is 2.5 eV for the oxygen poor O2 structure, and 2.0 eV for the O3 structure. Next we consider extended Frenkel pairs for each of the interfaces. To eliminate the spurious interaction between the EFPs we double the simulation cell in the lateral directions. Then we remove an oxygen atom from the interfacial layer and transfer it to the adjacent Rh layer, either close to the interface (short-EFP) or deeper inside the Rh layer (long EFP). The short EFP formation energies are 2.77 eV and 3.38 eV for

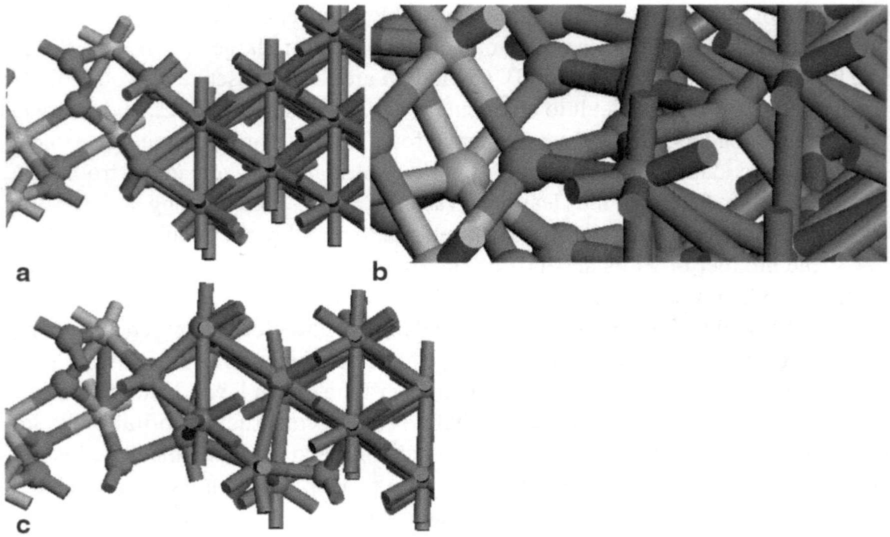

Fig. 4.21 **a** O4 interface structure between hafnia and Rh. There are four oxygen atoms at the interfacial layer per cell. **b** Short EFP for O4. **c** Long EFP for O4 [75] (Reprinted with permission. Copyright 2009 American Physical Society)

O4 and O3 respectively. The long EFP formation energies are 3.81 eV, 3.87 eV, and 4.12 eV. For the long EFPs the VBO increased by 0.3 eV, while for the short EFPs there is almost no effect. This suggests that Rh would be a stable contact metal.

These results can be understood as follows. When an oxygen vacancy is created in bulk HfO_2, two electrons will occupy the vacancy state in the gap about 3.8 eV above the valence band edge, resulting in high formation energy of the defect. However, the vacancy formation energy is lowered dramatically near an interface with a large work function metal. If the Fermi level of the metal lies below the vacancy state, the electrons will transfer to the metal lowering the energy. In addition, the system will actually gain energy by forming a metal oxide (MO_n). The FEP formation energy is the sum of four terms,

$$E_{EFP} = E_{form}(V^{\circ}) + E_{ox}(1/nMO_n) + 2(E(V^{\circ}) - E_{F,m}) + V_{es}(d) \tag{4.33}$$

E_{form} is the formation energy of the neutral vacancy with respect to molecular O_2, the zero chemical potential of oxygen. E_{ox} is the free energy of formation of the oxide MO_n per oxygen atom. The third term is the energy gained by the two electrons falling from the V° level to the metal Fermi energy, at infinite separation. V_{es} is the electrostatic contributions to the energy, including the image potential [75] between the charged vacancy and its negative image in the metal,

$$V_{\text{image}} = \frac{2N^2 ed}{\varepsilon_0 \kappa} \text{eV} \tag{4.34}$$

where N is the areal density of vacancies at distance d from the interface, +2 is their charge, and κ is the dielectric constant of HfO_2 (20–25). This term is of order 0.5 eV.

Caution should be used when choosing the dielectric constant of the oxide because screening is enhanced in the close proximity of the metal [76]. The third term will be reduced by the band bending in the oxide due to the other nearby charged defects. Equation (4.33) explains the much reduced formation energy for O vacancies in HfO_2 when in contact with metals of higher work function. The basic cost of the vacancy 6.3 eV is greatly reduced by the oxide formation and the charge transfer. It also explains the difference in EFP formation energies at O-rich and O-poor interfaces. The O-rich interface has a smaller valence band offset, so the energy gain is greater, and the formation energy is smaller. The charge transfer associated with forming an FEP creates a double layer that shifts the original band alignment.

4.6 Conclusions

Density functional theory is being rather successfully used to support materials development for high-k dielectric gate stacks in advanced CMOS technology despite its limitations with respect to the excited sates and computational expense associated with a large number of atoms in these systems. Many properties of bulk structures and thin films such as the atomic and electronic structure or linear dielectric response can be reliably calculated form first principles. First principles calculations are especially useful in situations where the exact positions of atoms are difficult to infer, such as surfaces and interfaces. The qualitative picture and trends predicted by first principles calculations can be used to guide the experimental effort.

Acknowledgements We wish to thank Jaekwang Lee, Gennady Bersuker, John Robertson, John Ekerdt, Alex Navrotsky and many others for insightful conversations we have had over the years. This work in part is supported by the National Science Foundation under grants DMR-0548182 and DMR-0606464, and by the Office of Naval Research under grant N000 14-06-1-0362. All calculations are performed at the Texas Advanced Computing Center (TACC).

References

1. M. Passlack, N. Medendorp, R. Gregory and D. Braddock, Appl. Phys. Lett. **83**, 5262 (2003).
2. P. Hohenberg and W. Kohn, Phys. Rev. **136**, 864 (1964).
3. W. Kohn and L. J. Sham, Phys. Rev. **140**, 1133 (1965).
4. J. R. Chelikowsky, N. Troullier and Y. Saad, Phys. Rev. Lett. **72**, 1240 (1994).
5. M. C. Payne, M. P. Teter, D. C. Allan, T. A. Arias and J. D. Joannopoulos, Rev. Mod. Phys. **64**, 1045 (1992).
6. J. C. Phillips and L. Kleinman, Phys. Rev. **116**, 287 (1959).
7. D. R. Hamann, M. Schlüter and C. Chiang, Phys. Rev. Lett. **43**, 1494 (1979).
8. N. Troullier and J. L. Martins, Phys. Rev. B **43**, 1993 (1991).
9. D. Vanderbilt, Phys. Rev. B **41**, 7892 (1990).
10. M. P. Allen and D. J. Tildesley, *Computer Simulation of Liquids*. (Clarendon Press, New York, 1988).
11. R. D. King-Smith and D. Vanderbilt, Phys. Rev. B **47**, 1651 (1993).

12. G. Kresse and J. Hafner, Phys. Rev. B **47**, 558 (1993).
13. G. Kresse and J. Furthmüller, Phys. Rev. B **54**, 11169 (1996).
14. D. M. Bylander and L. Kleinman, Phys. Rev. B **36**, 3229 (1987).
15. C. G. Van deWalle, Phys. Rev. B **39**, 1871 (1989).
16. J. Wang, H. P. Li and R. Stevens, J. Mater. Sci. **27**, 5397 (1992).
17. R. Ruh, H. J. Garrett, R. R. Domagala and N. M. Tallen, J. Am. Ceram. Soc. **51**, 23 (1968).
18. X. Luo, W. Zhou, S. Ushakov, A. Navrotsky and A. A. Demkov, Phys. Rev. B **80**, 134119 (2009).
19. J. Adam and M. D. Rodgers, Acta Crystallogr. **12**, 951 (1959).
20. R. Ruh and W. R. Corfield, J. Am. Ceram. Soc. **53**, 126 (1970).
21. R. Ruh and V. A. Patel, J. Am. Ceram. Soc. **56**, 606 (1973).
22. X. Y. Zhao and D. Vanderbilt, Phys. Rev. B **65**, 233106 (2002).
23. R. Stevens, *An Introduction to Zirconia*, 2nd ed. (Magnesium Electron, Twickenham, UK, 1986).
24. R. E. Hann, V. A. Suitch and J. L. Pentecost, J. Am. Ceram. Soc. C 285 (1985).
25. X. Y. Zhao, D. Ceresoli and D. Vanderbilt, Phys. Rev. B **71** (2005).
26. C. E. Curtis, L. M. Doney and J. R. Johnson, J. Am. Ceram. Soc. **37**, 458 (1954).
27. D. Taylor, Br. Ceram. Trans. **83**, 32 (1984).
28. V. B. Glushkova and M. V. Kravchinskaya, Ceram. Int. **11**, 56 (1985).
29. R. C. Weast, *CRC Handbook for Chemistry and Physics*, 65 ed. (CRC Press, 1985).
30. E. J. Little and M. M. Jones, J. Chem. Educ. **37**, 231 (1960).
31. P. E. Quintard, P. Barberis, A. P. Mirgorodsky and T. Merle-Mejean, J. Am. Ceram. Soc. **85**, 1745 (2002).
32. A. P. Mirgorodsky and P. E. Quintard, J. Am. Ceram. Soc. **82**, 3121 (1999).
33. G. K. Horton and A. A. Maradudin, *Dynamical Properties of Solids,* (American Elsevier Publishing Co., New York, 1974).
34. N. W. Ashcroft and N. D. Mermin, *Solid State Physics*. (Thomson Learning, 1975).
35. Y. Moriya and A. Navrotsky, J. Chem. Thermodyn. **38**, 211 (2006).
36. T. Tojo, T. Atake, T. Mori and H. Yamamura, J. Chem. Thermodyn. **31**, 831 (1999).
37. S. S. Todd, J. Am. Chem. Soc. **75**, 3035 (1953).
38. T. Tojo, T. Atake, T. Mori and H. Yamamura, J. Therm. Anal. Calorim. **57**, 447 (1999).
39. J. E. Bailey, Proc. R. Soc. A **27**, 359 (1964).
40. G. M. Wolten, Acta Crystallogr. **17**, 763 (1964).
41. R. N. Patil and E. C. Subbarao, Acta Crystallogr. **26**, 535 (1970).
42. M. Sternik and K. Parlinski, J. Chem. Phys. **122**, 064707 (2005).
43. J. Padilla and D. Vanderbilt, Surf. Sci. **418**, 64 (1998).
44. E. Chagarov, A. A. Demkov and J. B. Adams, Phys. Rev. B **71**, 075417 (2005).
45. M. Y. Ho, H. Gong, G. D. Wilk, B. W. Busch, M. L. Green, P. M. Voyles, D. A. Muller, M. Bude, W. H. Lin, A. See, M. E. Loomans, S. K. Lahiri and P. I. Raisanen, J. Appl. Phys. **93**, 1477 (2003).
46. J. Aarik, A. Aidla, H. Mandar, V. Sammelselg and T. Uustare, J. Cryst. Growth **220**, 105 (2000).
47. A. B. Mukhopadhyay, J. F. Sanz and C. B. Musgrave, Phys. Rev. B **73** (2006).
48. X. Luo, A. A. Demkov, D. Triyoso, P. Fejes, R. Gregory and S. Zollner, Phys. Rev. B **78**, 245314 (2008).
49. V. I. Anisimov, P. Kuiper and J. Nordgren, Phys. Rev. B **50**, 8257 (1994).
50. A. A. Demkov, Physica Status Solidi B **226**, 57 (2001).
51. A. A. Demkov, L. R. C. Fonseca, E. Verret, J. Tomfohr and O. F. Sankey, Phys. Rev. B **71**, 195306 (2005).
52. R. Q. Long, Y. P. Huang and H. L. Wan, J. Raman Spectrosc. **28**, 29 (1997).
53. V. V. Pushkarev, V. I. Kovalchuk and J. L. d'Itri, J. Phys. Chem. B **108**, 5341 (2004).
54. C. Li, Y. Sakata, T. Arai, K. Domen, K. I. Maruya and T. Onishi, J. Chem. Soc. Faraday Trans. I **85**, 1451 (1989).
55. A. Christensen and E. A. Carter, Phys. Rev. B **58**, 8050 (1998).

56. O. Sharia, *Theoretical Study of HfO_2 as a Gate Material for CMOS Devices,* Ph.D. thesis (University of Texas, Austin, 2008).
57. E. Bersch, S. Rangan, R. A. Bartynski, E. Garfunkel and E. Vescovo, Phys. Rev. B **78**, 085114 (2008).
58. G. Cerrato, S. Bordiga, S. Barbera and C. Morterra, Surf. Sci. **377**, 50 (1997).
59. S. V. Ushakov, A. Navrotsky, Y. Yang, S. Stemmer, K. Kukli, M. Ritala, M. A. Leskela, P. Fejes, A. Demkov, C. Wang, B. Y. Nguyen, D. Triyoso and P. Tobin, Physica Status Solidi B **241**, 2268 (2004).
60. A. Callegari, D. K. Sadana, D. A. Buchanan, A. Paccagnella, E. D. Marshall, M. A. Tischler and M. Norcott, Appl. Phys. Lett. **58**, 2540 (1991).
61. Y. Nakano, N. Negoro and H. Hasegawa, Jpn. J. Appl. Phys., Part 1 **41**, 2542.
62. S. J. Koester, E. W. Kiewra, Y. N. Sun, D. A. Neumayer, J. A. Ott, M. Copel, D. K. Sadana, D. J. Webb, J. Fompeyrine, J. P. Locquet, C. Marchiori, M. Sousa and R. Germann, Appl. Phys. Lett. **89**, 042104 (2006).
63. M. Ritala, M. Leskela, L. Niinisto, T. Prohaska, G. Friedbacher and M. Grasserbauer, Thin Solid Films **250**, 72 (1994).
64. R. Chau, S. Datta, M. Doczy, B. Doyle, J. Kavalieros and M. Metz, IEEE Electron Device Lett. **25**, 408 (2004).
65. J. Robertson and C. W. Chen, Appl. Phys. Lett. **74**, 1168 (1999).
66. W. Schottky, Zeit. Phys. **118**, 539 (1942).
67. J. Bardeen, Phys. Rev. **71**, 717 (1947).
68. J. Tersoff, Phys. Rev. B **30**, 4874 (1984).
69. O. Sharia, A. A. Demkov, G. Bersuker and B. H. Lee, Phys. Rev. B **75**, 035306 (2007).
70. A. A. Demkov, R. Liu, X. D. Zhang and H. Loechelt, J. Vac. Sci. Technol. B **18**, 2388 (2000).
71. O. Sharia, A. A. Demkov, G. Bersuker and B. H. Lee, Phys. Rev. B **77**, 085326 (2008).
72. L. Q. Zhu, N. Barrett, P. Jegou, F. Martin, C. Leroux, E. Martinez, H. Grampeix, O. Renault and A. Chabli, J. Appl. Phys. **105**, 024102 (2009).
73. H. B. Michaelson, J. Appl. Phys. **48**, 4729 (1977).
74. A. A. Demkov, Phys. Rev. B **74**, 085310 (2006).
75. O. Sharia, K. Tse, J. Robertson and A. A. Demkov, Phys. Rev. B **79**, 125305 (2009).
76. C. Berthod, N. Binggeli and A. Baldereschi, Phys. Rev. B **68**, 085323 (2003).

Chapter 5
Density Functional Theory Simulations of High-k Oxides on III-V Semiconductors

Evgueni A. Chagarov and Andrew C. Kummel

Abstract A comprehensive overview of density functional theory simulations of high-k oxide/III-V semiconductor interfaces is presented. The methodologies of realistic amorphous high-k oxide generation by hybrid classical-DFT molecular dynamics are compared. The simulation techniques, oxide/semiconductor model designs and rules for formation of unpinned high-k oxide/semiconductor interfaces are discussed. The density-functional theory molecular dynamics simulations of a-Al_2O_3/InGaAs and a-Al_2O_3/InAlAs/InGaAs stacks are presented and analyzed.

5.1 Introduction

5.1.1 High-k Oxides

The rapid scaling of complementary metal oxide semiconductor (CMOS) technology requires substituting the traditional gate oxide, silicon dioxide (SiO_2), with high-k dielectrics, which can maintain the same capacitance with much lower leakage current. Silicon dioxide was the major gate oxide material for decades. Since transistors have been rapidly decreasing their lateral sizes to increase surface density of microelectronic elements on the chip area, the gate oxide had to decrease its thickness to scale the capacitance, drive current, and device performance. However, decreasing SiO_2 thickness below 7–13 Å is not feasible since it leads to significant tunneling leakage, increased power consumption, and deterioration of device reliability [1, 2]. Replacement of SiO_2 oxide by high-k materials would allow an increase in gate oxide capacitance while diminishing the gate leakage.

The gate oxide in a Metal-Oxide-Semiconductor-Field-Effect-Transistor (MOSFET) can be considered as a parallel plate capacitor. Ignoring quantum

E. A. Chagarov (✉)
Department of Chemistry and Biochemistry, University of California, San Diego, La Jolla, California 92093, USA
e-mail: echagaro@ucsd.edu

S. Oktyabrsky, P. D. Ye (eds.), *Fundamentals of III-V Semiconductor MOSFETs*, DOI 10.1007/978-1-4419-1547-4_5, © Springer Science+Business Media LLC 2010 mechanical interaction and depletion effects at the oxide/semiconductor and oxide/electrode interfaces, the capacitance C of the gate oxide approximated as the parallel plate capacitor can be given by:

$$C = \frac{k\varepsilon_0 S}{d}, \qquad (5.1)$$

where S is a capacitor plate area, d is a distance between capacitor plates (equal to capacitor oxide material thickness), k is a relative dielectric constant, and ε_0 is the electrical permittivity of vacuum. As follows from Eq. (5.1), using a material with high dielectric constant k would allow an increase of oxide thickness while avoiding the problem of oxide current leakage and maintaining the same capacitance per unit area required for a high density of MOSFET devices on the chip surface.

Replacing SiO_2 by high-k gate oxide materials adds a whole plethora of new technological challenges to the device manufacturing process. Besides a high dielectric constant, the selected high-k gate oxide material should satisfy a whole range of additional requirements, such as proper band alignment to the semiconductor substrate, thermal stability, low interface roughness and associated with it high mobility of charge carriers, and low density of electrical defects in the oxide/semiconductor interface.

Currently, the most promising high-k gate oxide materials are hafnium dioxide (HfO_2), zirconium dioxide (ZrO_2), alumina (Al_2O_3), hafnium silicate ($HfSiO_4$), and zirconium silicate ($ZrSiO_4$) [3–6]. In the real-world oxide gate/semiconductor stacks, these materials are often found in amorphous phases due to deposition and post-deposition processing.

5.1.2 III-V Semiconductors

The III-V semiconductors are very promising materials for microelectronic and optoelectronic applications. They can provide much higher low field carrier mobility than Si-based devices and, therefore, are potentially beneficial for high-speed applications. The III-V semiconductors are compound semiconductors formed by chemical elements from groups III and V. In the crystal structure of III-V semiconductors, every atom of group III is bound to four group V atoms while every group V atom is bound to four group III atoms. The bonding model of III-V semiconductors is mainly covalent with moderate bond ionicity due to the modest electronegativity difference between group III and V elements. The III-V semiconductors can form binary, ternary, quaternary and even higher-order compounds mixing different number of III and V group elements. The mixing of various III and V group elements in one semiconductor compound provides wide possibilities for engineering of the semiconductor band-gap and the associated emission/adsorption wave-length, which are critically important properties for microelectronic and optoelectronic applications. Among binary III-V semiconductors, the most widely investigated are InAs, GaAs, InP, GaP, and their ternary alloys. Among III-V semiconductors, the typical technologically promising compounds include $In_xGa_{1-x}As$, $In_xAl_{1-x}As$, $Al_xGa_{1-x}As$, and $In_xGa_{1-x}P$.

5.1.3 Density-Functional Theory

The Density-Functional Theory (DFT) is an extremely successful quantum-mechanical first-principles technique capable of modeling the electronic structure of many-body atomic, molecular and condensed matter systems. DFT is based on the Kohn-Sham theorems and equations [7, 8], which reduce the many-body problem for N particle system to a one-parameter problem using the density-functional approach when the energy functional of the system can be described in terms of charge density as follows:

$$E[\rho] = T[\rho] + \int V_{ext}(r)\rho(r)dr + V_H[\rho] + E_{xc}[\rho], \tag{5.2}$$

where T is the kinetic energy of the system, V_{ext} is the external potential acting on the system (e.g., due to ions), E_{xc} is the exchange-correlation energy, and V_H is the Hartree (electron-electron interaction) energy given by:

$$V_H[\rho(r)] = \frac{1}{2}\int \frac{\rho(r)\rho(r')}{|r-r'|}drdr'. \tag{5.3}$$

In practical applications, DFT codes use the variational approach: the wavefunction representation is varied to minimize the system total energy until it is numerically converged to the ground state energy and charge density. The major problem of DFT is that the exchange-correlation functionals, E_{xc} are known exactly only for the free electron gas, but in all other more general cases, they have to be approximated. One of the classical approximations is the local-density approximation (LDA), where the exchange-correlation functional is determined only locally by the electron density at the given spatial point. The generalized gradient approximation (GGA) is the next refinement of LDA; while GGA is still a local approximation, it takes into account the gradient of the charge density at the given spatial point thereby making it more accurate for systems with steep spatial charge-density variations. Although for the majority of cases, DFT results show strong correlation to experimental data, standard DFT approaches (including LDA and GGA) still demonstrate problems with reproduction of the true band-gaps of solid phases, intermolecular interactions (especially van der Waals interaction), transition states and charge transfer excitations. To overcome these limitations, a set of improvements has been proposed for the standard DFT methodology. An excellent in-depth overview of DFT and other electronic structure methods can be found elsewhere [9].

5.2 Methodology of DFT Simulations of High-k Oxides on Semiconductor Substrates

Si has been the major semiconductor substrate for several decades which made it the focus of active DFT research for potential high-k oxide/semiconductor stack applications [3–5, 10–35]. Ge due to its high hole mobility and similarity to Si was also actively investigated as a potential substrate for high-k oxides [6, 36–38]. This chapter will give brief overview of DFT simulations of high-k oxides on Si or Ge concentrating mainly on methodology. The DFT simulations of oxide/semiconductor interfaces can be performed using different computational approaches to initial system design and different simulation algorithms.

5.2.1 Oxide Deposition Technique in DFT Simulations

The initial system configuration is extremely important factor which can significantly influence the final outcome of oxide/semiconductor simulations. There are several computational techniques for oxide deposition on the semiconductor substrate which can be employed to devise the initial interfacial structure.

1. Kinetic Monte-Carlo (KMC) simulations [39–41]. KMC simulations can model the time evolution of physical and chemical processes occurring with a given rate. These rates are input variables for KMC simulations and are often obtained from DFT simulations of the energy barriers for certain system transitions.
2. Density-Functional Theory Molecular Dynamics (DFT MD) simulations can be employed to simulate oxide molecules randomly bombarding the semiconductor surface simulating experimental molecular beam, e-beam, or sputter deposition. Since nearly all oxides including Al_2O_3 and ZrO_2 evaporate incongruently, this approach can only provide very approximate description of oxide molecular beam deposition.
3. Artificial layer-by-layer deposition of oxide atoms without any correlation to real deposition speed followed by molecular dynamics annealing and/or relaxation can be used to simulate reactive oxide formation on semiconductors. For example, to investigate thermal oxidation by O_2, DFT can be used to simulate a very rapid reaction with atomic oxygen.
4. DFT molecular dynamics simulations with previously prepared bulk oxide sample stacked to the semiconductor surface and relaxed or annealed-cooled-relaxed can be employed to simulate experimental postdeposition-annealed oxide/semiconductor interfaces. The oxide bulk sample can be thoroughly tested prior to main oxide-semiconductor simulations. This method is the most realistic for systems in which the oxide and semiconductor are weakly interacting.
5. Hybrid methods mixing molecular dynamics and Monte-Carlo techniques can be employed to simulate different processes with various rates during deposition by different approaches.

Each of these methods has its own advantages and disadvantages, which should meet certain strict criteria of applicability, such as realism, computational efficiency, and achievable simulated timescale.

Method A: Kinetic Monte-Carlo simulations are computationally efficient and can simulate atom-by-atom deposition at long timescale. However in comparison with DFT MD, KMC simulations provide lower accuracy for atomistic modeling of oxide-semiconductor interface evolution because KMC simulations replace the true atomic dynamics with statistically-equivalent MC kinetics. The KMC simulations are based on a set of energy barrier calculations and associated rates for various configurations and transitions between the configurations. The computer simulations of oxide/semiconductor interfaces often involve significant deformation in the substrate, interface, and oxide regions with a very large number of degrees of freedom. An attempt to reproduce the realism of such oxide-semiconductor simulations with 3D KMC would require taking into account an unrealistic number of atomic configurations, transition barriers between the numerous configurations, and multiple DFT-calculated activation energies.

Method B: DFT-MD simulations modeling oxide atoms randomly bombarding the semiconductor surface cannot accurately reproduce experimental oxide-semiconductor interface growth for three reasons: (a) Most gate oxides are deposited by atomic layer deposition (ALD), not molecular beam deposition. (b) The timescale needed to deposit 100 atoms of oxide at realistic experimental deposition rate requires many orders of magnitude longer timescale than picoseconds, which modern DFT MD can afford. The typical MBE deposition rate is ~1.0 ML/s [42], the fastest MOCVD is ~3 ML/s [43] and the fastest sputtering deposition is ~1 ML/s [44]. (c) With the exception of $LaAlO_3$, nearly all common gate oxides evaporate incongruently; therefore, to form stoichiometric films using molecular beam deposition, a second oxygen source or post-deposition annealing must be employed.

Method C: Artificial layer-by-layer deposition of oxide atoms without any correlation to real deposition speed followed by molecular dynamics annealing and/or relaxation somewhat circumvents the problem of deposition speed. The technique works best when certain general bonding rules for the given chemical species are employed. This approach was successfully used by Hakala et al. for modeling of HfO_2 growth on Si substrate [24].

Method D: DFT molecular dynamics simulations with a previously prepared bulk oxide sample stacked to the semiconductor surface and relaxed or annealed-cooled-relaxed, provides much more elaborate system of checks to verify high quality of the crystalline or amorphous oxide sample prior to bonding to the semiconductor. Although methods (A), (B), and (C) are able to simulate atom-by-atom deposition they raise significant concerns about oxide film realism, especially if amorphous oxide films are investigated. The random deposition of atoms for relatively small atomistic system and limited statistical ensemble can produce significant deviations in oxide sample properties. Because of the checks to verify the quality of the amorphous oxide samples, method (D) often provides a realistic affordable computational alternative to methods (A), (B), and (C) for simulating amorphous oxide-semiconductor interfaces when the interfacial reactions are limited. The technique has been successfully used by many groups for atomistic simulations of oxide-semiconductor interfaces. Tse et al. successfully utilized this approach for modeling of ZrO_2/Ge [6]. Peacock et al., Puthenkovilakam et al., Fonseca et al. and Dong et al. successfully applied it to simulations of ZrO_2/Si, $ZrSiO_4$/Si and HfO_2/Si interfaces [4, 13, 16, 18, 45, 46]. Zhang et al., Peacock et al., Robertson et al., and Forst et al. used it for modeling of $SrTiO_3$/Si interface [14, 20, 47, 48]. Broqvist et al. and Capron et al. used it for simulations of a-HfO_2/a-SiO_2/Si and HfO_2/SiO_2 stacks [11, 33, 34], Chagarov et al. used it for simulations of a-Al_2O_3/Ge, a-ZrO_2/Ge, a-Al_2O_3/InGaAs and a-Al_2O_3/InAlAs/InGaAs [36, 37, 49, 50].

Method E tries to combine best of molecular dynamics and Monte-Carlo approaches. In this approach, all processes in the system are classified by their characteristic times (rates). The low barrier processes, which typically involve bulk and surface relaxations are described by MD simulations (classical or first-principle), while high-barrier processes like surface reactions, activation diffusion and similar are treated by Monte-Carlo family approaches. Knizhnik et al. successfully applied such hybrid approach to simulations of ZrO_2 deposition on Si(100) surface [29]. In their study, the system relaxation was modeled by classical MD with empirical potentials, while high-barrier processes were simulated by Kinetic Monte-Carlo approach. The Ng et al. used a similar technique to model oxidation kinetics of Si(100)-SiO_2 interface [51]. In the latter case, the system relaxation processes were simulated by DFT while high-barrier processes were modeled using the Metropolis algorithm. Hybrid methods (E) can simulate much longer timescales than standard molecular dynamics and better handle systems with high lattice irregularities like amorphous films; however, they still inherit original problems of the underlying techniques such as replacement of the true atomic dynamics with statistically-equivalent MC kinetics which inevitably introduces additional error in the final configuration. Although it is able to provide high degree of realism, molecular dynamics with its typical timestep of $\sim$1 fs still has serious limitations on simulated timescale. The final choice of the oxide/semiconductor deposition technique remains a difficult compromise between accuracy, timescale, and computational efficiency.

5.2.2 Oxide-Semiconductor Stack Design

Another variation of oxide/semiconductor system simulations comes from different ways to arrange oxide and semiconductor slabs in periodic boundary condition (PBC) box. Figure 5.1 presents two possible designs for oxide/semiconductor stack arrangement with periodic-boundary conditions. The design represented in Fig. 5.1a has only one oxide-semiconductor interface with a vacuum spacer above oxide; conversely design in Fig. 5.1b has two oxide-semiconductor interfaces and no vacuum layer. Very often simulations incorporate interfacial layer between oxide and semiconductor which can be trivially incorporated into both Fig. 5.1 designs. Each of these two designs has its own advantages and disadvantages.

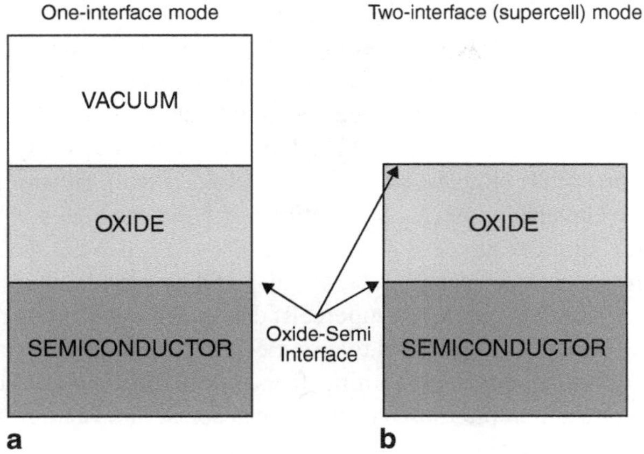

Fig. 5.1 One-interface and two-interface (supercell) designs of oxide/semiconductor simulated stack. Periodic-boundary conditions are assumed

The one-interface design (Fig. 5.1a) with a vacuum spacer layer provides freedom for interface height relaxation and release of possible internal vertical stresses which can be present after oxide/semiconductor stacking prior to DFT MD simulations. Artificial internal vertical stresses in the oxide/semiconductor stack can lead to significant errors in DFT simulations affecting system electronic structure and final atomic configuration. Although the sizes of the PBC box can vary in many DFT codes relaxing the system and minimizing the total energy, not every code can vary the PBC box size during MD runs at finite temperature. In case of crystalline oxide/semiconductor interfaces, the initial interface height can be roughly predicted from general bond lengths due to high interface regularity. However in case of amorphous oxides which have highly corrugated surfaces after oxide planar cutting and in case of interfaces with unclear bonding structure, it is difficult to estimate realistic interface height at the preliminary stage of system design. For these cases, the one-interface design can become a preferred solution because the vacuum spacer provides enough freedom to optimize the interface height and relax the oxide/semiconductor stack into the most energetically favorable interfacial bonding structure. Obviously the one-interface design is the best solution for film deposition simulations with layer-by-layer deposition or oxidation. Since this type of design includes only one interface, the major disadvantage of this design is that it requires chemical passivation of dangling bonds on the semiconductor and oxide interfaces with vacuum, which is often implemented by adding H atoms or OH ligands onto the dangling bonds. While this is a well established for the vacuum/semiconductor interface, it is not a trivial procedure for the vacuum/amorphous oxide interface. The vacuum spacer thickness is important; it should eliminate spurious interaction between semiconductor and oxide images through PBC translation and is usually about ~15 Å or thicker. The presence of relatively thick vacuum spacer increases the computational cost of DFT runs requiring larger internal grids to cover the whole space of the simulation box. Another challenge with the one-interface design (Fig. 5.1a) is that periodic-boundary conditions for semiconductor/oxide stack (which in general have different work-functions) create a spurious electric field perpendicular to interface, with the field mainly localized in the least-screened region (vacuum) and requiring compensation to avoid computational errors. Fortunately majority of modern DFT codes provides automatic correction/compensation of such artificial electric fields [52–54]. The one-interface design was successfully applied by Puthenkovilakam et al. for simulations of ZrO_2/Si and $ZrSiO_2$ stacks [4], by Monaghan et al. for simulations of Hf silicates on Si [23], by Hakala et al. for simulations of HfO_2 on Si [24], by Gavartin et al. for simulations of HfO_2/SiO_2/Si stacks [26], by Chagarov et al. for simulations of a-Al_2O_3/Ge, a-ZrO_2/Ge, a-Al_2O_3/InGaAs and a-Al_2O_3/InAlAs/InGaAs interfaces [36, 37, 49, 50], and by other authors.

The two-interface (supercell) design (Fig. 5.1b) has its own advantages and disadvantages. It provides two oxide-semiconductor interfaces in one model with one additional interface coming from periodic-boundary condition translation. It has reduced computational cost due to absence of a vacuum spacer and a smaller system size. It also avoids the need to passive the vacuum/oxide interface and does not create artificial electric fields. However, its fixed structure in direction perpendicular to interface does not provide the system with enough freedom to release internal interfacial stresses which can form either during initial system design or during DFT MD annealing and/or relaxation. Although majority of DFT codes can vary the box size during relaxation, not all of them can not do it during DFT MD annealing. In case of unclear bonding structure in interface region and highly corrugated oxide and/or semiconductor surface (like in case of amorphous oxides), this type of design can lead to presence of significant stresses in the system distorting electronic structure and final atomic geometry. However this type of design can be satisfactory in case of pre-determined bonding structure and previously well estimated interfacial bond-lengths. The two-interface (supercell) design was successfully used by Tse et al. for simulations of ZrO_2/Ge interfaces [6], by Peacock et al. and by Ha et al. for simulations of ZrO_2/Si and HfO_2/Si stacks [16, 27], by Broqvist et al. for modeling of HfO_2/SiO_2/Si and by other authors [11, 32, 33].

The different choice of the semiconductor/oxide stack design leads to different evaluation of some of its major properties. For example, for two-interface (supercell) design (Fig. 5.1b), the interface formation energy can be determined by:

$$E_{form} = \frac{E_{tot} - (nE_{oxide} + mE_{semi} + l\mu_N)}{2S}, \tag{5.4}$$

where E_{tot} is a total energy of the relaxed system, n and E_{oxide} are the number of oxide units and the total energy of such unit, m and E_{semi} are the number of semiconductor units and the total energy of each unit, l and μ_N are the number of additional atomic species (like extra O or H for example) and their chemical potential, and S is an interface area. In case of one-interface design with vacuum spacer (Fig. 5.1a), the Eq. (5.4) requires modification to take into account surface energies of passivated or unpassivated oxide and semiconductor surfaces.

The final choice of the oxide-semiconductor system design is determined by specific research goals, interface to be investigated, and the computational cost.

5.2.3 *Crystalline vs. Amorphous Oxides in DFT Simulations*

Experimentally, amorphous oxides for gate dielectrics are considered to be superior to crystalline/polycrystalline ones since grain boundaries and dislocations in polycrystalline oxide films can lead to increased current leakage through the oxide. In addition, highly-ordered polycrystalline oxide/crystalline semiconductor interfaces usually have lattice mismatch which results in a high density of dangling bonds acting as interfacial electrical defects. Some amorphous oxides, such as ZrO_2 and HfO_2, can crystallize at relatively low temperatures (<500 °C) [55]. Crystalline oxides tend to have higher dielectric constants so sometimes nanocrystalline materials are experimentally employed.

DFT simulations have different issues with modeling crystalline and amorphous oxide samples. Although for many perspective high-k oxide/semiconductor stacks such as HfO_2/Si or ZrO_2/Si, amorphous oxides are the most promising, there are many simulations of these stacks with crystalline oxides instead of amorphous ones [3, 4, 6, 13, 14, 16, 18, 27, 28]. The validity of approximation of amorphous oxides by crystalline systems in oxide/semiconductor stacks is not obvious; however, it has been very successful for simulating oxide defects [56–66].

One of the main reasons why crystalline oxide models are often used in DFT simulations of oxide/semiconductor stacks is that generation of realistic amorphous oxide models is a much more complicated task than building crystalline oxide samples from well-known crystallographic information. The overview of techniques for computational generation of amorphous oxides will be provided later. The substitution of amorphous oxides in oxide/semiconductor DFT simulations by crystalline systems can be problematic for modeling some interface properties. Crystalline oxide/semiconductor lattice mismatch can lead to artificial oxide compression/elongation in DFT models to match semiconductor area. Due to high computational cost of DFT runs, the typical atomic system size is usually limited to several hundred atoms thereby limiting the area of the oxide/semiconductor interface. The problem of lattice mismatch could be alleviated by employing larger interfacial areas and larger numbers of unit cells along the interfacial plane. However, this is not always feasible with typical computational resources. Since in DFT simulations both oxide and semiconductor share the same lateral sizes (parallel to the interface), the small interface area and lattice mismatch often require substantial oxide compression/elongation to match the semiconductor substrate area. This artificial oxide compression/elongation can lead to sensible electronic structure distortions such as deformation of band structure, introduction of gap states, changing of band offsets and bond lengths [4, 6, 16, 18]. The presence of vacuum spacer over the stack and system relaxation in vertical direction can provide only a partial reduction of these problems.

Another challenge of substituting amorphous oxides by crystalline ones comes from the enhanced risk of formation of partially filled dangling bonds in the interface region or distortion of substrate lattice positions during DFT-MD annealing. This can occur because of the lattice mismatch between the oxide and semiconductor. The amorphous oxide surface with its atomic long-range disorder in contact with semiconductor has more homogeneous sampling of the ordered semiconductor surface patterns minimizing the risk of dangling form formation and substrate lattice distortion if the bonding between the amorphous oxide and semiconductor is weak. However, if the bonding is strong between the oxide and semiconductor, there will be dangling bond formation and substrate lattice distortion at the interface for both types of oxides.

Besides purely technical problems such as oxide-semiconductor lattice mismatch and dangling bond formation in interface region, substitution of amorphous high-k oxides in simulations of oxide/semiconductor interfaces by crystalline systems changes some oxide properties particularly coordination numbers which in turn affects the band gap.

One of the major differences between the crystalline and amorphous high-k oxides comes from very different coordination distributions. Figure 5.2 presents atomic coordination distributions for crystalline Al_2O_3 (corundum) and amorphous Al_2O_3 generated by computer simulations [37, 67]. It can be seen that the crystalline and amorphous Al_2O_3 phases demonstrate very different coordination distributions. While crystalline Al_2O_3 has 4-coordinated O and 6-coordinated Al with no coordination broadening, the coordination distribution for amorphous Al_2O_3 introduces significant broadening, lowers Al coordination maximum to 4 and O coordination to 3 (Fig. 5.2). This comparison of coordination distributions for amorphous and crystalline Al_2O_3 samples (Fig. 5.2) indicates that despite chemical similarity, the amorphous phase can have a different chemical environment of atoms which may affect some physical properties of the oxide. Momida et al. performed theoretical investigation of dielectric response of amorphous alumina and calculated that the difference in band gap between amorphous Al_2O_3 in comparison with crystalline Al_2O_3 was 2.8–3.9 eV (DFT value), which is in good correlation with the experimental band gap difference of 1.9–3.7 eV with the amorphous having a smaller bandgap than crystalline oxide [68–70]. The more detailed theoretical research of Momida et al. indicated that the band-gap reduction in a-Al_2O_3 was caused by the local potential changes on Al and O sites due to their lower coordination in comparison with crystalline polymorphs correlating well with experimental measurements proposing roughly linear

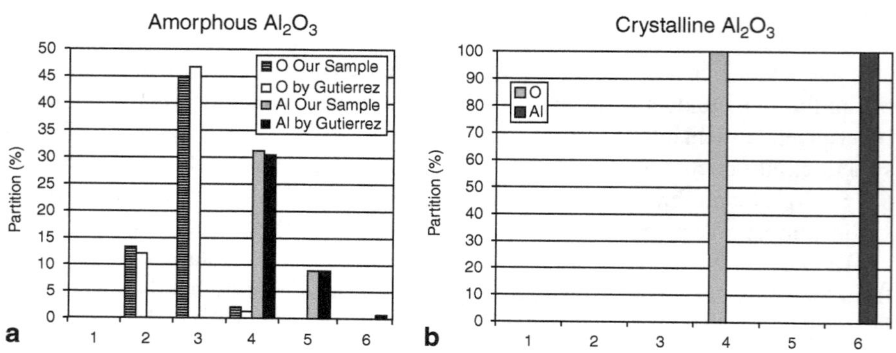

Fig. 5.2 Coordination distribution for **a** amorphous Al_2O_3, **b** crystalline Al_2O_3 (corundum) [37] (Reprinted with permission. Copyright 2009 American Institute of Physics)

relation between Al_2O_3 bandgap and Al average atomic coordination [68, 69]. Some amorphous oxides such as a-SiO_2 have the same coordination distribution as their crystalline polymorphs such as α-quartz due to their covalent bonding. However most high-k oxides such as a-Al_2O_3, a-ZrO_2, a-HfO_2 have ionic bonding and their amorphous phases have different coordination distributions than crystalline polymorphs, which can significantly affect physical properties such as band gaps.

5.2.4 The Oxide-Semiconductor Stack Simulation Techniques: DFT Relaxation vs. Molecular Dynamics

After high-k oxide is deposited on the semiconductor substrate (with or without an interfacial layer), there are two major computational approaches to simulate interface evolution to the final configuration: geometry optimization (relaxation) and molecular dynamics (annealing). Although there are other simulation techniques such as Monte-Carlo and hybrid MC-Molecular Dynamics approaches, the relaxation and molecular dynamics annealing remain to be the most widely used computational techniques for oxide/semiconductor interface research.

The goal of geometry optimization is to generate the lowest energy structure of an atomic system starting from an arbitrary initial configuration. Following the Born-Oppenheimer approximation, the motion of nuclei and electrons can be separated; therefore, geometry optimization typically includes a series of self-consistent single point energy calculations and adjustments of atomic positions. The forces on the nuclei can be calculated from the converged wavefunction using the Hellmann-Feynman theorem [71]:

$$F_{X_\gamma} = -\frac{\partial E}{\partial X_\gamma} = -\left\langle \psi \left| \frac{\partial \hat{H}}{\partial X_\gamma} \right| \psi \right\rangle, \tag{5.5}$$

where H–system Hamiltonian, ψ–system wavefunction, and λ–parameter corresponding to the coordinates of the nuclei. From the calculated forces acting on the atoms, the atomic coordinates are adjusted to minimize the system total energy by applying various relaxation algorithms. One of the most popular relaxation algorithms is Broyden–Fletcher–Goldfarb–Shanno (BFGS) method developed for solving an unconstrained nonlinear optimization problem [72, 73]. There are two sets of computational problems associated with geometry relaxation. The first set of problems includes Pulay force, which is an error introduced into the Hellman-Feynman forces if the basis set is incomplete, and a Pulay stress, which arises if the simulation cell shape changes during the simulation [74, 75]. The second set of problems results from the geometry optimization performed at zero kinetic (thermal) energy; this can lead to system relaxation into the closest configuration with zero forces, which can be a local, not the global minimum of the system. This feature makes geometry optimization very sensitive to the choice of initial configuration.

The common approaches to avoid such problem is either to carefully choose initial configuration (this is not always possible) or to perform an initial energy minimization using molecular dynamics simulations at finite temperature to obtain the vicinity of the global minimum and then to perform a full geometry relaxation.

Molecular dynamics (MD) is a simulation technique, which in the most classical form is described by the following algorithm:

1. Choose the initial atomic positions $\bar{R}_i^{(0)}$ and choose a short timestep Δt;
2. Calculate the forces acting on atoms using certain forcefield (first-principle or empirical);
3. Evolve system in time moving atoms applying various integration algorithms, which approximate the expansion: $R^{(j+1)} = R^{(j)} + V^{(j)}\Delta t + 1/2a\ \Delta t^2 + \cdots$
4. Update time: $t = t + \Delta t$;
5. Repeat steps (2)–(4) as long as necessary.

In case of DFT molecular dynamics simulations the forces in step (2) are often obtained by using the Hellmann-Feynman theorem (Eq. 5.5). There are various atom motion integration algorithms. The classical atom integration algorithms are Verlet and "leap-frog" (which is a modified version of Verlet algorithm) [76–80].

In Verlet algorithm, the atom positions are updated according to the following expression:

$$R(t + \Delta t) = 2 * R(t) - R(t - \Delta t) + a(t) * \Delta t^2. \tag{5.6}$$

The Verlet algorithm is straightforward, has modest memory demands, but moderate precision.

In the "leap-frog" algorithm, the atomic coordinate and velocities are updated according to different expressions:

$$R(t + \Delta t) = R(t) + v(t + \frac{1}{2}\Delta t) * \Delta t, \tag{5.7}$$

$$v(t + \frac{1}{2}\Delta t) = v(t - \frac{1}{2}\Delta t) + a(t) * \Delta t. \tag{5.8}$$

In this algorithm, the velocities *leap* over the positions; afterwards, the positions *leap* over the velocities. In the leap-frog algorithm, the velocities are calculated explicitly, however, they are not calculated at the same time as atomic coordinates. In this case, the velocities at any moment of time t can be approximated by the following relationship:

$$v(t) = \frac{1}{2} * \left[v(t - \frac{1}{2}\Delta t) + v(t + \frac{1}{2}\Delta t)\right]. \tag{5.9}$$

In modern classical MD simulations, old atom integration algorithms such as Verlet or "leap-frog" are often substituted by more sophisticated and more efficient algorithms such as r-RESPA, which implements a multi-timestep scheme applied differently to integration of short- and long-range interactions in the system providing significant boost to the computational efficiency [81–83].

The timestep in molecular dynamics simulations is chosen to provide satisfactory accuracy of atom motion integration and energy conservation in an adiabatic ensemble. If the timestep is too long, the violation of energy conservation becomes very severe. A typical timestep for molecular dynamics simulations is about 1 fs $= 1.0 \times 10^{-15}$ (s). However in case of very high temperatures or the presence of very light elements such as H, the timestep of MD simulations is often reduced to 0.1–0.5 fs to maintain proper level of energy conservation and integration accuracy. The relatively short typical timestep of ~1 fs and significant computational cost of DFT calculations for one atomic configuration limit the typical simulated time-span for DFT MD simulations to ~1–100 ps.

One of the major differences between geometry relaxation and molecular dynamic simulations is that many algorithms for geometry relaxation, such as BFGS or Conjugate Gradient (CG), are not bound to a real time frame and include only artificial atom movements, while the molecular dynamics and its system evolution are always logically connected to the real time frame. Another major difference is that many relaxation techniques, such as BFGS or Conjugate-Gradient, imply zero temperature (zero thermal energy) of the system and have an internal propensity to converge to the local minima. Conversely, molecular dynamics simulations are generally performed at finite temperature, which allows the system to pass over modest energy barriers and evolve to the global minimum. MD simulation can be performed as a function of temperature and annealing time to optimize the evolution to the global minimum. In general, finite-temperature molecular dynamics simulations of the oxide/semiconductor interfaces are finalized by cooling-relaxation or relaxation to converge to the ground state atomic configuration. Contrary to relaxation by GC and BFGS, DFT MD simulations can be performed as a function of temperature and annealing time to analyze system behavior. Although molecular dynamics simulations can be performed for different numbers of timesteps (simulated time span), in general, DFT geometry relaxations of oxide/semiconductor interfaces are computationally cheaper while DFT MD runs are generally much more computationally expensive.

The geometry relaxation and molecular dynamics approaches have been used for multiple simulations of high-k oxide/semiconductor interfaces. Peacock et al., Tse et al., Robertson et al., and Xiong et al. successfully applied DFT geometry relaxation approach to the investigation of ZrO_2/Si, HfO_2/Si, and ZrO_2/Ge interfaces [6, 13–16]. Giorgi et al., Kawamoto et al., Puthenkovilakam et al., Dong et al., and Zhang et al. applied geometry relaxation to the investigation of $SrTiO_3$/Si, ZrO_2/Si, HfO_2/Si, $ZrSiO_4$/Si, and $HfSiO_4$ interfaces [4, 5, 18, 20, 28]. In these cases, geometry relaxation was preferred over molecular dynamics since initial interfaces in these studies were designed to have very specific configurations; molecular dynamics with its kinetics at finite temperature would bring significant chaos to the atomic configurations. Broqvist et al. and Ha et al. successfully applied geometry relaxation to the investigation of defects in HfO_2/Si and a-HfO_2/Si systems creating defects and relaxing the interfaces [11, 27].

Although geometry relaxation can be a preferred technique when interface configuration is roughly known a priori, in situations when the oxide/semiconductor interface configuration is not known, molecular dynamics at finite temperature becomes a preferred technique since it can sample many atomic configurations both in local and global minima. Monaghan et al., Hakala et al., Gavartin et al., Broqvist et al., and Chagarov et al. successfully applied DFT molecular dynamics simulations to the investigation of $HfSiO_4$/Si, HfO_2/Si, HfO_2/SiO_2/Si, a-HfO_2/SiO_2/Si, a-Al_2O_3/Ge, a-ZrO_2/Ge, a-Al_2O_3/InGaAs and a-Al_2O_3/InAlAs/InGaAs interfaces [23, 24, 26, 32, 33, 36, 37, 49, 50].

The geometry relaxation and molecular dynamics techniques have their own advantages and disadvantages; therefore, the final choice of technique is usually dictated by specifics of the research goals and affordable computational cost.

5.3 DFT Simulations of High-k Oxides on Si/Ge Substrates

Although DFT simulations of high-k/III-V semiconductor stacks are different from simulations of high-k/Si or high-k/Ge interfaces, they can naturally inherit many useful methodological and research solutions previously developed and successfully employed for investigation of high-k/Si stacks. This chapter will provide overview of some DFT approaches and results on high-k/Si and high-k/Ge systems useful for modeling of the next generations of high-k/III-V oxide-semiconductor interfaces.

The main reason high-k oxides have become a necessity in commercial MOSFETs is that the scaling of the traditional SiO_2 oxide below 7–13 Å is fundamentally impossible due to the direct tunneling through the oxide [1, 2]. Furthermore, at oxides thickness of only a few atomic monolayers, the oxide properties begin to significantly deviate from the traditional bulk values. This problem is important even for high-k/Si devices because ultra-thin a-SiO_2 layers are often spontaneously formed between high-k dielectrics and Si substrates [84]. Although the dielectric constant of the bulk a-SiO_2 oxide ($\varepsilon_0 = 3.9$) is often used for estimations of the oxide electrical properties in the a-SiO_2/semiconductor stacks, there are significant concerns about validity of this assumption since experimental measurements report unusual electronic and structural properties for the first few angstroms of SiO_2 such as reduced band-gap and higher dielectric constant values [2, 84, 85].

To clarify this effect, Giustino et al. performed DFT simulations of Si-SiO_2 stacks varying the SiO_2 thickness from 0 to 20 Å and calculating the static (ε_0) and high-frequency (ε_∞) dielectric constants for the ultra-thin SiO_2 layers as a function of the film thickness [25]. The oxide was modeled by β-cristobalite using two interface models with various ratios of silicon to oxygen with all the oxygen in an O^{-2} state. In model I, the interfacial layer was formed by 1 ML of Si^{+1} and 1/2 ML of Si^{+2} [25, 86]. In model II, the interfacial layer was formed by equal amounts with 1/2 ML of Si^{+1}, Si^{+2}, and Si^{+3} atoms [25, 87, 88]. The static and high-frequency permittivities were calculated by the standard Berry-phase approach [89]. As shown in Fig. 5.3, the static and high-frequency dielectric permittivities of SiO_2 increase as the oxide thickness decreases, indicating significant deviation from the bulk-type properties due to different Si oxidation states in ultra-thin oxide layers. For model I, the static permittivity increases up to $\varepsilon_0 = 10.1$, while high-frequency permittivity reaches $\varepsilon_\infty = 5.1$. Model II has a smaller increase of permittivity reaching $\varepsilon_0 = 6.8$ and $\varepsilon_\infty = 3.7$. This study indicated that the ultra-thin SiO_2 interlayer between high-k oxide and Si has a less severe effect on the total stack capacitance than expected from bulk-like SiO_2 models with dielectric permittivity of $\varepsilon_0 = 3.9$.

To function properly, the gate in a MOSFET should sweep the Fermi level across the semiconductor band-gap. The perfect oxide-semiconductor interface should have no states in the semiconductor band-gap since these states can pin the Fermi level. For Si, mid-gap states are often associated with undercoordinated Si atoms having partially filled non-bonding orbital, i.e., dangling bonds. In the SiO_2/Si interface, the analysis of the bonding structure was relatively easy due to the covalent nature of oxide and semiconductor and could be performed by simple electron-pair bond counting. However this simple approach can not be applied to high-k oxide interfaces since high-k oxides usually have ionic bonding without fixed coordination numbers. In order to solve this problem and to predict high-k oxide-semiconductor

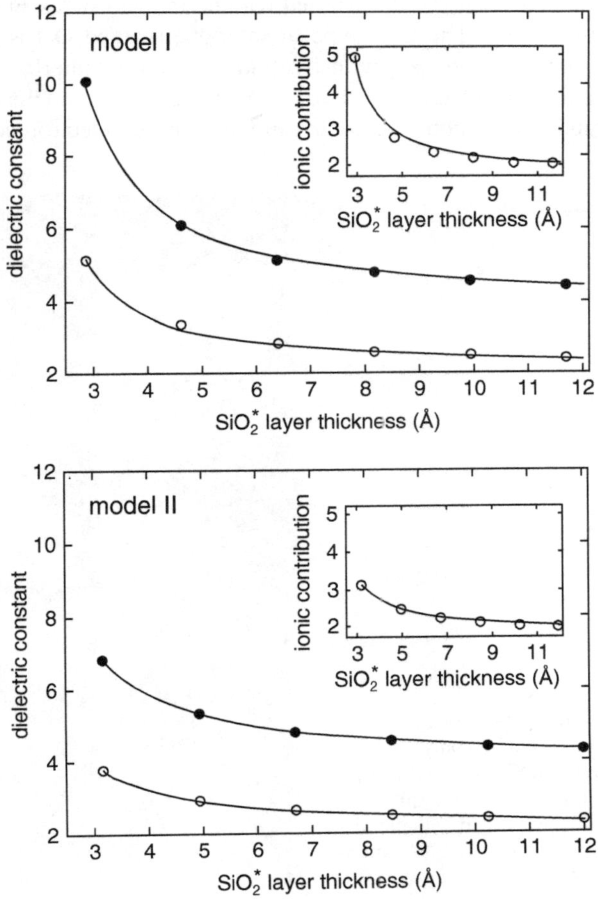

Fig. 5.3 Static ε_0 (*disks*) and high-frequency ε_∞ (*open circles*) dielectric constants of the oxide overlayer SiO_2^* plotted vs. its thickness, together with the behavior deduced from the simplified model (*solid lines*). In the *insets*, the ionic contributions to the static dielectric constants [25] (Reprinted with permission. Copyright 2004 Elsevier)

insulating interfaces, Peacock et al. formulated the following electron counting rules for ionic oxide/covalent semiconductor substrate interfaces [13]:

1. terminate with faces with enough excess oxygen so that the interfacial Si dangling bonds (DBs) are formally Si^{+} and empty, or
2. terminate with excess metal so the Si DBs are formally Si^{-} and filled.

These rules are quite general and can be applied to the bonding between high-k (crystalline or amorphous) oxide and III-V semiconductors.

To test and verify these set of electron counting rules, Peacock et al. performed set of DFT relaxation calculations for various O- and Zr-terminated ZrO_2/Si interfaces (Fig. 5.4) [13]. The O_{4V} configuration (Fig. 5.4a) has fourfold coordinated interfacial O atoms and leaves the Si dangling bonds to be half-filled predicting the interface to be metallic in full agreement with DFT calculations. The interface in Fig. 5.4b (O_4) in its initial state has twice as many interfacial oxygens with sixfold Si's and fourfold O's. After relaxation to the configuration presented in Fig. 5.4c with fourfold interfacial Si's, it is predicted to be an insulating interface confirmed by DFT calculations; by insulating Peacock et al. imply the valence and conduction bands are well separated with no midgap states, and, therefore consistent with unpinning. The third type of interface (named O_3) is presented in Fig. 5.4d and relaxes to the state with fivefold coordinated interfacial Si atoms, while half the interfacial O atoms bond to two Si's and one Zr, and the other half bonds to two Zr's and one Si atom. According to the proposed electron counting rules, this interface

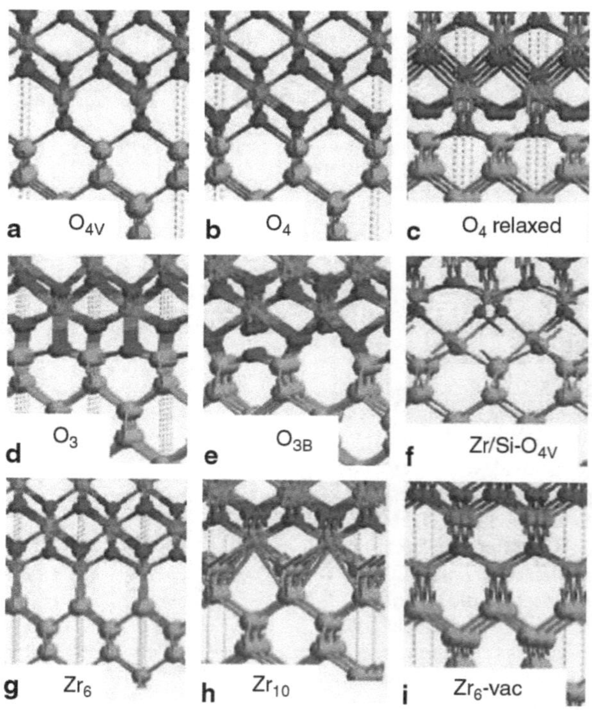

Fig. 5.4 Atomic configurations of various unrelaxed [(**b**) only] and relaxed Si:ZrO_2 interfaces, viewed in the (110) plane [13] (Reprinted with permission. Copyright 2004 American Physical Society)

should be insulating in complete agreement with DFT calculations [13]. The last O-terminated interface among investigated is presented in Fig. 5.4e (named O_{3B}). For this configuration one of the Si dangling bonds is accommodated in the Si–O–Si bridge, while the second dangling bond is bonded to the OZrO unit. However since OZrO unit has a closed shell, Si dangling bond needs an extra O half-monolayer to donate an electron resulting in configuration $Si^{+}(O^{2-})_{0.5}OZrO$. The formulated electron counting rules predict this interface to be insulating in complete agreement with performed DFT calculations. The O_3, O_{3B}, and O_4 interfaces have the same number of O atoms, and the O_3 interface (Fig. 5.4d) is the most stable configuration.

One of the possible Zr-terminated interfaces (Zr_6) is presented in Fig. 5.4g. In this interface, there are two Zr–Si bonds per interfacial Si, and Zr is in the Zr^{+4} state satisfying bonding rules and predicting the interface to be insulating. The DFT simulations show a band-gap in agreement with the prediction; however, there are gap states (i.e., pinned) due to the Zr electronegativity, which makes this interface not usable for microelectronics applications. The second Zr-terminated interface (Zr_{10}) is presented in Fig. 5.4h and has tenfold coordinated Zr atoms. According to the electron counting rules, this interface should be insulating, however DFT simulations find it to be metallic (i.e., pinned) due to the high Zr coordination number causing band overlap. In addition Peacock et al. investigated O_{4V} structure with half-monolayer of Zr substituted for Si in the interfacial layer (Fig. 5.4f) as suggested by Fiorentini et al. and found it to be metallic consistent with the electron counting rules [13, 19]. The two Zr-terminated interfaces (Zr_6 and Zr_{10}) emphasize the general validity of the electron counting predictions while demonstrating that they can be violated by the electron structure nuances of the particular oxide/semiconductor stacks. Nevertheless, the formulated electron counting rules proved to be valuable tool for predicting the high-k oxide/semiconductor stack properties and can be extrapolated to the amorphous high-k oxides and III-V semiconductors [32].

When applying the electron counting rules, two technical aspects should be taken into account: first, engineering of the initial interface to satisfy these rules does not guarantee that the final ground state bonding configuration will be insulating since geometry relaxation can distort the interface geometry resulting in unsatisfied bonding rules and a metallic interfacial electronic structure; secondly, the geometry relaxation of the manually designed interfaces performed at zero temperature have a significant propensity to converge to the local (not global) minima; this problem can be solved by applying molecular dynamics simulations at finite temperature capable of sampling many interface configurations and predicting the most energetically favorable one.

Another important aspect of high-k/Si DFT simulation heritage that is useful for future modeling of high-k/III-V stacks is a modeling of defects in high-k oxides. The problem of defects (vacancies or interstitials) in high-k dielectrics and their passivation has tremendous technological importance, since unpassivated defects can create pinning states in the bandgap region having detrimental effect on the total device performance. The DFT with its microscopic, atomistic approach can provide detailed insight into the major aspects of the defect problem in high-k oxides such as energy levels in the bandgap, formation energies, defect migration energy barriers and their passivation [11, 12, 26, 27, 32–35, 56, 90–93].

The defect formation energy in a charge state q can be expressed via the electron chemical potential μ referred to the valence band maximum ε_υ [94]:

$$E_f^q(\mu) = E_{tot}^q - E_{tot}^{bulk} - \sum_\alpha n_\alpha \eta_\alpha + q(\mu + \varepsilon_\upsilon), \quad (5.10)$$

where E_{tot}^q is a total energy of the defect system, E_{tot}^{bulk} the total energy of the unperturbed system, n_α the number of extra atoms of species α needed to create the defect, and η_α the corresponding atomic chemical potential.

The defects can demonstrate different properties in crystalline and amorphous polymorphs of high-k oxides. Broqvist et al., and Capron et al. performed comprehensive studies of oxygen vacancies in crystalline, amorphous HfO_2 systems and a-HfO_2/SiO_2/Si interfaces [11, 32–34]. These studies provided comparison between defect properties in crystalline and amorphous high-k oxides. In addition, defect tracking in an amorphous system with its more disordered network becomes a non-trivial issue, which makes these studies interesting from methodological point of view. For monoclinic HfO_2, the formation energies for three- and fourfold coordinated neutral oxygen vacancies are 9.1 eV and 8.9 eV (endothermic) respectively agreeing well with the previously published data [11]. Besides coordination, the vacancy formation energy depends on vacancy charge state. For monoclinic HfO_2, the threefold oxygen vacancy in positively charged states (+1 and +2) has a lower energy, while fourfold oxygen vacancy is a more favorable for the neutral and negatively charged states (−1 and −2) [11]. For amorphous HfO_2, the reported DFT-calculated vacancy formation energy has an average value of 8.85 eV (with standard deviation of 0.29 eV), which is lower than for the monoclinic phase [11]. The DFT MD annealing at 2000 K demonstrated the stability of the neutral oxygen vacancy in amorphous HfO_2 network and stability of its formation energy. However the DFT MD annealing of the doubly-positively charged vacancy V^{2+} in a-HfO_2 demonstrated different behavior with the formation energy decreasing by 2.6 eV making the system more stable [11].

The defect energy levels and their location relative to the band gap are very important for technological applications to avoid pinning states. For monoclinic HfO_2, the energy levels of the neutral (V^0) and doubly-positively charged vacancies (V^{2+}) for three- and fourfold coordinations were calculated; the neutral vacancies created occupied states in the middle of the HfO_2 band gap, while the V^{2+} vacancies created states in the band gap closer to the conduction band [11]. The energy level for the three-coordinated oxygen vacancy is closer to the conduction band than the four-coordinated vacancy by ~0.2 eV for the neutral vacancy and by ~0.3 eV for the V^{2+} charged vacancy. The vacancy energy levels in amorphous HfO_2 demonstrate certain similarity in comparison with the monoclinic phase: the neutral vacancy V^0 levels are located in the band gap middle (2.7 eV from the valence band), while V^{2+} vacancy levels are a little closer to the conduction band than the corresponding levels in the monoclinic phase. After DFT MD annealing of the amorphous HfO_2, the energy levels of the neutral V^0 and charged V^{2+} vacancies shift to the conduction band with V^{2+} level getting very close to the conduction band edge [11]. Although the annealing shifts these levels towards the conduction band, they still remain in the band gap thereby pinning it.

In the high-k oxide/semiconductor stacks, the vacancies can have different formation energies in different parts of the stack indicating propensity for vacancy migration in heterojunctions. To investigate this effect, Broqvist et al. modeled the $HfO_2/SiO_2/Si$ stack calculating the formation energies for the oxygen vacancy in the middle of the a-HfO_2 layer vs. the vacancy in the a-HfO_2/SiO_2 interfacial region [11]. While the vacancy in the middle of the a-HfO_2 layer had a formation energy (9.3 eV) close to the in-bulk a-HfO_2 value, the vacancy in the a-HfO_2/SiO_2 interface had formation energy of 8.1 eV, or by 1.2 eV lower than the in-bulk value clearly indicating the oxygen vacancy propensity to migrate from the a-HfO_2 bulk to the a-HfO_2/SiO_2 interface which was reported in previous theoretical studies [24, 95]. However, the computational propensity for vacancy migration is insufficient to insure experimental diffusion at finite temperature and should be complemented by the energy barriers low enough to be overcome at typical processing temperatures. Capron et al. performed DFT simulations of oxygen vacancy migration in neutral V^0 and charged V^{2+} states in monoclinic HfO_2 and through HfO_2/SiO_2 interface reporting that in monoclinic HfO_2 the activation barriers for the long-range diffusion are 2.4 eV and 0.7 eV for the neutral V^0 and charged V^{2+} vacancies respectively [34]. These DFT predictions of the vacancy thermodynamic driving force towards the SiO_2 interface and low V^{2+} vacancy migration barriers are in good agreement with reported experimental measurements [96–99].

The fact that defects in high-k oxides or their interfaces with semiconductors can form pinning states in the band gap region logically leads to the necessity of interface passivation to reduce pinning states. Ha et al. performed DFT simulations of HfO_2/SiO_2 interface suggesting several possible mechanisms of the interface passivation [27]. In these modeling studies, the interface was formed by crystalline (cubic) HfO_2 and crystalline SiO_2 (low-cristobalite). Although the ideal abrupt HfO_2/SiO_2 interface was simulated, the density of states revealed pinning states coming from under-coordinated interfacial Hf atoms. In the bulk, Hf has at least 7 oxygen neighbors, while in the simulated interface, Hf was bonded only to 5 or 6 oxygen atoms. As a result, nonbonding Hf d-orbital produced pinning state in the band gap. The similar oxygen under-coordination problem was found by Kralik et al. for ZrO_2 [100]. Ha et al. investigated usage of O interstitials to passivate Hf under-coordinated defects [27]. The DFT calculations indicate that O has a binding energy about 2 eV more favorable to segregate at the interface rather than to stabilize inside HfO_2 or SiO_2 regions [27]. When O interstitials are included into the interface region, they bind Hf nonbinding states, restore Hf normal coordination, and remove pinning states from the band gap. However, for practical applications, this method is not very useful since high concentration of oxygen will lead to Si subsurface oxidation decreasing gate capacitance and deteriorating stack properties. The chlorine (Cl) was investigated among other possible passivation chemicals [27]. Chlorine has a useful propensity to segregate in HfO_2/SiO_2 interface region where its binding energy is ~4 eV higher than in HfO_2 bulk [27]. While oxygen can accommodate two nonbonding electrons, Cl can accommodate just one electron of under-coordinated interfacial Hf. Similar to oxygen passivation, passivation by Cl requires very well controlled Cl dosing because excessive amounts of Cl will remain as interstitials defects, trapping electrons, and producing negative fixed charge. The oxygen and chlorine are not the only possible passivants. For a long time H has been used for passivation of defects in Si-based devices and has been a focus of active research [101–132].

5.4 Generation of Amorphous High-k Oxide Samples by Hybrid Classical-DFT Molecular Dynamics Computer Simulations

Experimentally, amorphous high-k oxide/semiconductor interfaces are expected to be superior to crystalline oxide-semiconductor counterparts due to the lack of lattice mismatch at amorphous oxide-semiconductor interfaces that can induce a high density of interface defects and the absence of grain-boundaries that can act as current leakage pathways. Despite their chemical composition similarity to crystalline polymorphs, amorphous high-k oxides demonstrate quite different microstructures, coordination distributions, and atomistic chemical environments.

The generation of realistic amorphous oxide samples is a non-trivial research task since contrary to crystalline samples, amorphous oxides do not have well-defined atomic structure. In case of high-k oxides, this problem is particularly challenging since, unlike covalent oxides, the high-k oxides very often have ionic bonding which gives broad coordination distributions. The one of the most robust computational techniques for generation of amorphous samples is a hybrid classical-DFT molecular dynamics "melt-and-quench" approach, which found many applications in generation of amorphous high-k oxides [11, 33, 36–38, 133–144]. This hybrid approach starts with classical molecular dynamics annealing after which the sample is transferred to DFT code and annealed for some additional time in the DFT forcefield. Sometimes sample transfer from classical to DFT code includes sample volume rescaling to reflect small difference of the sample density in the classical and DFT forcefields [36–38]. Usually the classical molecular dynamics has lower accuracy than DFT simulations, but classical MD is much more computationally efficient and can simulate much longer timescales than DFT MD. Conversely, DFT MD has higher accuracy (especially for irregular atomic systems as amorphous samples), but is much more computationally expensive and has limited affordable simulated timescales. The hybrid classical-DFT MD technique tries to combine the best features of the classical and DFT MD approaches. Figure 5.5 presents two possible hybrid classical-DFT MD approaches successfully applied to atomistic generation of realistic high-k oxides such as a-Al_2O_3, a-ZrO_2, and a-HfO_2 [11, 33, 36–38, 142–144].

The first sequence (Fig. 5.5a) starts from ordered system at artificially low classical density (such as $\rho_{norm}/\rho_{low} \sim 3.38$ [36–38]); annealing with classical MD at

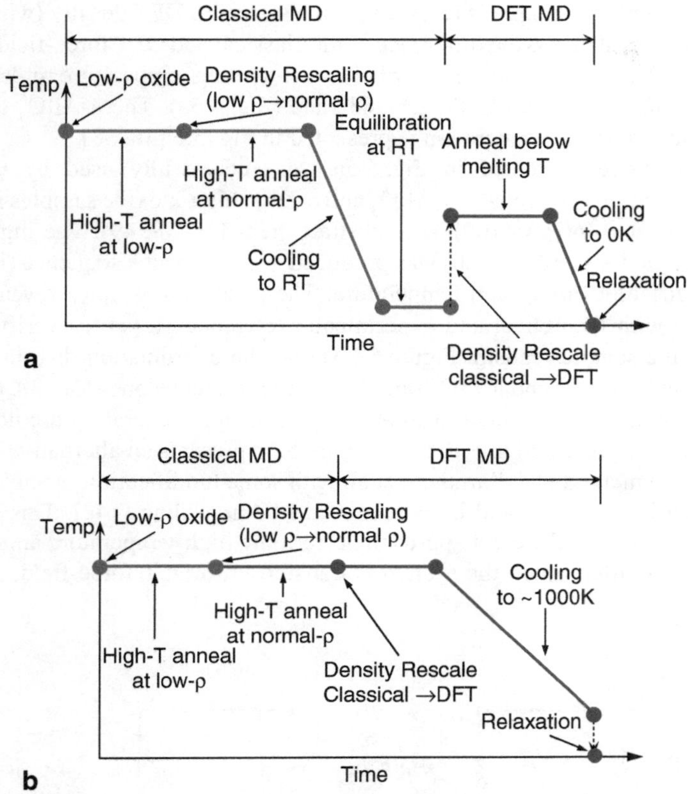

Fig. 5.5 Hybrid classical-DFT molecular dynamics generation of amorphous oxide samples

very high temperature (high-T) for a long time results in significant system intermixing which erases the original ordered geometry. Afterwards, the high-T melt is instantaneously rescaled from low to normal density, annealed at high temperature for an additional period of time to equilibrate it at the new normal classical density, and linearly cooled from high to room temperature (RT) passing through the amorphization stage. The cooled sample is equilibrated at RT finalizing the classical MD stage. Due to low computational cost, classical amorphous samples can be produced in batches varying annealing time and/or temperatures and choosing the most realistic sample [36–38, 145].

Published studies show that the classical molecular dynamics simulations with accurate interatomic potentials alone are able to produce amorphous high-k samples such as a-Al_2O_3 [133–138]. However, since DFT force-field generally provides a more accurate interaction model than empirical interatomic potentials of classical MD, the sequence in Fig. 5.5a introduces further sample refinement by performing subsequent DFT annealing, cooling, and relaxation to tune the sample to the more accurate DFT force-field. To bridge between classical and DFT MD, the selected classical amorphous sample is rescaled from classical to DFT density (which differ by several percent due to the difference in the classical and DFT force-fields), transferred to DFT code, annealed slightly below the melting temperature of the sample, cooled to 0 K and relaxed to the ground state (Fig. 5.5a). The a-Al_2O_3 sample at different stages of such generation is presented in Fig. 5.6 [36–38].

The technique presented in Fig. 5.5a was successfully used by Chagarov et al. for generation of realistic a-Al_2O_3 and a-ZrO_2 high-k oxide samples matching Ge(100)-2×1 and InGaAs(100)-4×2 substrate areas [36–38, 49]. The high-quality a-ZrO_2 sample for Ge(2×1)(100) was produced using the same sequence (Fig. 5.5a) but with different timing and temperature. The generated samples revealed good correlation with theoretical and experimental reference properties verifying high degree of the sample realism. Figure 5.7 shows the coordination distributions for the generated a-Al_2O_3 and a-ZrO_2 samples vs. the reference ones [36–38, 67, 140].

The techniques of computational amorphous sample generation are not limited to the sequence presented in Fig. 5.5a. Figure 5.5b presents an alternative approach similar to the method of J. Sarnthein et al. with some modifications such as low-to-normal density rescaling and initial classical MD annealing [142]. This sequence (Fig. 5.5b) follows a different approach performing high-temperature annealing by classical MD, after which the system is transferred to DFT force-field, annealed,

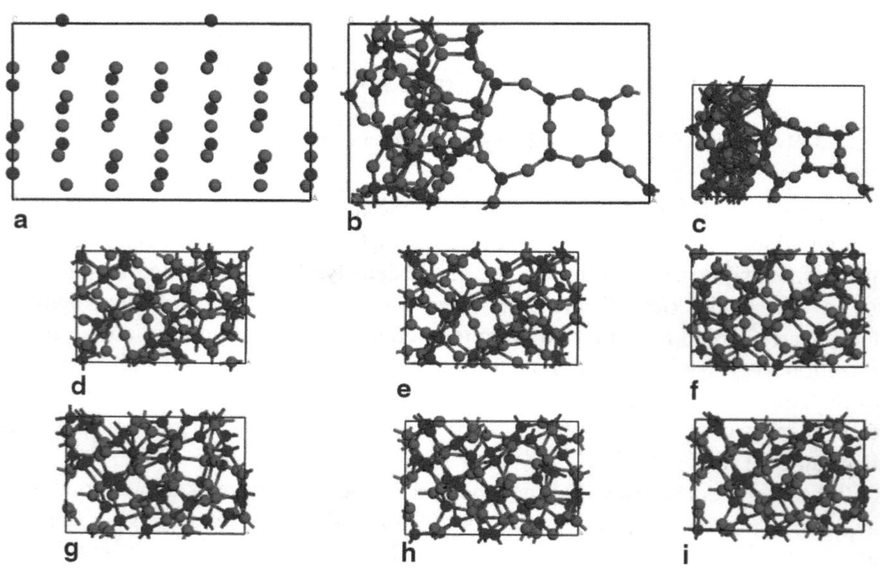

Fig. 5.6 The generation of a-Al_2O_3 sample. **a** initial system at low density, **b** low-density system after 5000 K annealing just before rescaling, **c** system just after rescaling to normal classical amorphous density, **d** after annealing at 5000 K, normal density, **e** after cooling to RT, **f** after equilibration at RT, **g** after DFT annealing at 1500 K, **h** after DFT cooling to 0 K, **i** after DFT relaxation. Stages **a**–**f** correspond to classical MD. Stages **g**–**i** correspond to DFT MD [38] (Reprinted with permission. Copyright 2008 The Electrochemical Society)

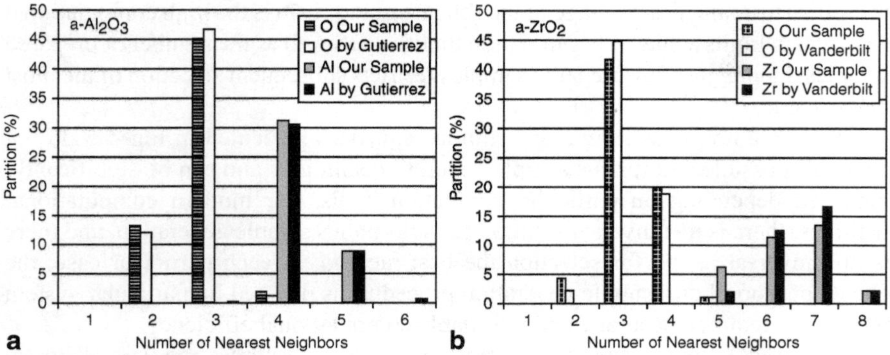

Fig. 5.7 Nearest neighbor distribution of: **a** classical a-Al_2O_3 sample from Ref. [37] vs. a larger scale classical sample from Ref. [67]. Cutoff radius is 2.2 Å. **b** DFT annealed a-ZrO_2 sample from Ref. [37] vs. DFT generated sample from Ref. [140]. Cutoff radius is 3 Å [37] (Reprinted with permission. Copyright 2009 American Institute of Physics)

cooled producing amorphization of the melt, and finally relaxed. The generation sequences presented in Fig. 5.5 have their own advantages and disadvantages.

The method in Fig. 5.5a is much more computationally efficient and can simulate much lower cooling rates than the Fig. 5.5b sequence since the Fig. 5.5a technique has higher fraction of the classical vs. DFT molecular dynamics. The experimental cooling rates are far lower than the rates affordable in DFT MD computer simulations; therefore, the Fig. 5.5a sequence can bring additional realism to the sample generation by its lower cooling rates of the melt. The empirical classical potentials are usually obtained by fitting to several crystalline system states in equilibrium and are less accurate than DFT forcefields. During MD simulations the atomistic system spends a significant amount of time in non-equilibrium states which introduces computational errors into the force and energy values. Therefore, the classical stage of generation sequence in Fig. 5.5a requires very accurate classical potentials which are not always available. The DFT annealing applied as a second stage of the sequence in Fig. 5.5a improves the sample quality by annealing in the more accurate DFT force-field and is able to produce amorphous samples in strong correlation with pure DFT-generated and experimental samples [36–38]. The high computational efficiency of the sequence in Fig. 5.5a allows multiple runs to generate large number of samples for further selection.

The generation sequence presented in Fig. 5.5b has a different set of advantages and disadvantages. For this sequence, the sample amorphization occurs during DFT cooling in the more accurate DFT force-field. However, the high computational cost of DFT runs often shifts affordable cooling rates to values higher by several orders of magnitude, which might have a negative effect on the sample realism. Since in this generation sequence high-temperature annealing, cooling and final relaxation are performed in the DFT force-field, this method does not require as accurate classical potentials as the method employed in Fig. 5a. As an example, Broqvist et al. and Chagarov et al. successfully employed schemes similar to sequence in Fig. 5.5b for generation of a-HfO_2 samples using a classical potential for crystalline ZrO_2

[11, 50]. The main disadvantage of the scheme in Fig. 5.5b is the high computational cost, which limits achievable simulation timescale as well as the number of prepared samples; having just few prepared samples restricts subsequent selection of the most realistic one.

The two amorphous sample generation sequences presented in Fig. 5.5 do not cover all possible amorphous sample generation solutions and can be significantly modified depending on particular simulation goals. For modern computational facilities, there is no universal solution for amorphous sample generation, and there is no universal recipe for selection the best method. In each particular case, the choice of amorphous sample generation procedure is dictated by particular system properties, required accuracy and affordable computational efficiency.

The computational generation of amorphous oxide samples requires elaborate tests to verify the quality of the samples. The choice of particular sample tests depends on the available reference properties and may be based upon both experimental and simulated data. Among the most frequently used sample quality tests are the calculations of radial-distribution functions (RDF), angular-distribution functions (ADF), coordination distributions, average coordination numbers, neutron scattering static structure factors and their comparison to the reference data.

The radial-distribution function (RDF) for atomic pair α-β is calculated according to Eq. (5.11), where $n_{\alpha,\beta}(r, r+\Delta r)$ is a number of β atoms within cutoff radius/shell $(r, r+\Delta r)$ from α-type atom, $\rho_\beta = N_\beta / V$ is the number density of atoms β, and N_β is the total number of β atoms. To make the RDF curves smoother, they are often calculated for a series of atomic configurations spanned by certain time interval (e.g., 10 fs), and finally averaged to minimize statistical noise. The typical RDF curves for classically generated a-Al_2O_3 sample are presented in Fig. 5.8.

$$g_{\alpha\beta}(r) = \frac{\langle n_{\alpha,\beta}(r, r+\Delta r)\rangle}{4\pi r^2 \rho_\beta \Delta r}. \tag{5.11}$$

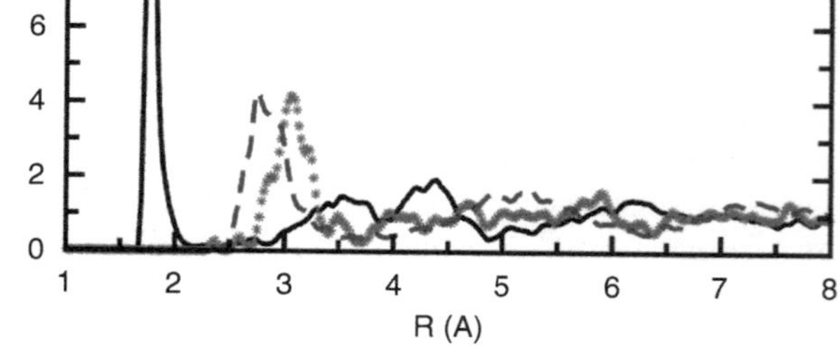

Fig. 5.8 RDF curves for a-Al_2O_3 sample generated by classical MD

The calculated RDF curve can be quantified via the main peak position and its full width at half maximum (FWHM).

Another useful computational metric, average nearest neighbor number $n_{\alpha\beta}(R)$ (Eq. 5.12), can be obtained by integrating the corresponding RDF curve up to the cutoff radius R, which is usually the position of the first minimum after the main RDF peak:

$$n_{\alpha\beta}(R) = 4\pi\rho_\beta \int_0^R g_{\alpha\beta}(r) r^2 dr. \tag{5.12}$$

One of the major metrics to characterize amorphous sample quality is the coordination distribution. In computer simulations the "coordination" value is determined by the number of nearest neighbors within a certain cutoff radius (Figs. 5.2 and 5.7). This can result in small differences in coordination number distributions determined by direct experimental imaging of electron density. Contrary to RDF curves, coordination distribution often demonstrates much greater variability from one simulated sample to another, providing clear venue for sample evaluation and selection.

The RDF curves provide radial distribution for specific α-β atomic pairs, ADF curves provide the angular distributions for specific α-β-χ atom type triples and can be useful for sample characterization.

Another useful metric to characterize generated amorphous sample is a neutron scattering static structure factor, which can be experimentally measured and theoretically calculated for atomistic models. The neutron scattering static structure factors ($S_N(q)$) (Eq. 5.13) are obtained from the partial static structure factors ($S_{\alpha\beta}(q)$) (Eq. 5.14), which are calculated from RDF curves $g_{\alpha\beta}$(r) (Eq. 5.11), where b_α is the coherent neutron scattering length of species α and $c_{\alpha(\beta)} = N_{\alpha(\beta)}/N$ is the concentration of $\alpha(\beta)$ species [38, 67].

$$S_N(q) = \frac{\sum_{\alpha\beta} b_\alpha b_\beta (c_\alpha c_\beta)^{1/2} \left[S_{\alpha\beta}(q) - \delta_{\alpha\beta} + (c_\alpha c_\beta)^{1/2} \right]}{\left(\sum_\alpha b_\alpha c_\alpha \right)^2} \tag{5.13}$$

$$S_{\alpha\beta}(q) = \delta_{\alpha\beta} + 4\pi\rho (c_\alpha c_\beta)^{1/2} \int_0^R r^2 \left[g_{\alpha\beta}(r) - 1 \right] \frac{\sin(qr)}{qr} \frac{\sin(\pi r/R)}{\pi r/R} dr \tag{5.14}$$

Although other metrics can be used for screening simulated amorphous samples, the described analysis based on RDF, ADF curves, coordination distributions, average coordination numbers, and neutron scattering static structure factor is able to provide a very detailed estimate of the generated amorphous sample quality and validity. Since for many amorphous systems, the available experimental data are very much limited, this can impose additional limitations on applicable sample tests.

5.5 The Current Progress in DFT Simulations of High-k Oxide/III-V Semiconductor Stacks

The III-V semiconductors due to their very promising properties have been a focus of active DFT research for a long time. The III-V's were theoretically investigated for bulk properties [146–150], growth mechanisms [151–158], surfaces and passivation [159–163], defects [164–166], and various III-V/III-V junctions [167–172]. There are certain methodological differences in modeling of high-k/III-V vs. high-k/Si: (a) III-V's do not tend to form thick interfacial layers with oxides but instead interdiffuse; (b) III-V's tend to have higher mobility and smaller bandgaps; (c) oxide deposition on III-V's is done at lower temperatures. Although amorphous high-k oxide/III-V interfaces are very important from technological point of view, the comprehensive DFT studies of these systems are just emerging and are very limited in number [49, 50].

The a-Al_2O_3 interfaces with InGaAs(4×2) and InAlAs(4×2)/InGaAs substrates were investigated by DFT molecular dynamics at 800 K which is a realistic post-deposition annealing temperature (PDA) [49]. To artificially accelerate atomic kinetics, elongate the DFT MD simulated timescale, and investigate the effect of temperature, the a-Al_2O_3/InGaAs interface was also investigated at the artificially high 1100 K annealing temperature. The a-Al_2O_3/InGaAs(4×2) stack was investigated in two designs discussed previously: one-interface (with vacuum spacer) (Fig. 5.9) and two-interface (supercell with no vacuum) (Fig. 5.10). For better statistics, the one-interface a-Al_2O_3/InGaAs stack was investigated with two different initial a-Al_2O_3 cuts in interface region (Fig. 5.9a, b).

These simulations were performed with high-quality amorphous Al_2O_3 samples generated by hybrid classical–DFT MD "melt and quench" approach discussed previously. The oxide samples perfectly matched $In_{0.5}Ga_{0.5}As$ and $In_{0.5}Al_{0.5}As$ substrate surface areas to avoid artificial internal stresses [36–38, 49]. The oxide sample high realism and quality were verified by a set of tests including the RDF main peak positions and FWHM's, coordination distributions, average nearest neighbor numbers, neutron scattering static structural factors, and band gaps demonstrating good correlation to previously simulated and experimentally measured reference properties [67, 68, 173]. In particular, the defect-free band gap of the a-Al_2O_3 samples (~3.7 eV) was in a good agreement with previously reported bandgaps of 3.8 eV and 3.77 eV [37, 68]. The $In_{0.5}Ga_{0.5}As$ substrate had the 4×2 surface reconstruction. To mimic continuous semiconductor bulk, the 3 bottom layers of InGaAs slab were fixed in their bulk positions and As dangling bonds at the bottom were passivated by 3/4 |e| charge H's [174]. The initial $In_{0.5}Ga_{0.5}As$ bulk unit cell had a DFT optimized lattice constant and was built from GaAs by substituting half of the Ga atoms by In following checkerboard pattern.

The InAlAs/InGaAs semiconductor stack was formed by 7 layers of $In_{0.5}Al_{0.5}As$ and 6 layers of $In_{0.5}Ga_{0.5}As$ with bottom As passivated by 3/4 |e| H atoms in order to simulate infinite InGaAs bulk [174]. The $In_{0.5}Al_{0.5}As$ bulk unit cell had a DFT optimized lattice constant and was formed from GaAs by substituting half of the Ga by Al and the other half by In following checkerboard pattern. The DFT optimized

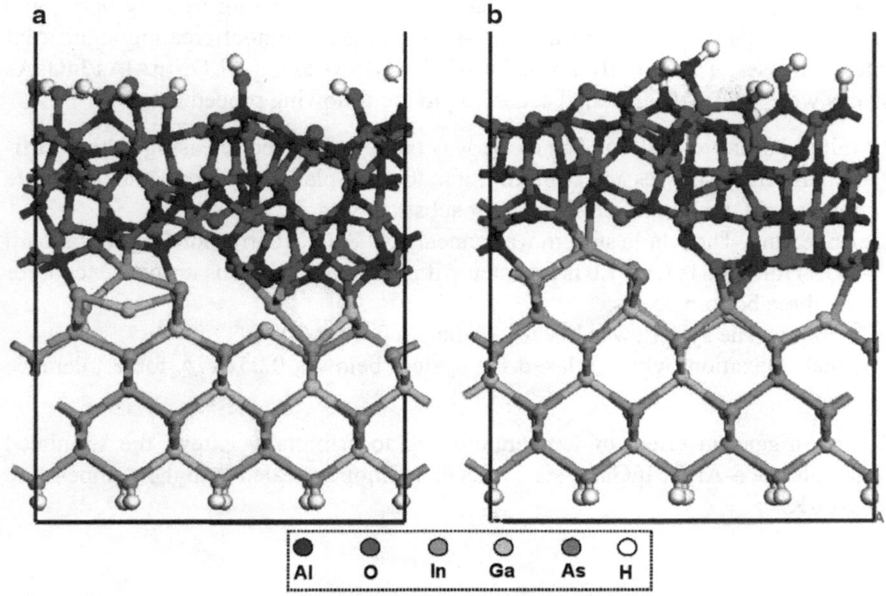

Fig. 5.9 DFT MD a-Al_2O_3/InGaAs interface annealed at 800 K. **a** a-Al_2O_3 Cut I. **b** a-Al_2O_3 Cut II [49] (Reprinted with permission. Copyright 2009 Elsevier)

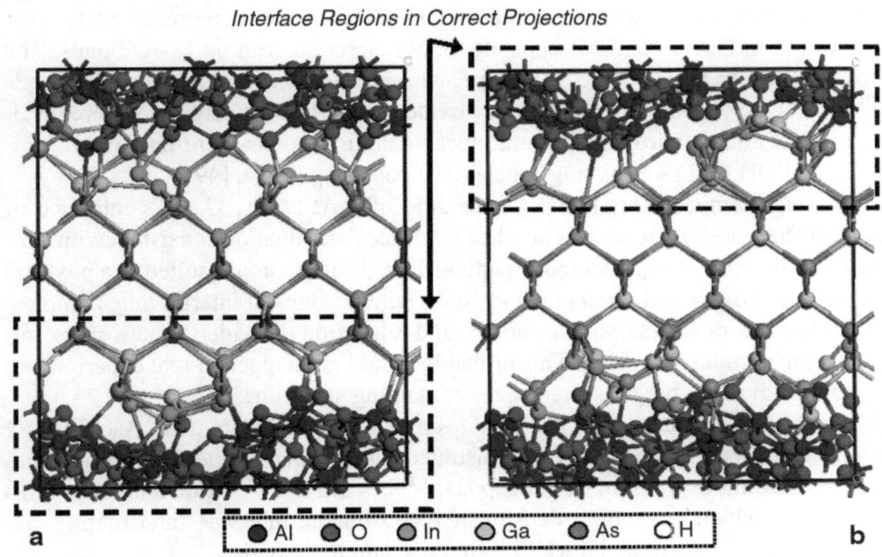

Fig. 5.10 a-Al_2O_3/InGaAs two-interface (supercell) model after final relaxation. Annealing temperature is 800 K. The model (**b**) is obtained from model (**a**) by 90° rotation around vertical axis. The dashed boxes indicate interface regions in correct spatial projection [49] (Reprinted with permission. Copyright 2009 Elsevier)

InAlAs and InGaAs unit cells have lattice constants differing by only 0.3%, and in the stack they share the same cross-sectional area without creating significant lateral stresses. The initially formed a-Al_2O_3/InGaAs and a-Al_2O_3/InAlAs/InGaAs stacks were DFT-MD annealed according to the following procedure:

1. Initial partial relaxation. The oxide was partially relaxed releasing initial artificial interface stresses and conforming to the non-planar semiconductor substrate topography while the semiconductor substrate was fixed in space;
2. Annealing. The whole system was annealed at 800 K (corresponding to standard PDA) for 1000 fs with 1.0 fs timestep with the substrate atoms unfixed except for the three bottom layers;
3. Cooling. The system was linearly cooled to 0 K for 200 fs;
4. Final relaxation, which relaxed the system below a 0.05 eV/Å force tolerance level.

To investigate an effect of temperature and to artificially extend the simulated timescale the a-Al_2O_3/InGaAs stack was in addition annealed at higher temperature of 1100 K.

5.5.1 Interfacial Bonding Structure

The DFT-MD simulated a-Al_2O_3/$In_{0.5}Ga_{0.5}As$(100)-4×2 interfaces annealed at 800 K for two cases of one-interface design (Fig. 5.9) and one case of two-interface design (Fig. 5.10) demonstrated strong similarity in interface bonding represented by polar As–Al and In/Ga–O bonds of opposite dipole direction with no O–As bonds. The a-Al_2O_3/InGaAs stack (Fig. 5.9a) after final relaxation had several dangling bonds at the a-Al_2O_3/vacuum interface (upper oxide surface) which originally were localized on Al under-coordinated atoms; these dangling bonds were passivated by 8 additional OH groups removing states in the band-gap region [49].

The high-temperature (1100 K) annealing of a-Al_2O_3/$In_{0.5}Ga_{0.5}As$ interface of one-interface design produced complete interface delamination consistent with generally weak oxide-semiconductor bonding. This delamination resulted in a physical separation of oxide and semiconductor slabs with no signs of intermixing, complete breaking of oxide-semiconductor bonds, and relaxation of oxide/vacuum and semiconductor/vacuum interfaces. This probably would not happen in real experimental systems with much bigger system sizes containing step-edges.

The bonding structure of DFT-MD simulated a-Al_2O_3/$In_{0.5}Ga_{0.5}As$(100)-4×2 interfaces (Figs. 5.9 and 5.10) demonstrated the absence of the three major phenomena leading to creation of midgap states between high-k oxide and III-V semiconductor, which are a high density of As–O bonds, interface intermixing, and disruption of the substrate lattice to form new dangling bonds [175].

The a-Al_2O_3/InAlAs interface of the a-Al_2O_3/InAlAs/InGaAs stack demonstrated different behavior with signs of intermixing during annealing when a few Al atoms were pulled out of InAlAs into the oxide forming single Al–In bonds. The interfacial

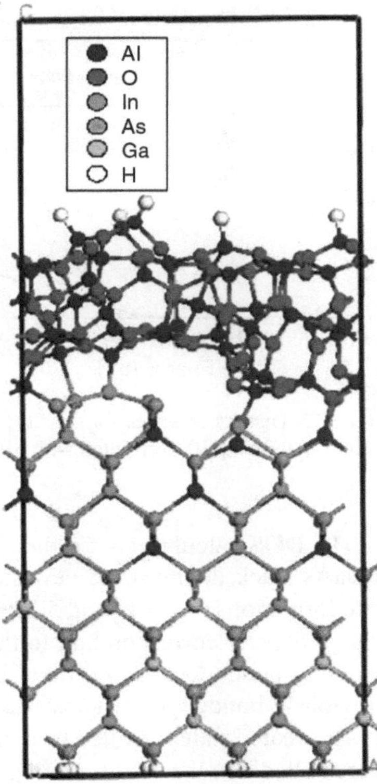

Fig. 5.11 DFT MD a-Al_2O_3/InAlAs/InGaAs interface annealed at 800 K [49] (Reprinted with permission. Copyright 2009 Elsevier)

bonding was formed by the low density of Al–O, Al–As, and O–In bonds with moderately compensating bond-dipoles (Fig. 5.11).

5.5.2 *Density of State Analysis*

The electronic structure of the a-Al_2O_3/InGaAs and a-Al_2O_3InAlAs interfaces was investigated by calculating the density of states (DOS) for the interface regions, semiconductor channels below interfaces, and clean semiconductor substrates without oxide (Fig. 5.12) [49]. Note that in comparison with experimental data, standard DFT underestimates calculated band gaps for a-Al_2O_3, InGaAs and InAlAs due to approximated nature of implemented exchange interaction.

The a-Al_2O_3/InGaAs interface demonstrates DOS free of pinning states with the Fermi level positioned in the middle of the band gap region (Figs. 5.9a and 5.12a). This result is consistent with the hypothesis than an unpinned interface between the highly ionic metal oxide and semiconductor channel can be formed if the interaction is weak similar to observations for gate oxides on carbon nanotubes [176].

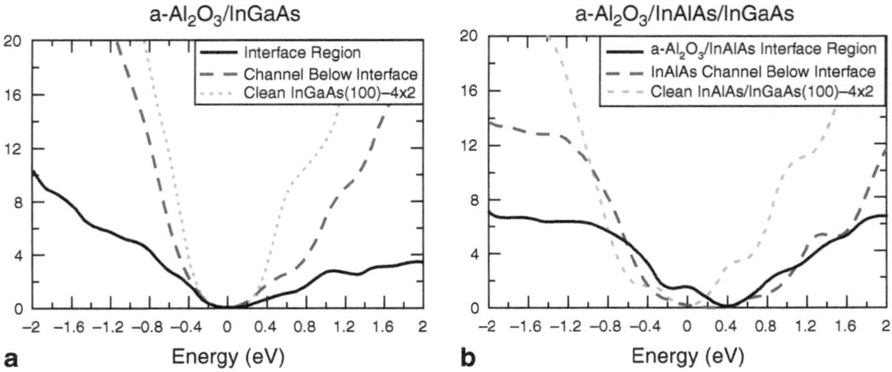

Fig. 5.12 Density of states for a-Al_2O_3/InGaAs and a-Al_2O_3/InAlAs/InGaAs stacks. The Fermi level is at 0.0 eV [49] (Reprinted with permission. Copyright 2009 Elsevier)

The DOS calculations for the a-Al_2O_3/InAlAs interface of the a-Al_2O_3/InAlAs/InGaAs stack demonstrate new states created in the band-gap region after interface formation (Fig. 5.11 and 5.12b). The 3D visualization of the band-decomposed charge density corresponding to these newly created states indicates that they are dangling bonds localized on two interfacial As atoms in InAlAs row which were previously bonded to the substrate Al atom pulled into the oxide. Therefore, these newly created states resulted from interface intermixing when the substrate Al atoms were pulled into the oxide creating In–Al metal-metal bonds (Fig. 5.11) [49]. In real processing with much longer annealing, this intermixing could be more extensive. The presented a-Al_2O_3/InAlAs interface is pinned due to dangling bonds produced by intermixing instead of large changes in the charge state of the interface atoms due to ionic bonding alone [49]. It is noted that due to the wider bandgap of the InAlAs compared to InGaAs and due to the local nature of the pinning states, the midgap defect states at the oxide/InAlAs interface induce almost no midgap states in the InGaAs channel; this may be very helpful in buried channel MOSFETs as predicted by Ayubi-Moak et al. [177].

5.5.3 Semiconductor Substrate Deformation

The intermixing and deformation are highly undesirable for oxide-semiconductor interfaces since they have significant negative impact on interface physical and electrical properties, such as formation of interfacial layer, decrease of carrier mobility, and creation of midgap pinning states.

To estimate average layer-by-layer deformation in semiconductor substrate after interface formation the following norm is employed:

$$\Delta \bar{R}_i = \frac{1}{N_i} \sum_j \left| \bar{R}_j - \bar{R}_{0j} \right|, \tag{5.15}$$

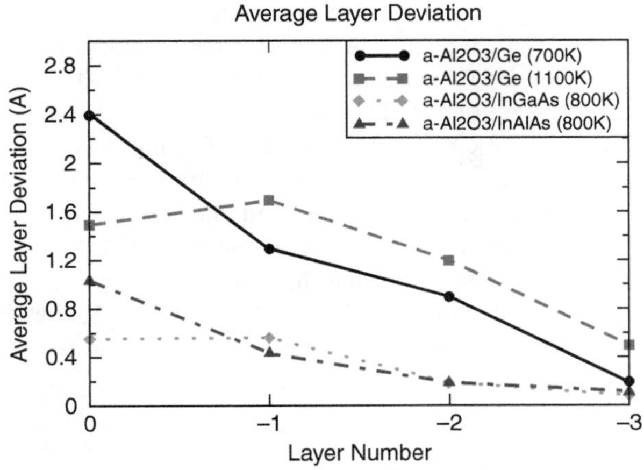

Fig. 5.13 Average layer deviation starting with uppermost semiconductor layer (layer 0) and down into the bulk (layer −3) [49] (Reprinted with permission. Copyright 2009 Elsevier)

where $\bar{R}_j$ and $\bar{R}_{0j}$ are coordinates of atom j belonging to the horizontal layer i after the interface relaxation and in the initial relaxed clean substrate slab, while index j goes along every substrate atom in horizontal layer i, and N_i is the number of atoms in horizontal layer i [49]. The average layer-by-layer deformation for the a-Al_2O_3/InGaAs (Fig. 5.9a), and a-Al_2O_3/InAlAs/InGaAs (Fig. 5.11) stacks is presented in Fig. 5.13.

The a-Al_2O_3/InGaAs interfaces for the investigated one-interface (Fig. 5.9) and two-interface models (Fig. 5.10) have practically no intermixing, and small displacements of InGaAs interface atoms relative to the initial atomic positions in the InGaAs slab (Figs. 5.9, 5.10, and 5.13). Conversely, the a-Al_2O_3/InAlAs interface of the a-Al_2O_3/InAlAs/InGaAs stack undergoes medium intermixing with an Al atom pulled from the InAlAs row into the oxide creating a metal-metal Al-In bond (Figs. 5.11 and 5.13). The a-Al_2O_3/Ge interface (annealed at 700 K and 1100 K) clearly demonstrates that the largest deformation and intermixing consistent with the group IV semiconductor surfaces being more reactive than III-V surfaces because of the half filled dangling bonds on surface tri-coordinated group IV atoms [37].

5.5.4 *Bader Charge Analysis*

The interface charge transfer and polarity are very important properties of oxide/semiconductor stacks since too high charge transfer through interface can have a negative impact on device performance. To investigate charge transfer and interface polarity on atomistic level, a Bader charge analysis was employed [178, 179].

For the a-Al_2O_3/InGaAs interface (Fig. 5.9a), the Bader charge analysis of the interfacial InGaAs atoms relative to the clean $In_{0.5}Ga_{0.5}As(100)$-4×2 surface demonstrated weakly polar bond formation with no ionic bonding. Interfacial semiconductor atoms in an unpinned interface should have near bulk-like charges. To verify this the Bader charge of the interfacial InGaAs atoms was compared with in-bulk InGaAs atoms indicating that the relative to bulk atoms, the As bonded to Al had excessive charge of ~0.26 |e|, the Ga atoms bonded to one O were depleted by ~0.08 |e|, the In atoms bonded to one O atom were depleted by ~0.18 |e|, and the In atoms bonded to two O's were depleted by ~0.40 |e|. The analysis of the total Bader charge transfer through the interface indicates that after interface formation the InGaAs substrate was depleted by ~1.23 |e| leading to a limited normalized charge transfer of -8.6×10^{-3} |e|/Å^2 consistent with the good interface for microelectronic applications [49].

The comparative analysis of the two investigated high-k/III-V semiconductor interfaces indicates that the a-Al_2O_3/InGaAs demonstrates the lowest absolute charge transfer through the interface (-8.6×10^{-3} |e|/Å^2), and the charge transfer through the a-Al_2O_3/InAlAs interface is -1.08×10^{-2} |e|/Å^2 or ~1.3 times higher in absolute value than in a-Al_2O_3/InGaAs [49].

5.5.5 Comparison to Experimental Data

The presented DFT-MD simulations of a-Al_2O_3/InGaAs and a-Al_2O_3/InAlAs/InGaAs stacks demonstrated good correlation to experimental measurements. Note that while computer simulations often consider ideal interfaces, experimental measurements usually introduce various collateral deviations due to oxide deposition and stack processing. In addition, DFT MD due to its high computational cost typically limits investigated system size to few hundreds of atoms and simulated timescale to ~1–100 ps. These two DFT MD limitations and experimental data non-ideality do not allow straightforward comparison between DFT MD results and experimental measurements.

5.5.5.1 a-Al_2O_3/InAlAs

The experimental data on a-Al_2O_3 bonding to Al-V semiconductor substrates are very limited. Yasuda et al. performed C-V studies of a-Al_2O_3 interfaces with GaAs(100), $In_{0.53}Ga_{0.47}As$(100), $In_{0.52}Al_{0.48}As$, and $Al_{0.5}Ga_{0.5}As$ substrates [180]. In their studies the oxide was grown by ALD with preliminary wet-cleaning in NH_3 solution. The studies demonstrated the best C-V curves (modest frequency dispersion and low hysteresis) for InGaAs samples, the higher dispersion and hysteresis for GaAs(100) and $In_{0.52}Al_{0.48}As$ substrates, while the $Al_{0.5}Ga_{0.5}As$ substrate demonstrated almost no capacitance modulation.

Although the presented DFT-MD simulations investigated ideal interfaces while experimentally measured stacks had some preexisting oxides, the intermixing observed during DFT-MD studies can be roughly compared with experimentally measured substrate oxide formation. Yasuda et al. observed that the amount of post-ALD As oxide was almost six times greater on InAlAs(100) substrate than on InGaAs, while for In and Ga post-ALD oxides the difference between the two substrates was very modest [180]. The enhanced interfacial intermixing was also reported for HfO_2 ALD deposition on InAlAs substrate in comparison with InGaAs [181]. These experimental data correlate very well with the presented DFT-MD simulations predicting higher degree of intermixing for a-Al_2O_3/InAlAs interface in comparison with a-Al_2O_3/InGaAs stack demonstrating no intermixing.

5.5.5.2 a-Al_2O_3/InGaAs

The a-Al_2O_3/$In_{0.2}Ga_{0.8}As$ interface bonding structure was investigated by XPS to analyze chemical shifts with high accuracy [182]. For XPS shifts, the bulk oxide and semiconductor bulk states were used as reference states; InGaAs would be assigned as In^{+0}, Ga^{+0}, and As^{+0} while Al_2O_3 would be assigned as Al^{+3} and O^{-2}. The samples were wet-cleaned in NH_4S and TMA, and although TMA reduces the Ga_2O_3, a remaining Ga^{+1} peak indicated either residual sulfur or Ga–O bonding [183]. The As was found in As^{+0} state indicating absence of arsenic oxides and Al peaks pointed to an O–Al–O atomic environment. A similar experiment was performed by Aquirre-Tostado et al. on samples using atomic hydrogen to remove native oxides instead of NH_4S wet cleaning [184]. A Ga^{+1} XPS peak found at the interface was assigned to Ga–O bonding since no sulfur was present. The Ga–O bonds at the interface can be assigned either to residual substrate oxides or bonding between a-Al_2O_3 and the substrate. This surface can not be directly compared to DFT-MD simulations since cleaning by atomic hydrogen depletes the surface of In and creates As–As bonds. However, these experiments correlate well with DFT-MD predictions that the bonding between a-Al_2O_3 and InGaAs is sufficiently weak that the As, Al, and O charge states are bulk-like and the Ga is slightly shifted to the loss of a partial charge [49].

Kim et al. have obtained a chemically abrupt interface of a-Al_2O_3/$In_{0.53}Ga_{0.47}As$(100) observed by HR-TEM, high-angle annular dark-field (HAADF) TEM and prepared by in-situ decapping of As_2-capped $In_{0.53}Ga_{0.47}As$(100) followed by a-Al_2O_3 ALD [185, 186].

The angle-resolved XPS spectra demonstrated no chemical shifts for the interfacial Ga, In, and As atoms. However that XPS spectrometer had lower resolution than previously employed by Shahrjerdi et al., therefore a small chemical shift could still be present [183]. These data correlate well with the presented DFT-MD simulations demonstrating abrupt interface with no intermixing, weak nearly covalent bonding between a-Al_2O_3 and InGaAs(100) with chemical shifts of less than one electron for all interfacial atoms.

Cheng and Fitzgerald performed TEM studies of Al_2O_3/GaAs MOCVD grown interfaces on As-rich substrate, which makes it different from DFT-MD simulations using In/Ga rich InGaAs(100)-4×2 surface [49, 187]. TEM images demonstrated a completely abrupt interface, while XPS showed the absence of any arsenic oxides. These TEM results are similar to the results by Shahrjerdi et al. and by Hong-Liang for ALD deposited a-Al_2O_3 on HF cleaned GaAs(100) [183, 188]. These experiments were performed on GaAs(100) substrate while DFT-MD investigated InGaAs(100); however, this comparison can be made due to chemical similarity between InGaAs and GaAs substrates. Huang et al. reported HR-TEM measured a sharp interface between ALD-grown a-Al_2O_3 and wet-cleaned InGaAs both for as-deposited and nitrogen annealed at 500°C stacks [180]. The C-V measurements of $Al_2O_3/In_{0.53}Ga_{0.47}As$ stacks demonstrated significant inversion at elevated temperatures, consistent with an unpinned Al_2O_3/InGaAs interface and correlating well with DFT density of states (Fig. 5.12) [185, 186].

The a-Al_2O_3/InGaAs interface implemented in transistors prepared by wet cleaning of the substrate demonstrated very high output current, low threshold voltages, reasonable subthreshold slopes, and reasonably low off current for submicron devices [189]. However, the better device performance was obtained with higher indium content [190]. These experimental device data correlate well with DFT calculated density of states for the a-Al_2O_3/InGaAs interface region consistent with unpinned interface (5.12). The better device performance might be obtained with decapped samples and in-situ oxide deposition after heterostructure growth since the studies cited above indicated that the wet cleaning leaves some gallium oxides/sulfites likely creating some interface states, and UHV MBE deposited oxides on in-situ regrown semiconductor surfaces have the lowest reported subthreshold swings [191].

5.6 Summary

In summary, the presented DFT MD simulations of a-Al_2O_3/InGaAs and a-Al_2O_3/InAlAs interfaces indicated different degree of interface intermixing, electronic structure and Bader charge transfer for 800 K annealing temperature. The a-Al_2O_3/InGaAs interfaces demonstrated no intermixing, good electronic structure consistent with unpinned Fermi level (Figs. 5.9a and 5.12a), low interface polarity, and low charge transfer. Conversely, the a-Al_2O_3/InAlAs interface of the a-Al_2O_3/InAlAs/InGaAs stack annealed at 800 K showed moderate interface intermixing with Al atoms pulled from the InAlAs substrate into the oxide creating dangling bonds on row As which contribute states in the band-gap region [49].

From the presented DFT-MD simulations, it follows that the weak bonding between the oxide and the semiconductor substrate is the best predictor of an unpinned interface between a highly ionic metal oxide and a compound semiconductor. In practical ULSI manufacturing this can be achieved only if the oxide deposition does not perturb the substrate and if the semiconductor surface has few partially filled dangling bonds which are far more reactive than filled or empty dangling bonds [49].

References

1. M. Schulz, Nature (London) 399, 729 (1999).
2. D. Muller, T. Sorsch, S. Moccio, F. Baumann, K. Evans-Lutterodt, G. Timp, Nature 399, 758 (1999).
3. P.W. Peacock and J. Robertson, J. Appl. Phys. 92, 4712 (2002).

4. R. Puthenkovilakam, E. Carter, J.P. Chang, Phys. Rev. B 69, 155329 (2004).
5. A. Kawamoto, K.J. Cho, R. Dutton, J. Comput. Aided Mater. Des. 8, 39 (2002).
6. K.-Y. Tse and J. Robertson, J. Appl. Phys. 100, 093713 (2006).
7. P. Hohenberg and W. Kohn, Phys. Rev. 136, 864 (1964).
8. W. Kohn and L.J. Sham, Phys. Rev. 140, 1133 (1965).
9. R.M. Martin, *Electronic Structure: Basic Theory and Practical Methods* (Cambridge University Press, 2004).
10. G. Lucovsky and J.C. Phillips, Thin Solid Films 486, 200 (2005).
11. P. Broqvist and A. Pasquarello, Microelectronics Eng. 84, 2022 (2007); A.S. Foster, F.L. Gejo, A.L. Shluger, R.M. Nieminen, Phys. Rev. B 65, 174117 (2002).
12. M. Houssa, G. Pourtois, M.M. Heyns, A. Stesmans, J. Phys. Condens. Matter 17, S2075 (2005).
13. P.W. Peacock and J. Robertson, Phys. Rev. Lett. 92, 057601 (2004).
14. J. Robertson and P.W. Peacock, Physica Status Solidi (b) 241, 2236 (2004).
15. K. Xiong, P.W. Peacock, J. Robertson, Appl. Phys. Lett. 86, 012904 (2005).
16. P.W. Peacock, K. Xiong, K. Tse, J. Robertson, Phys. Rev. B 73, 075328 (2006).
17. J. Robertson, J. Vac. Sci. Technol. B 18, 1785 (2000).
18. Y.F. Dong, Y.P. Feng, S.J. Wang, A.C.H. Huan, Phys. Rev. B 72, 045327 (2005).
19. V. Fiorentini and G. Gulleri, Phys. Rev. Lett. 89, 266101 (2002).
20. X. Zhang, A.A. Demkov, H. Li, X. Hu, Y. Wei, J. Kulik, Phys. Rev. B 68, 125323 (2003).
21. F. Sacconi, J.M. Jancu, M. Povolotskyi, A.D. Carlo, IEEE Trans. Electron Devices 54, 3168 (2007).
22. C.J. Forst, P.E. Blochl, K. Schwarz, Comput. Mater. Sci. 27, 70 (2003).
23. S. Monaghan, J.C. Greer, S.D. Elliott, Phys. Rev. B 75, 245304 (2007).
24. M.H. Hakala, A.S. Foster, J.L. Gavartin, P. Havu, M.J. Puska, R.M. Nieminen, J. Appl. Phys. 100, 043708 (2006).
25. F. Giustino, P. Umari, A. Pasquarello, Microelectronics Eng. 72, 299 (2004).
26. J.L. Gavartin and A.L. Shluger, Microelectronics Eng. 84, 2412 (2007).
27. J.-H. Ha, P.C. McIntyre, K.J. Cho, J. Appl. Phys. 101, 033706 (2007).
28. G. Giorgi, A. Korkin, K. Yamashita, Comput. Mater. Sci. 43, 930 (2008).
29. A.A. Knizhnik, A.A. Bagaturyants, I.V. Belov, B.V. Potapkin, A.A. Korkin, Comput. Mater. Sci. 24, 128 (2002).
30. C.J. Forst, C.R. Ashman, K. Schwarz, P.E. Blochl, Microelectronics Eng. 80, 402 (2005).
31. A. Alkauskas, P. Broqvist, F. Devynck, A. Pasquarello, Phys. Rev. Lett. 101, 106802 (2008).
32. P. Broqvist, A. Alkauskas, J. Godet, A. Pasquarello, J. Appl. Phys. 105, 061603 (2009).
33. P. Broqvist, A. Alkauskas, A. Pasquarello, Appl. Phys. Lett. 92, 132911 (2008).
34. N. Capron, P. Broqvist, A. Pasquarello, Appl. Phys. Lett. 91, 192905 (2007).
35. A. Alkauskas, P. Broqvist, A. Pasquarello, Phys. Rev. Lett. 101, 046405 (2008).
36. E. Chagarov and A.C. Kummel, Surf. Sci. 602, L74 (2008).
37. E. Chagarov and A.C. Kummel, J. Chem. Phys. 130, 124717 (2009).
38. E. Chagarov and A.C. Kummel, ECS Trans. 16(10), 773 (2008).
39. W. Young and E. Elcock, Proceedings of the Physical Society of London 89 (565P), 735 (1966).
40. D.R. Cox and H.D. Miller, *The Theory of Stochastic Processes* (Methuen, London, 1965).
41. A. Bortz, M. Kalos, J. Lebowitz, J. Comput. Phys. 17(1), 10 (1975).
42. Y. Liang, J. Kulik, Y. Wei, T. Eschrish, J. Curless, B. Craigo, S. Smith, Presented at the MRS Symposium Proceedings, (2004).
43. S. Pasko, A. Abrutis, L. Hubert-Pfalzgraf, Mater. Lett. 59(2, 3), 261 (2005).
44. Y. Zhou, N. Kojima, K. Sasaki, J. Phys. D Appl. Phys. 41(17), 175414 (2008).
45. R. Puthenkovilakam and J.P. Chang, J. Appl. Phys. 96, 2701 (2004).
46. L. Fonseca, A. Demkov, A. Knizhnik, Physica Status Solidi B 239, 48 (2003).
47. P. Peacock and J. Robertson, Appl. Phys. Lett. 83, 5497 (2003).
48. C. Forst, C. Ashman, K. Schwarz, P. Blochl, Nature 427, 53 (2004).
49. E.A. Chagarov and A.C Kummel, Surf. Sci. 603, 3191 (2009).
50. E.A. Chagarov and A.C. Kummel, Surf. Sci. (submitted).

51. K.-O. Ng and D. Vanderbilt, Phys. Rev. B 59, 10132 (1999).
52. G. Kresse and J. Furthmuller, Comput. Mater. Sci. 6(1), 15 (1996).
53. G. Kresse and J. Furthmuller, Phys. Rev. B 54(16), 11169 (1996).
54. J. Neugebauer and M. Scheffler, Phys. Rev. B 46(24), 16067 (1992).
55. A.A. Demkov and A. Navrotsky (editors), *Materials Fundamentals of Gate Dielectrics* (Springer, New York, 2004), p. 77.
56. K. Xiong, J. Robertson, M.C. Gibson, S.J. Clark, Appl. Phys. Lett. 87, 183505 (2005).
57. J.T. Ryan, P.M. Lenahan, J. Robertson, G. Bersuker, Appl. Phys. Lett. 92, 123506 (2008).
58. K. Xiong, Y. Du, K. Tse, J. Robertson, J. Appl. Phys. 101, 024101 (2007).
59. K. Tse and J. Robertson, Microelectronics Eng. 84, 663 (2007).
60. J. Robertson, O. Sharia, A.A. Demkov, Appl. Phys. Lett. 91, 132912 (2007).
61. J. Robertson, K. Xiong, S.J. Clark, Thin Solid Films 496, 1 (2006).
62. K. Tse and J. Robertson, Appl. Phys. Lett. 89, 142914 (2006).
63. K. Xiong, J. Robertson, S.J. Clark, Physica Status Solidi B 243, 2071 (2006).
64. J. Robertson, K. Xiong, B. Falabretti, IEEE Trans. Device Mater. Reliab. 5, 84 (2005).
65. P.W. Peacock and J. Robertson, Appl. Phys. Lett. 83, 2025 (2003).
66. J. Robertson, J. Appl. Phys. 93, 1054 (2003).
67. G. Gutierrez and B. Johansson, Phys. Rev. B 65, 104202 (2002).
68. H. Momida, T. Hamada, Y. Takagi, T. Yamamoto, T. Uda, T. Ohno, Phys. Rev. B 73, 054108 (2006).
69. S. Miyazaki, J. Vac. Sci. Technol. B 19, 2212 (2001).
70. A.M. Goodman, J. Appl. Phys. 41, 2176 (1970).
71. R.P. Feynman, Phys. Rev. 56, 340 (1939).
72. C.G. Broyden, J. Inst. Math. Appl. 6, 76–90 (1970).
73. R. Fletcher, Comput. J. 13, 317–322 (1970).
74. P. Pulay, Molec. Phys. 19, 197 (1969).
75. G.P. Francis and M.C. Payne, J. Phys. Condens. Matter 2, 4395 (1989).
76. L. Verlet, Phys. Rev. 159, 98 (1967).
77. L. Verlet, Phys. Rev. 165, 201 (1967).
78. M.P. Allen and D.J. Tildesley, *Computer Simulation of Liquids* (Oxford, 1987).
79. J.M. Haile, *Molecular Dynamics Simulation* (Wiley, 1992).
80. D.C. Rapaport, *The Art of Molecular Dynamics Simulation* (Cambridge Univ. Press, 1995).
81. M. Tuckerman, B. Berne, A. Rossi, J. Chem. Phys. 94, 1465 (1990).
82. S. Stuart, R. Zhou, B. Berne, J. Chem. Phys. 105, 1426 (1996).
83. P. Procacci and M. Marchi, J. Chem. Phys. 104, 3003 (1996).
84. D.A. Muller and G.D. Wilk, Appl. Phys. Lett. 79, 4195 (2001).
85. C.M. Perkins, B.B. Triplett, P.C. McIntyre, K.C. Saraswat, S. Haukka, M. Tuominen, Appl. Phys. Lett. 78, 2357 (2001).
86. N. Tit and M.W.C. Dharma-Wardana, J. Appl. Phys. 86, 387 (1999).
87. A. Pasquarello, M.S. Hybertsen, R. Car, Phys. Rev. B 53, 10942 (1996).
88. A. Pasquarello, M.S. Hybertsen, R. Car, Appl. Surf. Sci. 104, 317 (1996).
89. P. Umari and A. Pasquarello, Phys. Rev. Lett. 89, 157602 (2002).
90. J.M. Knaup, P. Deak, T. Frauenheim, A. Gali, Z. Hajnal, W.J. Choyke, Phys. Rev. B 72, 115323 (2005).
91. J.L. Gavartin, D. Muñoz Ramo, A.L. Schluger, G. Bersuker, B.H. Lee, Appl. Phys. Lett. 89, 082908 (2006).
92. S. Ismail-Beigi and S.G. Louie, Phys. Rev. Lett. 95, 156401 (2005).
93. M. Hedstrom, A. Schindlmayr, G. Schwarz, M. Scheffler, Phys. Rev. Lett. 97, 226401 (2006).
94. C.G. Van de Walle and J. Neugebauer, J. Appl. Phys. 95, 3851 (2004).
95. W.L. Scopel, A.J.R. da Silva, W. Orellana, A. Fazzio, Appl. Phys. Lett. 84, 1492 (2004).
96. M. Copel and M.C. Reuter, Appl. Phys. Lett. 83, 3398 (2003).
97. S.J. Wang, P.C. Lim, A.C.H. Huan, C.L. Liu, J.W. Chai, S.Y. Chow, J.S. Pan, Q. Li, C.K. Ong, Appl. Phys. Lett. 82, 2047 (2003).

98. G. Bersuker, J. Barnett, N. Moumen, B. Foran, C.D. Young, P. Lysaght, J. Peterson, B.H. Lee, P.M. Zeitzoff, H.R. Huff, Jpn. J. Appl. Phys. Part 1 43, 7899 (2004).
99. D.Y. Cho, S.J. Oh, Y.J. Chang, T.W. Noh, R. Jung, J.C. Lee, Appl. Phys. Lett. 88, 193502 (2006).
100. B. Kralik, E.K. Chang, S.G. Louie, Phys. Rev. B 57, 7027 (1998).
101. R.Q. Wu, M. Yang, Y.P. Feng, Y.F. Ouyang, Appl. Phys. Lett. 93, 082107 (2008).
102. P.J.H. Denteneer, C.G. Van de Walle, S.T. Pantelides, Phys. Rev. Lett. 62, 1884 (1989).
103. C.G. Van de Walle, P.J.H. Denteneer, Y. Bar-Yam, S.T. Pantelides, Phys. Rev. B 39, 10791 (1989).
104. C.S. Nichols, D.R. Clarke, C.G. Van de Walle, Phys. Rev. Lett. 63, 1090 (1989).
105. P.E. Blöchl, C.G. Van de Walle, S.T. Pantelides, Phys. Rev. Lett. 64, 1401 (1990).
106. C.G. Van de Walle, Physica B 170, 21 (1991).
107. C.G. Van de Walle and R.A. Street, Phys. Rev. B 49, 14 766 (1994) (RC).
108. C.G. Van de Walle and R.A. Street, Phys. Rev. B 51, 10 615 (1995).
109. N.H. Nickel, N.M. Johnson, C.G. Van de Walle, Phys. Rev. Lett. 72, 3393 (1994).
110. C.G. Van de Walle and N.H. Nickel, Phys. Rev. B 51, 2636 (1995) (RC).
111. J. Neugebauer and C.G. Van de Walle, Phys. Rev. Lett. 75, 4452 (1995).
112. C.G. Van de Walle and J. Neugebauer, Phys. Rev. B 52, R14 320 (1995) (RC).
113. J. Neugebauer and C.G. Van de Walle, Appl. Phys. Lett. 68, 1829 (1996).
114. C.G. Van de Walle, Phys. Rev. B 56, R10 020 (1997).
115. C.G. Van de Walle, J. Non Cryst. Solids 227–230, 111 (1998).
116. C.G. Van de Walle, Phys. Rev. Lett. 80, 2177 (1998).
117. C.G. Van de Walle, C. Stampfl, J. Neugebauer, J. Cryst. Growth 189/190, 505 (1998).
118. C.G. Van de Walle, J. Vac. Sci. Technol. A 16, 1767 (1998).
119. B. Tuttle, C.G. Van de Walle, J.B. Adams, Phys. Rev. B 59, 5493 (1999).
120. C.G. Van de Walle and J. Goss, Mater. Sci. Eng. B 58, 17 (1999).
121. B. Tuttle and C.G. Van de Walle, Phys. Rev. B. 59, 12884 (1999).
122. C.G. Van de Walle and B. Tuttle, IEEE Trans. Electron Devices 47, 1779 (2000).
123. C. Herring, N.M. Johnson, C.G. Van de Walle, Phys. Rev. B 64, 125209 (2001).
124. C.G. Van de Walle and J. Neugebauer, Phys. Rev. Lett. 88, 066103 (2002).
125. S. Limpijumnong and C.G. Van de Walle, Physica Status Solidi B 228, 303 (2001).
126. C.G. Van de Walle, Physica Status Solidi B 235, 89 (2003).
127. B. Szûcs, A. Gali, Z. Hajnal, P. Deák, C.G. Van de Walle, Phys. Rev. B 68, 085202 (2003).
128. C.G. Van de Walle and J. Neugebauer, Nature 423, 626 (2003).
129. C.G. Van de Walle, J. Vac. Sci. Technol. A 21, S182 (2003).
130. C.G. Van de Walle, Physica B 376–377, 1 (2006).
131. J. Weber, A. Janotti, P. Rinke, C.G. Van de Walle, Appl. Phys. Lett. 91, 142101 (2007).
132. C.G. Van de Walle, J. Alloys Compd. 446–448, 48 (2007).
133. G. Gutierrez and B. Johansson, Phys. Rev. B 65, 104202 (2002).
134. S.P. Adiga, P. Zapol, L.A. Curtiss, Phys. Rev. B 74, 064204 (2006).
135. H. Momida, T. Hamada, Y. Takagi, T. Yamamoto, T. Uda, Phys. Rev. B 73, 054108 (2006).
136. V.V. Hoang, Phys. Rev. B 70, 134204 (2004).
137. V.V. Hoang and N.H. Hung, Physica Status Solidi B 243, 416 (2006).
138. V.V. Hoang and S.K. Oh, Phys. Rev. B 72, 054209 (2005).
139. X. Zhao, D. Ceresoli, D. Vanderbilt, Phys. Rev. B 71, 085107 (2005).
140. D. Vanderbilt, X. Zhao, D. Ceresoli, Thin Solid Films 486, 125 (2005).
141. D. Ceresoli and D. Vanderbilt, Phys. Rev. B 74, 125108 (2006).
142. J. Sarnthein, A. Pasquarello, R. Car, Phys. Rev. Lett. 74, 4682 (1995).
143. P. Broqvist and A. Pasquarello, Appl. Phys. Lett. 90, 082907 (2007).
144. J. Godet, P. Broqvist, A. Pasquarello, Appl. Phys. Lett. 91, 262901 (2007).
145. E. Chagarov, J.B. Adams, J. Kieffer, Model. Simul. Mater. Sci. Eng. 12, 337 (2004).
146. I.N. Remediakis and E. Kaxiras, Phys. Rev. B 59, 5536 (1999).
147. S. Li and C. Ouyang, Phys. Lett. A 336, 145 (2005).
148. A.H. Reshak, Eur. Phys. J. B 47, 503 (2005).
149. D.C. Gupta and S. Kulshrestha, J. Phys. Condens. Matter 20, 255204 (2008).

150. Q.Y. Chen, M. Xu, H.P. Zhou, M.Y. Duan, W.J. Zhu, H.L. He, Physica B 403, 1666 (2008).
151. A. Palma, E. Semprini, A. Talamo, N. Tomassini, J. Cryst. Growth 150, 180 (1995).
152. K. Seino, A. Ishii, T. Aisaka, Prog. Theor. Phys. Suppl. 138, 164 (2000).
153. K. Seino, A. Ishii, T. Kawamura, Jpn. J. Appl. Phys. 39, 4285 (2000).
154. P. Kratzer, E. Penev, M. Scheffler, Appl. Phys. A 75, 79 (2002).
155. A. Ishii, K. Fujiwara, T. Aisaka, Appl. Surf. Sci. 216, 478 (2003).
156. P. Kratzer, C.G. Morgan, M. Scheffler, Prog. Surf. Sci. 59, 135 (1998).
157. P. Kratzer, E. Penev, M. Scheffler, Appl. Surf. Sci. 216, 436 (2003).
158. P. Kempisty and S. Krukowski, J. Cryst. Growth 303, 37 (2007).
159. G-X Qian, R.M. Martin, D.J. Chadi, J. Vac. Sci. Technol. B 5, 933 (1987).
160. J. Northrup and C.G.V. der Walle, Appl. Phys. Lett. 84, 4322 (2004).
161. S. Wang, Y. Song, X. Zhao, J. Li, J. Liu, Y. Jia, Mod. Phys. Lett. B 20, 1275 (2006).
162. B. Szucs, Z. Hajnal, Th. Frauenheim, C. Gonzalez, J. Ortega, R. Perez, F. Flores, Appl. Surf. Sci. 212, 861 (2003).
163. A.L. Rosa and J. Neugebauer, Phys. Rev. B 73, 205346 (2006).
164. A. Gulans, R.A. Evarestov, I. Tale, C.C. Yang, Physica Status Solidi C 2, 507 (2005).
165. A. Janotti, S.B. Zhang, S.-H. Wei, C.G. Van de Walle, Opt. Mater. 25, 261 (2004).
166. H. Williams, W.A. Hofer, E. Cavar, A. Mikkelsen, E. Lungren, Phys. Rev. B 78, 205309 (2008).
167. R.G. Dandrea and C.B. Duke, J. Vac. Sci. Technol. A 11, 848 (1993).
168. A. Kley and J. Neugebauer, Phys. Rev. B 50, 8616 (1994).
169. S.-F. Ren and J. Shen, J. Appl. Phys. 81, 1169 (1996).
170. C.-F. Shih, N.-C. Chen, P.-H. Chang, K.-S. Liu, Jpn. J. Appl. Phys. 44, 7892 (2005).
171. R. Choudhury, D.R. Bowler, M.J. Gillan, J. Phys. Condens. Matter 20, 235227 (2008).
172. B.K. Agrawal, S. Agrawal, R. Srivastava, Physica E 33, 17 (2006).
173. P. Lamparter and R. Kniep, Physica B 234, 405–406 (1997).
174. K. Shiraishi, J. Phys. Soc. Jpn. 59, 3455–3458 (1990).
175. D. Winn, M. Hale, T. Grassman, J. Sexton, A. Kummel, J. Chem. Phys. 127, 134705 (2007).
176. A. Javey, H. Kim, M. Brink, Q. Wang, A. Ural, J. Guo, P. McIntyre, P. McEuen, M. Lundstrom, H. Dai, Nat. Mater. 1, 241 (2002).
177. J. Ayubi-Moak, B. Benbakhti, G.W. Paterson, R. Hill, M. Passlack, I. Thayne, A. Senov, Microelectronic Eng. 86, 1564 (2009).
178. G. Henkelman, A. Arnaldsson, H. Jonsson, Comput. Mater. Sci. 36, 354–360 (2006).
179. E. Sanville, S. Kenny, R. Smith, G. Henkelman, J. Comput. Chem. 28, 899–908 (2007).
180. T. Yasuda, N. Miyata, H. Ishii, T. Itatani, O. Ichikawa, N. Fukuhara, M. Hata, A. Ohtake, T. Haimoto, T. Hoshii, M. Takenaka, S. Takagi, *SISC 2008* (San Diego, IEEE, 2008), p. 41.
181. S. Oktyabrsky, V. Tokranov, S. Koveshnikov, M. Yakimov, R. Kambhampati, H. Bakhru, R. Moore, W. Tsai, J. Cryst. Growth 311, 1950 (2009).
182. M. Milojevic, F.S. Aguirre-Tostado, C.L. Hinkle, H.C. Kim, E.M. Vogel, J. Kim, R.M. Wallace, Appl. Phy. Lett. 93, 202902 (2008).
183. D. Shahrjerdi, D.I. Garcia-Gutierrez, E. Tutuc, S.K. Banerjee, Appl. Phys. Lett. 92, 223501 (2008).
184. F.S. Aguirre-Tostado, M. Milojevic, B. Lee, J. Kim, R.M. Wallace, Appl. Phy. Lett. 93, 172907 (2008).
185. E. Kim, E.A. Chagarov, J. Cagnon, Y. Yuan, A.C. Kummel, P.M. Asbeck, S. Stemmer, K.C. Saraswat, P.C. McIntyre, J. Appl. Phys. (in press).
186. P.C. McIntyre, Y. Oshima, E. Kim, E. Chagarov, J. Cagnon, K.C. Saraswat, S. Stemmer, A.C. Kummel, *SISC 2008* (San Diego, IEEE, 2008).
187. C.-W. Cheng and E.A. Fitzgerald, Appl. Phy. Lett. 93, 031902 (2008).
188. L. Hong-Liang, L. Yan-Bo, X. Min, D. Shi-Jin, S. Liang, Z. Wei, W. Li-Kang, Chin. Phys. Lett. 23, 1929 (2006).
189. Y. Xuan, Y.Q. Wu, P.D. Ye, Electron Device Lett. 29, 294 (2008).
190. Y. Xuan, T. Shen, M. Xu, Y.Q. Wu, P.D. Ye, *IEDM 2008* (San Fransisco, 2008), 371.
191. R.J.W. Hill et al., IEEE Electron Device Lett. 28, 1080 (2007).

Chapter 6
Interfacial Chemistry of Oxides on III-V Compound Semiconductors

Marko Milojevic, Christopher L. Hinkle, Eric M. Vogel and Robert M. Wallace

Abstract This chapter reviews the interfacial chemistry observed for III-V semiconductors, with particular attention to native and deposited oxide gate dielectrics. The chemical states observed at the III-V surface as studied by x-ray photoelectron spectroscopy are presented, and the suppression of oxide formation with deposition techniques, particularly atomic layer deposition, is discussed. The resultant electrical properties of these interfaces are then reviewed, with attention to the interpretation of capacitance-voltage and related transistor measurements.

6.1 Introduction

The prospect of enhanced device performance from III-V materials has been recognized for at least 50 years, and yet, relative to the phenomenal size of the Si-based IC industry, these materials fulfilled only specific niches and were often referred to as "the material of the future" [1]. A key restriction enabling widespread use of III-V materials is the lack of a high quality, natural insulator for III-V substrates like that available for the SiO_2/Si materials system [2]. The prospect of impending scaling challenges for technologies based on silicon metal oxide semiconductor field effect transistor (MOSFET) devices has brought renewed focus on the use of alternate surface channel materials from the III-V compound semiconductor family. The performance of the traditional MOSFET device structure is dominated by defects at the semiconductor/oxide interface, which in turn requires a high quality semiconductor surface. In this chapter, we summarize recent progress in the understanding of the dielectric/III-V interface, particularly in regard to the interfacial chemistry that impacts the resultant electrical behavior observed. The first section summarizes the current understanding of the structural and chemical properties of

R. M. Wallace (✉)
Department of Materials Science and Engineering, University of Texas at Dallas, Richardson, Texas 75044, USA
e-mail: rmwallace@utdallas.edu

S. Oktyabrsky, P. D. Ye (eds.), *Fundamentals of III-V Semiconductor MOSFETs*,
DOI 10.1007/978-1-4419-1547-4_6,

III-V semiconductor surfaces. Next, the nature of the oxidation states of surface oxides are described. We then summarize the deposition of such oxides on the III-V surface in view of the interfacial chemical reactions employed [3]. Finally we examine the resultant electrical properties of these oxides on III-V surfaces.

6.2 Surfaces of III-V MOSFET Semiconductor Candidates

Despite the fact that semiconductor materials are largely attributed the title as the most engineered and defect free materials known, one has to keep in mind that the MOSFET device architecture relies, by definition, on the control of the largest structural defect possible; a crystal surface. In the best case scenario the material is terminated in a single crystal plane offering a perfectly flat and homogenous surface. Regardless of the method used to obtain such an "ideal" surface, the atoms in the terminating plane are inherently in a chemically unstable environment due to the bond scission and resulting formation of unsatisfied "dangling" bonds. Thermodynamic concerns drive these unpaired electrons to form covalent bonds between adjacent atoms in a process termed dimerization. It is not unexpected, due to the extreme degree of long range order of the bulk crystal, that these surface dimers tend to arrange in highly ordered rows in order to reduce the free energy of the system.

The simplest example of this process is readily observable on a silicon (100) surface where, when hydrogen terminated, a (1×1) reconstruction is observed [4]. In this case, each surface atom offers one dangling bond which is then a bonding site for a hydrogen atom. Chemically, this is a very stable structure which can withstand even atmospheric conditions for tens of minutes. Heating in ultra high vacuum (UHV) conditions to above ~600 °C desorbs the hydrogen from the surface leading to the formation of dimer rows. These rows are then observed as a (2×1) reconstruction in diffraction experiments.

Naturally, the general structure of the surface is dictated by the bulk crystal, which in the case of Si is a diamond cubic structure obtained by interlocking two FCC lattices. These sublattices are shifted to each other by half the diagonal of the FCC cube. This means that each atom is connected to four adjacent neighbors in a tetragonal arrangement characteristic of highly covalent bonding. III-V compound semiconductors follow a similar arrangement with the distinction that one of the FCC sublattices is composed of group III atoms such as gallium or indium while the other is occupied by group V atoms like arsenic, antimony or phosphorus. This leads to each group III atom to be bonded to four group V atoms and vice versa in the desirable tetragonal arrangement. This crystal structure is termed cubic sphalerite or zinc blende and is depicted in Fig. 6.1.

Despite the bulk structural similarities between the two material systems, III-V compound semiconductor surfaces have drastically different behavior. Dangling bond density is minimized analogously to the silicon system by using wafers oriented in the (100) direction with the notable difference that a wafer oriented with a (100) surface can potentially be terminated with a plane made up entirely of group III

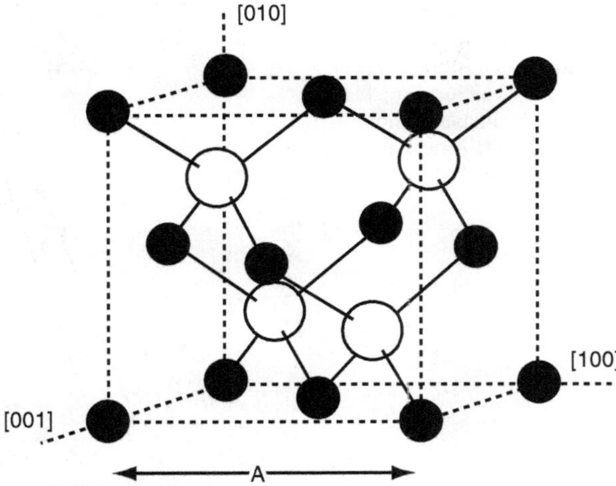

Fig. 6.1 The cubic zinc blende structure for III-V compound semiconductors

or alternatively group V atoms. This offers a plethora of possibilities for the overall composition and structure of the near surface region, which are explored in detail for several prominent compounds in the following sections.

6.2.1 GaAs

GaAs has a long history of study [5]. The lattice constant of the zinc blende unit cell is 5.65 Å resulting in a spacing between As and Ga layers in the [001] direction of 1.41 Å and a monolayer 2.82 Å thick that is composed of both As and Ga atoms. When not reconstructed, a square (1 × 1) lattice is visible on the surface regardless of the terminating species [6].

The observed reconstructions vary from gallium rich (4 × 2) terminated through a (2 × 4) As-rich reconstruction. Under extremely As-rich conditions a (4 × 4) reconstruction has been observed as well [7–9]. The arsenic rich surface is the one most relevant for MOSFET applications and appears in a variety of surface reconstructions that can be controlled reversibly by varying the deposition temperature between 400 and 700 °C and the beam equivalent pressure (BEP) [10] of the As and Ga beam fluxes between 0.1 and 10 Torr during epitaxial growth [11]. The dependence of the surface structure on these deposition parameters is illustrated in Fig. 6.2, with the most interesting structure for device applications being the (2 × 4) obtained under "medium" temperature and As-flux conditions. This surface is typically used as the starting surface in the epitaxial growth of GaAs layers.

Figure 6.3 illustrates a schematic of the possible surface phases that display this reconstruction. The most important feature is that the (2 × 4) surface indeed is terminated with arsenic dimers aligned along the [1–10]. direction, as opposed to the (4 × 2) surface, which has Ga dimers along the [−1–10] direction [9].

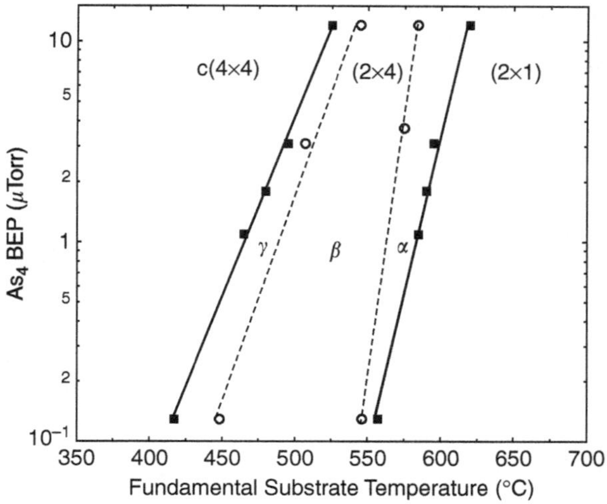

Fig. 6.2 The surface reconstructions for GaAs as a function of deposition temperature and arsenic pressure [11] (Reprinted with permission. Copyright 2001 American Institute of Physics)

A hypothesis for the structure was that the 2× periodicity was due to the dimerization of surface As atoms while the 4× results from surface vacancies [12]. These results, along with subsequent evidence collected using reflection high energy electron diffraction (RHEED) [13–16], low energy electron diffraction (LEED) [15, 17, 18], scanning tunneling microscopy (STM) [19–25], and theoretical simulations [13, 14, 24, 26–30] results in a consensus that the surface structure responsible for the (2 × 4) reconstruction resembles the "β2" model. LaBella et al. [6, 24] summarizes this debate concisely and also point to the fact that the onset of the α phase is associated with spontaneous island formation and surface roughening.

The appearance of this reconstruction is associated with a relatively lower surface concentration of arsenic at the surface as it appears at higher substrate temperatures and lower As BEP, as was observed by Tersoff et al. initially [31, 32]. LaBella et al. [11] argue that since arsenic leaves the surface under lower substrate temperatures than gallium, a gallium rich surface is left behind. These residual gallium atoms form a highly diffusive species that are responsible for the island formation

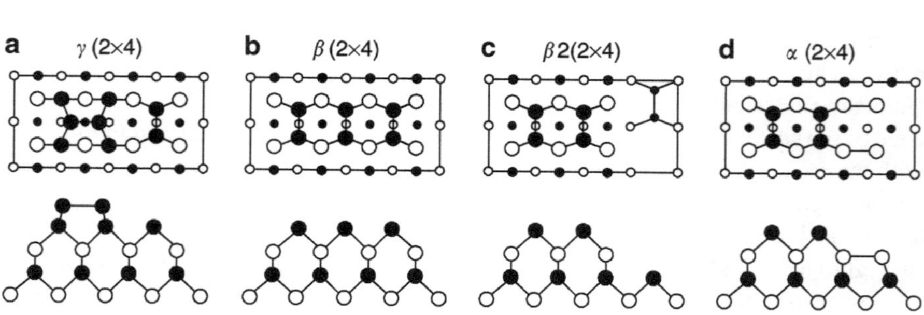

Fig. 6.3 The (2 × 4) surface reconstruction models of GaAs. *Filled circles* indicate As atoms (from Ref. [28])

observed at high temperature and low arsenic BEP. This behavior can then be modeled similarly to single element semiconductors.

Like silicon, the GaAs surface is quite susceptible to chemical reactions with atmospheric gases, which leads to unprotected wafers forming moderately thick native oxides and some carbonaceous adsorbents. The reaction mechanisms for the oxidation of III-V surfaces will be discussed in detail in subsequent sections. The propensity for reaction with oxygen has major implications for MOSFET performance as dielectric thickness is one of the most aggressively scaled transistor dimensions. Fortunately, III-V surfaces can be protected by an arsenic cap [19, 33] for weeks. The cap is formed by cooling an epitaxially grown surface in an arsenic flux. This arsenic layer can then be thermally desorbed in UHV preserving the ideal (2 × 4) reconstructed surface (Fig. 6.4) ready for further processing. Molecular beam epitaxy (MBE) and metal-organic chemical vapor deposition (MOCVD) processes offer a high surface quality sufficient for MOS applications, with MOCVD most likely to enable sufficient throughput for large scale production applications. It is noted however that the III-V integrated circuit industry has also relied on MBE production tools for substrates and devices for some time. Thus there

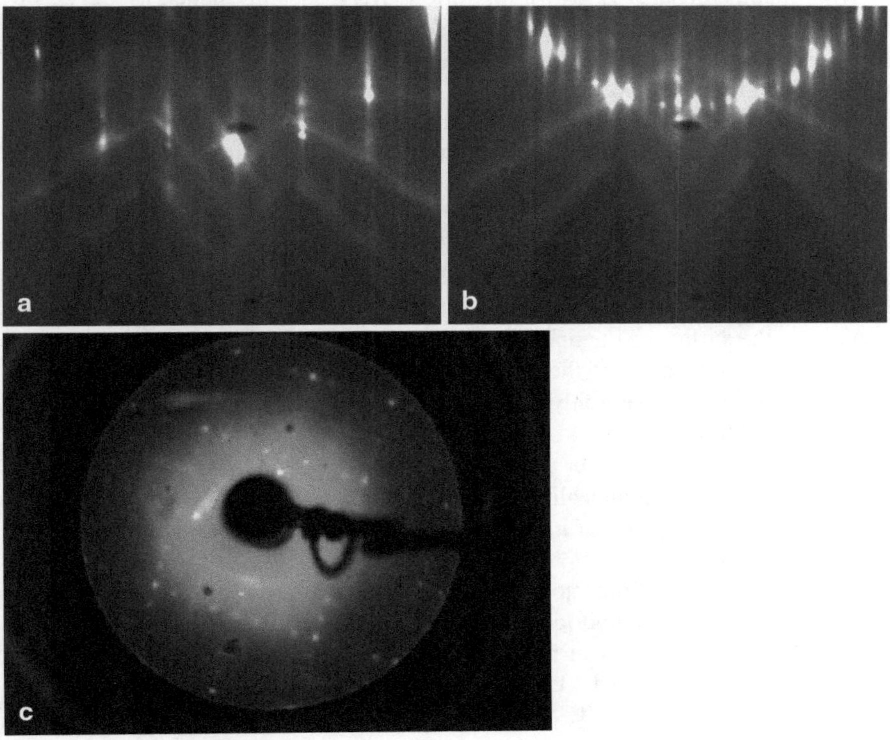

Fig. 6.4 RHEED pattern from the decapped $In_{0.53}Ga_{0.47}As$ surface along. **a** [110] and **b** [–110] directions demonstrating the (2 × 4) reconstructed surface. **c** Associated LEED pattern acquired at 70 eV

is ample experience in the growth of these materials. The challenge is really the growth of sufficiently high quality material on more mechanically stable substrates, such as Si. Progress has been made in this area as well but is beyond the scope of this chapter.

6.2.2 GaAs/InAs Alloys: InGaAs

The surfaces of GaAs have been studied extensively and the acquired concepts prove to be essential to understanding the behavior of other III-V semiconductors. Since elements of a chemical group tend to share many properties, it is not surprising that replacing some gallium with indium results in a compound with very similar physical properties. Due to the superior electrical performance, InGaAs has surpassed GaAs as a potential compound semiconductor for future MOSFET applications. Stable InGaAs exists in a variety of indium concentrations allowing for control of both physical properties, and therefore electrical properties such as mobility and band gap.

High quality InGaAs is grown on various III-V substrates such as GaAs, InP and InAs using epitaxial techniques. The desired indium concentration dictates the substrate choice due to defects such as misfit dislocations. Since the indium atom is larger than gallium, the maximum indium concentration that can be grown on GaAs is limited by strain arising from lattice constant mismatch between the substrate and the InGaAs layer. Replacing 20% of the gallium with indium allows for the growth of a high quality, 20 nm thick $In_{0.2}Ga_{0.8}As$ channel relatively free of misfit dislocations and other defects.

Regardless of the relative InAs and GaAs concentrations, the alloy exhibits similar surface reconstruction behavior, albeit scaled to a slightly lower temperature regime. Since the In–As bond is slightly weaker, free indium and arsenic is readily created on the surface. At temperatures higher than 500 °C, preferential desorption of As_2 is observed leading to the (2×4) reconstruction changing to a (4×2) indium rich surface. The (4×4) highly As-rich surface reconstruction, occurs at temperatures below 380 °C.

Even at temperatures lower than 500 °C, the surface composition can be manipulated through vacuum annealing as well as exposure to atomic hydrogen under UHV conditions. The loss of indium is reported using in-situ x-ray photoelectron spectroscopy (XPS) analysis [34]. Figure 6.5 shows the Ga 3*d*/In 4*d* region of the XPS spectrum with various atomic hydrogen exposures. This loss is attributed to the formation of volatile hydrides at these conditions. It is seen that As-oxides and Ga-oxides on the surface can be readily reduced from vacuum annealing and/or exposure to atomic hydrogen [35]. The loss of surface indium is accompanied by the narrowing of In features in the XPS spectrum. Figure 6.5c illustrates this change as the In 3*d* peak, obtained after a protective As-cap is desorbed, narrows following 30 and 90 minute exposures to atomic hydrogen. The exact chemical state responsible

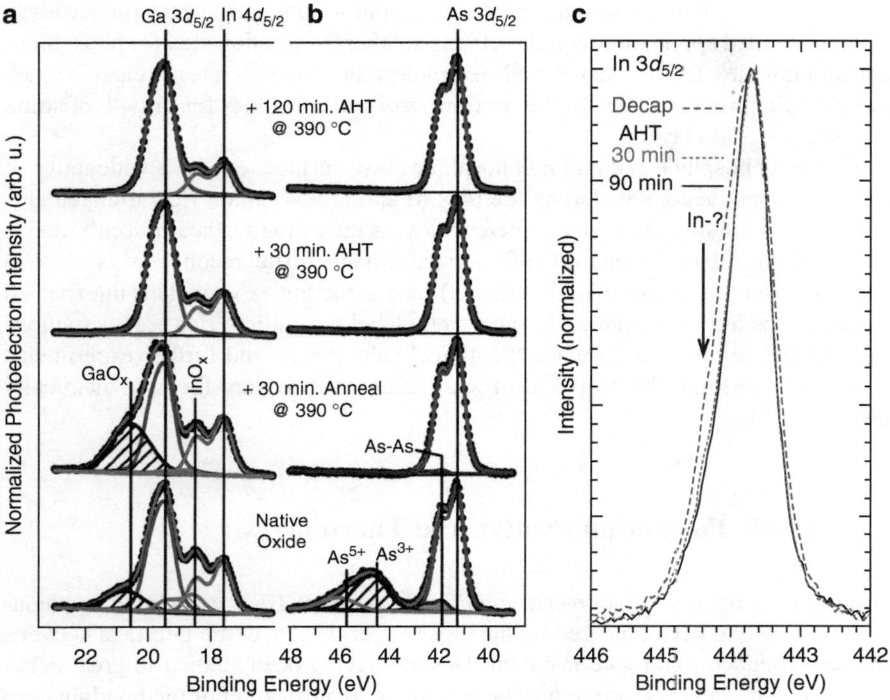

Fig. 6.5 XPS spectra of $In_{0.2}Ga_{0.8}As$ after annealing/exposure to atomic hydrogen produced by thermal cracking of H_2. **a** From the Ga and In features as well as the **b** As spectra, removal of surface oxides as well as In is detected. **c** The narrowing of the asymmetric In $3d_{5/2}$ line is also evident. "AHT" = atomic hydrogen treatment [34] (Reprinted with permission. Copyright 2008 American Institute of Physics)

for this shoulder feature is not clear, but it does make the correct In oxide peak assignment challenging as will be illustrated in later sections.

In addition to the improvement in bulk electrical properties obtained by alloying GaAs with InAs, the InGaAs surface offers distinct advantages in chemical behavior. Brammertz et al. [36] have shown through comparison of photoluminescence (PL) intensity measurements taken on GaAs vs. InGaAs interfaces with their respective native oxides that InGaAs surfaces should be easier to chemically passivate. Their results indicate that replacing just 15% of Ga with In results in an reduction of the surface recombination velocity by an order of magnitude.

6.2.3 Phosphides and Antimonides

III-V semiconductors arising from utilizing the group V anions P and Sb are also of interest. As seen from the previous section, the group III cations may be readily exchanged from gallium to indium in order to tailor bulk and surface properties to suit an intended application. Similar effects are therefore anticipated for phosphides and antimonides. Of the two possibilities, phosphides have been researched in much more detail particularly due to the usefulness of InP substrates for growth of strain free $In_{0.53}Ga_{0.47}As$ layers.

Gallium phosphide and indium phosphide (100) surfaces exhibit a wide range of surface reconstructions including the (4 × 4) and (2 × 4) anion rich arrangements observed for the InGaAs system. Interestingly, as the cation surface concentration is increased, the mixed dimer and Ga-dimerized surfaces do not reconstruct as a (4 × 2) reconstruction but rather exhibit a (2 × 4) reconstruction as well. This unexpected tendency has led earlier papers to incorrectly label the gallium rich reconstructions on GaP and InP as (4 × 2) when theoretical calculations and further experiments later confirmed that the proposed (4 × 2) reconstructions are thermodynamically unstable [37].

6.3 Oxide Formation (Native and Thermal)

The viability of field effect electronic devices made of III-V compound semiconductors has long been hindered by the presence of defects at the interface between the semiconductor and gate insulator. Despite having been studied in great detail for more than 35 years,[1] these interfaces and the identification of the bonding configurations that cause the defects has remained challenging [38–41]. In contrast, this issue has been acceptably solved in well established Si-based field effect transistors (FETs) since the 1960s, and is largely due to the superior natural oxide for Si, viz. SiO_2. Problems with metal-oxide-semiconductor (MOS) devices on GaAs and InGaAs, including frequency dispersion of capacitance and sub-optimal electron mobility, have been attributed to a number of different native surface species including Ga–O bonds, As–O bonds, elemental As, and As and Ga anti-sites [42–44]. An understanding of the specific bonding configurations that arise on III-V semiconductors has great implications for the electronics industry contemplating the move away from Si-channel devices to those with higher mobility, such as GaAs and InGaAs.

6.3.1 *Stable Oxidation States on $In_xGa_{(1-x)}As$*

The stable oxides of $In_xGa_{(1-x)}As$ include As_2O_3, As_2O_5, Ga_2O_3, Ga_2O, In_2O_3, $GaAsO_4$, and $InAsO_4$ [45]. The relative stability of these native oxides in this sys-

[1] For an excellent review of the developments of gate dielectric studies for GaAs up until ~1985, see: *Physics and Chemistry of III-V Compound Semiconductor Interfaces*, edited by C. Wilmsen, (Plenum, New York, 1985).

Table 6.1 List of stable III-V oxides and their bulk oxide Gibbs free energies

Oxide	Gibbs free energy, ΔG (kcal/mol)	XPS core-level binding energy (eV)
Ga_2O	−75.3 [50]	1117.3 (2*p*3/2) [47], 19.6 (3*d*) [47]
Ga_2O_3	−238.6 [45]	1117.9 (2*p*3/2) [47], 20.2 (3*d*) [47]
$GaAsO_4$	−212.8 [50]	1118.8 (2*p*3/2) [45], 21.1 (3*d*) [45]
$GaSbO_4$	−218.4 [50]	N/A
$GaPO_4$	−310.1 [50]	1119.3 (2*p*3/2) [51], 21.6 (3*d*)13
In_2O_3	−198.6 [45]	444.7 (3*d*) [45]
$InAsO_4$	−209.4 [45]	445.3 (3*d*) [45]
$InSbO_4$	−198.4 [50]	N/A
$InPO_4$	−287 [50]	445.7 (3*d*) [52]
P_2O_3	−176 [50]	Volatile [53]
P_2O_5	−322.4 [50]	135.6 (2*p*) [51]
As_2O_3	−137.7 [50]	1325.4 (2*p*3/2) [46], 43.7 (3*d*) [46]
As_2O_5	−187 [50]	1326.7 (2*p*3/2) [46], 45.0 (3*d*) [46]
Sb_2O_3	−151.5 [50]	~530.5 (3*d*5/2) [54]

Also shown are XPS core-level binding energies of the stable oxides. All binding energies measured by XPS are measured with respect to the semiconductor Fermi level. As such, exact peak positions must be adjusted to take doping concentration, charging, and band bending into consideration. Relative shifts such as the energy difference from the C 1*s* peak or shifts with respect to the bulk semiconductor peak are commonly used to determine chemical states rather than absolute binding energies.

tem is easily explained in terms of bulk thermodynamic values of the individual oxides. Table 6.1 shows the Gibbs free energies for each of the relevant native oxides of $In_xGa_{(1-x)}As$. The As-oxides are the most unstable of the oxides and are easily removed or converted to the more stable Ga oxides, in particular Ga_2O_3. The stability of this oxidation state makes it difficult to remove during typical chemical processing or relatively low temperature thermal desorption [46]. An in-depth study of the critical Ga-oxides [47] is presented in the next section.

For GaAs ($x = 0$), the low temperature grown native oxide (Fig. 6.6) consists primarily of As 5+, As 3+, Ga 3+, and Ga 1+ oxidation states in addition to a small amount of elemental As [48]. When the oxides are grown at higher temperature, or the low temperature oxides are subjected to an increased thermal budget, the composition of the oxide layer changes dramatically. As the Gibbs free energies predict, the less stable As-oxides become less prevalent as temperature increases eventually either evaporating away or being converted to the most stable oxide in this system Ga_2O_3 [46]. As–O bonding can then be further reduced by a post-etch, in-situ anneal in vacuum up to 450 °C. Analysis of the corresponding O1*s* region for the annealed surface reveals that the O concentration (intensity) for the etched surface and the annealed surface is comparable, indicating no significant loss of oxygen and therefore suggesting that bond conversion from As–O bonding to Ga–O bonding is favored with such treatments. For temperatures above 500 °C, As-oxide is below the level of detection as the oxide is almost completely Ga_2O_3. As the reduction equation shows [45],

$$As_2O_3 + 2GaAs \rightarrow Ga_2O_3 + 4As \tag{6.1}$$

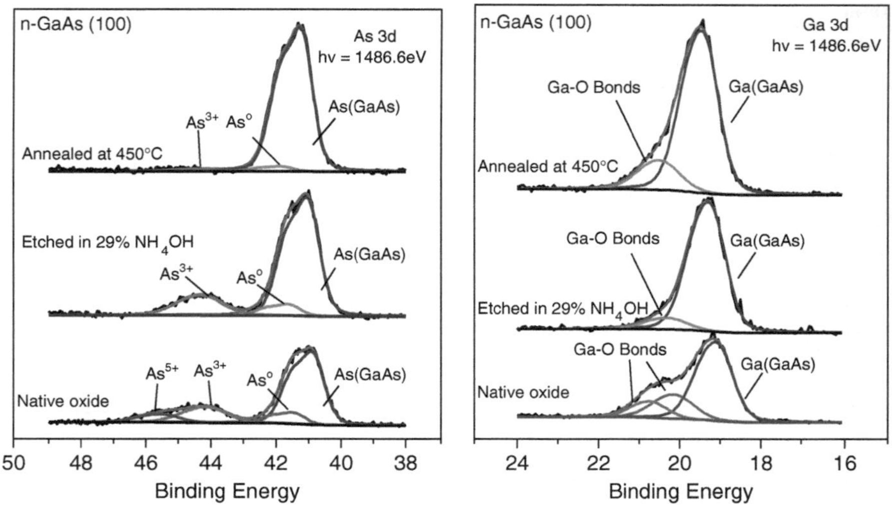

Fig. 6.6 XPS core-level spectra showing the oxidation states of GaAs native oxide. Two states are observed for both As and Ga. Chemically etching the native oxide removes much of the oxide including all of the As 5+ oxidation state. Anneals at temperatures greater than 350 °C convert all remaining As-oxides to the most thermodynamically stable oxide in the system, Ga_2O_3 [48] (Reprinted with permission. Copyright 2009 The Electrochemical Society)

there should also be elemental As created for these higher temperature oxides which is reported in most of the literature. As noted above, this elemental As is located primarily at the interface although some of it may subsequently diffuse through the Ga_2O_3 to form $GaAsO_4$-like regions. Any As-oxides that may be formed via this elemental As diffusion to the surface is likely to be highly unstable and would therefore evaporate [49].

InAs ($x = 1$), despite having fundamentally the same phase diagram [55] as GaAs has a different oxide growth structure primarily due to the relative thermodynamic stabilities of the possible native oxides (Fig. 6.7). Whereas for GaAs, Ga_2O_3 was the most stable of the possible oxides, for InAs the most stable compound is $InAsO_4$ [45]. Therefore, for thermally grown oxides above 350 °C, the oxide consists almost entirely of ternary $InAsO_4$ (although $InAsO_3$ has been suggested as a possibility as well) with minimal elemental As at the interface. For air-grown native oxides, the structure is similar to that of GaAs, with As 5+, As 3+, and In 3+ species present in XPS spectra.

The compounds in between $x = 0$ and $x = 1$ of $In_xGa_{(1-x)}As$ exhibit oxides that are once again consistent with thermodynamic values of stable bulk oxides. Low temperature (air-grown) native oxides consist of a mixture of oxides of each of the substrate elements, As 5+, As 3+, In 3+, Ga 3+ and Ga 1+ as well as a small amount of elemental As. Increasing the thermal budget either by anneal or thermal oxidation removes the As-oxides while increasing the amount of Ga 3+ and In 3+ oxidation states. During this bond reconfiguration, the amount of oxygen that gets transferred to Ga appears to be greater than the amount transferred to In [56]. It is noted that

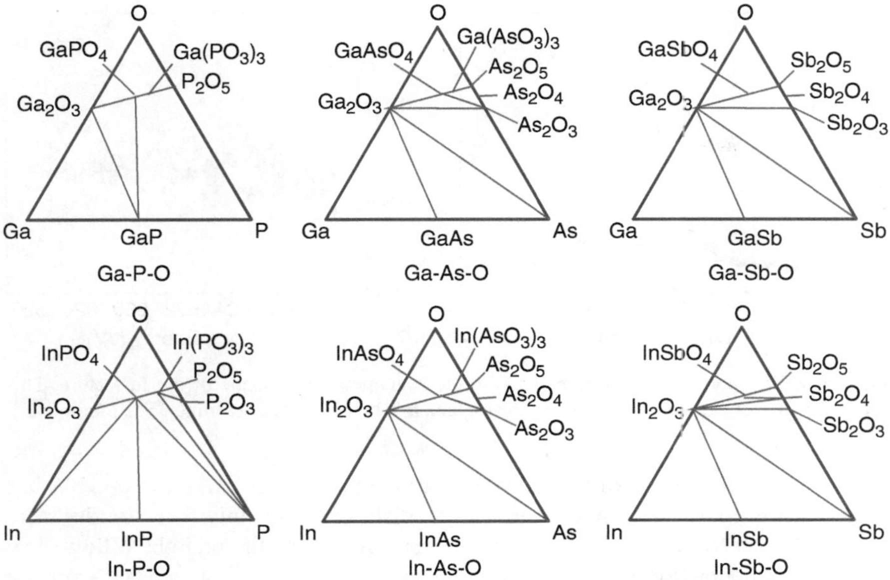

Fig. 6.7 Phase diagrams for III-V based systems. The similarities between the arsenides and the antimonides is noted [55] (Reprinted with permission. Copyright 1983 Elsevier)

since there is no detectable As oxide at higher temperatures on $In_xGa_{(1-x)}As$ (up to at least $x = 0.65$), $InAsO_4$ is ruled out as a possible interfacial oxide in these systems. Also of note is that as In content increases, the amount of air-grown Ga_2O_3 decreases even when normalized to the lower Ga content of these compounds (i.e., Ga_2O_3:GaAs decreases with increasing In content) [57]. This could be due to the increased size of the In atom reducing the available space for O to bond to the interfacial Ga atoms.

It is important to point out one of the major complications with using XPS to determine the presence of In-oxides. The line-shape for In, even for a surface that is oxide-free, is asymmetric (Fig. 6.8) due to core-hole screening by conduction electrons [58]. A useful curve fitting for this type of phenomenon is achieved by using a Doniach-Sunjic line-shape [59]. Using an incorrect line-shape for these spectra can lead to the misinterpretation of the detection of sub-monolayer In-oxides.[2] Also as noted previously, Ga–O formation is also more stable than In–O under equilibrium conditions.

Exposure of the native oxides of $In_xGa_{(1-x)}As$ to chemical etches (notably HF and NH_4OH) appears to result in the reduction of the As–O bond concentration as well. Particularly of note is the complete removal by this chemical treatment of the As 5+ oxidation state leaving the As 3+ state as the only As–O surface bond (Fig. 6.1) [60–62]. These etch solutions also reduce the amount of Ga-oxide, but much less efficiently than for the As-oxides.

[2] See: C. L. Hinkle, M. Milojevic, E. M. Vogel and R. M. Wallace, *Appl. Phys. Lett.* **95**, 151905 (2009).

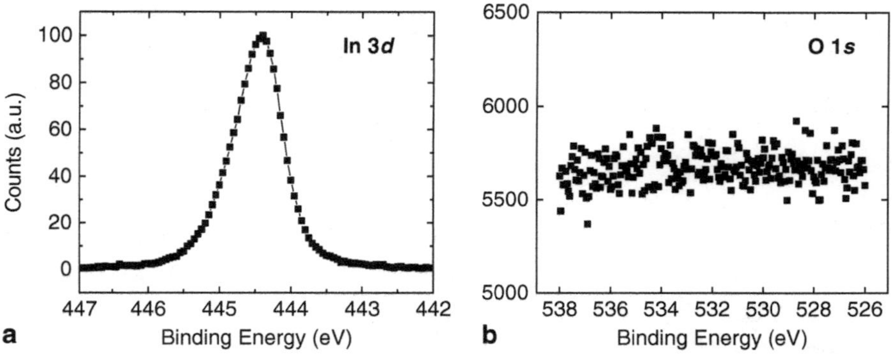

Fig. 6.8 **a** XPS core-level spectrum showing the asymmetric line-shape of the In feature. This asymmetry is not caused by an oxide as the oxygen **b** is below detectable limits [See footnote 2.]

A critical comparison of the native oxide chemistry on n-GaAs and p-GaAs has been recently conducted where chemically indistinguishable interfacial oxides were observed upon NH_4OH etching and exposure to laboratory air and light, followed by Al_2O_3 ALD deposition [63]. The background-subtracted, normalized XPS spectra of the two companion GaAs substrates with chemical treatment followed by 12 min atmospheric exposure fully overlap, clearly showing that the surfaces exhibit the same bonding environment to within detectible limits (Fig. 6.9). The extent of oxidation on these samples is chemically identical to each other for an n- and p-type substrate. This shows that there is no difference detected in the oxidation rate or chemical state of the n- or p-type substrates as long as they are exposed to identical conditions. Al_2O_3 deposition on these same samples continues to show that n- and p-type GaAs interfaces oxidize in identical manners. It is noted that the As *2p* and Ga *2p* spectra are very surface sensitive due to the binding energies of those core level electrons and are therefore particularly useful in analyzing interfacial oxides. All other spectral ranges analyzed (Al *2p*, As *3d*, Ga *3d*, C *1s*) also show identical chemical states for both n-type and p-type substrates.

6.3.2 Ga-Suboxides

Since the As-oxides can easily be removed with chemical or thermal treatment, the role of the most stable bound Ga native oxides, GaO_x, is of particular interest. Throughout the analysis described above, it is clear that residual Ga-oxides appear to be present at some level. To clarify the role of Ga-oxides, several experiments were conducted to control the Ga oxidation states on GaAs and $In_xGa_{(1-x)}As$ [47, 48]. The accurate identification of the oxidation states of Ga is often thought to be limited to high photon intensity synchrotron x-ray photoelectron spectroscopy (XPS) data and is usually only applied to the Ga 3*d* spectrum. The study of this spectrum with typical laboratory-based x-ray sources results in energetic photoelectrons (photoelectron kinetic energy KE ≈ 1465 eV for an Al Kα x-ray source)

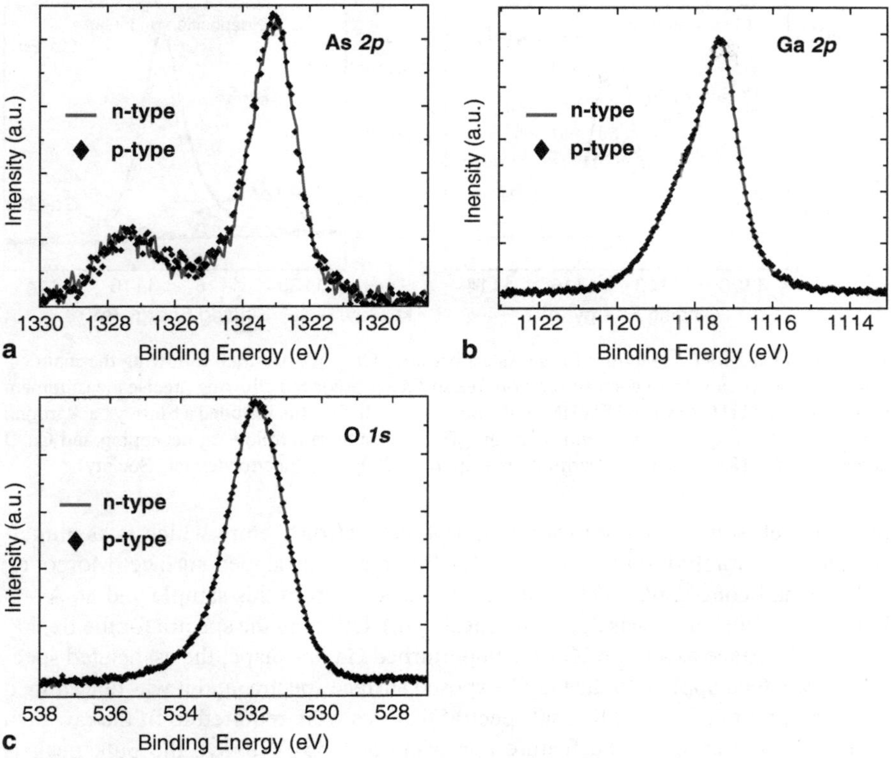

Fig. 6.9 XPS core-level spectra showing that oxide growth on GaAs does not depend on the semiconductor dopant [63] (Reprinted with permission. Copyright 2008 American Institute of Physics)

resulting in a significant decrease in sensitivity to photoelectrons originating from the near surface region. In contrast, the Ga 2*p* spectrum (KE ≈ 369 eV) exhibits considerable surface sensitivity, but is notoriously difficult to deconvolve into individual chemical (oxidation) states due to the broad Gaussian widths of the peaks in conjunction with the proximity of the chemical state peak positions. Moreover, the Ga 2*p* extreme surface sensitivity makes ex-situ studies difficult due to the ultrathin films necessary for photoelectron transparency and the susceptibility of the film to spurious contamination and oxidation due to environmental considerations. However, the identification of these bonding states is crucial in terms of the relation to potential defects at the interface with the III-V semiconductor.

As seen in Fig. 6.10a, in-situ analysis of a thermally decapped InGaAs surface can provide a model binding energy reference (i.e., chemicals state) and an associated linewidth for the Ga 2*p* spectra of the Ga–As bond. Analysis of other relevant photoelectron regions for this sample (viz. As, In, O, C—not shown) indicate that no oxidized species or spurious C is detected [48]. Similar results are obtained for hydrogen-cleaned or decapped GaAs surfaces as well.

Subsequent in-situ room temperature exposure of the decapped InGaAs surface to a flux of molecular Ga_2O (Fig. 6.10b) from a Ga_2O_3 powder charge in an Ir

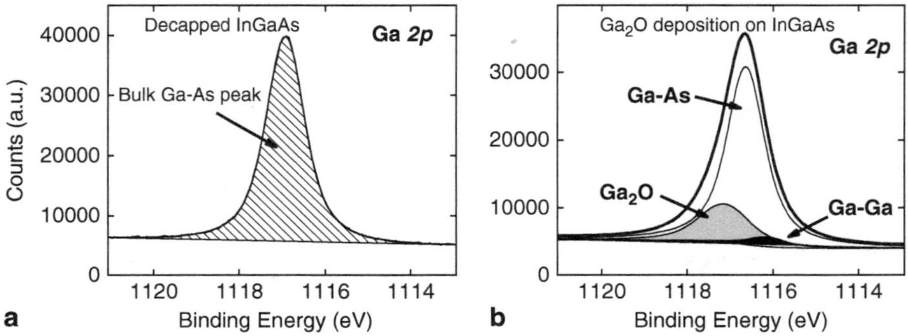

Fig. 6.10 **a** Ga 2*p* XPS spectrum for an As capped $In_{0.53}Ga_{0.47}As$ substrate following thermal desorption of the As cap. An oxygen and carbon free surface is produced allowing precise measurement of the position (1116.7 eV) and FWHM of the bulk peak. All XPS fits included a Shirley background subtraction. **b** Ga 2*p* XPS spectrum of an $In_{0.53}Ga_{0.47}As$ substrate following decapping and Ga_2O deposition [48] (Reprinted with permission. Copyright 2009 The Electrochemical Society)

effusion cell source [64–66] results in a photoelectron feature which is asymmetric and measurably broader than that for the freshly decapped surface. Moreover, only a small concomitant O1*s* feature was detected from this sample and no As–O bonding or spurious C was detected (not shown). Utilizing the spectra for the freshly decapped surface as a model for the unperturbed Ga lineshape, the associated spectral fit was then applied to the Ga_2O-exposed surface spectra, and it was determined that a minimum of two additional spectral features were required to fit the raw data with similar precision. The feature appearing at 0.55 eV above the bulk peak is therefore attributed to the Ga 1+ oxidation state, originating from Ga_2O present on the surface, while a barely detectable feature at ~1116.1 eV is attributed to Ga–Ga bonding originating from the powder (rather than polycrystalline) source used in this experiment. Again, similar results are obtained for GaAs.

As noted above, higher oxidation states are seen on $In_xGa_{(1-x)}As$ surfaces as well. Ga–O bonds are known to exist in chemical states that manifest themselves with binding energies around 1–2 eV above the bulk Ga–As peak. Chemically treating a native oxide with NH_4OH leaves a surface oxide that exhibits a large shoulder on the high binding energy side of the bulk peak. The accurate analysis of this spectrum requires two oxidation states of Ga, the Ga 1+ oxidation state (Ga_2O) and the Ga 3+ oxidation state (Ga_2O_3) as shown in Fig. 6.11. Using the previously determined parameters for the bulk peak and the Ga_2O surface oxide, the peak position for the Ga 3+ oxidation state is determined and is found to be 1.2 eV above the bulk peak, precisely where photoelectron studies of Ga_2O_3 powders place it [45]. There is also a higher binding energy state that is sometimes seen in the Ga 2*p* spectra which arises at 1.6 eV above the bulk peak and is seen only on samples with relatively thick oxides. Possible identities for this oxidation state include $GaAsO_4$, $Ga(AsO_3)_3$ as well as $Ga(OH)_3$ [45]. This state is removed easily with chemical treatment.

Deconvolution of the spectra after careful analysis of a number of in-situ and ex-situ surface preparations suggests that Ga-suboxides (Ga 1+: Ga_2O) are likely stable on *all* substrates that have had any exposure to oxygen, hydroxyls, etc. despite

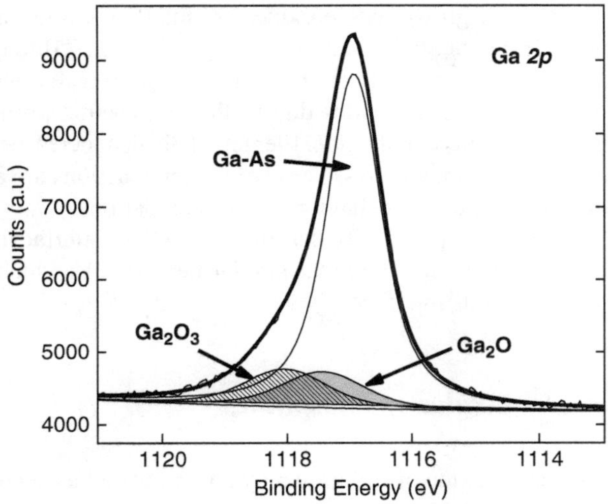

Fig. 6.11 XPS core-level spectrum showing the two interfacial oxidation states of Ga. The suboxide is present for any surface exposed to oxygen or hydroxyls

its relatively high (less stable) Gibbs free energy. For example, the HF-last GaAs surface has residual Ga oxides on the surface, mainly Ga 3+ and Ga 1+. These states were deconvoluted using the control sample procedure outlined above. The suboxides are found specifically at the substrate interface [47] suggesting that the formation energies in this regime are not governed by bulk oxide values for up to one monolayer but rather are perturbed due to the other elements nearby. Such suboxide species have been previously identified as crucial for low defect densities in MBE-grown dielectrics on GaAs using STM methods [64, 67], but have been difficult to measure conclusively using complementary characterization methods conducted under ex-situ conditions such as HRTEM [66, 68], HREELS [69, 70] or XPS [71]. These oxidation states play a critical role in device characteristics as is presented in detail in a later section of this chapter [47, 57].

6.3.3 Oxides of Antimonides

The oxides of antimonide compounds (InSb and GaSb) behave similarly to each other and notably similar to those of GaAs. These results are unsurprising considering the similarity in the phase diagrams of these compounds (Fig. 6.7) [50, 55].

Specifically, native oxides of InSb tend to be comprised of approximately 15–20 Å of a mixture of In_2O_3 and Sb_2O_3 with possibly some Sb_2O_5 as well. There is also elemental Sb detectable [72]. The Sb oxides are easily removable with thermal treatments as the thermodynamic stability of these oxides is considerably less than that of In_2O_3 [52, 73]. Vasquez and Grunthaner [74] found that thermal oxidation of InSb resulted in an In rich native oxide (In_2O_3) with elemental Sb piled up at the interface of the substrate and oxide.

GaSb air-grown native oxides are made of Ga_2O_3 and Sb_2O_3 and Sb_2O_5 along with elemental Sb found at the interface [52, 73, 75]. Increasing the thermal budget through either anneals or thermal oxidation results once again in a reduction or removal of the Sb-oxides due to thermodynamic properties that result in Ga_2O_3 being the preferential oxide. Elemental Sb also increases at the interface for the case of elevated temperatures. These oxidation reactions are all completely analogous to the aforementioned behavior of GaAs oxidation.

It may be a possibility that In_2O and Ga_2O interfacial oxides are also present in these Antimonides due to the similarities to GaAs which contains the Ga 1+ oxidation state at its interface.

6.3.4 Oxides of Phosphides

The native oxide of InP grown by air exposure has been shown to be comprised of primarily $InPO_4$, particularly when the oxide layer is less than 3 nm thick [50, 51, 76, 77]. As the native oxide layer increases in thickness, In_2O_3 is the likely oxide species away from the interface due to the diffusion of In through the oxide layer. The diffusion rate of P is much smaller than that of In, especially at temperatures <650 °C, which limits the growth of $InPO_4$. As the thickness of this oxide increases, a concomitant elemental P build up at the interface occurs as the In diffusion proceeds for top surface In_2O_3 growth.

The same story is true for the growth of thermal oxides on InP for low temperature conditions. However, for temperatures greater than 650 °C, P diffusion is much more likely and the thermally grown native oxide continues as $InPO_4$. At these elevated temperatures, severe degradation of the substrate occurs due to the diffusion of the constituent atoms as well as "blistering" due to evaporation of P making for a poor quality interface between the InP and the oxide.

For GaP, both air-grown and thermal native oxides grow uniformly as $GaPO_4$ as the thermal budget necessary for diffusion of all the elements simultaneously is much lower than for InP [50, 78, 79]. Elemental P does not pile up at the interface due to this. As a result, the oxidation of GaP proceeds near equilibrium for most temperatures resulting in the most thermodynamically stable oxide. Unlike InP, the interface does not roughen for GaP up to 950 °C.

6.4 Oxide Deposition on III-V Substrates[3]

Work on gate dielectrics for GaAs MOSFETs extends back into the 1960s [80], with the relative complexity of the capacitance-voltage behavior recognized early on for a number of substrate orientations [81]. The early work on gate dielectrics

[3] Portions of this section have been previously published in [3].

for GaAs entailed studies of surface oxidation through chemical, thermal, anodic, and sputter deposition methods [82]. Fundamental studies on the effects of GaAs surface oxidation and the resultant interface states were also reported, highlighting the effects of Fermi level pinning observed [83]. The prospects and challenges for InGaAs MOSFETs have also been previously summarized [84]. However, with the introduction of ALD into Si-based IC fabrication for high-k gate dielectric applications[4], a new vista of gate dielectric material choices are now available for consideration in III-V devices.

For MOSFET surface channels beyond the 16 nm node, the candidate III-V channel materials primarily include $In_xGa_{1-x}As$ ($0 \leq x \leq 1$) alloys as the bandgap for these materials ($0.36 \leq E_g \leq 1.42$ eV) is $E_g \geq 14$ kT, making them potentially suitable for many applications with power supply operating voltages ≤0.5 V envisioned for short gate lengths ($L_g \leq 22$ nm), such as low stand-by power applications. Alloys with an In content greater than 50% are of particular interest for NMOS applications as the bulk electron mobility is very high ($\sim 10^4\, cm^2/V\, s$) relative to Si ($\sim 10^3\, cm^2/V\, s$), thus potentially enabling high performance surface channel devices. With the possibility of leveraging existing Si-fabrication investments, interest in the (100) orientation of these materials is particularly acute, as lattice-matched epitaxial growth schemes are possible with an underlying Si(100) substrate. For these reasons, we will confine our review to $In_xGa_{1-x}As$ alloy results reported to date. We note however that there have also been a few ALD dielectric studies of higher E_g (3.14 eV) GaN surface channel FETs [85, 86] and low E_g (0.17 eV) InSb surface channel FETs [87], as well as buried quantum well transistor work also on InSb [88].

6.4.1 Preparation of III-V Substrate Surfaces

It is well known that surface preparation [89] (sometimes also described as "passivation" in the context of device technologies) of any substrate affects the subsequent ALD process for dielectric deposition. A number of surface treatments have been examined on III-V (100) surfaces (GaAs, in particular), and are now under examination for dielectrics deposited by ALD. Previous (typically wet) chemical surface preparation studies include the effects of chalcogenides [90] (e.g., sulfur: $(NH_4)_2S_x$ solutions [91–94], Na_2S solutions [95, 96], H_2SO_4 solutions [97], H_2S/S_2 exposures [98], selenium [99], and tellurium [100]), etchants such as, HCl solutions [101, 102], HF solutions [61, 103], hydroxide solutions [60, 104, 105], as well as combinations of these [106, 107]. Alternative treatments (typically performed in vacuum) include vacuum annealing [17], As-flux passivation (capping) and desorption (de-capping) [108], hydrogen plasmas [109, 39], atomic (thermally dissociated) hydrogen [110, 111], as well as the deposition of overlayers such as silicon and germanium on the III-V surface [112–118].

[4] For an early review of high-k gate dielectrics, see the *MRS Bulletin* March 2002 issue.

At this point in time, systematic surface preparation studies on $In_xGa_{1-x}As$ are relatively less complete, and thus the subject of current research. Many of these investigations utilize similar surface treatments to that employed for GaAs surfaces. The addition of In generally results in the formation of In-oxides on the surface, with a formation energy reported to be intermediate to that of AsO_x and GaO_x [45]. Thus thin native oxides on this surface generally exhibit oxides associated with all three of the constituent elements, and so similar oxide-etch chemistries may be anticipated. However, relative to GaAs, the details on the surface preparation of InGaAs are less established, and so various surface preparation methods are currently under investigation on this surface including wet chemical approaches [119, 120], thermal vacuum annealing [35], As-capping and decapping [121], exposure to H-plasmas [120, 122, 123] and thermally-cracked atomic H [34], exposure to molecular species including $As[N(CH_3)_2]_3$ [124], and H_2S [125].

6.4.2 *Atomic Layer Deposition (ALD) on III-V Substrate Surfaces*

The ALD process parameters (reactant delivery and deposition temperatures, purge and exposure times, etc) are frequently decided by the precursor characteristics, such as vapor pressure, decomposition behavior, sticking coefficient and reaction chemistry, etc. The ALD deposition process should not cause unintentional thermal instability of III-V substrates. Elemental As desorption occurs from GaAs surfaces [126] in vacuum at $T > 400\,°C$ and from $T > 300\,°C$ for InAs [127]. Indium segregation at the surface has been reported at 480 °C for InGaAs [128]. Metal precursors utilized on III-V surfaces can be divided into two main groups: metalorganic and inorganic precursors. Halide precursors, such as $HfCl_4$, are the most common inorganic precursors used for CVD and ALD applications. They are typically highly reactive and thermally stable (up to 750 °C). However halides are typically solid phase with relatively low volatility (except $TiCl_4$). HfO_2 films grown by ALD using $HfCl_4$ and water below 300–350 °C was reported to have high residual chlorine and hydrogen content (2–5 at%) [129]. Alkyls, such as trimethyl-aluminum (TMA), are ideal metalorganic precursors containing direct bonding between the metal ion and carbon, while alkoxides and amides have oxygen and nitrogen bonding between the metal and alkyl groups, respectively. They are highly volatile and very reactive with water through hydrolysis. On the other hand, they often decompose at a relatively low temperature. For example, both TMA and Tetrakis(ethylmethylamid o)hafnium (TEMA-Hf) decompose at temperatures higher than 275 °C [130, 131]. The chelation of C (β-diketonates), O (cyclopentadienyls) and N (amidinates) with alkyls to a metal enhances thermal stability compared to single bond precursors, while they frequently have low vapor pressures at deposition temperatures due to their bulky ligands.

Water is the most commonly used oxygen source providing an hydroxyl terminated surface in metal oxide ALD. Water frequently needs a sufficiently long purge time due to its sticking coefficient with surfaces, while ozone is easily purged out of the chamber and it potentially improves the throughput of an ALD process. It was also reported that an ozone process with Tetrakis(dimethylamino)hafnium (TDMA-Hf) at 300 °C on Si enhances electrical characteristics of HfO_2 films and reduces C contamination in the films as well [136]. However, there are potential concerns on O_3 with III-V substrates regarding undesirable interface oxide formation due to its strong oxidation power, and possible residual carbonate formation at low temperature.

A number of precursors have been reported for ALD deposition of dielectrics on III-V surfaces. Representative examples, typically the first published reports, are summarized in Table 6.2 with the associated dielectric produced, precursors reported to date, oxidizing agent, deposition temperature, and maximum post-deposition temperature.

As of this writing (mid-2009), Al_2O_3, HfO_2 and their alloys constitute the majority of reports for ALD on III-V substrates. Given the available precursors developed, additional ALD metal oxides will likely be evaluated for gate dielectrics of III-V substrates soon. Particularly, the recent development of amidinates and cyclopentadienyls precursors enables La_2O_3, Gd_2O_3 and their nanolaminate oxides using ALD [139–141].

6.4.2.1 ALD on GaAs

The first reported depletion mode MOSFET work utilizing ALD dielectrics directly on GaAs(100) is by Ye and coworkers [142, 143] with 8–16 nm thick Al_2O_3. The $(CH_3)_3Al$ (trimethyl-aluminum: "TMA")/water chemistry utilized to obtain the Al_2O_3 layer was apparently conducted on a GaAs surface exposed to the laboratory ambient after MBE growth, thus initially having a thin native oxide layer (likely with spurious organic contamination from the air exposure) consisting of Ga- and As-oxides. It was noted in this work that the ALD process removes the native oxide and excess As on the GaAs surface, resulting in ~0.6 nm Ga-oxide interfacial layer. Capacitor measurements indicated interface state densities D_{it} ~$10^{12}\,cm^{-2}\,eV^{-1}$ for this gate stack, which should be compared to the benchmark stack for GaAs: Gd_2O_3/Ga_2O_3/GaAs grown by molecular beam epitaxy (MBE) methods which yields D_{it} ~$5 \times 10^{10}\,cm^{-2}\,eV^{-1}$ [144].

A subsequent more detailed study by Frank et al., examining HfO_2 as well as Al_2O_3 deposition by ALD, indicated that the surface oxides are indeed affected by the ALD process [145]. A small reduction of the GaAs surface native oxide content was reported for "vacuum pre-annealing" (i.e., annealing at 300 °C prior to ALD deposition) indicating that little native oxide "thinning" from such annealing was detected from ex-situ analysis, consistent with prior reports. A native oxide thickness of ~2.5 nm was reported for the GaAs surface. In the case of TMA/water chemistry ALD at 300 °C, an amorphous, 4 nm Al_2O_3 layer was observed from deposition on either the native oxide or an HF-last surface. An interfacial layer, reported to contain significant Ga_2O content, was observed to be ~1 nm thick for

Table 6.2 ALD precursors and oxidizers studied on III-V surfaces. ${}^{i}Pr$ = isopropyl

Dielectric/substrate	Precursor	Vapor pressure*			Oxidizer/Nitridizer	T_{dep} °C	T_{max} °C	Refs.
		20 °C	100 °C	200 °C				
Al_2O_3/n-GaAs	$Al(CH_3)_3$ [TMA]	8.6	315	–	H_2O	300	600 O_2	[142, 143]
Al_2O_3/n-InGaAs	$Al(CH_3)_3$ [TMA]	8.6	315	–	H_2O	300	550 O_2	[132]
Al_2O_3/p-GaAs	$Al(CH_3)_3$ [TMA]	8.6	315	–	$(CH_3)_2CHOH$	400	–	[120]
AlN/n-, p-InGaAs	$Al(N(CH_3)_2)_3$ [TDMA-Al]				NH_3	250	550 N_2	[133]
HfO_2/p-InGaAs	$HfCl_4$	10^{-6}	0.006	5.1	H_2O	320	500 O_2, N_2	[134]
HfO_2/n-InGaAs	$Hf(NCH_3C_2H_5)_4$ [TEMA-Hf]	0.005	1.7	155	H_2O	200	–	[135]
HfO_2/n-, p-GaAs	$Hf(N(CH_3)_2)_4$ [TDMA-Hf]	0.06	15	1074	H_2O	200	500 N_2	[136]
Hf-aluminate/	$Al(CH_3)_3$	8.6	315	–	H_2O	300	600	[137, 138]
n-, p-GaAs	$HfCl_4$	10^{-6}	0.006	5.1			N_2	
La-aluminate/	$Al(CH_3)_3$	8.6	315	–	H_2O	200	–	[141]
n-InGaAs	$La({}^{i}PrNCHN{}^{i}Pr)_3[({}^{i}Pr_2\text{-fmd})3\text{-La}]$	0.008	0.06	0.256				

* See: http://www.safchitech.com

the Al_2O_3 deposition, while a thicker interfacial layer (~2–2.5 nm) was observed from the HfO_2 deposition using a $HfCl_4$/water chemistry ALD. These investigators note relatively thinner interfacial layer from the Al_2O_3 deposition suggests that volatile interfacial layer products may be formed or conversion of interfacial oxides to Al_2O_3 occurs during the ALD process. The difference in the behavior of the ALD precursors was attributed to the enhanced reactivity of $Al(CH_3)_3$ compared to $HfCl_4$ based upon formation enthalpies.

This "self cleaning" interfacial oxide reaction has been subsequently observed by others on GaAs [137, 146–148]. Dalapati and coworkers also examined capacitors using Al_2O_3 (TMA/water), HfO_2 ($HfCl_4$/water) and nanolaminated mixtures by ALD on HCl + $(NH_4)_2S$ treated GaAs (100) surfaces [137]. The capacitance-voltage (C–V) behavior was studied at room temperature, demonstrating higher frequency dispersion for capacitors on n-GaAs vs. p-GaAs, consistent with reports published nearly 20 years ago on anodic native oxides [149, 150]. Generally, the maximum capacitance ("C_{ox}") observed in a C–V measurement is observed to decrease as the measurement frequency is increased. Such behavior has also been more recently observed by others as well utilizing PVD [43] and ALD [46, 151] dielectrics. The utilization of Si (or Ge) interfacial "passivation" layers noted above, and/or post-deposition annealing, reduces the observed dispersion behavior on GaAs [152].

Dalapati et al. [137] also examined the chemical nature of the interface using ex-situ x-ray photoelectron spectroscopy in conjunction with thin (~1.5 nm) Al_2O_3, HfO_2 and Al_2O_3/HfO_2 ("HfAlO") nanolaminates. It was found that Ga- and As-oxides were detected at the interface, and the C–V behavior for all three stacks investigated was attributed to the interfacial oxide layer, with HfAlO exhibiting the better behavior and a thinner interfacial layer for p-GaAs substrates. In the case of n-GaAs substrates, they report that interfacial oxidation is relatively suppressed for all dielectrics investigated, with the interfacial layer associated with the HfAlO stack essentially indistinguishable from the p-GaAs case. Taken together, the results suggest a dopant–dependent oxidation process [153, 154] in conjunction with the ALD chemistry employed [138]. However, as noted in Sect. 6.3.1, recent studies of chemically identical oxides on n-GaAs and p-GaAs, followed by Al_2O_3 deposition by ALD, suggests that the difference in C–V behavior stems from the differences in the capture time constants for electrons and holes rather than any dopant–dependent interface state effects [63].

However, there can be significant differences in the chemical species identified at the GaAs interface by XPS dependent upon the surface preparation or whether the interface studies are conducted ex-situ or in-situ. As can be seen in Fig. 6.12, in-situ studies of surface native oxides after subsequent ALD processing indicate that such oxides can be actually reduced *below* the limit of XPS detection by the ALD process depending upon the oxidation state/precursor combination employed [62]. The results are even more dramatic for surfaces chemically treated prior to ALD, where the weaker bonded, higher-oxidation states are initially removed completely by wet cleans (e.g., NH_4OH shown in Fig. 6.12). In contrast, previous literature also indicates that surface oxide species remain after ALD [137, 148, 155].

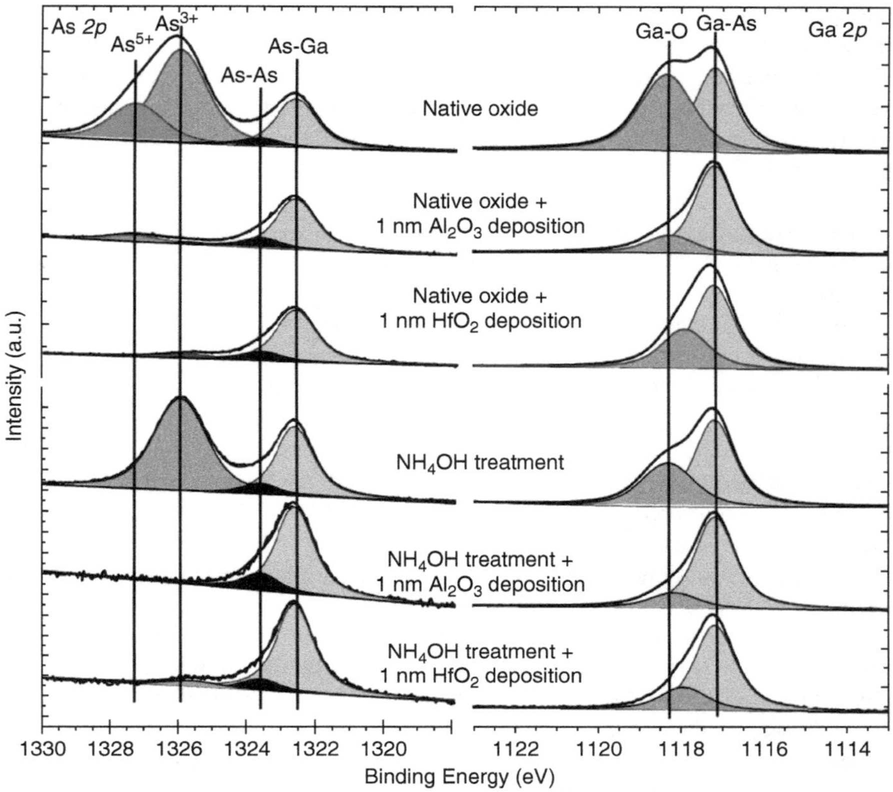

Fig. 6.12 X-ray photoelectron spectroscopy of the interfacial reactions after atomic layer deposition of Al_2O_3 and HfO_2. The reactions with the surface oxides exhibit precursorspecific and oxidation state–specific behavior [179] (Reprinted with permission. Copyright 2009 Elsevier)

This apparent discrepancy in the literature is not surprising as the thin films required for typical XPS analysis must be photoelectron transparent (≤6 nm). As discussed above, the photoelectron kinetic energy associated with the features (e.g., 3*d* or 2*p* lines) of interest using laboratory x-ray sources are ~1480 eV or much less. Thus, very thin films and the interfaces will likely oxidize upon extended exposure to the ambient prior to surface analysis, resulting in oxidized interfacial species not a priori associated with the dielectric growth process itself. However, such results indicate that considerable caution must be exercised when drawing conclusions on correlations of physical characterization results to those obtained electrically such as C–V curves. Device processing often entails steps which result in exposure of the gate stack to the cleanroom ambient, and thus ex-situ studies certainly have relevance in this context. In contrast, in-situ studies enable a better understanding of the film interface behavior during growth and the impact of controlled oxidation on electrical performance where the devices are constructed under carefully controlled conditions to limit spurious interfacial oxidation.

The data [62] presented in Fig. 6.12 indicates a reaction mechanism that is consistent with a ligand exchange process [156], whereby Al from the TMA preferentially reacts with the As 3+ and Ga 3+ oxidation states, resulting in bond conversion to Al-oxide. In contrast, the reaction with the 3+ oxidation state is less efficient for Hf originating from the TEMA-Hf precursor, yet effective for the 5+ state and indicative of a more complex process. In either case, it is evident that the weakly bonded oxides may be reduced from the ALD process, with the stronger Ga–O bonding, including potential Ga sub-oxide species, remaining at the interface. It therefore seems possible to control a significant portion of the interfacial oxidation through such precursor-mediated reactions, and thus impact detrimental electrical behavior from defects induced by uncontrolled oxidation such as Fermi level pinning and C–V frequency dispersion [57, 157]. Recently, the use of isopropanol as an oxidizer has been investigated in this context as well for Al_2O_3 deposition [120].

Careful inspection of the Ga–O feature binding energy in the Ga2*p* spectra shown in Fig. 6.12 indicates that a general shift *toward* the bulk Ga–As peak is observed upon ALD film growth. Such chemical shifts may be consistent with M–O–Ga bonding (where M=Al or Hf), the presence of Ga sub-oxides species (such as O–Ga–O) in addition to Ga 3+ (viz. Ga_2O_3) as noted in Sect. 6.3.2, as well as band bending effects [63, 107].

6.4.2.2 ALD on InGaAs

The higher bulk mobility and potentially more favorable surface passivation behavior [36] associated with $In_xGa_{1-x}As$ alloys has stimulated recent research on MOSFETs with ALD dielectrics as well. As of mid-2009, a number of Al_2O_3 ALD studies, most emphasizing device characteristics, have been conducted with various In content including $x = 15\%$ In [158, 176], 20% In [132, 159–161], 53% In [162, 163], 65% In [164], and 100% In [165]. Other high-k dielectrics deposited by ALD including HfO_2 [133, 134, 166, 167], ZrO_2 [168, 169]. Hf-aluminates [170, 171] and La-aluminates [141, 172] have also been recently examined on InGaAs. The interest in aluminates [173, 174] stems from a potentially higher gate dielectric permittivity with a minimal (low-k) interfacial layer—an essential aspect when device scaling is taken into account [2, 175].

For ALD on $In_{0.2}Ga_{0.8}As$ (as well as InSb [73]), both "self cleaning" [176–178] and predeposition annealing [161] have been recently examined in view of the desire to control interfacial oxidation. Figure 6.13 demonstrates from in-situ half-cycle ALD studies that the reaction with the *initial* TMA pulse at 300 °C results in most of the oxide reduction on the $(NH_4)_2S$-treated n-$In_{0.2}Ga_{0.8}As$ surface [178, 179]. In addition to the suppression of the 3+ oxidation state consistent with a ligand exchange mechanism, it is also seen that As-S bonding is rendered below the limit of detection by this reaction, while Ga-S (and possibly Ga-suboxide) remains throughout the growth process of the resultant 1 nm Al_2O_3 film. Further reduction of the residual oxides and As-As bonding has been recently reported by vacuum annealing chemically treated n- and p-$In_{0.2}Ga_{0.8}As$ surfaces to 380–390 °C prior to

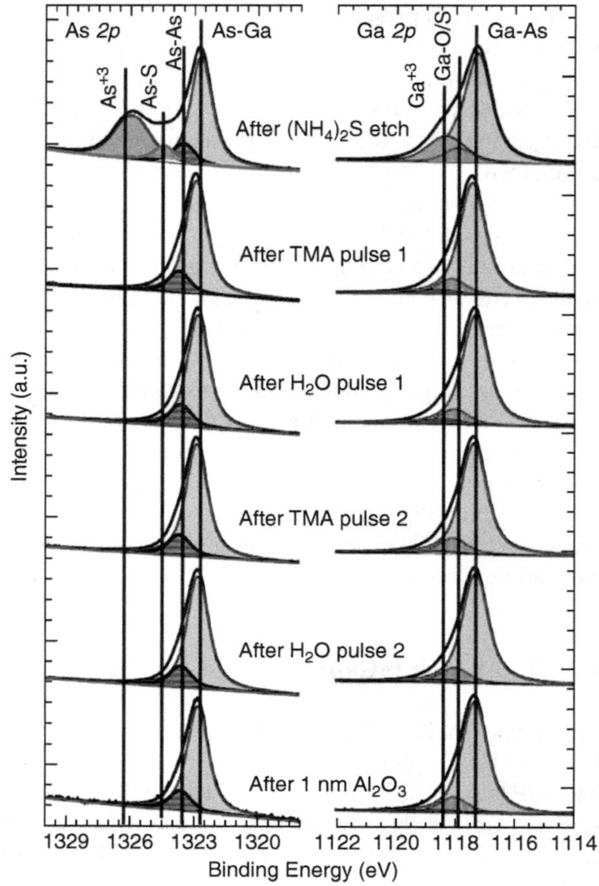

Fig. 6.13 In-situ x-ray photoelectron spectroscopy analysis of atomic layer deposition halfcycle reactions for Al_2O_3 on $In_{0.2}Ga_{0.8}As$. TMA, trimethyl-aluminum [178] (Reprinted with permission. Copyright 2008 American Institute of Physics)

initiating ALD deposition [34, 161]. Exposure of the dielectric stack to forming gas (N_2:H_2) anneals at 450 °C results in the reduction of hydroxyls and a concomitant negative flatband voltage shift [161]. The effect of ozone as an oxidant has also been recently examined, where extensive oxidation is observed [180].

From a surface/interface chemistry perspective, the $In_{0.53}Ga_{0.47}As$ surface (on InP substrates) appears to be among the most studied, and is particularly interesting from a potential transistor performance point of view. Chang et al. examined HfO_2 on this surface produced from TEMA-Hf/H_2O ALD at 200 °C using synchrotron angle-resolved XPS [166]. They report that no detectable In, Ga, nor As species is seen within the HfO_2 film, and that only Ga_2O_3, In_2O_3, and $In(OH)_3$ are detected at the HfO_2/$In_{0.53}Ga_{0.47}As$ interface. A conduction-band offset of $\Delta E_C = 1.8 \pm 0.1$ eV and a valence-band offset of $\Delta E_V = 2.9 \pm 0.1$ eV were determined in this work, and a large midgap interface state density $D_{it} \approx 1 \times 10^{12}\,cm^{-2}\,eV^{-1}$ was deduced and attributed to the presence of the interfacial oxides. Lee et al. examined the effects of air-exposure after MBE growth of $In_{0.53}Ga_{0.47}As/InP$ and ALD deposited HfO_2 [167]. It was found that limiting exposure to air to 10 min. results in thinner In- and Ga-oxides, while As-oxide was below the limit of detection by XPS. A similar $D_{it} \approx 1 \times 10^{12}\,cm^{-2}\,eV^{-1}$ was deduced from capacitor measurements, and capacitance equivalent oxide thickness (CET) of 1 nm was reported. In contrast, Oh et al. [171] reported that MOCVD studies of HfO_2 films on $In_{0.53}Ga_{0.47}As$ results in no detectable interfacial oxides and cite the deposition approach as a potential cause. A comparison to MOCVD Hf-aluminate formation on this surface also indicated a lower interface state density ($\sim 5 \times 10^{11}\,cm^{-2}\,eV^{-1}$) without detectable interfacial oxides compared to the HfO_2 case ($\sim 9 \times 10^{12}\,cm^{-2}\,eV^{-1}$). Similar band offsets were reported for the $HfO_2/In_{0.53}Ga_{0.47}As$ interface, while $\Delta E_C = 2.4 \pm 0.1$ eV and $\Delta E_V = 3.3 \pm 0.1$ eV for the $HfAlO_x/In_{0.53}Ga_{0.47}As$ interface. Shahrjerdi et al. [133], examined $HfO_2/AlN/ In_{0.53}Ga_{0.47}As$ where small, but detectable, As-oxides were observed, in conjunction with Ga- and In-oxides as well. A high $D_{it} \approx 8 \times 10^{12}\,cm^{-2}\,eV^{-1}$ was also deduced from the capacitors fabricated, and the self-aligned transistors fabricated indicated room for further optimization. Taken together these reports suggest that even small amounts of detectable surface oxides results in a significant interface state density.

Control of the interfacial oxides on $In_{0.53}Ga_{0.47}As$ upon gate stack formation has been recently explored along two avenues: precursor selection and interfacial Si layers. Xuan et al. have examined transistor performance using ALD Al_2O_3 directly on $In_{0.53}Ga_{0.47}As$ reporting $D_{it} \sim 10^{12}\,cm^{-2}\,eV^{-1}$ and corrected mobilities as high as 2200 $cm^2/V\,s$ [162]. Recent work on La-aluminate deposition by ALD on $In_{0.2}Ga_{0.8}As$ using tris(*N-N'*-diisopropylformamidinato)La indicates that formation of higher oxidation states may be controlled effectively through the ligand-exchange reaction mechanism [141]. Only suboxides of Ga were detected in these films. Alternatively, the use of a thin Si layer has been shown to react with $In_xGa_{1-x}As$ surface oxides, as well as subsequent dielectrics deposited by PVD [181] and ALD [57] methods. Essentially, the Si film serves as a "getter" layer for relatively weaker bound surface oxides resulting in more stable SiO_x bond formation. The effect of employing Si interfacial layers is explicitly discussed in the next section.

Scaling of surface channel transistors beyond the 16 nm node (~2020) will necessarily require a suitable high-k dielectric with an equivalent SiO_2 thickness below 1.0 nm which discourages thick low-k interface layers, and so the precursor reaction chemistry route to controlling interfacial defects and their resultant trap states will likely be pursued vigorously. It should also be noted that alternative device architectures, which rely on buried channels for example, are also under investigation [182]. In general, the performance of all III-V field effect transistor architectures to date have yet to be scaled to establish the performance at the 16 nm node and, although buried channel high electron mobility transistors (HEMT) are making significant strides [183], several challenges remain for this technology to be widely adopted for high performance logic applications.

6.5 Electrical Behavior of Oxides on III-V and Interfacial Chemistry

For over 30 years, there has been a significant effort on development of MOS devices with various oxides on compound semiconductors. The electrical behavior of these MOS devices intimately depends on details of processing conditions (e.g., substrate type, surface preparation, dielectric deposition technique, post-deposition anneal). Therefore, it is impossible to review in a cohesive manner all of the literature available regarding the electrical behavior of oxides on III-V semiconductors. Instead, the intent of this section is to describe a few salient features of the electrical behavior of oxides on III-V semiconductors with respect to interfacial chemistry and that will be critical for the ongoing development of $In_xGa_{1-x}As$ MOSFETs.

6.5.1 C–V Measurements and Issues

Capacitance–Voltage (C–V) measurements are a staple in the traditional characterization of MOS devices and materials [184]. However, application of the technique to III-V systems requires considerable care, as described extensively by Passlack [185, 186] and Brammertz et al. [187–191]. In this section, we discuss the issues associated with electrical measurements obtained in view of the interfacial chemistry described in previous sections.

6.5.1.1 Frequency Dispersion

One of the commonly observed anomalous phenomena is that of strong frequency dispersion of the C–V characteristics in maximum capacitance as shown in Fig. 6.14.

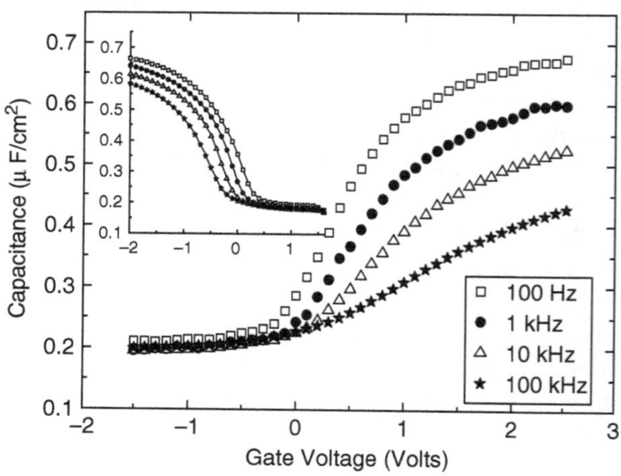

Fig. 6.14 Commonly observed frequency dispersion in accumulation for TaN/Al_2O_3 stack on both n- and p-type (*inset*) GaAs MOS capacitor structures. The capacitor area is 7.85×10^{-5} cm^2

In this case, both n-type (2–4 × 10^{17} cm^{-3} Si doped) and p-type (2–4 × 10^{17} cm^{-3} Zn doped) GaAs substrates received a standard degrease (acetone, methanol and IPA for 1 min each) followed by cleaning in 29% NH_4OH for 3 min. A 10 nm Al_2O_3 film was deposited using a Cambridge Nanotech Savannah-100 Atomic Layer Deposition (ALD) chamber using trimethylaluminum (TMA) and H_2O at 300 °C as the gate oxide. A post deposition anneal (PDA) was performed at 600 °C in N_2 for 60 s. TaN (150 nm) was rf sputtered through a shadow mask as the gate metal to form MOS capacitors. An electron-beam evaporated Ni/Au/Ge alloy annealed at 425 °C was used as a back side Ohmic contact. The frequency dispersion of maximum capacitance is more prominent in n-type GaAs compared to p-type GaAs MOS capacitors. This behavior has been observed on numerous III-V semiconductors for over 30 years [150, 192, 193]. For $In_xGa_{1-x}As$, the effect is primarily observed for low In concentration and is typically not observed for $x = 0.53$ and above.

One of the possible causes for frequency dispersion in maximum capacitance is that of series resistance (contact, substrate, cabling) altering the measured C–V [194]. However, there are several reasons that series resistance cannot explain the dispersion behavior observed here. The device capacitance (C_c), parallel conductance (G_c) and series resistance (R_s), and the measured capacitance (C_m) has the following dependence with measurement frequency (ω)

$$C_m = \frac{C_c}{(G_c R_s + 1)^2 + \omega^2 (C_c R_s)^2}. \tag{6.2}$$

Frequency dispersion due to series resistance depends on ω^{-2} which is not observed in the measured results. To provide an independent measure of series resistance, the gate dielectric of the MOS capacitor was broken down by going to extremely high positive bias. The series resistance was measured to be approximately 20 Ω. This value is consistent with separate measurements of resistivity of the TaN, resistivity of the GaAs, and contact resistance measurements of the AuGeNi backside contact. Using this value of series resistance, a parallel conductance (obtained from the derivative of the tunneling current–voltage characteristic) and a reasonable estimate for the C_c–V_g relationship, no dispersion is obtained in modeled C_m–V_g behavior for a frequency of 1 MHz. Therefore, series resistance cannot explain the observed dispersion.

Previous researchers have ascribed this anomalous dispersion behavior to a high density of interface states and associated Fermi level pinning [43, 150, 185, 192, 193, 195, 196]. Figure 6.15 shows modeled n-GaAs C–V characteristics including classical interface state capacitance with an extremely high interface state density of 1×10^{13} cm^{-2} eV^{-1} uniformly distributed in energy [57]. C–V characteristics were simulated using a classical model of the total semiconductor charge [197]. Using the surface potential from this solution, the frequency dependent capacitance associated with interfacial defects (C_{it}), averaged over band bending, and for a p-type substrate, was calculated numerically using the traditional approach [184] yielding:

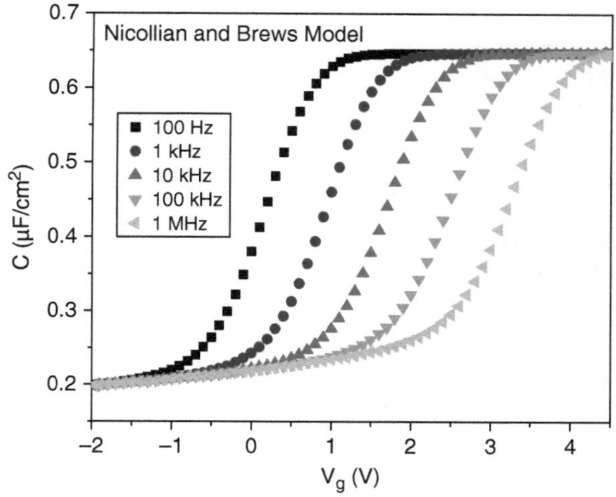

Fig. 6.15 Modeled GaAs C–V characteristics including classical interface state capacitance (Eq. 6.2)

$$C_{it} = \frac{qD_{it}\left(2\pi\sigma_s^2\right)^{-1/2}}{2\omega\tau_p}\int_{-\infty}^{\infty}\exp\left(\frac{-\upsilon^2}{2\sigma_s^2}\right)\exp\left(-\upsilon\right)\tan^{-1}\left(2\omega\tau_p\exp(\upsilon)\right)\mathrm{d}\upsilon, \quad (6.3)$$

with

$$\tau_p = \frac{1}{\bar{v}\sigma_p p_s} \quad (6.4)$$

where σ_s^2 is the variance of band bending in units of kT/q, ω is the measurement frequency in radians, τ_p is the characteristic capture time constant for holes, v is band bending, $\bar{v}$ is the thermal velocity of the carriers (typically 10^7 cm/s in silicon at room temperature), σ_p is the capture cross-section for holes, and p_s is the density of free holes at the substrate surface. The total capacitance for the MOS capacitor (C_{tot}) at a given gate voltage (V_g) is then calculated using

$$C_{tot} = \left[(C_{it} + C_{sub})^{-1} + C_{ox}^{-1}\right]^{-1}. \quad (6.5)$$

The results show a frequency-dependent kink in the depletion portion of the curve similar to that of silicon but quite different than the measured GaAs behavior. In accumulation, the free electron density (n_s) at the semiconductor surface is large which implies that the trapping time constant given by Eq. (6.4) will be very small. Therefore, all of the interface states can respond to the frequencies of interest, and C_{it} is approximately equal to qD_{it}. In strong depletion, n_s is very small so that τ_n is very large. Therefore, no interface trap response occurs for the frequencies of interest and C_{it} approaches zero. In the gate bias range between strong accumulation and strong depletion, there is limited interface trap response, and the value of C_{it} is between these two extreme cases. Even the use of physically unrealistic parameters (e.g., capture cross-section of $10^{-4}\,cm^2$) or different energy distributions within the classical interface state capacitance framework cannot reproduce the measured behavior of many compound semiconductors.

6.5.1.2 Hasegawa and Sawada C_{it} Model

Hasegawa and Sawada have previously developed a model that can explain the behavior [192, 193]. It is assumed in the classical treatment of interface state capacitance that defects distributed away from the interface (into the dielectric) cannot respond to the small signal associated with the capacitance measurement. However, Hasegawa and Sawada performed a time domain analysis of Deep Level Transient Spectroscopy (DLTS) measurements of interface states which suggests trapping time constants not consistent with this assumption. The trapping time constants suggest an interfacial region with a conduction band 0.33 eV lower than that associated with crystalline GaAs. They suggest that this region is associated with a thin disordered interfacial layer at the interface of III-V semiconductors and the related disorder-induced gap states (DIGS) where the defects are distributed in both energy and space. A similar low resistivity interfacial region has also been suggested by Passlack to explain the dispersion behavior [185]. Hasegawa and Sawada found that the dispersion results can be reproduced if one assumes an exponentially decaying spatial distribution of traps into the dielectric,

$$N_T(x) = N_{TO} \exp(-\alpha x), \tag{6.6}$$

where $N_T(x)$ is the trap density as a function of position and α is the decay constant. Assuming tunneling into these defects (after Preier [198]), the following relationship was obtained for the interface state capacitance,

$$C_{it} = \frac{q^2 N_{TO}}{2\kappa_0} (\omega\tau_0)^{(\alpha/2\kappa_0)} \int_0^{1/\omega\tau_0} z^{(\alpha/2\kappa_0)} \tan^{-1}\left(z^{-1}\right) dz \tag{6.7}$$

assuming,

$$\tau(x) = \tau_0 \exp(2\kappa_0 x) \tag{6.8}$$

where κ_0 is the quantum-mechanical decay constant of electron wave function, and τ_0 is the time constant of the trap located at the interface.

Figure 6.16 shows the simulated interface state capacitance plots of an n- and p-type GaAs MOS capacitor using this Hasegawa-Sawada model. Donor type D_{it} was assumed to be in the lower half of the gap and acceptor type D_{it} in the upper

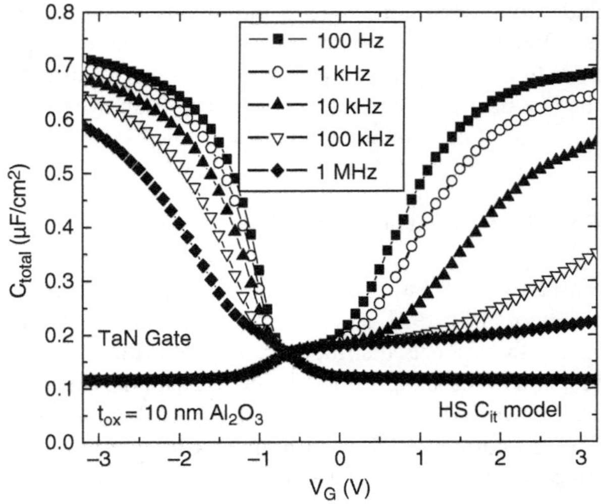

Fig. 6.16 Simulated C–V characteristics of MOS capacitors on both n- and p-type GaAs using the Hasegawa-Sawada C_{it} model. The model capacitance shows a frequency dispersion similar to the experimentally observed phenomena

half of the bandgap. The difference between the classical model and the tunneling model is immediately apparent. The C_{it} has a frequency dependence that varies with each decade frequency change. The value of $\alpha = 50$ Å and $k_0 = 2$ Å were used in these calculations. As the total capacitance is dominated by C_{it} rather than C_{sub}, the C_{total}–V_g plot shows a frequency dependence of the maximum capacitance. The frequency dependent capacitance characteristics shown in Fig. 6.16 are quite similar to the experimentally observed frequency dispersion of the GaAs MOS capacitors shown in Fig. 6.14. The disparity in dispersion of n-type vs. p-type GaAs is primarily related to the difference in trapping time constants for n-type vs. p-type (differences in effective density of states for electrons and holes) and the energy distribution of interface states [63].

6.5.1.3 Detection of Free Carriers

An important factor necessary to observe dispersion in the capacitance is that the substrate capacitance must be small as compared to the interface state capacitance. With low to moderate interface state density, the total capacitance for all frequencies merges in accumulation and, potentially, inversion due to the substrate capacitance becoming much larger than interface state capacitance. To observe frequency dispersion, the interface state capacitance in Eq. (6.3) must be larger than the substrate capacitance. This means that the observation of maximum capacitance in a low frequency capacitance-voltage curve does not necessarily indicate the presence of free carriers [46, 150, 187, 204]. This statement is valid for both majority and minority carrier response. For majority carriers, the substrate capacitance does not have a dependence on measurement frequency (F ≈ 10^2 to 10^6 Hz). For minority carriers, the substrate capacitance of a capacitor is limited by the time constant for minority carrier generation. Therefore, the minority carrier (inversion) substrate capacitance depends on frequency as well as sweep rate [188]. The maximum capacitance in inversion as a function of frequency is sometimes used to infer the presence of inversion. However, the interface trap time constants and associated capacitance response as a function of frequency demonstrates that this methodology cannot be used to necessarily infer inversion.

6.5.2 *Interface States of* $In_xGa_{1-x}As$

The interface state density of $In_xGa_{1-x}As$ is extremely important from a technological point of view. Although there are subtle differences in interface states with different dielectrics and interfacial cleans, the most salient differences are observed by applying an interfacial "passivation" layer such as amorphous Si and by changing In concentration. The following will provide a brief review of the impact of these experimental parameters.

6.5.2.1 Effect of Silicon Interfacial Passivation Layer (IPL)

Interfacial passivation layers (e.g., amorphous Silicon) between the dielectric and III-V semiconductor have been explored previously [199, 200] and has been reinvestigated for thin (~1–2 nm) films recently [43, 46, 181, 196]. Figure 6.17b reproduces the C–V characteristics shown in Fig. 6.14 for GaAs MOS capacitors, while those obtained by similar fabrication conditions except for the presence of an amorphous Si IPL deposited using PECVD (100 sccm of 2% SiH_4/He, 400 sccm of He, 50 W, 200 °C) are shown in Fig. 6.17d. This deposition condition results in approximately 1.2 nm of amorphous Silicon. It is clear that the dispersion effect is reduced substantially due to the presence of this interfacial layer. Measurements at elevated temperatures show a similar reduction [48].

The associated Ga 2*p* XPS spectra for these interfaces are also presented in Fig. 6.17 [47, 48, 201]. In addition to a complete absence of detectable As-oxides at these interfaces (not shown here) due to either the ALD "self-cleaning" phenomenon [62] or the gettering reaction of the Si IPL with surface oxides to form Si-O species [201], it is seen that the presence of the silicon interlayer dramatically reduces the presence of the Ga 3+ oxidation state, while leaving a detectable Ga 1+ oxidation state. It is therefore proposed that the higher oxidation states of Ga (and not As) are related to the species that cause high D_{it} for devices similarly fabricated, and hence Fermi level pinning (low substrate capacitance), either from a direct removal of defect states induced from Ga 3+ or from a resultant bonding reconfiguration, such as the formation of undimerized As, when Ga_2O_3 is present [67]. It is again noted that the Ga_2O bonding arrangement remains for *all* interfaces that have been exposed to oxidizing species at some point in the fabrication process, suggesting that the Ga 1+ oxidation state is *not* the species primarily responsible for

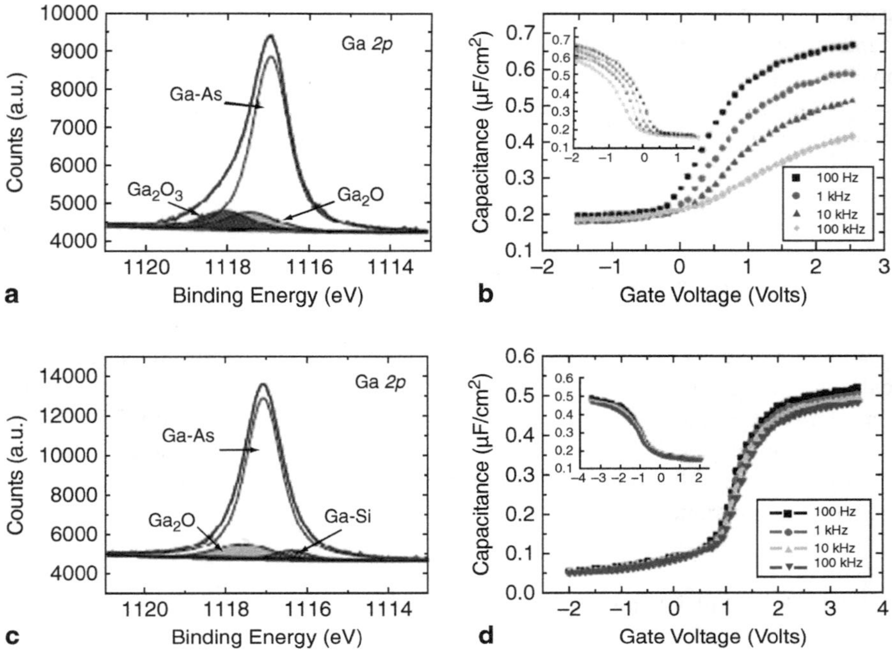

Fig. 6.17 Capacitance-voltage characteristics of GaAs MOS capacitors with ~1.2 nm PECVD amorphous Silicon interlayer and ~11 nm of ALD Al_2O_3 [201] (Reprinted with permission. Copyright 2009 Elsevier)

Fermi level pinning. This observation is consistent with prior reports utilizing Ga_2O deposition on GaAs by MBE methods with improved electrical characterics [64, 68, 202]. Transport characteristics of $In_xGa_{1-x}As$ MOSFETs with and without a silicon interlayer are consistent with these C–V results. Drastically improved MOSFET performance can be achieved using an amorphous silicon passivation layer [57]. An extracted peak mobility >3600 cm^2/Vs is achieved for $In_{0.53}Ga_{0.47}As$ with an amorphous silicon interlayer upon correction of interface state capacitance [203].

Although a lower defect density is one of the requirements necessary to reduce frequency dispersion of the maximum capacitance, another critical requirement is to alter the time constant of these defects. The Hasegawa-Sawada model alters the typical interface state time constant by permitting trapping and detrapping in a thin disordered interfacial layer and the related DIGS. Figure 6.18 shows the experimental C–V characteristics for GaAs MOS capacitors with conditions similar to that of Fig. 6.17 except the amorphous Silicon interlayer is formed using in-situ MBE methods [46]. Although the interface state density is very high and results in a shift in the transition region with frequency, the maximum capacitance shows vastly reduced dispersion. This behavior can be reproduced using the classical interface state model as shown in Fig. 6.15. The results suggest that the formation of the Ga 3+ oxidation state occurs in conjunction with a disordered interfacial layer. The disorder induced gap states (DIGS) have time constants which permit dispersion

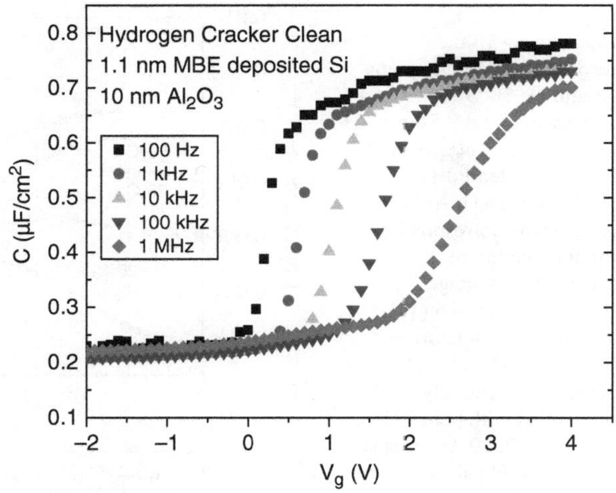

Fig. 6.18 Capacitance–voltage characteristics of GaAs MOS capacitors with ~1.1 nm MBE deposited Silicon interlayer and ~10 nm of ALD Al_2O_3

of the maximum capacitance. The silicon interlayer reduces the DIGS, but typical interface states are still possible.

6.5.3 Effect of Indium Concentration

The In concentration also dramatically influences the measured interface state density. Figure 6.19 shows C–V characteristics for ALD HfO_2 on $In_{0.53}Ga_{0.47}As$ and GaAs from O'Conner et al. [204]. It is important to note that the frequency dispersion observed for the ALD HfO_2 film on GaAs is very similar to the results of Fig. 6.14 for ALD Al_2O_3. This provides further evidence that the most important parameter controlling interfacial defect density is the oxygen bonding and that details associated with the specific dielectric used are a second-order effect. It is noted in that work that the frequency dispersion of maximum capacitance for GaAs, $In_{0.15}Ga_{0.85}As$ and $In_{0.30}Ga_{0.70}As$ are very similar suggesting very high interfacial defect density. However, the frequency dispersion of maximum capacitance for $In_{0.53}Ga_{0.47}As$ is dramatically reduced as observed in Fig. 6.19a. The C–V results as a function of In concentration are consistent with transport data from $In_xGa_{1-x}As$ MOSFETs. The maximum drive current for $In_{0.53}Ga_{0.47}As$ is ~5×10^7 higher than $In_{0.20}Ga_{0.70}As$ with no silicon interlayer [57]. This difference is related to the extremely small inversion charge density of $In_{0.20}Ga_{0.70}As$ due to pinning of the Fermi level by the DIGS.

There are two possible explanations for this behavior. The first is that the primary defect responsible for the dispersion in maximum capacitance is in the upper half of the bandgap of GaAs such that the defect is within the conduction band for $In_{0.53}Ga_{0.47}As$. Figure 6.20 shows the conduction band minimum and valence band maximum of $In_xGa_{1-x}As$ as a function of In concentration. The electron affinity

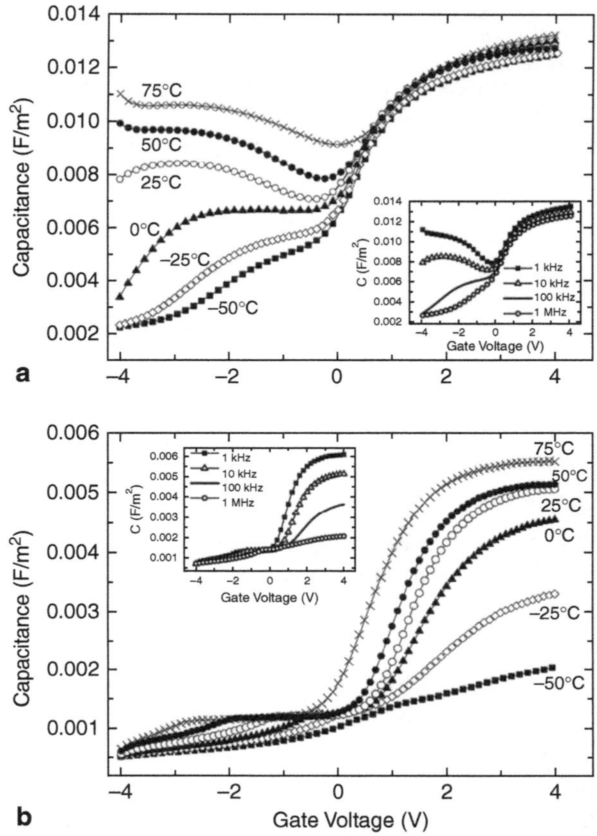

Fig. 6.19 10 kHz capacitance–voltage response with varying temperature (−50 to 75 °C) of **a** unpassivated Pd/9.5 nm ALD HfO_2 on $In_{0.53}Ga_{0.47}As$ and **b** unpassivated Pd/11.4 nm ALD HfO_2 on GaAs. The *insets* show corresponding room temperature capacitance–voltage frequency variation (1kHz to 1 MHz) in unpassivated $In_{0.53}Ga_{0.47}As$ and GaAs devices, respectively [204] (Reprinted with permission. Copyright 2009 American Institute of Physics)

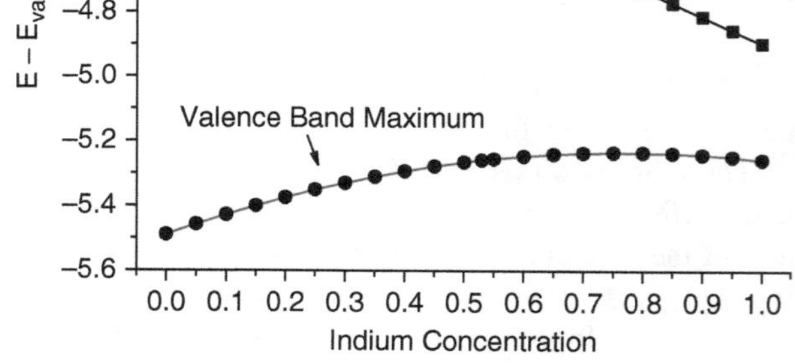

Fig. 6.20 $In_xGa_{1-x}As$ conduction band minimum and valence band maximum referenced to vacuum as a function of Indium concentration (x)

(conduction band minimum) for $In_{0.53}Ga_{0.47}As$, $In_{0.30}Ga_{0.70}As$, and GaAs is ~4.51 eV ~4.32 eV, and ~4.07 eV [205, 206]. Assuming that the defect energy with respect to vacuum is independent of In concentration, the defect responsible for frequency dispersion of the maximum capacitance would be in this energy range (~4.07 to ~4.51 eV). The second possible explanation is that the density of interfacial defects decreases with increasing In concentration. XPS results shown in the previous section indicate that the density of the Ga 3+ oxidation state is indeed reduced with increasing In concentration, as the gallium concentration is concomitantly reduced. The likely associated decrease in DIGS would result in reduced frequency dispersion of the maximum capacitance. The influence of interfacial In-oxides (which as noted above appears to be a minority interfacial species) on such states remains a topic of further investigation.

6.6 Conclusions

Despite having been studied in great detail for more than 35 years, the dielectric/III-V semiconductors interfaces and the identification of the bonding configurations that cause the defects has remained challenging. Problems with metal-oxide-semiconductor devices on GaAs and InGaAs, including frequency dispersion of capacitance and sub-optimal electron mobility, have been attributed to a number of different defects. Recent research indicates that avoiding surface and interfacial defect formation is critical in every step of device fabrication. This includes the right surface reconstruction as well as the formation of particular species of interfacial oxides, either through direct deposition or alternatively through targeted reduction of native oxides during ALD of high-k dielectrics.

Acknowledgements The authors gratefully acknowledge the discussions with our colleagues actively working in this field: G. Brammertz, R.A. Chapman, K.J. Cho, K.J. Choi, L. Colombo, A. Craven, N. Goel, P. Hurley, G. Hughes, H.C. Kim, J. Kim, A. Kummel, P. Mahji, P. Longo, P.C. McIntyre, S. Oktyabrsky, M. Passlack, F.S. Tostado-Agurirre, I. Thayne and P.D. Ye. The hard work and dedication of our student colleagues is also acknowledged: B. Brennan, R. Conteras, R. Galatage, M. Jivani, B. Lee, S. McDonnell, E. O'Conner, A. Sonnet, and W. Wang. This work is supported by the MARCOSRC Focus Center on Materials, Structures, and Devices, and the NIST Semiconductor Electronics Division.

References

1. H. C. Gatos, J. Lagowski, T. E. Kazior, *Jpn. J. Appl. Phys. Suppl.* **22**, 11 (1983), and refs. therein.
2. G. D. Wilk, R. M. Wallace, J. M. Anthony, *J. Appl. Phys.* **89**, 5243 (2001).
3. R. M. Wallace, P. C. McIntyre, J. Kim, Y. Nishi, *Mat. Res. Bull.* **34**(7), 493 (2009).
4. R. E. Schlier, H. E. Farnsworth, *J. Chem. Phys.* **30**, 917 (1959).
5. H. Welker, *Z. Naturforsch. A* **7**, 744 (1952).

6. V. P. LaBella, M. R. Krause, Z. Ding, P. M. Thibado, *Surf. Sci. Rep.* **60**, 1–53 (2005).
7. L. L. Chang, L. Esaki, W. E. Howard, R. Ludeke, *J. Vac. Sci. Technol.* **10**, 11 (1973).
8. L. L. Chang, L. Esaki, W. E. Howard, R. Ludeke, G. Schul, *J. Vac. Sci. Technol.* **10**, 655 (1973).
9. A. Y. Cho, *J. Appl. Phys.* **47**, 2841 (1976).
10. R. A. Kubiak, S. M. Newstead, P. Sullivan, in *Moleculaar Beam Epitaxy*, R. Farrow, Ed., Noyes Publications, Park Ridge, NJ (1995) 77.
11. V. P. LaBella, D. W. Bullock, C. Emery, Z. Ding, P. M. Thibado, *Appl. Phys. Lett.* **79**, 3065 (2001).
12. J. A. Appelbaum, G. A. Baraff, D. R. Hamann, *J. Vac. Sci. Technol.* **13**, 751 (1976).
13. P. K. Larsen, J. H. Neave, B. A. Joyce, *J. Phys. C: Solid State Phys. (UK)* **14**, 167 (1981).
14. P. K. Larsen, J. F. van der Veen, A. Mazur, J. Pollmann, J. H. Neve, B. A. Joyce, *Phys. Rev. B* **26**, 3222 (1982).
15. H. Tsuda, T. Mizutani, *Appl. Phys. Lett.* **60**, 1570 (1992).
16. W. Barvosa-Carter, R. S. Ross, C. Ratsch, F. Grosse, J. H. G. Owen, J. J. Zinck, *Surf. Sci.* **499**, 129 (2002).
17. J. R. Arthur, *Surf. Sci.* **43**, 449 (1974).
18. J. Massies, P. Etienne, F. Dezaly, N. T. Linh, *Surf. Sci.* **99**, 121 (1980).
19. M. D. Pashley, K. W. Haberern, W. Friday, J. M. Woodall, P. D. Kirchner, *Phys. Rev. Lett.* **60**, 2176 (1988).
20. D. K. Biegelsen, R. D. Bringans, L. E. Swartz, *Proc. SPIE* **1186**, 136 (1990).
21. D. K. Biegelsen, R. D. Bringans, J. E. Northrup, L. E. Swartz, *Phys. Rev. B* **41**, 5701 (1990).
22. M. D. Pashley, K. W. Haberern, R. M. Feenstra, *J. Vac. Sci. Technol. B* **10**, 1874 (1992).
23. M. D. Pashley, K. W. Haberern, *Ultramicroscopy* **42–44**, 1281 (1992).
24. V. P. LaBella, H. Yang, D. W. Bullock, P. M. Thibado, P. Kratzer, M. Scheffler, *Phys. Rev. Lett.* **83**, 2989 (1999).
25. M. D. Pashley, K. W. Haberern, R. M. Feenstra, P. D. Kirchner, *Phys. Rev. B* **48**, 4612 (1993).
26. D. J. Chadi, C. Tanner, J. Ihm, *Surf. Sci.* **120**, 425 (1982).
27. B. A. Joyce, J. H. Neave, P. J. Dobson, P.K. Larsen, *Phys. Rev. B* **29**, 814 (1984).
28. D. J. Chadi, *J. Vac. Sci. Technol. A* **5**, 834 (1987).
29. D. J. Frankel, C. Yu, J. P. Harbison, H. H. Farrell, *J. Vac. Sci. Technol. B* **5**, 1113 (1987).
30. W. G. Schmidt, F. Bechstedt, *Phys. Rev. B* **54**, 16742 (1996).
31. M. D. Johnson, K. T. Leung, A. Birch, B. G. Orr, J. Tersoff, *Surf. Sci.* **350**, 254 (1996).
32. J. Tersoff, M. D. Johnson, B. G. Orr, *Phys. Rev. Lett.* **78**, 282 (1997).
33. W. M. Lau, R. N. S. Sodhi, S. Jin, S. Ingrey, N. Puetz, A. Spring Thorpe., *J. Appl. Phys.* **67**, 768 (1990).
34. F. S. Aguirre-Tostado, M. Milojevic, C. L. Hinkle, E. M. Vogel, R. M. Wallace, S. McDonnell, G. J. Hughes, *Appl. Phys. Lett.* **92**, 171906 (2008).
35. S. I. J. Ingrey, W. M. Lau, R. N. S. Sodhi, *J. Vac. Sci. Technol. A* **7**, 1554 (1989).
36. G. Brammertz, M. Heyns, M. Meuris, M. Caymax, D. Jiang, *Microelectronic Eng.* **84**, 2154 (2007).
37. W. G. Schmidt, *Appl. Phys. A* **75**, 89 (2002).
38. T. Sawada, H. Hasegawa, *Electron. Lett.* **12**, 471 (1976).
39. A. Callegari, P. D. Hoh, D. A. Buchanan, D. Lacey, *Appl. Phys. Lett.* **54**, 332 (1989).
40. G. Sixt, K. H. Ziegler, W. R. Fahrner, *Thin Solid Films* **56**, 107 (1979).
41. M. Passlack, M. Hong, J. P. Mannaerts, *Appl. Phys. Lett.* **68**, 1099 (1996).
42. W. E. Spicer, N. Newman, C. J. Spindt, Z. Lilientalweber, E. R. Weber, *J. Vac. Sci. Tech. A* **8**, 2084 (1990).
43. S. Koveshnikov, W. Tsai, I. Ok, J. C. Lee, V. Torkanov, M. Yakimov, S. Oktyabrsky, *Appl. Phys. Lett.* **88**, 022106 (2006).
44. D. L. Winn, M. J. Hale, T. J. Grassman, A. C. Kummel, R. Droopad, M. Passlack, *J. Chem. Phys.* **126**, 084703 (2007).
45. G. Hollinger, R. Skheytakabbani, M. Gendry, *Phys. Rev. B* **49**, 11159 (1994).

46. C. L. Hinkle, A. M. Sonnet, E. M. Vogel, S. McDonnell, G. J. Hughes, M. Milojevic, B. Lee, F. S. Aguirre-Tostado, K. J. Choi, J. Kim, R. M. Wallace, *Appl. Phys. Lett.* **91**, 163512 (2007).
47. C. L. Hinkle, M. Milojevic, B. Brennan, A. M. Sonnet, F. S. Aguirre-Tostado, G. J. Hughes, E. M. Vogel, R. M. Wallace, *Appl. Phys. Lett.* **94**, 162101 (2009).
48. C. L. Hinkle, M. Milojevic, A. M. Sonnet, H. C. Kim, J. Kim, E. M. Vogel, R. M. Wallace, *ECS Trans.* **19**, 387 (2009).
49. G. P. Schwartz, G. J. Gualtieri, G. W. Kammlott, B. Schwartz, *J. Electrochem. Soc.* **126**, 1737 (1979).
50. A. J. Bard, R. Parsons, J. Jordan, *Standard Potentials in Aqueous Solution*, International Union of Pure and Applied Chemistry.
51. H. He, C.-F. Cheng, S. Seal, T. L. Barr, J. Klinowski, *J. Phys. Chem.* **99**, 3235 (1995).
52. G. Hollinger, E. Bergignat, J. Joseph, Y. Robach, *J. Vac. Sci. Technol. A* **3**, 2082 (1985).
53. P. G. Hofstra, D. A. Thompson, B. J. Robinson, M. P. Besland, M. Gendry, P. Regreny, G. Hollinger, *J. Appl. Phys.* **77**, 5167 (1995).
54. E. Weiss, O. Klin, S. Grossman, S. Greenberg, P. C. Klipstein, R. Akhvlediani, R. Tessler, R. Edrei, A. Hoffman, *J. Vac. Sci. Technol. A* **25**, 736 (2007).
55. G. P. Schwartz, *Thin Solid Films* **103**, 3 (1983).
56. M. Milojevic, C. L. Hinkle, E. M. Vogel, R. M. Wallace (unpublished).
57. A. M. Sonnet, C. L. Hinkle, M. N. Jivani, R. A. Chapman, G. P. Pollack, R. M. Wallace, E. M. Vogel, *Appl. Phys. Lett.* **93**, 122109 (2008).
58. C. L. Hinkle, M. Milojevic, E. M. Vogel, R. M. Wallace (unpublished).
59. S. Doniach M. Sunjic, *J. Phys. C* **3**, 285 (1970).
60. M. V. Lebedev, D. Ensling, R. Hunger, T. Mayer, W. Jaegermann, *Appl. Surface Sci.* **229**, 226 (2004).
61. S. Adachi, D. Kikuchi, *J. Electrochem. Soc.* **147**, 4618 (2000).
62. C. L. Hinkle, A. M. Sonnet, E. M. Vogel, S. McDonnell, G. J. Hughes, M. Milojevic, B. Lee, F. S. Aguirre-Tostado, K. J. Choi, H. C. Kim, J. Kim, R. M. Wallace, *Appl. Phys. Lett.* **92**, 071901 (2008).
63. C. L. Hinkle, A. M. Sonnet, M. Milojevic, F. S. Aguirre-Tostado, H. C. Kim, J. Kim, R. M. Wallace, E. M. Vogel, *Appl. Phys. Lett.* **93**, 113506 (2008).
64. R. Droopad, M. Passlack, N. England, K. Rajagopalan, J. Abrokwah, A. Kummel, *Microelectronic Eng.* **80**, 138 (2005).
65. R. Droopad, K. Rajagopalan, J. Abrokwah, L. Adams, N. England, D. Uebelhoer, P. Fejes, P. Zurcher, M. Passlack, *J. Cryst. Growth* **301–302**, 139 (2007).
66. M. Holland, P. Longo, G. W. Paterson, W. Reid, A. R. Long, C. R. Stanley, A. J. Craven, I. Thayne, R. Gregory, *Microelectronic Eng.* **86**, 244 (2009).
67. M. J. Hale, S. I. Yi, J. Z. Sexton, A. C. Kummel, M. Passlack, *J. Chem. Phys.* **119**, 6719 (2003).
68. Z. Yu, C. D. Overgaard, R. Droopad, M. Passlack J. K. Abrokwah, *Appl. Phys. Lett.* **82**, 2978 (2003).
69. I.-J. Ok, H. Kim, M. Zhang, F. Zhu, S. Park, J. Yum, H. Zhao, D. Garcia, P. Majhi, J. C. Lee, *Appl. Phys. Lett.* **92**, 202908 (2008).
70. P. Longo, A. J. Craven, M. C. Holland, D. A. J. Moran, I. G. Thayne, *Microelectronic Eng.* **86**, 214 (2009).
71. M. Passlack, M. Hong, R. L. Opila, J. P. Mannaerts, J. R. Kwo, *Appl. Surf. Sci.* **104/105**, 441 (1996).
72. A. J. Rosenberg, *J. Phys. Chem.* **64**, 1143 (1960).
73. A. H. Hou, M. C. Chen, C. H. Chang, T. B. Wu, C. D. Chiang, J. J. Luo, *J. Electrochem. Soc.* **155**, G180 (2008).
74. R. P. Vasquez F. J. Grunthaner, *J. Vac. Sci. Technol.* **19**, 431 (1981).
75. K. Loschke, G. Kuhn, H. J. Bitz, G. Leonhardt, *Thin Solid Films* **48**, 229 (1978).
76. J. F. Wager, K. M. Geib, C. W. Wilmsen, L. L. Kazmerski, *J. Vac. Sci. Technol. B* **1**, 778 (1983).

77. J. F. Wager, D. L. Ellsworth, S. M. Goodnick, C. W. Wilmsen, *J. Vac. Sci. Technol.* **19**, 513 (1981).
78. M. Rubenstein, *J. Electrochem. Soc.* **113**, 540 (1966).
79. R. Nishitani, H. Iwasaki, Y. Mizokawa, S. Nakamura, *Jpn. J. Appl. Phys*. **17**, 321 (1978).
80. T. Ikoma, H. Yokomizo, H. Tokuda, *Jpn. J. Appl. Phys. Suppl.* **18**, 131 (1979).
81. T. Sawada, H. Hasegawa, *Thin Solid Films* **56**, 183 (1979).
82. H. H. Wieder, *J. Vac. Sci. Technol. B* **11**, 1331 (1993); and refs. therein.
83. W. E. Spicer, I. Lindau, P. Pianetta, P. W. Chye, C. M. Garner, *Thin Solid Films* **56**, 1 (1979).
84. K. Heime, *InGaAs Field Effect Transistors,* Research Studies Press Ltd., Taunton, Somerset, England (1989), ISBN: 0 86380 083 1.
85. Y. Q. Wu, P. D. Ye, G. D. Wilk, B. Yang, *Mater. Sci. Eng. B* **135**, 282 (2006), and refs. therein.
86. Y. C. Chang, W. H. Chang, H. C. Chiu, L. T. Tung, C. H. Lee, K. H. Shiu, M. Hong, J. Kwo, J. M. Hong, C. C. Tsai, *Appl. Phys. Lett.* **93**, 053504 (2008).
87. T. Ashley, A. B. Dean, C. T. Elliott, R. Jefferies, F. Khaleque, T. J. Phillips, "High-speed, Low-Power InSb Transistors," *Proc. Int. Elec. Dev. Meet.* **97** (30.4.1), 751 (1997). This work actually utilized sputter deposited SiO_2.
88. S. Datta, *Micrelectronic Eng.* 84, 2133 (2007). This work actually utilizes ALD Al_2O_3 to isolate the gate for the buried QW structure, and is not in direct contact with the InSb.
89. M. V. Lebedev, *Prog. Surf. Sci.* **70**, 153 (2002).
90. V. N. Bessolov, M. V. Lebedev, *Semiconductors* **32**, 1141 (1998).
91. J.-F. Fan, H. Oigawa Y. Nannichi, *Jpn. J. Appl. Phys., Part 2* **27**, L1331 (1988).
92. Y. Nannichi, J. Fan, H. Oigawa, A. Koma, *Jpn. J. Appl. Phys., Part 2* **27**, L2367 (1988).
93. H. Oigawa, J.-F. Fan, Y. Nannichi, K. Ando, K. Saiki, A. Koma, *Jpn. J. Appl. Phys., Part 2*, **28**, L340 (1989).
94. E. Yablonovitch, B. J. Skromme, R. Bhat, J. P. Harbison, T. J.Gmitter, *Appl. Phys. Lett.* **54**, 555 (1989).
95. C. J. Sandroff, R. N. Nottenburg, J.-C. Bischoff, R. Bhatt, *Appl. Phys. Lett.* **51**, 33 (1987).
96. E. Yablonovitch, C. J. Sandroff, R. Bhat, T. Gmitter, *Appl. Phys. Lett.* **51**, 439 (1987).
97. Z. Liu, Y. Sun, F. Machuca, P. Pianetta, W. E. Spicer, R. F. W. Pease, *J. Vac. Sci. Technol. A* **21**, 212 (2003).
98. S. M. Mokler, P. R. Watson, *J. Vac. Sci. Technol. A* **9**, 1374 (1991).
99. B. A. Kuruvilla, S. V. Ghaisas, A. Datta, S. Banerjee, S. K. Kulkarni, *J. Appl. Phys.* **73**, 4384 (1993).
100. S. A. Chambers, V. S. Sundaram, *J. Vac. Sci. Technol. B* **9**, 2256 (1991).
101. C. C. Chang, P. H. Citrin, B. Schwartz, *J. Vac. Sci. Technol.* **14**, 943 (1977).
102. Y. Ishikawa, H. Ishii, H. Hasegawa, T. Fukui, *J. Vac. Sci. Technol. B* **12**, 2713 (1994).
103. M. S. Abrahams, C. J. Buiocchi, *J. Appl. Phys.* **36**, 2855 (1965).
104. E. Yablonovitch, H. M. Cox, T. J. Gmitter, *Appl. Phys. Lett.* **52**, 1002 (1988).
105. G. C. DeSalvo, C. A. Bozada, J. L. Ebel, D. C. Look, J. P. Barrette, C. L. A. Cerny, R. W. Dettmer, J. K. Gillespie, C. K. Havasy, T. J. Jenkins, K. Nakano, C. I. Pettiford, T. K. Quach, J. S. Sewell, G. D. Via, *J. Electrochem. Soc.* **143**, 3652 (1996).
106. P. T. Chen, Y. Sun, E. Kim, P. C. McIntyre, W. Tsai, M. Garner, P. Pianetta, Y. Nishi, C. O. Chui, *J. Appl. Phys.* **103**, 034106 (2008).
107. F. S. Aguirre-Tostado, M. Milojevic, K. J. Choi, H. C. Kim, C. L. Hinkle, E. M. Vogel, J. Kim, T. Yang, Y. Xuan, P. D. Ye, R. M. Wallace, *Appl. Phys. Lett.* **93**, 061907 (2008).
108. N. J. Kawai, T. Nakagawa, T. Kojima, K. Ohta, M. Kawashima, *Electron. Lett.* **20**, 47 (1984).
109. R. P. H. Chang, S. Darack, *Appl. Phys. Lett.* **38**, 898 (1981).
110. S. Sugata, A. Takamori, N. Takado, K. Asakawa, E. Miyauchi, H. Hasimoto, *J. Vac. Sci. Technol. B* **5**, 1087 (1988).
111. T. Suguya, M. Kawabe, *Japan. J. Appl. Phys.* **30**(3A), L402 (1991).
112. R. W. Grant, J. R. Waldrop, *J. Vac. Sci. Technol. B* **5**, 1015 (1987).

113. G. G. Fountain, S. V. Hattangady, D. J. Vitkavage, R. A. Rudder, R. J. Markunas, *Electron. Lett.* **24**, 1134 (1988).
114. J. L. Freeouf, J. A. Silberman, S. L. Wright, S. Tawari, J. Batey, *J. Vac. Sci. Technol. B* **7**, 854 (1989).
115. S. Tawari, S. L. Wright, J. Batey, *IEEE Electron Device Lett.* **9**, 488 (1988).
116. A. Callegari, D. K. Sadana, D. A. Buchanan, A. Paccagnella, E. D. Marshall, M. A. Tischler, M. Norcoft, *Appl. Phys. Lett.* **58**, 2540 (1991).
117. D. G. Park, M. Tao, J. Reed, K. Suzue, A. E. Botchkarev, Z. Fan, G. B. Gao, S. J. Chey, J. Van Nostrand, D. G. Cahill, H. Morkoç, *J. Crys. Growth* **150**, 1275 (1995).
118. J. Ivanco, T. Kubota, H. Kobayashi, *J. Appl. Phys*. **97**, 073712 (2005).
119. O. E. Tereshchenko, D. Paget, P. Chiaradia, J. E. Bonnet, F. Wiame, A. Taleb-Ibrahimi, *Appl. Phys. Lett.* **82**, 4280 (2003).
120. Y. Sun, P. Pianetta, P.-T. Chen, M. Kobayashi, Y. Nishi, N. Goel, M. Garner, W. Tsai, *Appl. Phys. Lett.* **93**, 194103 (2008).
121. W. M. Lau, R. N. S. Sodhi, S. Jin, S. Ingrey, N. Puetz, A. Springthorpe, *J. Appl. Phys.* **67**, 768 (1990).
122. J. H. Thomas, III, G. Koganowicz, J. W. Robinson, *J. Electrochem. Soc.* **135**, 1201 (1988).
123. S. J. Pearton, F. Ren, C. R. Abernathy, W. S. Hobson, T. R. Fullowan, R. Esagui, J. R. Lothian, *Appl. Phys. Lett.* **61**, 586 (1992).
124. J. M. Orton, Y. Cordier, J. C. Garcia, D. Adam, C. Grattepain, *Jpn. J. Appl. Phys.* **35**, 5964 (1996).
125. E. O'Connor, R. D. Long, K. Cherkaoui, K. K. Thomas, F. Chalvet, I. M. Povey, M. E. Pemble, P. K. Hurley, B. Brennan, G. Hughes, S. B. Newcomb, *Appl. Phys. Lett.* **92**, 022902 (2008).
126. B. Banse, J. Creighton, *Appl. Phys. Lett.* **60**, 856 (1992).
127. C. Sasaoka, Y. Kato, A. Usui, *Appl. Phys. Lett.* **62**, 2338 (1993).
128. K. Evans, R. Kaspi, J. Ehret, M. Skowronski, C. Jones, *J. Vac. Sci. Tech. B* **13**, 1820 (1995).
129. K. Kukli, J. Aarik, M. Ritala, T. Uustare, T. Sajavaara, J. Lu, J. Sundqvist, A. Aidla, L. Pung, A. Harsta, M. Leskela, *J. Appl. Phys.* **96**(9), 5298 (2004).
130. R. Puurunen, M. Lindblad, A. Root, A. Krause, *Phys. Chem. Chem. Phys*. **3**, 1093 (2001).
131. K. Kukli, M. Ritala, T. Sajavaara, J. Keinonen, M. Leskela, *Chem. Vap. Dep.* **8**, 199 (2002).
132. P. D. Ye, G. D. Wilk, B. Yang, J. Kwo, H.-J. L. Gossmann, M. Hong, K. K. Ng, J. Bude, *Appl. Phys. Lett.* **84**, 434 (2004).
133. D. Shahrjerdi, T. Rotter, G. Balakrishnan, D. Huffaker, E. Tutuc, S. K. Banerjee, *IEEE Electron Device Lett.* **29**, 575 (2008).
134. N. Goel, P. Majhi, C. O. Chui, W. Tsai, D. Choi J. S. Harris, *Appl. Phys. Lett.* **89**, 163517 (2006).
135. C. H. Chang, Y. K. Chiou, Y. C. Chang, K. Y. Lee, T. D. Lin, T. B. Wu, M. Hong, J. Kwo, *Appl. Phys. Lett.* **89**, 242911 (2006).
136. M. Cho, D. Jeong, J. Park, H. Park, S. Lee, T. Park, C. Hwang, G. Jang, J. Jeong, *Appl. Phys. Lett.* **85**, 5953 (2004).
137. G. K. Dalapati, Y. Tong, W. Y. Loh, H. K. Mun, B. J. Cho, *IEEE Trans. Electron Device* **54**, 1831 (2007).
138. T. Yang, Y. Xuan, D. Zemlyanov, T. Shen, Y. W. Wu, J. M. Woodall, P. D. Ye, F. S. Aguirre-Tostado, M. Milojevic, S. McDonnell, R. M. Wallace, *Appl. Phys. Lett.* **91**, 142122 (2007).
139. K. Kim, D. Farmer, J.-S. M. Lehn, P. V. Rao, R. Gordon, *Appl. Phys. Lett.* **89**, 133512 (2006).
140. W. He, D. S. H. Chan, S. Kim, Y.-S. Kim, S.-T. Kim, B. J. Choc, *J. Electrochem. Soc.* **155**(10), G189 (2008).
141. F. S. Aguirre-Tostado, M. Milojevic, B. Lee, J. Kim, R. M. Wallace, *Appl. Phys. Lett.* **93**, 172907 (2008).
142. P. D. Ye, G. D. Wilk, B. Yang, J. Kwo, S. N. G. Chu, S. Nakahara, H.-J. L. Gossmann, J. P. Mannaerts, M. Hong, K. K. Ng, J. Bude, *Appl. Phys. Lett.*, **83**, 180 (2003).

143. P. D. Ye, G. D. Wilk, J. Kwo, B. Yang, H.-J. L. Gossmann, M. Frei, S. N. G. Chu, J. P. Mannaerts, M. Sergent, M. Hong, K. K. Ng, J. Bude, *IEEE Electron Device Lett.* **24**, 209 (2003).
144. M. W. Hong, J. R. Kwo, P. J. Tsai, Y. C. Chang, M. L. Huang, C. P. Chen, T. D. Lin, *Jpn. J. Appl. Phys., Part 1* **46**, 3167 (2007); and refs. therein.
145. M. M. Frank, G. D. Wilk, D. Starodub, T. Gustafsson, E. Garfunkel, Y. J. Chabal, J. Grazul, D. A. Muller, *Appl. Phys. Lett.* 86, 152904 (2005).
146. D. Shahrjerdi, E. Tutuc, S.K. Banerjee, *Appl. Phys. Lett.* **91**, 063501 (2007).
147. C. Y. Kim, S. W. Cho, M.-H. Cho, K. B. Chung, C.-H. An, H. Kim, H. J. Lee, D.-H. Ko, *Appl. Phys. Lett.* **93**, 192902 (2008).
148. J. C. Hackley, J. D. Demaree, T. Gougousi, *Appl. Phys. Lett.* **92**, 162902 (2008).
149. T. Sawada H. Hasegawa, *Thin Solid Films* **56**, 183 (1979).
150. H. Hasegawa T. Sawada, *IEEE Trans. Electron Device* **27**, 1055 (1980).
151. D. Shahrjerdi, D. I. Garcia-Gutierrez, T. Akyol, S. R. Bank, E. Tutuc, J. C. Lee, S. K. Banerjee, *Appl. Phys. Lett.* **91**, 193503 (2007).
152. J. P. de Souza, E. Kiewra, Y. Sun, A. Callegari, D. K. Sadana, G. Shahidi, D. J. Webb, J. Fompeyrine, R. Germann, C. Rossel, C. Marchiori, *Appl. Phys. Lett.* **92**, 153508 (2008).
153. M. D. Pashley, K. W. Haberern, R. M. Feenstra, P. D. Kirchner, *Phys. Rev. B* **48**, 4612 (1993).
154. D. Yan, E. Look, X. Yin, F. H. Pollak, J. M. Woodall, *Appl. Phys. Lett.* **65**, 186 (1994).
155. H.-L. Lu, L. Sun, S.-J. Ding, M. Xu, D. Wei Zhang, L.-K. Wang, *Appl. Phys. Lett.* **89**, 152910 (2006).
156. R. L. Puurunena, *J. Appl. Phys.* **97**, 121301 (2005).
157. M. J. Hale, J. Z. Sexton, D. L. Winn, A. C. Kummel, M. Erbudak, M. Passlack, *J. Chem. Phys*. **120**, 5745 (2004), and refs. therein.
158. M. L. Huang, Y. C. Chang, C. H. Chang, T. D. Lin, J. Kwo, T. B. Wu, M. Hong, *Appl. Phys. Lett.* **89**, 012903 (2006).
159. Y. Xuan, P. D. Ye, H. C. Lin, G. D. Wilk, *Appl. Phys. Lett.* **89**, 132103 (2006).
160. Y. Xuan, P. D. Ye, H. C. Lin, G. D. Wilk, *Appl. Phys. Lett.* **88**, 263518 (2006).
161. B. Shin, D. Choi, J. S. Harris, P. C. McIntyre, *Appl. Phys. Lett.* **93**, 052911 (2008).
162. Y. Xuan, Y. Q. Wu, H. C. Lin, T. Shen, P. D. Ye, *IEEE Electron Device Lett.* **28**, 935 (2007).
163. T. D. Lin, H. C. Chiu, P. Chang, L. T. Tung, C. P. Chen, M. Hong, J. Kwo, W. Tsai, Y. C. Wang, *Appl. Phys. Lett.* **93**, 033516 (2008).
164. Y. Xuan, P. D. Ye, T. Shen, *Appl. Phys. Lett.* **91**, 232107 (2007).
165. N. Li, E. S. Harmon, J. Hyland, D. B. Salzman, T. P. Ma, Y. Xuan, P. D. Ye, *Appl. Phys. Lett.* **92**, 143507 (2008).
166. Y. C. Chang, M. L. Huang, K. Y. Lee, Y. J. Lee, T. D. Lin, M. Hong, J. Kwo, T. S. Lay, C. C. Liao, K. Y. Cheng, *Appl. Phys. Lett.* **92**, 072901 (2008).
167. K. Y. Lee, Y. J. Lee, P. Chang, M. L. Huang, Y. C. Chang, M. Hong, J. Kwo, *Appl. Phys. Lett.* **92**, 252908 (2008).
168. S. Koveshnikov, N. Goel, P. Majhi, H. Wen, M. B. Santos, S. Oktyabrsky, V. Tokranov, R. Kambhampati, R. Moore, F. Zhu, J. Lee, W. Tsai, *Appl. Phys. Lett.* **92**, 222904 (2008).
169. N. Goel, D. Heh, S. Koveshnikov, I. Ok, S. Oktyabrsky, V. Tokranov, R. Kambhampati, M. Yakimov, Y. Sun, P. Pianetta, C. K. Gaspe, M. B. Santos, J. Lee, P. Majhi, W. Tsai, *IEEE Int. Electron Device Meet. Tech. Dig.* 363 (2008).
170. Y. Xuan, Y. Q. Wu, T. Shen, T. Yang, P. D. Ye, *IEEE Int. Electron Device Meet. Tech. Dig.* 637 (2007).
171. H. J. Oh, J. Q. Lin, S. J. Lee, G. K. Dalapati, A. Sridhara, D. Z. Chi, S. J. Chua, G. Q. Lo, D. L. Kwong, *Appl. Phys. Lett.* **93**, 062107 (2008).
172. H. Li, D. V. Shenai, R. Pugh, J. Kim, *Mater. Res. Soc. Symp. Proc.* **1036** (2008).
173. D. Choia J. S. Harris, M. Warusawithana, D. G. Schlom, *Appl. Phys. Lett.* **90**, 243505 (2007).
174. S. Koveshnikov, C. Adamo, V. Tokranov, M. Yakimov, R. Kambhampati, M. Warusawithana, D. G. Schlom, W. Tsai, S. Oktyabrsky, *Appl. Phys. Lett.* **93**, 012903 (2008).

175. R. M. Wallace, G. D. Wilk, *MRS Bull.* **27**(3), 192 (2002); and associated articles in that issue.
176. M. L. Huang, Y. C. Chang, C. H. Chang, Y. J. Lee, P. Chang, J. Kwo, T. B. Wu, M. Hong, *Appl. Phys. Lett.* **87**, 252104 (2005).
177. C. H. Chang, Y. K. Chiou, Y. C. Chang, K. Y. Lee, T. D. Lin, T.B. Wu, M. Hong, J. Kwo, *Appl. Phys. Lett.* **89**, 242911 (2006).
178. M. Milojevic, F. S. Aguirre-Tostado, C. L. Hinkle, H. C. Kim, E. M. Vogel, J. Kim, R. M. Wallace, *Appl. Phys. Lett.* **93**, 202902 (2008).
179. C. L. Hinkle, M. Milojevic, E. M. Vogel, R. M. Wallace, *Microelectronic Eng.* **86**, 1544 (2009).
180. B. Brennan, M. Milojevic, H. C. Kim, P. K. Hurley, J. Kim, G. Hughes, R. M. Wallace, *Electrochem. Sol. St. Lett.* **12**, H205 (2009).
181. S. Oktyabrsky, V. Tokranov, M. Yakimov, R. Moorea, S. Koveshnikov, W. Tsai, F. Zhuc, J. C. Lee, *Mater. Sci. Eng. B* **135**, 272 (2006).
182. R. J. W. Hill, D. A. J. Moran, X. Li, H. Zhou, D. Macintyre, S. Thoms, A. Asenov, P. Zurcher, K. Rajagopalan, J. Abrokwah, R. Droopad, M. Passlack, I. G. Thayne, *IEEE Electron Device Lett.* **28**(12), 1080 (2007).
183. D. H. Kim, J. A. del Alamo, *IEEE Electron Device Lett.* **29**(8), 830 (2008).
184. E. H. Nicollian, J. R. Brews, *MOS (Metal Oxide Semiconductor) Physics and Technology*. Wiley-Interscience, New York (1982).
185. M. Passlack, in *Materials Fundamentals of Gate Dielectrics*, edited by A. A. Demkov and A. Navrotsky (Springer, Dordrecht, 2005), p. 403.
186. M. Passlack, R. Droopad, Z. Yu, Member, N. Medendorp, D. Braddock, Senior Member, X. W. Wang, T. P. Ma, T. Büyüklimanli, *IEEE Electron. Device. Lett.* **29**, 1191 (2008).
187. K. Martens, W. Wang, K. De Keersmaecker, G. Borghs, G. Groeseneken, H. Maes, *Microelectron. Eng.* **84**, 2146 (2007).
188. G. Brammertz, K. Martens, S. Sioncke, A. Delabie, M. Caymax, M. Meuris, M. Heyns, *Appl. Phys. Lett.* **91**, 133510 (2007).
189. G. Brammertz, H.-C. Lin, K. Martens, D. Mercier, S. Sioncke, A. Delabie, W. E. Wang, M. Caymax, M. Meuris, M. Heyns, *Appl. Phys. Lett.* **93**, 183504 (2008).
190. G. Brammertz, H. C. Lin, K. Martens, D. Mercier, C. Merckling, J. Penaud, C. Adelmann, S. Sioncke, W. E. Wang, M. Caymax, M. Meuris, M. Heyns, *J. Electrochem. Soc.* **155**, H945 (2008).
191. G. Brammertz, H. C. Lin, K. Martens, D. Mercier, C. Merckling, J. Penaud, C. Adelmann, S. Sioncke, W. E. Wang, M. Caymax, M. Meuris, M. Heyns, *ECS. Trans.* **16**, 507 (2008).
192. H. Hasegawa, T. Sawada, *Surf. Sci.* **98**, 597 (1980).
193. H. Hasegawa, L. He, H. Ohno, T. Sawada, T. Haga, Y. Abe, H. Takahashi, *J. Vac. Sci. Technol. B* **5**, 1097 (1987).
194. E. M. Vogel, V. Misra, in *Handbook of Silicon Semiconductor Metrology*, edited by A. C. Diebold, p. 59, Marcel Dekker, Austin (2001).
195. P. D. Ye, G. D. Wilk, B. Yang, J. Kwo, H. J. L. Gossmann, M. Frei, J. P. Mannaerts, M. Sergent, M. Hong, K. K. Ng, J. Bude, *J. Elec. Matl.* **33**, 912 (2004).
196. H. S. Kim, I. Ok, M. Zhang, T. Lee, F. Zhu, L. Yu, J. C. Lee, *Appl. Phys. Lett.* **89**, 222903 (2006).
197. E. M. Vogel, C. A. Richter, B. G. Rennex, *Sol. St. Elecs.* **47**, 1589 (2003).
198. H. Preier, *Appl. Phys. Lett.* **10**, 361 (1967).
199. H. Hasegawa, M. Akazawa, H. Ishii, K. Matsuzaki, *J. Vac. Sci. Technol. B* **7**, 870 (1989).
200. J. L. Freeouf, D. A. Buchanan, S. L. Wright, T. N. Jackson, B. Robinson, *Appl. Phys. Lett.* **57**, 1919 (1990).
201. C. L. Hinkle, M. Milojevic, E. M. Vogel, R. M. Wallace, *Microelectron Eng.* **86**, 1544 (2009).
202. Z. Yu, C. D. Overgaard, R. Droopad, M. Passlack, J. K. Abrokwah, *Appl. Phys. Lett.* **82**, 2978 (2003).
203. C. L. Hinkle, A. M. Sonnet, R. A. Chapman, E. M. Vogel, *IEEE Electron Device Lett.* **30**, 316–319 (2009).

204. É. O'Connor, S. Monaghan, R. D. Long, A. O' Mahony, I. M. Povey, K. Cherkaoui, M. E. Pemble, G. Brammertz, M. Heyns, S. B. Newcomb, V. V. Afanas'ev, P. K. Hurley, *Appl. Phys. Lett.* **94**, 102902 (2009).
205. M. E. Levinshtein, S. L. Rumyantsev, in *Handbook Series in Semiconductor Parameters, Vol 1*, World Scientific, Singapore (1996), p. 77.
206. Y. A. Goldberg, N. M. Shmidt, in *Handbook Series in Semiconductor Parameters, Vol 2*, World Scientific, Singapore (1999), p. 62.

Chapter 7
Atomic-Layer Deposited High-k/III-V Metal-Oxide-Semiconductor Devices and Correlated Empirical Model

Peide D. Ye, Yi Xuan, Yanqing Wu and Min Xu

Abstract Si CMOS scaling is reaching its physical limit at the 15 nm technology node and beyond. III-V compound semiconductor is one of the leading candidates to replace main-stream Si as n-channel material due its much higher electron mobility. Lacking a suitable gate insulator, practical III-V metal-oxide-semiconductor field-effect transistors (MOSFETs) remain all but a dream for more than four decades. The physics and chemistry of III-V compound semiconductor surfaces or interfaces are problems so complex that even after enormous research efforts understanding is still limited. Most of the research is focused on surface pretreatments, oxide formation and dielectric materials. Less attention is given to the III-V substrate itself. In this chapter, the history and present status of III-V MOSFET research is briefly reviewed. An empirical model for high-k/III-V interfaces is proposed based on the experimental works we performed on III-V MOSFETs using ex-situ atomic-layer-deposited high-k dielectrics and also reported works in the literature using in-situ molecular beam expitaxy grown $Ga_2O_3(Gd_2O_3)$ as gate dielectric. The results show that physics related to III-V substrates is as important as surface chemistry and gate oxide properties for realizing high-performance III-V MOSFETs. The central concept of this empirical model is that the band alignment between trap neutral level (E_0) and conduction band minimum (CBM) or valence band maximum (VBM) and the magnitude of interface trap density governs the device performance of inversion-mode III-V MOSFETs.

7.1 Introduction

GaAs metal-oxide-semiconductor field-effect transistors (MOSFETs) have been a subject of study for more than four decades. The renewed research thrust is for advanced ultra-large-scale-integration (ULSI) digital applications or complementary

P. D. Ye (✉)
School of Electrical and Computer Engineering and Birck Nanotechnology Center, Purdue University, West Lafayette, Indiana 47907, USA
e-mail: yep@purdue.edu

S. Oktyabrsky, P. D. Ye (eds.), *Fundamentals of III-V Semiconductor MOSFETs*,
DOI 10.1007/978-1-4419-1547-4_7, © Springer Science+Business Media LLC 2010

metal-oxide-semiconductor (CMOS) technology beyond the 22-nm node by using III-V compound semiconductors as conduction channels to replace traditional Si or strained Si, integrating novel high-k dielectrics with these high mobility materials, and heterogeneously incorporating them on Si or silicon-on-insulator (SOI). The lack of high-quality, thermodynamically stable gate dielectric insulators on GaAs or III-V that can match device criteria similar to SiO_2 on Si, remains the main obstacle to realizing a III-V MOSFET technology with commercial value. The understanding of the interface physics and chemistry of the III-V's is still quite limited, though enormous research efforts have been invested in this field. The literature on this subject is spread over the last 40 years and appears in many journals and conference proceedings. Understanding these literatures requires considerable efforts by the seasoned researchers, and even more for those who are new in the field. The first part of this chapter provides the readers with a brief overview on history and current status of III-V MOSFET research. The second part of this chapter proposes an empirical model based on trap neutral level (E_0) concept at oxide/semiconductor interface to explain all experimental works on III-V MOSFETs using ex-situ atomic-layer-deposited (ALD) high-k dielectrics and also in-situ molecular beam expitaxy (MBE) grown $Ga_2O_3(Gd_2O_3)$ gate dielectric. The third part of the chapter provides more detailed experimental results on different ALD high-k dielectrics and different III-V substrates to verify the validation of the empirical mode. The investigated III-V substrates include $In_xGa_{1-x}As$ (x = 0, 0.2, 0.53, 0.65, 0.7, 0.75, 1), InP, InSb, GaN, and GaSb. The studied ALD high-k dielectrics include Al_2O_3, HfO_2, and their HfAlO nanolaminates.

7.2 History and Current Status

The advantage of GaAs MOSFET over its Si counterpart has long been recognized because the bulk electron mobility in GaAs (8800 cm^2/Vs) is five or more times higher than that in Si. The electron mobility advantage for InSb is astonishing as high as 77000 cm^2/Vs. The GaAs MOSFET research has its own phenomenal cycles coincidently with the well-know ten-year semiconductor industrial business cycles. The first GaAs MOSFET work was reported by Becke and White by the Radio Corporation of America in 1965 [1, 2]. Although deposited SiO_2 is used as the gate dielectric with large amount of interface traps, the devices are operated successfully at several-hundred-megahertz frequency range showing the feasibility of this approach. It was quickly realized that SiO_2 is not the right gate dielectric for GaAs, which ignited an enormous research effort in the following decades searching for a low-defect, thermo-dynamically stable gate dielectric for GaAs. The efforts could be divided into two scenarios: (a) deposited oxide, and (b) native oxide.

A variety of dielectrics and techniques have been investigated. The well studied dielectrics on GaAs include pyrolytically deposited silicon dioxide [1], silicon nitride [2], silicon oxynitride [3], and aluminum oxide [4]. All these processes require relatively high temperatures ranging from ~350 to ~600 °C. The chemical reaction between GaAs and oxygen in the gas ambient is expected to form Ga-oxide, As-oxide, remaining elemental As and sometimes even a large amount of vacancies due to the high volatility of As. A low-temperature process is believed to be essential for obtaining a high-quality interface [5]. Plasma-enhanced deposition with the process temperature lower than 200 °C was attempted [6], though it could potentially induce more defects or fixed charges on GaAs surface or interface due to the plasma damage effects.

Except for the conventional physical vapor deposition (PVD) and chemical vapor deposition (CVD), the molecular beam epitaxy (MBE) approach was also used to form insulating films on GaAs after it started to be widely used to grow III-V compound semiconductor heterostructures at the end of 1970s. For example, highly resistive AlGaAs [7] or low-temperature grown GaAs [8] was used as insulating layer for GaAs channel. But the relatively high gate leakage current limits the wide application of these approaches. To improve the barrier heights at the heterojunction interface, thermal oxidation of AlAs epi-layers grown on GaAs using MBE was also investigated [9]. The difficulty of this approach is to control the oxidation process to terminate at the AlAs-GaAs interface, though the large difference in the rate of oxide formation of AlAs and GaAs exists. The research in this direction is still active currently [10].

Motivated by the success of thermal oxidized SiO_2 on Si, using native oxides on GaAs as gate dielectrics was also intensively studied at the very beginning. Some representative approaches include thermal oxidation [11], wet chemical anodization [12–13], dc and RF plasma oxidation [14–18], laser-assisted oxidation [19], vacuum ultraviolet photochemical oxidation [20] and photo-wash oxidation [21]. Any of these is not optimistic as a feasible approach leading to a commercial GaAs MOSFET technology. One general observation is that the native oxide is not stable, mostly leaky with low dielectric breakdown strength, and cannot be forward biased beyond a few volts. All these studies are essential to enrich our understanding of chemistry and physics properties of III-V interfaces. Although there are some controversies within a large amount of experimental data, a consensus emerges that a significant amount of As_2O_3, As_2O_5 and elemental As present in native oxides pin the Fermi-level of GaAs and are not favorable for an ideal gate dielectric. The book "Physics and Chemistry of III-V Compound Semiconductor Interfaces" edited by C.W. Wilmsen and published in 1985 summaries most of experimental work in 1970s and the beginning of 1980s [22]. Recent work reveals that controlling the oxidation state of Ga at the interface (Ga_2O vs. Ga_2O_3), formed from native oxide or ALD processes, could also play a critical role on unpinning of Fermi level on GaAs [23, 24].

A variety of models on semiconductor interfaces have also been developed at the same period of time beyond the Mott-Schottky model in 1938, J. Bardeen's surface states model in 1947 and P.W. Anderson's electron affinity model in 1962. Some representative works include Cowley and Sze's interfacial layer model in 1965 [25], Heine's metal induced gap state (MIGS) model in 1965 [26], Tejedor and Flores' model on line-up at charge neutrality level (CNL) in 1978 [27], Spicer's unified defect model (UDM) in 1980 [28], Hasegawa and Ohno's disorder-induced gap state (DIGS) model in 1986 [29], and Tersoff's dielectric midgap energy model in 1984 and 1985 [30, 31]. All these works have strong influences on semiconductor device research including III-V MOS. For examples, W.E. Spicer's extensive experiments, published in 1980, concludes that the Schottky-barrier formation on III-V semiconductors is due to defects formed near the interface by deposition of metals or any chemisorption of oxygen [32]. This model also applies to formation of states at the III-V oxide interface. Fermi-level position in GaAs is pinned at the midgap no matter whether it is metal-semiconductor or oxide-semiconductor interface. This UDM doesn't persist with the modern oxide deposition techniques such as MBE and ALD with selective oxides. Very recently, J. Robertson presented a generalized model of the density of interface states at III-V oxide interfaces [33]. In this chapter, Ye (one of the authors) proposes an empirical model, refined from Refs. [34–36], in the second portion of this chapter. This empirical model is supported by a large amount of experimental work done in Ye's group at Purdue and also other groups worldwide. Some of the original ideas of this model could be tracked back to Hasegawa's DIGS' model proposed two decades ago [29].

Fermi-level pinning in III-V semiconductors stymies the enthusiasm among III-V researchers to compete with Si in large-scale integrated circuit front. But significant progress was made on high-performance microwave GaAs metal-semiconductor field-effect transistors (MESFETs) pioneered by Hooper and Lehrer [37] using GaAs Schottky barrier directly. The development of MBE and metal-organic CVD technologies in the 1970s made heterojunctions, quantum-wells, and superlattices practical. At Bell Labs, Dingle et al. first demonstrated the enhancement of mobility in the AlGaAs/GaAs modulation-doped superlattice in 1978 [38]. Stormer et al. subsequently reported similar effect using a single AlGaAs/GaAs heterojunction [39] which leads to the discovery of fractional quantum Hall effect in 1981. In 1980, Mimura et al. applied a similar concept and invented high electron mobility transistors (HEMT) [40]. Similar work was also reported later in the same year by Delagebeaudeuf et al [41]. GaAs HEMT eventually becomes a commercialized technology and finds its wide applications in communication, military and aerospace industry today. GaAs MOSFET research was not continued at a large scale after the successful introduction of GaAs HEMT.

The research work on search for suitable dielectrics or passivation layers on GaAs continues. In 1987, researchers at Bell Labs discovered that a class of sulfides [Li_2S, $(NH_4)_2S$, $Na_2S{\cdot}9H_2O$, etc.] are able to passivate GaAs surface and provide excellent electronic properties to GaAs surfaces [42, 43]. The work generated new interests in GaAs MOSFETs using sulfur passivation before dielectric deposition. Hundreds of papers were published related with GaAs sulfur passivation. But sulfur passivation didn't become a widespread technology since sulfur is not a stable material with very low thermal budget. Nevertheless, it is still very interesting scientifically. Sulfur, as a VI element next to oxygen, has the same electron number in its outer shell as oxygen, but less chemically reactive. It could passivate pristine GaAs surface with a few monolayers of sulfur in certain conditions and prevent GaAs from oxidation to form As-oxides. Sulfur passivation is still used in today's research. However, sulfur layers could serve as protective layers for III-V surface oxidization in ALD process. Other works on effective passivation of GaAs using hydrogen or nitrogen plasma were also reported by IBM group [44]. Another interesting approach is to use a very thin amorphous or crystalline Si layer as an interfacial control layer (ICL) between GaAs and SiO_2 or Si_3N_4 [45–48]. This approach has been intensively studied in 90s [49–51] and recently also adopted to integrate with high-k HfO_2 gate dielectrics on GaAs [52–55]. Promising results are obtained on GaAs MOSFETs using SiH_4 passivation [56, 57], which is believed to form 1 to 2 nm interficial SiO_2 layer also except oxide reduction by hydrogen environment. The details of this work are reviewed in Chaps. 8 and 11 in this book.

In 1995, M. Passlack and M. Hong at Bell Labs reported that in-situ deposition of $Ga_2O_3(Gd_2O_3)$ dielectric film on GaAs surface by electron beam evaporation from single-crystal $Ga_5Gd_3O_{12}$ produced MOS structures on GaAs with a low D_{it} [58–60]. Since the experiment is realized in an ultra-high-vacuum (UHV) multi-chamber MBE system, it is referred to as MBE grown $Ga_2O_3(Gd_2O_3)$ most of the time. A series of device work, mainly led by M. Hong and F. Ren, were carried out at Bell Labs after this breakthrough in material science. It includes GaAs depletion-mode (D-mode) and enhancement (E-mode) MOSFETs [61, 62], InGaAs enhancement-mode MOSFETs [63], GaAs complementary MOSFETs [64] and GaAs power MOSFETs [65]. This $Ga_2O_3(Gd_2O_3)$ dielectric device work continues at Agere Systems [66], a spin-off from Bell Labs and Lucent Technologies, and National Tsinghua University (NTHU) in Taiwan. Inversion-mode InGaAs NMOSFETs with outstanding on-state performance are reported by Hong's group at NTHU in 2008 [67]. In 2003, Passlack et al. at Motorola/Freescale started to report a series of works by modifying the previous $Ga_2O_3(Gd_2O_3)$ dielectric process and using Ga_2O_3 template in $Gd_xGa_{0.4-x}O_{0.6}$/Ga_2O_3 dielectric stacks on GaAs [68]. An implant-free enhancement-mode device concept was introduced and good device performance was demonstrated in 2006 to eliminate the difficulty to realize the inversion-operation due to the relative low thermal budget of III-V and gate dielectric stacks [69]. The work continues as a collaborative effort at the University of Glasgow [70]. UHV scanning tunneling microscopy study reveals that the unpinning of GaAs Fermi level results from Ga_2O restoring the surface arsenic and gallium atoms to near-bulk charge [71].

ALD is an ex-situ, robust, and manufacturable process, which attracts wider interests in academia and industry. At the end of 2001, Ye and Wilk at Bell Labs/Agere Systems, started to work on ALD high-k Al_2O_3 and HfO_2 on GaAs and other III-V materials. A series of D-mode MOSFETs on GaAs, InGaAs and GaN using ALD Al_2O_3 as gate dielectrics were demonstrated [72–76]. The work on ALD integration with high-mobility channel materials continues and enhances in Ye's group at Purdue University. Detailed interface studies were carried out to demonstrate the unpinning of Fermi-level in InGaAs and GaAs (111)A surface [24, 35, 77–82] including the fundamental understanding of ALD chemical process on GaAs reported by other groups [83–86]. The "self-cleaning" effect on III-V by ALD is believed important to clean up As-oxide and provide high-quality high-k/III-V interfaces. High-performance inversion-mode III-V MOSFETs were demonstrated with unprecedented drain current as high as 1.1 A/mm and transconductance as high as 1.3 S/mm [87–93]. This is the first surface channel device in III-V with drain current beyond 500 mA/mm by reporting time and with transconductance beyond 1 S/mm among any reported III-V MOSFETs with oxide as a gate dielectric. Even further, InGaAs FinFETs with 40 nm fin width and 100 nm gate-length are demonstrated experimentally using ALD Al_2O_3 [94]. In-rich InGaAs is identified as the potential channel material for future 15 nm technology node or beyond with higher effective mobility and manageable bandgap for low drain voltage. The fundamental understanding on this successful demonstration is explained by the proposed empirical model. Device process and results are briefly described in the third portion of this chapter. Currently, many research groups including Intel, IBM, IMEC, SEMATECH, UCSB III-V CMOS Center, Stanford, MIT, UT Austin, UT Dallas, NTHU, and many others are working on inversion-mode ALD high-k/InGaAs MOSFETs [95–100]. Many of the results are reviewed elsewhere in this book.

The new cycle of interest in III-V MOSFETs was initiated by Intel in 2005 for alternative device technologies beyond Si CMOS. III-V is one of its main focuses with the excellent publications on InSb-based quantum well transistors in collaborations with QinetiQ [101–102]. P-channel InSb quantum well transistors, another hassle in III-V field due to low hole mobility, with outstanding performance was also reported in IEDM 2008 [103]. Some fine experiments based on In-rich InGaAs and InAs heterostructures were also carried out at MIT J. del Alamo's group to investigate the ultimate transport properties of III-V for logic applications [104, 105]. We are hoping that the current III-V research in Si CMOS community is at the similar stage as high-k concept was introduced at the end of 1990s. After collective efforts in academia and industry, we are able to make this long-standing GaAs MOSFET dream become a real commercial technology as the successful story of high-k in Si CMOS.

7.3 Empirical Model for III-V MOS Interfaces

ALD is based on the self-limiting chemical reactions to form ultra-thin, uniform, conformal, and pin-hole free films. The ALD or Atomic-layer epitaxy (ALE) concept was invented in 1970s. Strong interest in non-native oxides for Si CMOSFETs began in the mid-1990s. High-k dielectric research, especially the development of ALD high-k dielectrics for Si MOSFETs, has since flourished [106]. In 2007, Intel claimed its successful integration of ALD Hf-based high-k dielectric and metal gate process into its 45 nm node technology as one of the biggest technical leaps in Si CMOS development, after the introduction of poly-silicon gate in the 1960s. The success of ALD high-k dielectrics on Si has created much more research on ALD itself and other applications beyond Si CMOS.

ALD shows its uniqueness in high-k/III-V integration. First, the ALD dielectric process, in particular, trimethylaluminum (TMA)-related Al_2O_3, enables unpinning the Fermi levels on most of III-V semiconductors so far studied. Al_2O_3 is a highly desirable gate insulator from both physical and electrical standpoints. Al_2O_3 has a wide bandgap (~9 eV), high breakdown field (~10 MV/cm), high thermal stability

(>1000 °C), and remains amorphous under typical IC processing conditions. Al_2O_3 can be easily wet-etched, yet robust against interfacial reactions and moisture absorption. The uniqueness of ALD on III-V is so-called "self-cleaning effect", which is described in great details in Chap. 6.

The pre-cleaning or surface preparation before ALD is a very important process step to have a high-quality high-k/III-V interface, in particular, for GaAs. The simplified surface preparation for ALD on III-V developed by Ye et al. is following [79]. Before surface treatment, all wafers underwent standard degrease process using acetone, methanol and iso-propanol. The NH_4OH treatment is carried out by soaking the samples in NH_4OH (29%) solution for 3 min to remove native oxide and rinsing in flowing de-ionized (DI) water followed by gently drying the surface using N_2 blow-gun. The NH_4OH etching step removes most of arsenic and gallium oxides from the surface. The further ALD process results on the disappearance of arsenic oxides and self-cleaning of GaAs surface [73, 83–86]. The interface quality could be further improved sometimes by $(NH_4)_2S$ treatment. The process is to soak the sample in $(NH_4)_2S$ for 10 min in room temperature and dry using N_2 gun. The effectiveness of $(NH_4)_2S$ passivation is mostly related with mono-layers of sulfur preventing further oxidation of III-V in ex-situ process. Most of sulfur is puffed off from III-V surface during the ALD process when the ALD chamber is heated to growth temperature of 300 °C. With these appropriate surface pretreatments, the interface trap density in the range of high 10^{11}-low 10^{12}/cm^2-eV can be realized and the inversion-type enhancement-mode MOSFETs can be demonstrated on GaAs, InGaAs and InP. During the past decades, the research community focused mainly on dielectric materials and III-V surface chemistry study, and paid less attention to device physics of III-V substrates. We emphasize that substrate is also extremely important as the second determinant for realizing a high inversion current as the proposed empirical mode describes below. This simple trap neutral level (TNL)-based empirical model can explain all experimental work on III-V MOSFETs using ex-situ ALD high-k dielectrics and also in-situ MBE-grown $Ga_2O_3(Gd_2O_3)$ gate dielectric.

A difficult problem with III-V MOSFETs results from the unpassivated dangling bonds on III-V free surfaces. The energy locations of the dangling bonds are directly related to the bulk band structures of different III-V semiconductors. But in the tight-binding formalism, the conduction and valence bands are formed as bonding and anti-bonding combinations of the atomic sp^3 hybrids. That is why the dangling-bond energy is typically located in the middle of the bandgap [107]. There are many experimental studies that quantitatively measure the midgap energies of various semiconductors. There are also dozens of theoretical models (such as metal induced gap state model, unified defect model, and disorder-induced gap state model) to quantitatively calculate the electronic energies. The historical evaluation of interface models is briefly summarized in Ref. [108]. The empirical model is rooted from the unified disorder induced gap state model (DIGS) proposed by Hasegawa and Ohno in 1986, which explains the striking correlation between the energy location E_{min} for the minimum interface state density at the insulator-semiconductor interface and the Fermi level pinning position E_{pin} of the metal-semiconductor interface. The central concept

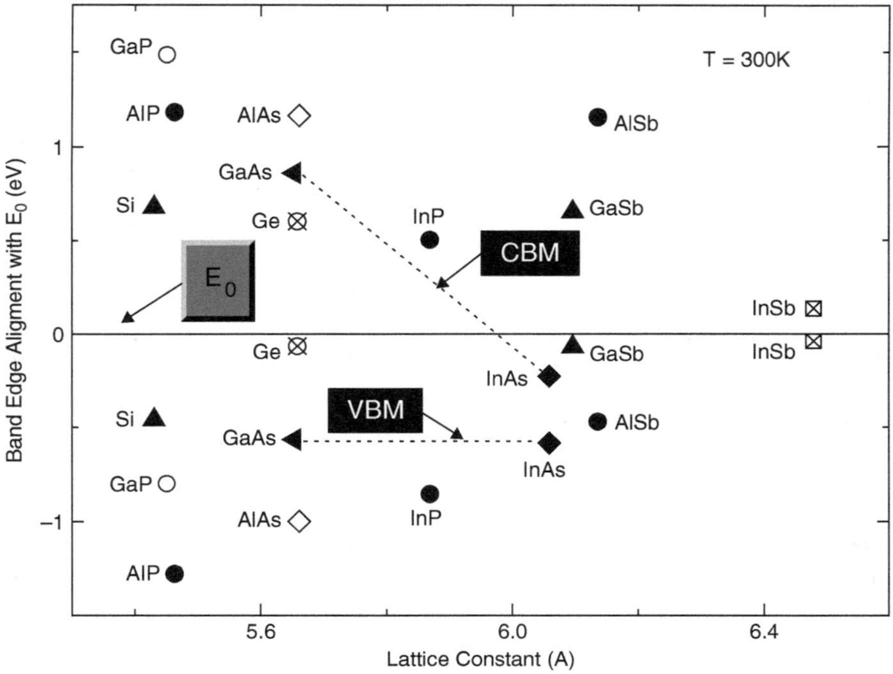

Fig. 7.1 The energy alignment of the trap neutral level E_0 with the band edges of the representative semiconductors at MOS interfaces. The band edge values are obtained from Tiwari and Frank's Applied Physics Letters 60, 630 (1992). The *dashed lines* represent the rough CBM and VBM positions of $In_xGa_{1-x}As$

is that there is an energy level called trap neutral level E_0 at the high-k/GaAs or III-V interface,above which the trap states are of acceptor type or electron traps and below which are of donor type or hole traps. E_0 is at the same or similar energy level as E_{min} and E_{HO} in Ref. [29] and E_{pin} at metal-semiconductor interfaces.

Figure 7.1 is the plot of several semiconductors' band edge alignment with the so-called trap neutral level (E_0) or Fermi level stabilization energy. In Si case, the state-of-the-art SiO_2/Si has low interface trap density of 10^9–10^{10}/cm^2-eV. Although the energy separation between valence band maximum (CBM) and E_0 is ~0.6 eV and E_0 and valence band maximum (VBM) is ~0.5 eV, the density of trap states is low thus the Fermi-level is fully unpinned. Outstanding NMOSFETs and PMOSFETs are demonstrated and CMOS technology is widely applied. In Ge case, significant interface traps exist at various dielectrics/Ge interfaces though good progress was made in the past several years. PMOSFET in Ge is much easier to be realized compared to NMOSFET because E_0 is much closer to the valence band maximum (VBM) and far away from the conduction band minimum (CBM) [109, 110].

In the case of GaAs, E_0 is far away from both CBM and VBM so that both NMOSFET and PMOSFET in GaAs are difficult to realize if significant interface traps exhibit. If conduction band edge points of GaAs and InAs are connected, the CBM of

In-rich InGaAs is very near E_0. That's why inversion-mode InGaAs NMOSFETs could have high performance such as $G_m > 1.0$ S/mm and $I_{DS} > 1.0$ A/mm [90, 93]. The position of VBM of In-rich InGaAs is almost the same as that for GaAs or InAs. InGaAs PMOSFETs are expected to be difficult. In the case of GaSb, whose band alignment is so like Ge, a GaSb PMOSFET must be very easy to realize with good device performance according to the model.

7.4 Experiments on High-k/III-V MOSFETs

7.4.1 High-k/GaAs MOSFETs

GaAs is the most studied III-V semiconductor in particular on (100) surface. A variety of devices are demonstrated and manufacturable on GaAs (100) surface such as GaAs HEMTs, LEDs and lasers. However, it is extremely difficult to realize Fermi level unpinning on this most useful surface with deposited oxide. GaAs inversion-mode or minority carrier devices on (100) surface mostly show minuscule drain current, indicating Fermi-level cannot be moved near or into CBM. We systematically study the electrical properties of inversion-mode NMOSFETs and PMOSFETs on GaAs (111)A, (111)B, (110) and (100) surfaces with ALD Al_2O_3 as gate dielectrics. (111)A is a pure Ga polar surface in contrast to (100) and (110) Ga-As non-polar surfaces, whilst (111)B is a pure As polar surface. The device work confirms that Fermi-level of GaAs (111)A surface is unpinned at the mid-gap with direct ALD Al_2O_3. The results obtained on GaAs (111)A surface are astonishingly different from those on GaAs (100), (110) and (111)B surfaces. The above proposed empirical model based on trap neutral level can explain the experimental observation. The XPS studies, in collaborations with UT Dallas, also confirm that Ga-oxide at Al_2O_3/GaAs (111)A interface is at detection limit in great contrast to what observed at GaAs (111)B, (110) and (100) surfaces [24].

MOSFET fabrication starts with 2-inch semi-insulating GaAs (111)A, (111)B, (110) or (100) substrates. After surface degreasing and ammonia-based native oxide etching, the wafers were transferred via room ambient to an ASM F-120 ALD reactor. A 30 nm thick Al_2O_3 layer was deposited at a substrate temperature of 300 °C as an encapsulation layer. Source and drain regions were selectively implanted with Si for NMOSFETs and Zn for PMOSFETs with the same dose of 5×10^{14} cm^{-2} at 40 keV through the 30 nm thick Al_2O_3 layer. Implantation activation was achieved by rapid thermal anneal (RTA) at 820 °C for 15 s in nitrogen ambient for NMOSFETs and at 750 °C for 15 s for PMOSFETs. An 8-nm Al_2O_3 film was regrown by ALD after removing the encapsulation layer by buffered oxide etch (BOE) solution and soaked in ammonia sulfide for 10 minutes for surface preparation. After 600 °C post-deposition anneal (PDA) in N_2 ambient, the source and drain ohmic contacts were made by an electron beam evaporation of a combination of AuGe/Ni/Au for NMOSFETs or Pt/Ti/Pt/Au for PMOSFETs and a lift-off process, followed by a RTA process at 400 °C for 30 s also in a N_2 ambient. The gate electrode was defined by electron beam evaporation of Ni/Au and a lift-off process. It was found during the process that GaAs (111)A surface is more hydrophilic like In-rich InGaAs surface and GaAs (100), (110) and (111)B surfaces are more hydrophobic. A hydrophilic surface is believed to be favorable for ALD two-dimensional growth and good interface properties.

A well-behaved I-V characteristic of a 4 μm-gate-length inversion-mode GaAs NMOSFET on (111)A surface is demonstrated in Fig. 7.2(c) with maximum drain current of 30 μA/μm, which is *a factor of 85000 or 25000 larger* than that obtained on (100) or (110) as shown in Fig. 7.2(a) and (b). Similar low inversion currents on GaAs (100) or (110) and even zero-current on GaAs (111)B lead to the conclusion of Fermi-level pinning in past reports. The GaAs (111)A surface is astonishingly different and Fermi-level is unpinned resulting in a large drain current. This astonishing difference can be partially explained by the empirical model as following.

By photoemission and other experiments, Spicer et al. discovered that E_{pin} in GaAs is 0.75 and 0.5 eV above the valence band maximum (VBM)[111]. The first energy given is associated with a missing anion (As) and the second with a missing cation (Ga). Ignoring the complications of surface reconstructions, GaAs (111)A surface is a Ga-terminated polar surface, which can be regarded as a missing anion (As) surface with $E_0 = 0.75$ eV above VBM [111]. GaAs (100) is a Ga-As terminated surface which might be more related with missing cations with $E_0 = 0.5$ eV above VBM as shown in Fig. 7.3. The drain current strongly depends on the energy separation between E_0 and CBM for NMOSFET or VBM for PMOSFET. With the measured near mid-gap interface trap density D_{it} of 2×10^{12}/cm^2 eV, the smaller the separation is, the less traps are filled in to prevent further Fermi level movement for strong inversion, the more inversion charge and drain current can be achieved. This model explains why NMOSFET on GaAs (111)A outperforms that on (100), and PMOSFET on GaAs (100) outperforms that on (111)A as reported in Ref. [35] that

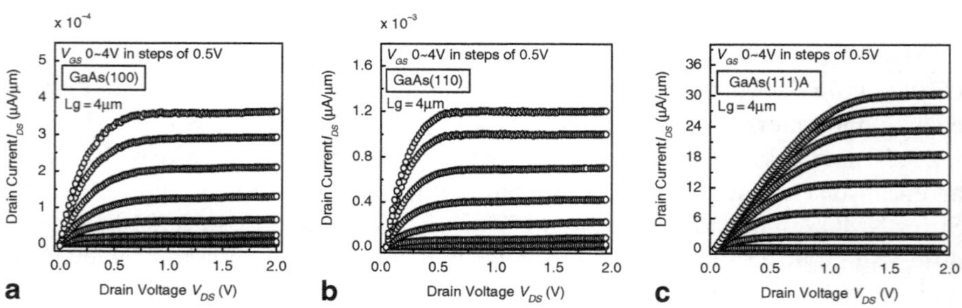

Fig. 7.2 **a** Output characteristic (I_{DS}~V_{DS}) for Al_2O_3/GaAs(100) NMOSFET with 4 μm gate length. The maximum drain current is 3.5×10^{-4} μA/μm. **b** Output characteristic (I_{DS}~V_{DS}) for Al_2O_3/GaAs(110) NMOSFET with 4 μm gate length. The maximum drain current is 1.2×10^{-3} μA/μm. **c** Output characteristic (I_{DS}~V_{DS}) for Al_2O_3/GaAs(111)A NMOSFET with 4 μm gate length. The maximum drain current is 30 μA/μm

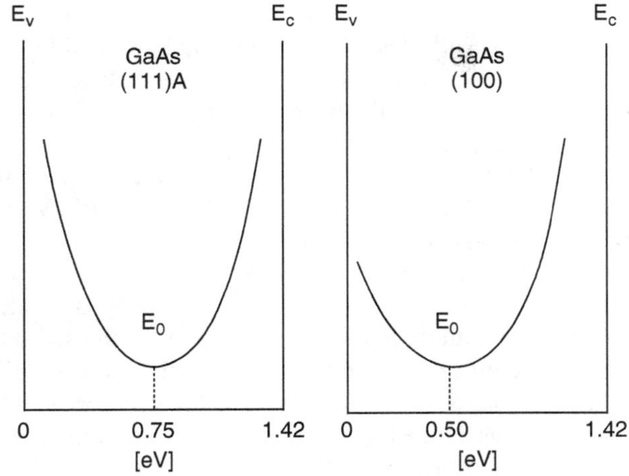

Fig. 7.3 The Empirical model on GaAs. 0.75 eV is associated with a missing anion due to Ga (111)A surface; 0.5 eV is associated with a missing cation, which is the case most likely for (100) or (110). The minimum D_{it} and U-shape curvature depends on processing conditions, while the location of E_0 remains constant for each semiconductor with the same crystal facet

shows that maximum drain current of 0.17 and 0.8 mA/mm are obtained on (111)A and (100) surfaces, respectively. Fermi-level pinning at the mid-gap of GaAs as proposed by the unified defect model (UDM) can be overcome by the appropriate surface preparation and the suitable dielectric deposition technique.

7.4.2 High-k/In_xGa_{1-x}As MOSFETs

This empirical model is even valid in explaining the experimental data on high-k/In_xGa_{1-x}As MOSFETs as shown in Fig. 7.4. The device structures and process are following. The channel is 15–20 nm thick $1 \times 10^{17}/cm^3$ doped p-type $In_{0.53}Ga_{0.47}$As

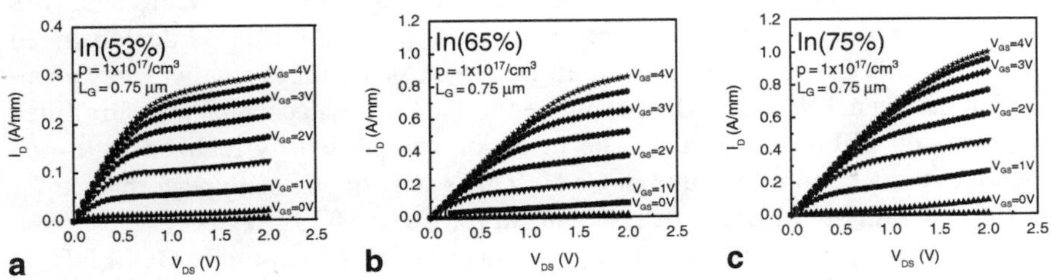

Fig. 7.4 **a** Drain current (I_D) versus drain bias (V_{DS}) as a function of gate bias (V_{GS}) for Al_2O_3(8 nm)/$In_{0.53}Ga_{0.47}$As NMOSFETs with 0.75-μm gate length. The maximum drain current is 0.3 A/mm. **b** Drain current versus drain bias as a function of gate bias for Al_2O_3 (10 nm)/$In_{0.65}Ga_{0.35}$As NMOSFETs with 0.75-μm gate length. The maximum drain current is 0.86 A/mm. **c** Drain current versus drain bias as a function of gate bias for Al_2O_3(10 nm)/$In_{0.75}Ga_{0.25}$As NMOSFETs with 0.75 μm gate length. The maximum drain current is 1.0 A/mm [91] (Reprinted with permission. Copyright 2008 IEEE)

or $In_{0.65}Ga_{0.35}As$ or $In_{0.75}Ga_{0.25}As$ channel layer, which is MBE epitaxially grown on $In_{0.53}Ga_{0.47}As$/InP substrate. 8–10 nm thick ALD Al_2O_3 is used as gate dielectric and Ni or Al is used as gate electrodes. After surface degreasing and ammonia-based native oxide etching, the wafers were transferred via room ambient to an ASM F-120 ALD reactor. A 30 nm thick Al_2O_3 layer was deposited at a substrate temperature of 300 °C as an encapsulation layer. Source and drain regions were selectively implanted with a Si dose of 1×10^{14} cm^{-2} at 30 keV and 1×10^{14} cm^{-2} at 80 keV through the 30 nm thick Al_2O_3 layer. Implantation activation was achieved by RTA at 700–800 °C for 10 s in a N_2 ambient. An 8–10 nm Al_2O_3 film was then re-grown by ALD after removing the encapsulation layer by BOE etching and ammonia sulfide surface preparation. After 400–600 °C PDA, the source and drain ohmic contacts were made by an electron beam evaporation of a combination of AuGe/Ni/Au and a lift-off process, followed by a RTA at 400 °C for 30 s also in N_2 ambient. The gate electrode was defined by electron beam evaporation of Ni/Au and a lift-off process. The fabricated MOSFETs have a nominal gate length varying from 0.40 to 40 μm and a gate width of 100 μm. From transmission electron microscopy (TEM) images, no visible interfacial layer between Al_2O_3/$In_{0.75}Ga_{0.25}As$ interface and relaxation of $In_{0.75}Ga_{0.25}As$ on $In_{0.53}Ga_{0.47}As$ are observed. The native oxide of III-V material has been effectively removed by HCl etching, NH_4OH and $(NH_4)_2S$ pretreatment and the ALD "self-cleaning" process.

Well-behaved I-V characteristic of 0.75-μm gate length inversion-type E-mode $In_{0.53}Ga_{0.47}As$, $In_{0.65}Ga_{0.35}As$ and $In_{0.75}Ga_{0.25}As$ NMOSFETs are demonstrated in Fig. 7.4 with the I_{DMAX} of 0.3, 0.86 and 1.0 A/mm, respectively. The gate leakage current (I_G) is less than 10^{-4} A/cm^2 at 4.0 V gate bias (V_G) for all devices. The extrinsic G_m, the intrinsic G_m, and the threshold voltage V_T for $In_{0.75}Ga_{0.25}As$ NMOSFETs are 0.43 S/mm, 0.52 S/mm, and 0.5 V, respectively, with 0.75-μm gate length. The I_{DMAX} and G_m increase with increasing indium content in InGaAs not only due to the increase of mobility and saturation velocity and reduced contact resistance. More importantly, according to Fig. 7.1, TNL level E_0 in $In_{0.75}Ga_{0.25}As$ is much near CBM than $In_{0.53}Ga_{0.47}As$. The inversion charge and inversion current on $In_{0.75}Ga_{0.25}As$ MOSFET is much larger than those on $In_{0.53}Ga_{0.47}As$ MOSFET as demonstrated experimentally in Fig. 7.4. In general, In-rich InGaAs has much smaller energy separation between TNL and CBM, compared to GaAs. Less acceptor traps are filled by inversion in In-rich InGaAs. The inversion charge density realized in In-rich InGaAs is much more than that in GaAs. That's why the on-state performance for In-rich InGaAs MOSFETs is better than that for GaAs MOSFETs.

With more demonstrated on-state performance of inversion-mode MOSFETs on In-rich InGaAs channels, more work are needed to study the fundamental limitation of narrow energy gap of In-rich InGaAs and the off-state performance related with exhibiting interface trap densities. For example, recent works unveil that the interface traps at ALD Al_2O_3/InGaAs interface are mostly donor-type with so far measured lowest D_{it} from 8.0×10^{11}/cm^2 eV to 2.0×10^{12}/cm^2 eV near the conduction band edge and increases continuously to $\sim 10^{13}$/cm^2 eV level at the valence band edge [112]. Further reducing the interface traps to the required device quality level is essential and remains a big challenge in III-V MOS field.

7.4.3 High-k/InP MOSFETs

Although InP is a commonly used compound semiconductor with wide applications in electronic, optoelectronic, and photonic devices, high-k dielectric integration on InP is largely unexplored. Compared to GaAs, InP is widely believed to be a more forgiving material with respect to Fermi level pinning and has a higher electron saturation velocity (2×10^7 cm/s) as well. Detailed Monte-Carlo simulations of deeply scaled n-MOS devices indicate that an InP channel could enable high-field transconductance ~60% higher than either Si, Ge, or GaAs at equivalent channel length [113]. It could be a viable material for high-speed logic applications if a high-quality, thermodynamically stable high-k dielectric could be found. In the few reported works on InP MOSFETs since the 1980s, SiO_2 was primarily used gate dielectric and devices suffered from significant current and threshold voltage drift due to the poor semiconductor-dielectric interface [114, 115]. Although Fermi-level unpinning was achieved through applying appropriate surface treatment before SiO_2 deposition, current and effective channel mobility remained low and interface trap density was far from applicable [116–119]. By implementing ALD high-k dielectrics on InP, we are able to revisit this historically unsolved problem and demonstrate Fermi level unpinning of InP surface with ALD high-k dielectrics. More importantly, InP as a different material from $In_xGa_{1-x}As$ (x from 0 to 1 including GaAs and InAs) series is an appropriate test for the validity of the proposed empirical model.

The starting material for an ALD high-k/InP MOSFET fabrication is an InP semi-insulating substrate with Fe as deep level traps [120]. After surface degreasing and $(NH_4)_2S$-based pretreatment, the wafers were transferred via room ambient to an ASM F-120 ALD reactor. A 30 nm thick Al_2O_3 layer was deposited at a substrate temperature of 300 °C, using alternately pulsed chemical precursors of $Al(CH_3)_3$ (the Al precursor) and H_2O (the oxygen precursor) in a carrier N_2 gas flow. Source and drain regions were selectively implanted with a Si dose of 1×10^{14} cm^{-2} at 140 keV through the 30 nm thick Al_2O_3 layer. Implantation activation was achieved by RTA at 720 °C for 10 s in a nitrogen ambient. The Al_2O_3 encapsulation layer was removed by HF and Al_2O_3 or HfO_2 gate dielectrics were grown by ALD again with the thickness between 4 and 10 nm. The source and drain ohmic contacts were made by an electron beam evaporation of a combination of AuGe/Pt/Au and a lift-off process, followed by a RTA process at 500 °C for 30 s also in a N_2 ambient. The gate electrode was defined by electron beam evaporation of Ti/Au and a lift-off process. The fabricated MOSFETs have a nominal gate length varying from 0.75 to 40 μm and a gate width of 100 μm.

A well-behaved I-V characteristic of an E-mode InP NMOSFET with 10 nm HfO_2 as gate dielectric is demonstrated in Fig. 7.5(a) with maximum drain current over 100 mA/mm at V_{ds} = 3V and V_{gs} = 4V. The device was simply fabricated on semi-insulating InP (100) substrate. The device is off at V_{gs} = 0V without significant drain-source leakage current or substrate current due to the high bandgap of InP. In general, it is surprised to obtain 100 mA/mm level drain current on InP substrate since only a few hundreds of μm/mm drain current was observed on GaAs (100)

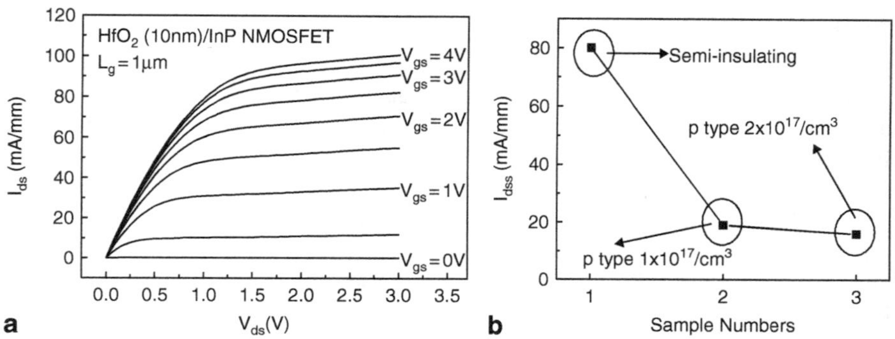

Fig. 7.5 **a** *I–V* characteristic of a 1.0 μm gate length InP MOSFET with a 30 nm ALD HfO_2 as a gate dielectric. **b** Comparison of maximum drain current with different InP channel doping concentration with Al_2O_3 as gate dielectric. Current in Sample 2 and Sample 3 is lower than that in Sample 1

without ICL layer. We ascribe it to the fact that the energy separation between CBM and TNL for InP is 0.5 eV instead of 0.8 eV for GaAs (100). This makes it easier for InP to realize inversion, compared to GaAs (100). Meanwhile, the energy separation between CBM and TNL for $In_{0.53}Ga_{0.47}As$ is ~0.2 eV and for $In_{0.75}Ga_{0.25}As$ is <0.1 eV. That's why $In_{0.53}Ga_{0.47}As$ MOSFET and $In_{0.75}Ga_{0.25}As$ MOSFET deliver much more drain currents than InP MOSFET. InP data further proves the validity of the empirical model for high-k/III-V MOSFETs. Similar devices were also fabricated on MOCVD grown InP epi-wafers with p-type doping concentration of $1 \times 10^{17}/cm^3$ and $2 \times 10^{17}/cm^3$. The maximum drain current drops to 17 mA/mm for p type $2 \times 10^{17}/cm^3$ channel and 20 mA/mm for $1 \times 10^{17}/cm^3$ channel as shown in Fig. 7.5(b). The result is consistent with what is observed in InGaAs MOSFETs [89]. The qualitative understanding of this observation is that more surface bending is required for high doping channels to realize strong inversion condition. In other words, for p-type doped InP, Fermi-level is located near the valence band maximum. If the Fermi-level and TNL is aligned at the first place, more negative charges are built in at the interface which makes further inversion more difficult.

7.4.4 High-k/GaSb MOSFETs

The modern microelectronic industry is built up on a Si-based CMOS technology in order to control the power assumption. CMOS technology requires high-performance NMOSFET and PMOSFET. Although significant progress was made in the past years on high-k/III-V integration, there are few breakthrough results on III-V PMOSFETs, compared to III-V NMOSFETs. One exception is Intel's p-channel InSb quantum well transistors [103]. However, this approach is hard to integrate in large scale due to the large gate current leakage. The issues and progress on III-V PMOSFET are reviewed in Chap. 12 in this book [121, 122]. An ultimate CMOS approach, yet to be demonstrated, is to use InGaAs as NMOSFET channel and

Ge as PMOSFET channel, and integrate both on Si CMOS platform. The cross contamination between III-V and Ge might be a big hassle for this approach. Here, we discuss the fundamental advantage of using GaSb as p-channel material for the potential full III-V CMOS technology in accordance with the empirical model. The principle of physics is supported by the preliminary results on ALD Al_2O_3/GaSb material system with C-V characterization of MOSCAPs and I-V characterization of PMOSFETs.

GaSb has a bandgap of 0.7 eV. GaSb is lattice-matched to InAs. GaSb could be epitaxially grown on In-rich InGaAs with controllable strain and eventually heterogeneously integrated on Si CMOS platform. Compared to Ge, GaSb doesn't have cross contamination issue with In-rich InGaAs. Hole mobility of GaSb is ~800 cm^2/Vs much higher than most of other III-V materials though lower than that of Ge (1900 cm^2/Vs). However, the limitation of current in inversion-mode alternative channel MOSFETs is not the mobility but the inversion charge density. For example, GaAs (100) has 8000 cm^2/Vs electron mobility in bulk and its inversion-mode MOSFETs show very low drain current with directly deposited high-k dielectrics. That's because it's hard to realize strong inversion charge density in GaAs with interface traps. This important understanding of non-Si MOS interfaces has been highlighted in Fig. 7.1. In the case of GaSb, whose band alignment is so like Ge, GaSb PMOSFET must be very easy to realize with good device performance.

N-type GaSb substrate with doping concentration of $5 \times 10^{17}\,cm^{-3}$ are used for MOSCAPs and MOSFETs study. After degreasing by acetone and methanol and pre-cleaning by HCl:H_2O = 1:1 for 30 s to etch away native oxide and NH_4OH for 60 seconds to remove elemental Sb [123], 8 nm Al_2O_3 was deposited at 300 °C in ASM F-120 ALD reactor. Ni is used as metal electrode for MOSCAPs. GaSb PMOSFETs were fabricated as following. 30 nm ALD Al_2O_3 served as an encapsulation layer and source and drain were Zn implanted at 40 KeV and $1 \times 10^{14}/cm^2$. 8 nm Al_2O_3 was re-grown, followed by 600 °C 45 s PDA in N_2 ambient for dopant activation. Ti/Pt/Au were used as Ohmic metal contacts and annealed at 400 °C by RTA. Ni/Au were used as metal gate. From the multi-frequency CV plots from 1 KHz to 1 MHz on Ni/Al_2O_3/GaSb MIS structure, although large frequency dispersion at accumulation side may indicate significant interface traps exhibiting, nevertheless, it does not inhibit the approach of inversion, seen in Fig. 7.6(a). As illustrated in Fig. 7.1 for GaSb, the E_0 is close to VBM and only traps located between E_0 and VBM decide the final inversion. Hole inversion is easy for GaSb, similar to Ge. Figure 7.6(b) shows the I-V characteristics of a 0.75 μm-gate-length GaSb PMOSFET with drain current 1.7 μA/μm. It proves the inversion can be achieved in GaSb at transistor level, though the drain current is much smaller than its counterpart Ge. The newest result shows that drain current of a 0.75 μm GaSb PMOSFET increases to 70 μA/μm with refined Si implanted source and drain. These on-state performances should be possible further improved with optimizing ALD growth condition and device fabrication process. The preliminary results on GaSb further confirms that the empirical model works quite general on all III-V semiconductors including GaSb.

The empirical model is also valid for other III-V materials such as InAs [124], InSb [102, 125] and GaN [126–128]. Our experiments, along with the results from

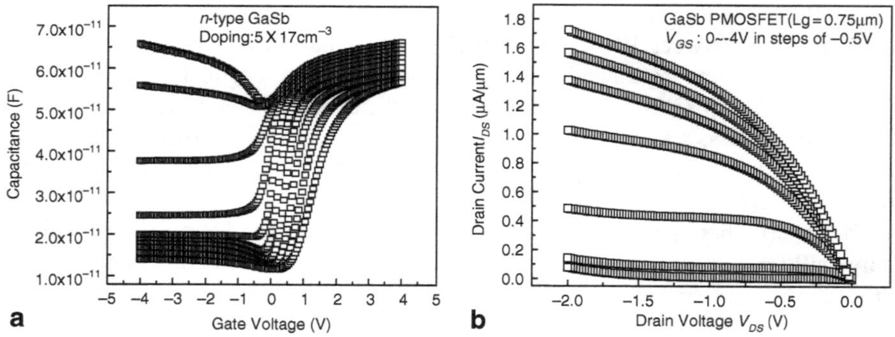

Fig. 7.6 **a** C–V plots on Ni/Al_2O_3/n-GaSb structure from 1 KHz to 1 MHz with 10 points. Large frequency-dispersion is observed at accumulation side, however, inversion characteristics can still be achieved due to that E_0 is close to E_V or VBM as illustrated in Fig. 7.1. **b** Output characteristic (I_{DS}~V_{DS}) for GaSb PMOSFET with 0.75 μm gate length. The maximum drain current is ~1.7 μA/μm

other leading groups in the field, verify this statement. The conclusion is that with decent interface quality as the pre-condition, the energy separation between TNL and CBM is the most important determinant for high-performance inversion-mode NMOSFETs and the energy separation between TNL and VBM for PMOSFET.

7.5 Conclusion

The quest for practical GaAs MOSFETs started in early 1960's has been continued in the 21st century with many work focused on ALD high dielectrics. We briefly reviewed the history over the past 40 years and current status of III-V MOSFET research to help those new in the field to have easy access to the many existing publications. We use an empirical model based on the TNL concept to explain all reported experimental observations on inversion-type enhancement-mode ALD high-k/III-V MOSFETs and MBE $Ga_2O_3(Gd_2O_3)$/III-V MOSFETs reported by M. Hong et al. Although the origin of this empirical model can be tracked back to the Hasegawa's DIGS model, this is the first time that we have systematically surveyed various III-V substrates and demonstrated the validity of the model. It is based on experimental demonstration of inversion-mode MOSFETs so that we call it "empirical model". The studied high-k/III-V interfaces include GaAs, $In_xGa_{1-x}As$, InAs, InP, InSb, GaN and GaSb. With ALD high-k dielectrics, in particular the most studied Al_2O_3, the interface quality with D_{it} between low 10^{11}-low 10^{12}/cm^2 eV is significantly improved from previous work in 1970s and 1980s. Fermi-levels on GaAs (111)A, In-rich InGaAs, InP, GaSb, and others are not pinned. Inversion-mode MOSFETs on different III-V substrates can be demonstrated as shown in Sect. 7.4. However, the validity of this empirical model also tells us the interface quality is far from ideal such as D_{it} of 10^9–10^{10}/cm^2 eV as SiO_2/Si interface. Much work is still needed to make high-k/III-V research become a manufacturable III-V CMOS technology.

Acknowledgments The authors would thank National Science Foundation, SRC FCRP MSD Focus Center, Army Research Office, and SRC CSR Program for supporting this work. The contributions from T. Shen, O. Koybasi, R. Wang, and H.C. Lin are greatly appreciated. The authors would like also to thank H. Hasegawa, J. Robertson, R.M. Wallace, G.D. Wilk, M. Hong, K.K. Ng, B. Yang, M.M. Frank, J. del Alamo, D.A. Antoniadis, A. Kummel, S. Oktyabrsky, M.S. Lundstrom, J.M. Woodall, M.A. Alam, and J.C.M. Hwang for the valuable discussions.

References

1. H. Becke, R. Hall, and J. White, *Solid-State Electron.*, **8**, 813 (1965).
2. H.W. Becke and J.P. White, *Electronics*, **40**, 82 (1967).
3. L. Messick, *Solid-State Electron.*, **22**, 71 (1979).
4. K. Kamimura and Y. Sakai, *Thin Solid Films*, **56**, 215 (1979).
5. T. Mimura and M. Fukuta, *IEEE Trans. Electron Devices*, **27**, 1147 (1980).
6. B. Bayraktaroglu, W.M. Theis, and F.L. Schuermeyer, in *37th Device Research Conference, Abstracts*, p. WP-A3, 1979.
7. H.C. Casey, Jr., A.Y. Cho, and E.H. Nicollian, *Appl. Phys. Lett.*, **32**, 678 (1978).
8. M.R. Melloch, N. Otsuka, J.M. Woodall, A.C. Warren, and J.L. Freeouf, *Appl. Phys. Lett.*, **57**, 1531 (1990).
9. W.T. Tsang, *Appl. Phys. Lett.*, **33**, 429 (1979).
10. X. Li, Y. Cao, D.C. Hall, P. Fay, B. Han, A. Wibowo, and N. Pan, *IEEE Electron Device Lett.*, **25**, 772 (2004).
11. D.N. Butcher and B.J. Sealy, *Electron. Lett.*, **13**, 558 (1977).
12. H. Hasegawa, K.E. Forward, and H.L. Hartnagel, *Appl. Phys. Lett.*, **26**, 567 (1975).
13. R.A. Logan, B. Schwartz, and W.J. Sundburg, *J. Electronchem. Soc.*, **120**, 1385 (1973).
14. O.A. Weinreich, *J. Appl. Phys.*, **37**, 2924 (1966).
15. K. Yamasaki and T. Sugano, *Jap. J. Appl. Phys.*, **17**, 321 (1978).
16. N. Yokoyama, T. Mimura, K. Odani, and M. Fukuta, *Appl. Phys. Lett.*, **32**, 58 (1978).
17. L.A. Chesler and G.Y. Robinson, *Appl. Phys. Lett.*, **32**, 60 (1978).
18. R.P.H. Chang and A.K. Sinha, *Appl.Phys. Lett.*, **29**, 56 (1976).
19. V.M. Bermudez, *J. Appl. Phys.*, **54**, 6795 (1983).
20. C.F. Yu, M.T. Schmidt, D.V. Podlesnik, E.S. Yang, and R.M. Osgood, Jr., *J. Vac. Sci. Technol.*, A **6**, 754 (1988).
21. S.D. Offsey, J.M. Woodall, A.C. Warren, P.D. Kirchner, T.I. Chappell, and G.D. Pettit, *Appl. Phys. Lett.*, 48, 475 (1986).
22. C.W. Wilmsen, *Physics and Chemistry of III-V Compound Semiconductor Interfaces*. New York: Plenum, 1985.
23. C.L. Hinkle, M. Milojevic, B. Brennan, A.M. Sonnet, F.S. Aguirre-Tostado, G.J. Hughes, E.M. Vogel, and R.M. Wallace, *Appl. Phys. Lett.*, **94**, 162101 (2009).
24. M. Xu, K. Xu, R. Contreras, M. Milojevic, T. Shen, O. Koybasi, Y.Q. Wu, R.M. Wallace, and P.D. Ye, in *IEDM Tech. Dig.*, 2009 (to be published).
25. A.M. Cowley and S.M. Sze, *J. Appl.Phys.*, **36**, 3212 (1965).
26. V. Heine, *Phys. Rev.*, **138**, 1689 (1965).
27. C. Tejedor and F. Flores, *J. Phys. C-Solid State Phys.*, **11**, L19 (1978).
28. W.E. Spicer, P.W. Chye, P.R. Skeath, C.Y. Su, and I. Lindau, *J. Vac. Sci. & Technol.*, **16**, 1422 (1979).
29. H. Hasegawa and H. Ohno, *J. Vac. Sci. & Technol.*, B **4**, 1130 (1986).

30. J. Tersoff, *Phys. Rev. Lett.*, **52**, 465 (1984).
31. J. Tersoff, *Phys. Rev. B,* **32**, 6968 (1985).
32. W.E. Spicer, I. Lindau, P. Skeath, C.Y. Su, and P. Chye, *Phys. Rev. Lett.*, **44**, 420 (1980).
33. J. Robertson, *Appl. Phys. Lett.*, **94**, 152104 (2009).
34. P.D. Ye, *J. Vac. Sci. Technol.*, A **26**, 697 (2008).
35. Y. Xuan, P.D. Ye, and T. Shen, *Appl. Phys. Lett.*, **91**, 232107 (2007).
36. M. Xu, Y.Q. Wu, O. Koybasi, T. Shen, and P.D. Ye, *Appl. Phys. Lett.*, **94**, 212104 (2009).
37. W.W. Hooper and W.I. Lehrer, *Proceedings of IEEE.*, **55**, 1237 (1967).
38. R. Dingle, H.L. Stormer, A.C. Gossard, and W. Wiegmann, *Appl. Phys. Lett.*, **33**, 665 (1978).
39. H.L. Stormer, R. Dingle, A.C. Gossard, W. Wiegmann, and M.D. Sturge, *Solid State Commun.*, **29**, 705 (1979).
40. T. Mimura, S. Hiyamizu, T. Fujii, and K. Nanbu, *Jpn. J. Appl. Phys.*, **19**, L225 (1980).
41. D. Delagebeaudeuf, P. Delescluse, P. Etienne, M. Laviron, J. Chaplart, and N.T. Linh, *Electron. Lett.*, **16**, 667 (1980).
42. B.J. Skromme, C.J. Sandroff, E. Yablonovitch, and T. Gmitter, *Appl. Phys. Lett.*, **51**, 2022 (1987).
43. E. Yablonovitch, C.J. Sandroff, R. Bhat, and T. Gmitter, *Appl. Phys. Lett.*, **51**, 439 (1987).
44. A. Callegari, P.D. Hoh, D.A. Buchanan, and D. Lacey, *Appl. Phys. Lett.*, **54**, 332 (1989).
45. H. Hasegawa, M. Akazawa, K.I. Matsuzaki, H. Ishii, and H. Ohno, *Jpn. J. Appl. Phys., Part 2* **L27**, L2265 (1988).
46. S. Tiwari, S.L. Wright, and J. Batey, *IEEE Electron Device Lett.*, **9**, 488 (1988).
47. G.G. Fountain, S.V. Hattangady, D.J. Vitkavage, R.A. Rudder, and R.J. Markunas, *Electron. Lett.*, **24**, 1134 (1988).
48. G.G. Fountain, R.A. Rudder, S.V. Hattangady, R.J. Markunas, and J.A. Hutchby, in *IEDM Tech. Dig.*, Dec. 1989, pp. 887–889.
49. M. Akazawa, H. Ishii, and H. Hasegawa, *Jpn. J. Appl. Phys.*, **30**, 3744 (1991).
50. D.S.L. Mui, H. Liaw, A.L. Demirel, S. Strite, and H. Morkoc, *Appl. Phys. Lett.*, **59,** 2847 (1991).
51. A. Callegari, D.K. Sadana, D.A. Buchanan, A. Paccagnella, E.D. Marshall, M.A. Tischler, and M. Norcott, *Appl. Phys. Lett.*, **58**, 2540 (1991).
52. S. Koveshnikov, W. Tsai, I. Ok, J.C. Lee, V. Torkanov, M. Yakimov, and S. Oktyabrsky, *Appl. Phys. Lett.*, **88**, 022106 (2006).
53. I.J. Ok, H.S. Kim, M.H. Zhang, C.Y. Kang, S.J. Rhee, C.W. Choi, S.A. Krishnan, T. Lee, F. Zhu, G. Thareja, and J.C. Lee, *IEEE Electron Device Lett.*, **27**, 145 (2006).
54. S.J. Koester, E.W. Kiewra, Y. Sun, D.A. Neumayer, J.A. Ott, M. Copel, D.K. Sadana, D.J. Webb, J. Fompeyrine, J.–P. Locquet, C. Marchiori, M. Sousa, and R. Germann, *Appl. Phys. Lett.*, **89**, 042104 (2006).
55. D. Shahrjerdi, M.M. Oye, A.L. Holmes, and S.K. Banerjee, *Appl. Phys. Lett.*, **89**, 043501 (2006).
56. J.P. de Souza E. Kiewra, Y. Sun, A. Callegari, D.K. Sadana, G. Shahidi, D.J. Webb, J. Fompeyrine, R. Germann, C. Rossel, and C. Marchiori, *Appl. Phys. Lett.*, **92**, 153508 (2008).
57. H.C. Chin, M. Zhu, C.H. Tung, G.S. Samudra, and Y.C. Yeo, *IEEE Electron Device Lett.*, **29**, 553 (2008).
58. M. Passlack, M. Hong, and J.P. Mannaerts, *Appl. Phys. Lett.*, **68**, 1099 (1996).
59. M. Passlack, M. Hong, J.P. Mannaerts, R.L. Opila, S.N.G. Chu, N. Moriya, F. Ren, and J.R. Kwo, *IEEE Electron Device Lett.*, **44**, 214 (1997).
60. M. Hong, J. Kwo, A.R. Kortan, J.P. Mannaerts, and A.M. Sergent, Science, **283**, 1897 (1999).
61. F. Ren, M. Hong, W.S. Hobson, J.M. Kuo, J.R. Lothian, J.P. Mannaerts, J. Kwo, S.N.G. Chu, Y.K. Chen, and A.Y. Cho, *Solid-State Electron.*, **41**, 1751 (1997).
62. Y.C. Wang, M. Hong, J.M. Kuo, J.P. Mannaerts, J. Kwo, H.S. Tsai, J.J. Krajewski, Y.K. Chen, and A.Y. Cho, *IEEE Electron Devices Lett.*, **20**, 457 (1999).
63. F. Ren, J.M. Kuo, M. Hong, W.S. Hobson, J.R. Lothian, J. Lin, H.S. Tsai, J.P. Mannaerts, J. Kwo, S.N.G. Chu, Y.K. Chen, and A.Y. Cho, *IEEE Electron Device Lett.*, **19**, 309 (1998).

64. M. Hong, J.N. Bailargeon, J. Kwo, J.P. Mannaerts, and A.Y. Cho, in *Proceeding of IEEE 27th International Symposium on Compound Semiconductors*, pp. 345–350, (2000).
65. Y.C. Wang, M. Hong, J.M. Kuo, J.P. Mannaerts, H.S. Tsai, J. Kwo, J. J. Krajewski, Y. K. Chen, and A. Y. Cho, *Electron. Lett.,* **35**, 667 (1999).
66. B. Yang, P.D. Ye, J. Kwo, M.R. Frei, H.J.L. Gossmann, J.P. Mannaerts, M. Sergent, M. Hong, and K.N.J. Bude, *J. Cryst. Growth.,* **251**, 837 (2003).
67. T.D. Lin, H.C. Chiu, P. Chang, L.T. Tung, C.P. Chen, M. Hong, J. Kwo, W. Tsai, and Y.C. Wang, *Appl. Phys. Lett.,* **93**, 033516 (2008).
68. M. Passlack, N. Medendorp, R. Gregory, and D. Braddock, *Appl. Phys. Lett.,* **83**, 5262 (2003).
69. K. Rajagopalan, J. Abrokwah, R. Droopad, and M. Passlack, *IEEE Electron Device Lett.,* **27**, 959 (2006).
70. R.J.W. Hill, D.A.J. Moran, X. Li, H. Zhou, D. Macintyre, S. Thoms, A. Asenov, P. Zurcher, K. Rajagopalan, J. Abrokwah, R. Droopad, M. Passlack, and L.G. Thayne, *IEEE Electron Device Lett.,* **28**, 1080 (2007).
71. M.J. Hale, S.I. Yi, J.Z. Sexton, A.C. Kummel, and M. Passlack, *J. Chem. Phys.,* **119**, 6719 (2003).
72. P.D. Ye, G.D. Wilk, J. Kwo, B. Yang, H.-J.L. Gossmann, M. Frei, S.N.G. Chu, J.P. Mannaerts, M. Sergent, M. Hong, K.K. Ng, and J. Bude, *IEEE Electron Device Lett.,* **24**, 209 (2003).
73. P.D. Ye, G.D. Wilk, B. Yang, J. Kwo, S.N.G. Chu, S. Nakahara, H.-J.L. Gossmann, J.P. Mannaerts, M. Hong, K.K. Ng, and J. Bude, *Appl. Phys. Lett.,* **83**, 180 (2003).
74. P.D. Ye, G.D. Wilk, B. Yang, J. Kwo, H.-J.L. Gossmann, M. Hong, K.K. Ng, and J. Bude, *Appl. Phys. Lett.,* **84**, 434 (2004).
75. P. D. Ye, G.D. Wilk, B. Yang, S.N.G. Chu, K.K. Ng, and J. Bude, *Solid-State Electron.,* **49**, 790 (2005).
76. P.D. Ye, B. Yang, K.K. Ng, J. Bude, G.D. Wilk, S. Halder, and J.C.M. Hwang, *Appl.Phys. Lett.,* **86**, 063501 (2005).
77. Y. Xuan, H.C. Lin, P.D. Ye, and G.D. Wilk, *Appl. Phys. Lett.,* **89**, 132103 (2006).
78. Y. Xuan, H.C. Lin, and P.D. Ye, *ECS Transactions,* **3**, 59 (2006).
79. Y. Xuan, H.C. Lin, and P.D. Ye, *IEEE Trans. Electron Devices,* **54**, 1811 (2007).
80. Y.Q. Wu, T. Shen, P.D. Ye, and G.D. Wilk, *Appl. Phys. Lett.,* **90**, 143504 (2007).
81. T. Yang, Y. Xuan, D. Zemlyanov, T. Shen, Y.Q. Wu, J.M. Woodall, P.D. Ye, F.S. Aguirre-Tostado, M. Milojevic, S. McDonnell, and R.M. Wallace, *Appl. Phys. Lett.,* **91**, 142122 (2007).
82. T. Yang, Y. Liu, P.D. Ye, Y. Xuan, H. Pal, and M.S. Lundstrom, *Appl. Phys. Lett.,* **92**, 252105 (2008).
83. M.M. Frank, G.D. Wilk, D. Starodub, T. Gustafsson, E. Garfunkel, Y.J. Chabal, J. Grazul, and D.A. Muller, *Appl. Phys. Lett.,* **86**, 152904 (2005).
84. M.L. Huang, Y.C. Chang, C.H. Chang, Y.J. Lee and P. Chang, J. Kwo, T.B. Wu, and M. Hong, *Appl. Phys. Lett.,* **87**, 252104 (2005).
85. C.H. Chang, Y.K. Chiou, Y.C. Chang, K.Y. Lee, T.D. Lin, T.B. Wu, M. Hong, and J. Kwo, *Appl. Phys. Lett.,* **89**, 242911 (2006).
86. C.L. Hinkle, A.M. Sonnet, E.M. Vogel, S. McDonnell, G.J. Hughes, M. Milojevic, B. Lee, F.S. Aguirre-Tostado, K.J. Choi, H.C. Kim, J. Kim, and R.M. Wallace, *Appl. Phys. Lett.,* **92**, 071901 (2008).
87. Y. Xuan, Y.Q. Wu, H.C. Lin, T. Shen, and P.D. Ye, *in Proceeding of 65th Device Research Conference,* Notre Dame, USA, 2007.
88. Y. Xuan, Y.Q. Wu, H.C. Lin, T. Shen, and P.D. Ye, *IEEE Electron Devices Lett.,* **28**, 935 (2007).
89. Y. Xuan, Y.Q. Wu, T. Shen, T. Yang, and P.D. Ye, in *IEDM Tech. Dig.,* Dec. 2007, pp. 637–640.
90. Y. Xuan, Y.Q. Wu, and P.D. Ye, *IEEE Electron Device Lett.,* **29**, 294 (2008).
91. Y. Xuan, T. Shen, M. Xu, Y.Q. Wu, and P.D. Ye, in *IEDM Tech. Dig.,* Dec. 2008, pp. 371–374.

92. Y.Q. Wu, W.K. Wang, O. Koybasi, D.N. Zakharov, E.A. Stach, S. Nakahara, J.C.M. Hwang, and P.D. Ye, *IEEE Electron Device Lett.,* **30**, 700 (2009).
93. Y.Q. Wu, M. Xu, R. Wang, O. Koybasi, and P.D. Ye, in *IEDM Tech. Dig.,* Dec. 2009, to be published.
94. Y.Q. Wu, R. Wang, T. Shen, J.J. Gu, and P.D. Ye, in *IEDM Tech. Dig.,* Dec. 2009, to be published.
95. M. Zhu, C. H. Tung, and Y. C. Yeo, *Appl. Phys. Lett.,* **89**, 202903 (2006).
96. H.L. Lu, L. Sun, S.J. Ding, M. Xu, D.W. Zhang, and L.K. Wang, *Appl. Phys. Lett.,* **89**, 152910 (2006).
97. G.K. Dalapati, Y. Tong, W.Y. Loh, H.K. Mun, and B.J. Cho, *IEEE Trans. Electron Devices,* **54**, 1831 (2007).
98. D. Shahrjerdi, E. Tutuc, and S. Banerjee, *Appl. Phys. Lett.,* **91**, 063501 (2007).
99. A.M. Sonnet, C.L. Hinkle, M.N. Jivani, R.A. Chapman, G.P. Pollack, R.M. Wallace, and E.M. Vogel, *Appl. Phys. Lett.,* **93**, 122109 (2008).
100. U. Singisetti, M.A. Wistey, G.J. Burek, A.K. Baraskar, J. Cagnon, B.J. Thibeault, S. Stemmer, A.C. Gossard, M.J.W. Rodwell, E. Kim, B. Shin, P.C. McIntyre, and Y.J. Lee, in *Proceeding of IEEE 67th Devcie Research Conference*, 2009, pp. 253–354.
101. R. Chau, S. Datta, and A. Majumdar, in *Proc. IEEE CSIC Dig.,* 2005, pp. 17–20.
102. S. Datta, T. Ashley, J. Brask, L. Buckle, M. Doczy, M. Emeny, D. Hayes, K. Hilton, R. Jefferies, T. Martin, T.J. Phillips, D. Wallis, P. Wilding, and R. Chau, in *IEDM Tech. Dig.,* 2005, pp. 783–786.
103. M. Radosvljevic, T. Ashley, A. Andreev, S.D. Coomber, G. Dewey, M.T. Emeny, M. Fearn, D.G. Hayes, K.P. Hilton, M.K. Hudait, R. Jefferies, T. Martin, R. Pillarisetty, W. Rachmady, T. Rakshit, S.J. Smith, M.J. Uren, D.J. Wallis, P.J. Wilding, and R. Chau, in *IEDM Tech. Dig.,* pp. 727–730, Dec. 2008.
104. D. H. Kim, J. A. del Alamo, J. H. Lee, and K. S. Seo, *IEEE Trans. Electron Devices,* **54**, 2606 (2007).
105. D. H. Kim, and J. A. del Alamo, *IEEE Electron Device Lett.,* **29**, 830 (2008).
106. G.D. Wilk, R.M. Wallace, and J.M. Anthony, *J. Appl. Phys.,* **89**, 5243 (2001).
107. D.D. Nolte, *Solid-State Electron.,* **33**, 295 (1990).
108. H. Hasegawa and M. Akazawa, *Appl. Surf. Sci.,* **254**, 8005 (2008).
109. J. Robertson and B. Falabretti, *J. Appl. Phys.,* **100**, 014111 (2006).
110. A. Dimoulas, P. Tsipas, A. Sotiropoulos, and E.K. Evangelou, *Appl. Phys. Lett.,* **89**, 252110 (2006).
111. W.E. Spicer, P.W. Chye, P.R. Skeath, C.Y. Su, and I. Lindau, *J. Vac. Sci. Tech.,* **16**, 1422 (1979).
112. D. Varghese, Y. Xuan, Y.Q. Wu, T. Shen, P.D. Ye, and M.A. Alam, *in IEDM Tech. Dig.,* 2008, pp. 379–382.
113. M.V. Fischetti and S.E. Laux, *IEEE Trans. Electron Devices,* **38**, 650 (1991).
114. D. L. Lile and M.J. Taylor, *J. Appl. Phys.,* **54**, 260 (1983).
115. D. L. Lile, *Solid-State Electron.,* **21**, 1199 (1978).
116. D. L. Lile, D. A. Collins, L. G. Meiners, and L. Messick, *Electronics Lett.,* **14**, 657 (1978).
117. T. Kawakami and M. Okamura, *Electronics Lett.,* **15**, 502 (1979).
118. Y. Shinoda and T. Kobayashi, *Solid State Electron.,* **25**, 1119 (1982).
119. W.F. Tseng, M.L. Bark, H.B. Dietrich, A. Christou, R.L. Henry, W.A. Schmidt, and N.S. Saks, *IEEE Electron Device Lett.,* **2**, 299 (1981).
120. Y.Q. Wu, Y. Xuan, T. Shen, P.D. Ye, Z. Cheng, and A. Lochtefeld, *Appl. Phys. Lett.,* **91**, 022108 (2007).
121. S. Oktyabrsky, V. Tokranov, S. Koveshnikov, M. Yakimov, R. Kambhampati, H. Bakhru, R. Moore, and W. Tsai, *J. Cryst. Growth.,* **311**, 1950 (2009).
122. N. Goel, D. Heh, S. Koveshnikov, I. Ok, S. Oktyabrsky, V. Tokranov, R. Kambhampati, M. Yakimov, Y. Sun, P. Pianetta, C.K. Gaspe, M.B. Santos, J. Lee, S. Datta, P. Majhi, and W. Tsai, *in IEDM Tech. Dig.,* pp. 363–366 (2008).

123. P.S. Dutta, H.L. Bhat, and V. Kumar, *J. Appl. Phys.,* **81**, 5821 (1997).
124. N. Li, E.S. Harmon, J. Hyland, D.B. Salzman, T.P. Ma, Y. Xuan, and P.D. Ye, *Appl. Phys. Lett.,* **92**, 143507 (2008).
125. T. Yang, Y. Xuan, P.D. Ye, W. Wang, J.C.M. Hwang, D. Lubyshev, J.M. Fastenau, W.K. Liu, T.D. Mishima, and M.B. Santos, in *Proceeding of TMS 2007 Electronic Materials Conferences*, Notre Dame, USA, 2007.
126. Y.Q. Wu, T. Shen, P.D. Ye, and G.D. Wilk, *Appl. Phys. Lett.,* **90**, 143504 (2007).
127. Y.C. Chang, W.H. Chang, H.C. Chiu, L.T. Tung, C.H. Lee, K.H. Shiu, M. Hong, J. Kwo, J.M. Hong, and C.C. Tsai, *Appl. Phy. Lett.,* **93**, 053504 (2008).
128. W. Huang, T. Khan, and T.P. Chow, IEEE *Electron Devices Lett.,* **27**, 796 (2006).

Chapter 8
Materials and Technologies for III-V MOSFETs

Serge Oktyabrsky, Yoshio Nishi, Sergei Koveshnikov, Wei-E Wang, Niti Goel and Wilman Tsai

Abstract The paper contains an overview of progress and challenges of group III-V MOSFETs. It begins with comparison of well-established high-electron mobility transistors for logic applications to MOSFET technology. Further, the results on improvement of current transport in buried modulation doped quantum well channels similar to HEMTs are presented. Next, the progress in interface passivation is reviewed, and detailed with descriptions of different technologies including atomic layer deposition with its property to self-clean a III-V surface, amorphous Si and Ge passivation, and use of in-situ high-k oxide deposition.

8.1 Introduction

After four decades of successful scaling of transistors, the trend is reaching fundamental material limits. Much interest has been generated in exploring high mobility materials such as III-V, Ge, graphene, as replacements for Si channel to sustain future CMOS nodes. The intrinsic properties of these materials may add to the challenge of successful device realization. Since III-Vs have been a work horse in communications and optoelectronics industries, they have been heavily researched. The success of III-V in potential CMOS technology will depends on heterogeneous integration on silicon with thinner buffer layers; compatible, low leakage and thermally stable gate dielectric with low interface state density, as well as defect free junctions with low external or access resistance. In addition, it is key to develop and orient various physical and electrical characterization techniques to probe and evaluate the interface and bulk characteristics effectively and correctly at the atomic level. The paper reviews various gate dielectric deposition techniques in addition to current understanding of the obstacles in implementing high-k oxides.

The chapter begins with comparison of well-established high-electron mobility transistors for logic applications to MOSFET technology. It follows with results on

W. Tsai (✉)
International SEMATECH, Austin, Texas 78741, USA
e-mail: wilman.tsai@intel.com

S. Oktyabrsky, P. D. Ye (eds.), *Fundamentals of III-V Semiconductor MOSFETs*,
DOI 10.1007/978-1-4419-1547-4_8, © Springer Science+Business Media LLC 2010

improvement of current transport in buried modulation doped quantum well channels similar to HEMTs. Next, the progress in interface passivation is reviewed, and detailed with descriptions of different technologies including atomic layer deposition with its property to self-clean a III-V surface, amorphous Si and Ge passivation, and use of in-situ high-k oxide deposition.

8.2 III-V HEMTs for Digital Applications

It is worth emphasizing challenges of group III-V MOSFETs by comparison to a mature technology of FETs with Schottky gate, such as metal-semiconductor FET (MESFET) or high-electron mobility transistor (HEMT).

III-V compound semiconductors are not novel materials for digital circuits. GaAs-based digital IC utilizing mainly implanted n-channel MESFET were in production since the 1980s. These ICs demonstrated almost an order of magnitude faster clocking speeds and 2–3 times intrinsic speed advantage over Si MOSFETs. The ICs with over 100 k directly-coupled GaAs MESFET gates per chip were commercially available in early 1990s and were employed in supercomputers, such as Convex C3800 and Cray-3 (with more modest gate-per-chip count). In fact, many thought that GaAs digital ICs would quickly replace silicon. However, the functional complexity of silicon chips was well ahead the GaAs ICs mainly due to low static power dissipation and simple circuitry of Si CMOS circuits. Digital GaAs ICs were pushed out of mainstream computing applications to data- and telecomm chips with the clock frequencies of 1–10 Gb/s and later 40 Gb/s and less-demanding complexity.

Of many device designs, technologies and circuits [1], enhancement mode (e-mode) MESFET (Fig. 8.1a) was most widely used in digital ICs due to simple resistive-load direct-coupled logic circuitry. Another modification of this device is the junction FET (JFET) which utilizes implanted p^+-type region under the gate, and thus, using a p-n junction rather than a Schottky junction to control depletion of the channel. These devices have a normally (at zero bias) depleted channel under the gate, which is determined by the Schottky (or p-n junction) barrier height and doping level of the channel as illustrated in the band diagram of the MESFET in Fig. 8.1a. To turn on the transistor, one has to apply a direct bias to the gate diode. The swing is therefore, limited by the barrier height, and in the case of GaAs MESFET is less than 0.5 V (~1.2 V in JFETs) resulting in small noise margin, and also relatively high gate leakage current.

Development of epitaxial heterostructure technologies, namely molecular beam epitaxy (MBE) and metal-organic chemical vapor deposition (MOCVD) allowed for a further significant improvement of III-V FETs. Two major changes were introduced as indicated in the Fig. 8.1b. Firstly, the channel was confined by a heterostructure potential to form a two-dimensional (2D) electron gas. The confinement can be achieved either at a heterointerface (e.g. AlGaAs/GaAs) as shown in the figure or using a quantum well (QW) potential with a thin layer of narrow band-

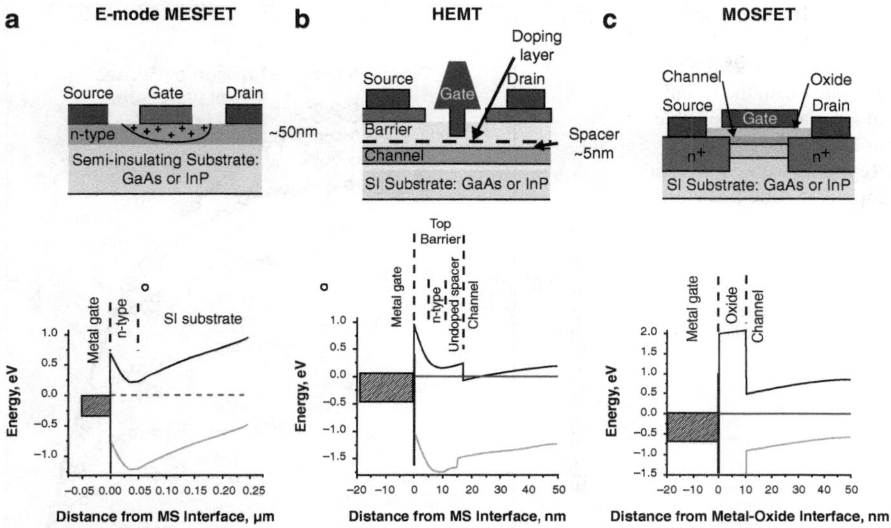

Fig. 8.1 Cross-sections and band diagram across the gate regions of **a** enhancement mode (e-mode) GaAs MESFET; **b** HEMT; and **c** MOSFET

gap semiconductor (InGaAs) sandwiched between wider bandgap (InAlAs) barriers. Secondly, the carriers in the channel were separated from the dopants reducing the Coulomb scattering due to ionized impurities. This was done using modulation doping into the barrier layer and introducing an undoped spacer layer between the doping layer and the channel. These devices are usually called high-electron mobility transistor (HEMTs), but historically also have been referred to as heterojunction FETs (HFETs) or modulation doped FETs (MODFETs).

Major advantages of HEMTs over MESFETs include higher mobility and velocity that results in higher speed of the circuits, and reduced gate-channel separation that leads to better gate control (higher gate capacitance) and device scalability. As shown in Fig. 8.2, improvements in technologies (mostly MBE) have resulted in 2D HEMT channels with mobilities limited by phonon scattering from room temperature down to ~60 K in GaAs QWs, with values as high as 6×10^6 cm^2/V s at 4.2 K [2] and a record 35×10^6 cm^2/V s below 1 K [3].

The III-V HEMT technology has been driven mostly by millimeter wavelength analog applications and has not been focused on the mainstream digital integration circuit requirements, such as high density and low power. However, HEMTs have been used in numerous digital integrated circuits with relatively low density [4, 5]. Although the mobility of p-type III-V materials is much less than that of electrons, the simplicity and low power of CMOS circuitry gave rise to development of these complimentary FET technologies based on HEMTs [6–8].

The physics, technology and applications of HEMTs have been covered previously [1, 4, 9–12] and we will not go into detail here, but it is worth while understanding physics-related benefits and limitations of HEMTs in comparison with

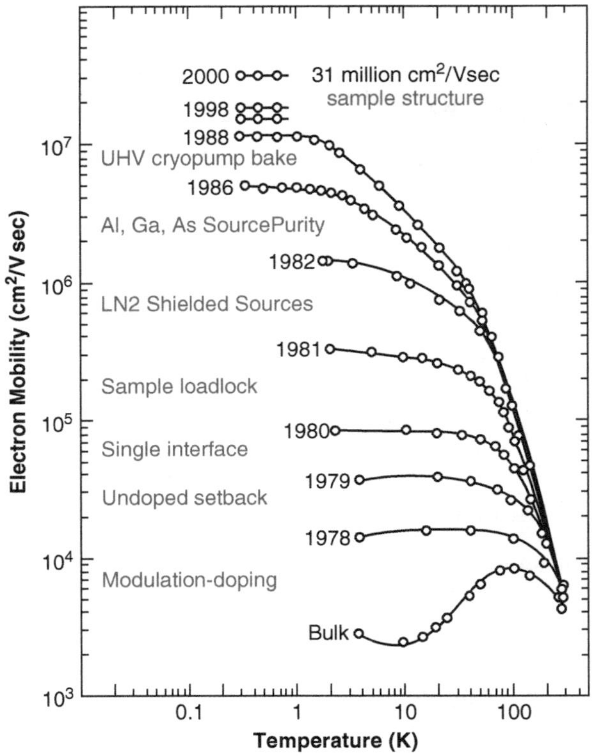

Fig. 8.2 History of improvements in the mobility of electrons in GaAs 2DEG, annotated with the technical innovation responsible for the Improvement [2] (Reprinted with permission. Copyright 2009 Elsevier)

MOSFETs. A typical III-V MOSFET design and bandstructure is shown in Fig. 8.1c. The major difference between a HEMT and a MOSFET is the MOSFET's significantly higher band offset (conduction band as shown in Figs. 8.1b, c) at the interface with the 2DEG. This interface is typically with one of high-k oxides (HfO_2, ZrO_2, Al_2O_3, etc.) which typically are materials with wide bandgaps, $E_g = 5$–9 eV, as compared to semiconductor barrier materials (AlGaAs, InAlAs, InP, etc.) with $E_g = 1.5$–2 eV. This rather quantitative difference results in drastic reduction of the gate current in MOSFETs. On the other hand, the interface with oxide is a source of excess carrier scattering in the channel as will be discussed later in this chapter. The interface with epitaxially grown wide-bandgap semiconductor barrier material can be made almost perfect with negligible scattering and defect density (see Fig. 8.2). This means that the transport in the HEMT channel is a good model and an ultimate goal for that in the MOSFET channel.

Most of the HEMTs for microwave and digital applications utilize GaAs or InGaAs channel on either GaAs or InP substrates (Table 8.1). The GaAs/AlGaAs interface is known to be the most perfect to localize the 2DEG and, if grown properly, is a source of negligible to scattering that allows for extremely high mobilities at low temperature as shown in Fig. 8.2. However, the effective mass reduces with increase of the indium content in bulk InGaAs, therefore, at room temperature, the

Table 8.1 Materials for HEMTs and highest cut-off frequencies, f_t

Channel/substrate	Barrier material	f_t, GHz (gate length, nm)
GaAs/GaAs, lattice-matched	AlGaAs	125 (100) [16]
$In_xGa_{1-x}As/GaAs$, x = 0.2–0.3, pHEMT	AlGaAs or InGaP	140 (120) [17]
$In_{0.53}Ga_{0.47}As/InP$, lattice-matched	$In_{0.52}Ga_{0.48}As$ or InP	396 (25) [18]
$In_xGa_{1-x}As/InP$, x = 0.7–1, InP pHEMT	$In_{0.52-y}Ga_{0.48+y}As$ or InP	628 (35) [19]
$In_xGa_{1-x}As/GaAs$, x = 0.8–1, mHEMT	$In_xAl_{1-x}As$, y ~ 0.7	610 (15) [20]
InSb/Si	InAlSb	305 (85) [15]

phonon-scattering-limited mobilities and electron velocities (and thus device speed) are higher in high-In content channels. Another advantage of high-In-content channels is a larger conduction band offset with the barrier and larger separation to the high-effective-mass L- and X-valleys which allows for higher number of states in the low mass Γ-valley and consequently higher current of electrons with low effective mass.

The InGaAs HEMTs may have either lattice-matched or biaxial compressively-strained channels (pseudomorphic—pHEMT) channels. The lattice-matched HEMTs typically have relatively thick channel layer to localize 2DEG at the top heterointerface. On the contrary, the pHEMTs contain a thin (6–12 nm) channel and a 2DEG confined by the quantum well (QW) potential. Higher In content in the channel results in increased strain and hence, limits the channel layer thickness by the critical thickness for plastic relaxation.

More recently, the metamorphic HEMT (mHEMT) was developed. It contains a high-In-content channel fabricated on a GaAs substrate. The mHEMT's specific feature is a relatively thick strain-release layer with graded composition, which contains high density of misfit dislocations and suited to increase the lattice constant from GaAs to $In_yAl_{1-y}As$ with x ~ 0.7. Although the mHEMT was considered previously as an inexpensive substitute for InP HEMTs, recently they demonstrated significantly improved performance [13]. This technology also gives rise to III-V HEMTs on Si, which are of great importance to digital electronics [14, 15].

For the fair comparison, it is important to use an adequate performance metric. For high-performance logic, this metric is determined by the two major requirements: increased speed and reduced power (consisting of dynamic and static power) while integration density keeps increasing (further scaling).

8.2.1 Intrinsic Delay

The device speed is usually described by the intrinsic delay time τ_i in linear approximation:

$$\tau_i \approx \frac{Q}{I_{Dsat}} \approx C_G \frac{V_D}{I_{Dsat}}, \tag{8.1}$$

which assumes switching the channel charge Q on the gate capacitance C_G by a constant "ON" drain current I_{Dsat} between two logic states with a voltage swing across the channel equal to power supply voltage V_D, or in CMOS circuits $V_d = V_{dd}$, the drain-to-drain voltage across the CMOS pair. The current is measured in saturation with equal gate and drain voltages, $V_{GS} = V_{DS} = V_D$.

In microwave applications, the τ_i is usually determined from a small-signal analysis the unity-current-gain cutoff frequency, f_t, obtained by extrapolating $|h_{21}|^2$ current gain roll-off at 6 dB/octave from typically measured 0–40 or 50 GHz frequency range to higher frequencies. This procedure is shown in Fig. 8.3 for microwave Si CMOS, InGaAs p-HEMT and InAs HEMT. Then the transistor intrinsic delay time can be extracted [21]:

$$\tau_i = \frac{1}{2\pi f_t} - \tau_p \tag{8.2}$$

The parasitc charging time is due to finite source and drain resistances (R_s, R_d):

$$\tau_p = C_{gd}\,(R_s + R_d), \tag{8.3}$$

and should be much lower than τ_i to have reliable estimates. The gate-to-drain capacitance C_{gd}, can be evaluated from measured S-parameters, and source and drain resistances from TLM measurements. However, special precautions should be made to evaluate and subtract the measuring circuit parasitics, in particular capacitance of the probes and pads [18].

The delay value (2) estimated from small-signal RF characteristics is expected to be close to the large-signal CV/I delay time (1), because the latter is approximated by the linear charging of the gate capacitance and is shorter than the real large-signal delay. This provides good basis for analogue and digital device comparison. However, it should be noted that many authors do not detail the procedures used to extract intrinsic parameters of the transistors. The comparison of gate delay scaling

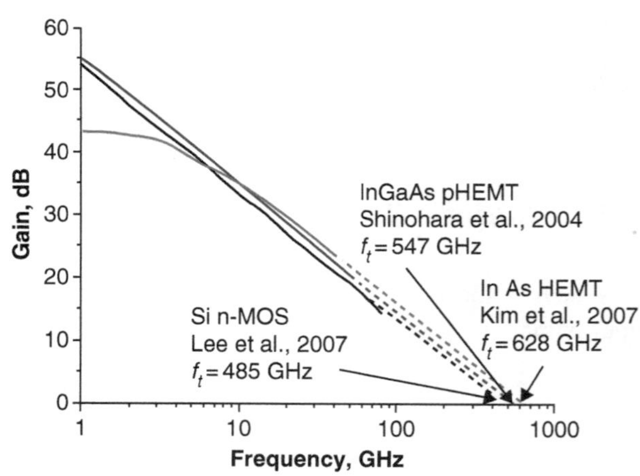

Fig. 8.3 Current gain $|h_{21}|^2$ vs. frequency for various FETs: $In_{0.7}Ga_{0.3}As$ 30-nm gate pHEMT biased at drain-source voltage $V_{ds} = 0.9$ V, and gate-source voltage $V_{gs} = -0.25$ V [22]; InAs 30-nm pHEMT with $V_{ds} = 0.6$ V, and gate-source voltage $V_{gs} = 0.1$ V; Si 45-nm nMOS with $V_{ds} = 1$ V, and gate-source voltage $V_{gs} = 0.6$ V [23]

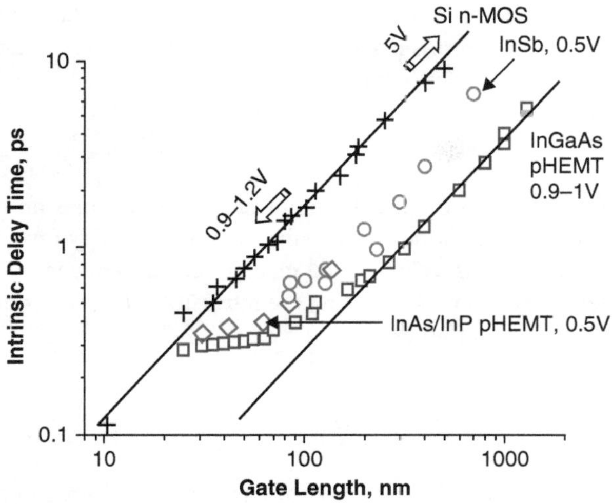

Fig. 8.4 Comparison of intrinsic gate delay scaling in different materials systems: Si n-MOS InGaAs pHEMT best delay times are adapted from Ref. [25], InSb data are from Ref. [26], and InAs PHEMT from Ref. [19]. Note the different operating voltages of the FETs

in FETs using different materials systems is shown in Fig. 8.4. For the long cannel devices, the III-V materials offer 4–6 times improvement over Si, but this improvement is reduced for $L_G < 80$ nm. The reason for the delay saturation is increased relative device parasitics mostly due to the fringing capacitance and possibly underestimated source and drain resistances in (2) in the deeply scaled FETs. The Si n-MOS data for $L_G < 60$ nm in Fig. 8.4 are for strained Si channels with improved electron transport. This improved transport is essential to keep the scaling of delay, otherwise, it would also show saturation at longer gates [24].

The intrinsic delay time is an important metric parameter for CMOS and for a long time it has scaled at 17% per year rate but will be likely slow down [27, 28]. The intrinsic delay can be also used to assess properties of different channel materials. It is inversely proportional to the average velocity v_{avr} of carriers in the channel $\tau_i \approx L_G/v_{avr}$, where L_G is the gate length. The v_{avr} is approaching the saturation velocity at high longitudinal fields when the L_G is reduced but drift transport is still dominant. In the deeply scaled channels the carrier velocity can be significantly higher than the saturation velocity [29, 30]. In this case, the performance limit is determined by the source injection velocity, v_{inj}, which is typically close to thermal velocity of carriers. To evaluate the delay time from the v_{inj}, one should take into consideration backscattering to the source [30, 31] and calculate the charge and current in the short-channel device as well as parasitics mainly due to effects of fringing and overlap capacitances and drain-induced barrier lowering (DIBL) [24]. For comparison of channel materials, it is good enough to approximate the intrinsic delay time as

$$\tau_i \approx \frac{L_G}{v_{inj}}. \tag{8.4}$$

Due to device and interconnect parasitics, the total circuit (extrinsic) delay time is longer than intrinsic one. The delay time can be estimated as

$$\tau_{ext} \approx C_{ext}\frac{V_D}{I_{Dsat}}, \tag{8.5}$$

and if $C_{ext} >> C_g$, the drain current drive becomes an important metric for circuit design. It is instructive to estimate how the delay time depends on materials properties, such as velocity and channel carrier concentration, n, for the same device geometry. As we have seen, the intrinsic and extrinsic delay times (through materials-related variables):

$$\tau_i \propto \left(v_{inj}\right)^{-1} \propto \sqrt{m^*} \quad \text{and} \quad \tau_{ext} \propto \frac{V_d}{nv_{inj}} \propto \frac{V_d}{\sqrt{m^*}}, \tag{8.6}$$

where m^* is effective mass of carriers. The source injection velocity ($\propto 1/\sqrt{m^*}$) in InAs and InGaAs was measured ~2.5–3 × 10^7 cm/s and is at least 2 times higher than in strained Si n-MOSFETs, Fig. 8.5 [19, 32]. On the other hand, the electron concentration (proportional to 2D density of states or m^*) in the channel is determined by the material bandstructure (through effective mass, non-parabolicity, higher-effective-mass valleys) and is lower in III-V's. Therefore, when parasitic capacitance is high, the speed benefits, if any, of III-V channels get reduced and the benefit is primarily the ability to reduce supply voltage, which is discussed in the following section.

8.2.2 Dynamic Power

The power metrics of FETs become increasingly important as the IC's approaching power dissipation level of 200 W/cm^2 [27], which is close to the physical limit for power dissipation at room temperature. MOS transistors have mostly capacitive

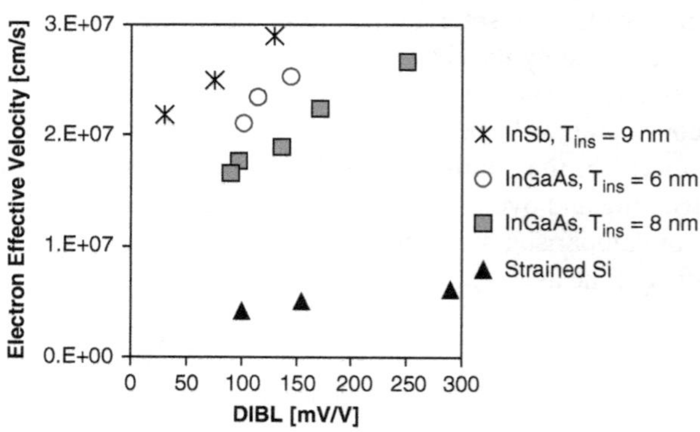

Fig. 8.5 Effective electron velocity versus DIBL for $In_{0.7}Ga_{0.3}As$ with T_{ins} of 6 and 8 nm, InSb with T_{ins} of 9 nm and strained Si nMOSFET at V_{ds} = 0.5 V and a gate overdrive of $(V_g - V_t)$ = 0.3 V. At a constant DIBL of 150 mV/V, $In_{0.7}Ga_{0.3}As$ (8 nm), $In_{0.7}Ga_{0.3}As$ (6 nm), and InSb (9 nm) show 3.7×, 4.1×, and 5× increase in v_{eff} over strained Si, respectively [32] (Reprinted with permission. Copyright 2008 IEEE)

input impedance and a certain energy is required to recharge the gate capacitance. The exact charging energy depends on distribution of carriers in the channel and capacitance change below threshold, therefore for metric purpose, it is assumed that the dynamic energy is proportional to $C_g V_d^2$ per bit and both dynamic metrics (energy and speed) are combined into intrinsic energy-delay product (the smaller—the better):

$$EDP_i = \frac{C_g V_d{}^2 \tau_i}{W}. \tag{8.7}$$

where W is the channel width. In HEMTs, direct measurement of gate capacitance is unreliable due to relatively high gate leakage, and can be extracted from high-frequency S-parameters [19], or estimated [25]: $C_g = \varepsilon_0 \varepsilon_b L_g W / (d_b + \Delta d)$, where ε_b and d_b are the dielectric constant and thickness of the top barrier layer, respectively, and Δd is the position of the maximum of the wavefunction of the 2DEG in the QW. Gate capacitance can be also evaluated from Eq. (8.1) if τ_i is available from cut-off frequency measurement (3). The scaling of the energy-delay product for different channels is shown in Fig. 8.6 [25].

It is again educational to evaluate the improvements related to material properties. The switching energy is proportional to Q^2/C_g or just $\propto n^2$ for the intrinsic power, if only the channel materials-related parameters are considered, and $C_{ext} V_d^2$ if large extrinsic capacitance is charging. Then, the intrinsic and extrinsic energy-delay products can be expressed through the materials-related parameters

$$EDP_i \propto \frac{n^2}{v_{inj}} \propto m^{*5/2} \quad \text{and} \quad EDP_{ext} \propto \frac{V_d{}^3}{n v_{inj}} \propto \frac{V_d{}^3}{\sqrt{m^*}}. \tag{8.8}$$

The energy-delay product metric shows that a III-V channel is considerably more favorable than Si-based even in the architectures with large parasitics because of V_d^3

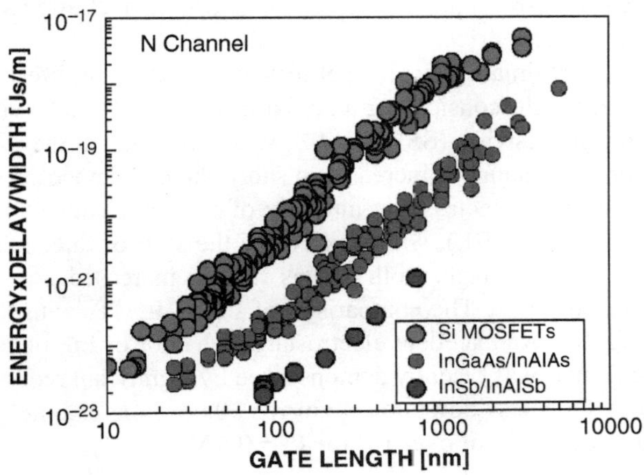

Fig. 8.6 Comparison of intrinsic energy-delay product in different materials systems [25] (Reprinted with permission. Copyright 2005 IEEE)

dependence. In the ballistic regime with large parasitics, the EDP improvements result from the ability to obtain high drain current at low voltage overdrive V_d–V_t, where V_t is the threshold voltage. The necessary voltage overdrive is proportional to the n/C_g, and therefore, scales roughly as 2D DOS or just effective mass, $V_d - V_t \propto m^*$. The Fig. 8.6 demonstrates more than an order of magnitude improvement of EDP over silicon transistors with the main benefit from the 2× reduction of V_d.

The benefits of the III-V channels (both in HEMT and future MOSFETs) can be realized in the circuits by reduction of power supply voltage, V_d (or V_{dd} in CMOS). Strictly speaking, the dynamic figures of merit depend on voltage overdrive V_d–V_t (with threshold voltage, V_t) if only intrinsic capacitance is charging, but depend on the entire voltage swing, V_d, if the parasitic capacitance is larger than C_g. In previous CMOS generations, the supply voltage were scaled down with the channel length to maintain more or less constant electric field in the channel and high (approaching saturation) carrier velocity. Scaling down the V_{dd} has become a difficult problem in short-channel devices, mainly because the threshold voltage has to be kept relatively high to maintain low subthreshold leakage current, and therefore, low static power. In other words, the power supply voltage can be reduced in III-V channels due to the lower voltage overdrive, V_d–V_t needed to achieve high average velocity in the channel.

8.2.3 Static Power

Static power in CMOS circuits, once considered negligible, approaches the dynamic power in deeply scaled devices and should be assessed and reduced. A standard metric for the static power is the transistor "OFF" current, I_{OFF}, or I_{ON}/I_{OFF} ratio. The I_{OFF} includes channel leakage (subthreshold) current and gate leakage current and depends on gate and drain voltages. As the gate length is scaled, the separation between the gate metal and the channel, which is typically referred to as equivalent oxide thickness—EOT (for dielectric constant k = 3.9 as in SiO_2), should be ideally scaled correspondingly to control short-channel effects and to increase the saturation current drive.

To maintain low channel leakage current, the threshold voltage should be relatively high (considering unipolar power supply and swing 0-to-V_d) with steep subthreshold swing (SS). The SS, which is close to the ideal value SS = 60 mV/dec for long channels, increases in short-channel devices. Another important parameter describing electrostatic integrity of a short channel FET is drain-induced barrier lowering (DIBL), which expresses the shift of threshold voltage due to change in the drain voltage. Both effects require increased power supply voltage to maintain low I_{OFF}. The comparison of III-V HEMTs with Si n-MOSFETs is shown in Fig. 8.7. An excellent electrostatic behavior of InP pHEMTs with InGaAs or InAs channels was recently demonstrated by significant reduction of channel-to-gate barrier thickness down to ~4 nm for 30 nm-long channel [33–35]. These resulted in I_{ON}/I_{OFF} ratio of over 10^3 for $V_d = 0.5$ V.

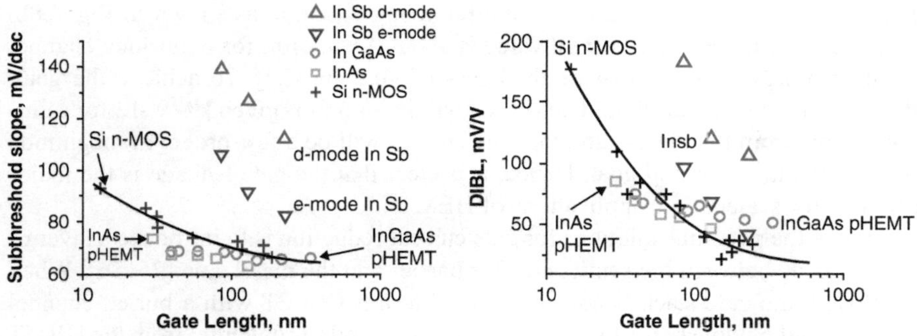

Fig. 8.7 Comparison of the subthreshold swing (*left*) and DIBL (*right*) of FETs with different channels: Si n-MOS and InAs PHEMT data from Ref. [19], InSb data are from Ref. [26]

This excellent scaling behavior is due to the thin top barrier semiconductor layer with relatively high dielectric constant (~12) that gives equivalent oxide thickness of 1.3 nm for 4 nm barrier. However, the Schottky gate with semiconductor top layer provides relatively low barrier height as compared to MOS gate (Fig. 8.1b, c). This leads to a significantly higher gate leakage current in Schottky gate. The dependences of the gate current and subthreshold characteristics are illustrated in Fig. 8.8. The gate current is mostly due to tunneling through the semiconductor barrier and

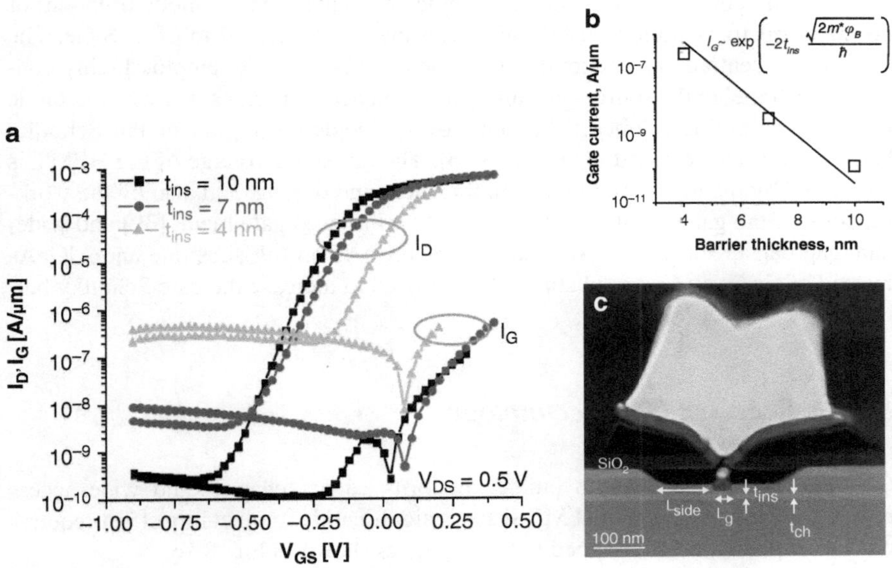

Fig. 8.8 **a** Leakage gate current and subthreshold characteristics of 30 nm InGaAs pHEMT (Courtesy of J. A. del Alamo, 2008); **b** scaling of the gate current with barrier thickness, t_{ins}, from **a** and exponential fit with barrier height $\phi_B = 0.5$eV; **c** TEM image of 30 nm InGaAs HEMT with recessed gate [19] (**c** reprinted with permission. Copyright 2008 IEEE)

is reasonably well fitted with a simple tunneling equation as shown in Fig. 8.8b. The target for the leakage current value is about 0.5 µA/µm for 6 nm-long channel which requires equivalent oxide thickness of ~0.5 nm [27]. To achieve the goal, the Schottky metal gate should be placed within ~1.5 nm (given k ~ 9–12 for semiconductor) from the 2D gas and the gate current will be a few orders of magnitude higher than the expected goal. Hence, it is clear that the gate leakage is the major problem for scaled logic applications of HEMTs.

One of the potential solutions for gate current reduction is to introduce a layer of high-k oxide between the semiconductor barrier and the metal gate [26, 36]. When employed, the gate stack becomes identical to a MOSFET with a buried channel [36, 37], and obviously the two gate technologies tend to converge: from the HEMT side by introducing a thin high-k oxide layer under the metal gate, and from MOSFET side by utilizing the high-mobility buried channel with modulation doping.

Further gate scaling opens up another mechanism contributing to channel leakage current that is band-to-band tunneling (BTBT). The narrow bandgap channel materials (with better electron transport properties) start suffering from the BTBT at longer channels. However, the channel design and supply voltage strongly affect the BTBT that may keep benefits of III-V's in deeply scaled FETs [38].

8.2.4 Enhancement Mode HEMTs

Both directly coupled and CMOS-type logic need enhancement mode (e-mode) or normally-off transistors built typically with inversion channel in Si or SiGe. The HEMT is essentially an accumulation mode device—it has to employ highly conductive channel in the source-gate and gate-drain access regions. However, e-mode HEMTs are well-developed. They employ the depletion region of the Schottky barrier under the recessed gate (Fig. 8.1b). The threshold voltage of the HEMT is controlled by the top barrier bandgap, thickness and doping, and also by the work-function of the gate metal. In e-mode HEMTs, Pt-based gate metal [39] and wide-bandgap barrier material InAlAs instead of InGaAs on InP substrate and AlGaAs instead of GaAs on GaAs substrate [40] are used to increase the gate Schottky barrier above 0.9–1.1 V.

8.2.5 Recessed Gate Technology

Contrary to the HEMT with built-in majority carrier channel and wide access regions, the inversion channel MOSFET is significantly more compact but requires contact implantation overlapped with the gate as shown in Fig. 8.1c.

Non-self-aligned recessed gate technology is a great approach which is responsible for HEMT progress due to relative ease of short gate fabrication and low source/drain resistances formed on highly-doped semiconductor layer as shown in the Fig. 8.1b. On the other hand, this technology is hardly useful for digital VLSI owing to large footprint of HEMT unacceptable for real estate-hungry VLSI and also resulting in higher parasitic capacitance. There are significant efforts to adopt and improve self-aligned gate technologies for HEMT [41–43] to reduce the length of the access regions. Recently, sub-100 nm self-aligned InGaAs pHEMTs were demonstrated with impressive parameters—extrinsic transconductance of 1.3 S/mm for 90 nm-long channel [44] and 1.7 S/mm and cut-off frequency of 520 GHz for 35 nm channel [45].

As outlined previously, the channel performance of HEMTs in ON state can be considered as an ultimate goal for MOSFETs, but high gate leakage current in HEMTs requires substitution of part or entire semiconductor barrier with large-barrier material, such as high-k oxide. However, placing the III-V/oxide interface in the device still faces significant challenges.

8.3 Challenges for III-V MOSFETs

Based on the comparison with mature HEMTs, it is now easy to obtain a list of essential requirements for III-V MOSFETs technology to be employed in logic (preferably CMOS) circuits (Table 8.2). Naturally for any technology, improvement of one

Table 8.2 Challenges for III-V CMOS Technology

Properties	Issues	Approaches and state-of-the-art
High drain current	Need of low mass, high concentration, low scattering	Buried channel, mobility above 5000 $cm^2/V\ s$ [46], drain current above 1 A/mm [47]
Low EOT and low gate leakage	Scalability, electrostatic integrity. Buried channel increases EOT	High-k gate oxides, 1 nm EOT [48, 49]
Thermal stability of gate stack	~750–800 °C for implant activation	Oxides preventing interdiffusion [50–53]
Low interface trap density (D_{it})	High D_{it} results in Fermi-level pinning, instability, large subthreshold swing	Interface passivation. D_{it} below $10^{11}\ cm^{-2}\ eV^{-1}$ [54]
Low source/drain resistance	Low-area ohmic contacts and access region resistance should be improved for further scaling	Implanted source-drain [52] or highly-modulation-doped channel [46], semiconductor regrowth [55–57] may give $<10^{-8}\ \Omega\ cm^2$, or $\sim 0.15\ \Omega$ mm
p-channel	Low hole mobility, large hole mass	Strained channel to split heavy/light holes, InSb channel with mobility over 1000 $cm^2/V\ s$ [58], or Ge channel
Integration with Si	Large lattice and thermal expansion mismatch	Direct growth of III-V's on Si with thick buffer [59, 60], epitaxial oxide interlayer [61], wafer bonding [62]
High device density	Need of low defect density material, currently ~100 cm^{-2}	Improvements of substrate quality and epi growth, no known solution
3D gate	D_{it} passivation on non-planar surface	Schottky gates [63, 64]; surface passivation [65]

parameter (for example, channel mobility in a buried channel) results in degradation of other parameters (correspondent increase of equivalent oxide thickness), therefore choice of design and technology is a subject of trade-offs and optimization.

Superior electron transport properties are a major reason why III-V's are considered for advanced CMOS circuits. However, placing the semiconductor channel in close proximity to high-k gate oxide reduces mobility significantly in a similar way as in Si channel due to interface charges, roughness and soft phonon scattering. Reduction of these scattering mechanisms may be achieved by inserting a 3–10 nm thick wide-bandgap semiconductor barrier spacer between the channel and oxide. But this approach adds up to the equivalent oxide thickness and reinstate the scaling issues as in HEMTs. Quite obviously optimization of the spacer layer can be obtained for a particular FET design and generation.

8.4 Mobility in Buried Quantum Well Channel

Excess scattering of carriers in MOSFET channel due to interface with high-k gate oxide interface significantly lowers the benefits of III-V high-mobility channel materials. The mobility is typically evaluated from electrical characteristics of a MOSFET, but in III-V's with relatively high interface trap density, D_{it} and source-drain resistances, determination of the channel charge and hence the channel mobility is quite a challenging problem. It can be obtained directly from Hall mobility measurements. Although the Hall mobility measurement does not exactly match the drift or effective channel mobilities due to non-unity Hall factor, it allows for more reliable measurements of the channel transport.

Figure 8.9 shows an example of electron Hall mobility dependence on separation with major source of scattering—HfO_2 high-k oxide [66]. The sheet concentration was maintained in the range (2–4) $\times 10^{12}$ cm^{-2}. The graph shows the degradation of the maximum mobility from 11400 to 1470 cm^2/V s in the surface channel due to remote scattering associated with the presence of HfO_2 gate oxide.

The temperature dependence of mobility (Fig. 8.10a) can be used to further analyze the contributions of different scattering mechanisms. The sample with deeply buried ($d = 50$ nm) QW channel shows $\mu \sim T^{-1.2}$ trend which is typical for InGaAs QWs in the 80–300 K temperature range [67] with the main contribution from polar optical phonon scattering and some role of almost temperature-independent alloy disorder and hetero-interface roughness scattering. The samples with thinner top barriers show the reduced slope of the curves and eventually the reversed slopes in the samples with $d < 3$ nm. More accurately, positive μ–T slope was observed in the samples with RT mobility below ~3000 cm^2/V s and is close to $\mu \sim T^{+1.0}$ in the surface QW channels indicating the dominant role of remote Coulomb scattering (RCS) due to charges in the oxide and at the interface.

The net positive μ–T slope is a quite unexpected result that indicates considerably stronger contribution of scattering due to the HfO_2 interface with InGaAs than that with Si [68–70]. This assumption is confirmed by significant improvement of

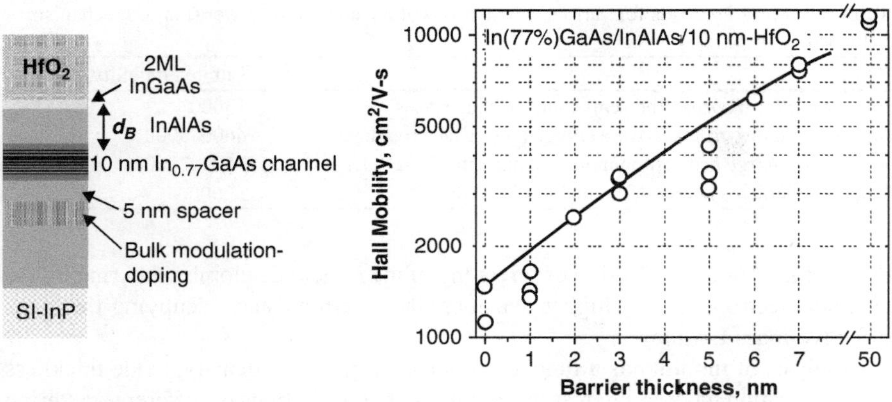

Fig. 8.9 Room temperature Hall electron mobility in $In_{0.77}Ga_{0.23}As$ QW channel as a function of top barrier layer thickness. Electron sheet density is in the range of $(2–4) \times 10^{12}$ cm^{-2}. *Solid line* is to guide an eye. Schematic cross-section of the structures used for Hall measurements is shown on the *left*

channel mobility if 1–2 nm-thick amorphous Si interface passivation layers (IPL) is introduced between the semiconductor surface and high-k (as discussed further in this chapter). The IPL is oxidized giving rise to SiO_x/InGaAs interface which is responsible for a significant increase in mobility.

Figure 8.10b contains room temperature Hall mobility results as a function of electron density for 3 and 5 nm-thick top barrier layers. Here the data from samples with different doping level and annealing are shown. The trend for the highest-mobility QWs is quite obvious: the mobility has a maximum in the range $(1–4) \times 10^{12}$ cm^{-2} and reduces at lower and higher carrier densities. The low-density mobility

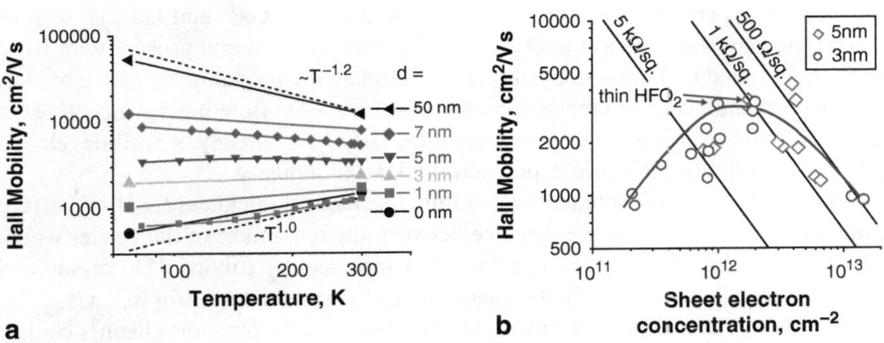

Fig. 8.10 **a** Temperature dependence of Hall mobility in QWs at various top barrier layer thickness and carrier density $(2–4) \times 10^{12}$ cm^{-2}; and **b** room temperature mobility as a function of electron density for 3 nm and 5 nm-thick top barrier layers. Trend for 3 nm thick barrier is shown by *solid line* [66]

Table 8.3 Mobility limits for various mechanisms at RT and 3 nm barrier (major mechanisms italicized)

Process	Limited mobility $cm^2/V\,s$
Phonons, alloy disorder and semi conductor interfaces	13000
Interface dipoles, trapped oxide charge and surface roughness	4600–5000
Bulk oxide charge and soft phonons (10 nm annealed HfO_2)	8500–20000
Trapped interface charge $\sim10^{12}$ cm^{-2}	>20000

drop is likely due to reduction of screening of the remote Coulomb scattering by the channel electrons, and at higher densities, the electrons start occupying the high-effective-mass L-valley.

Analysis of mobility as a function of annealing, carrier density, oxide thickness and passivation gave an interesting estimates of contributions of different scattering mechanisms at room temperature as summarized in Table 8.3.

The major scattering mechanism of electrons in QW with 3 nm-thick top barrier and HfO_2/InGaAs interface (Hall mobility is limited to ~3500 $cm^2/V\,s$) is remote Coulomb scattering (RCS) by interface dipoles and trapped oxide charge and, possibly, interface roughness. Strong (almost proportional) dependence of mobility on temperature implies a stronger contribution of the RCS over the surface/interface roughness scattering, which has weaker temperature dependence.

8.5 Interface Passivation Technologies

For sub-16 nm nodes the highly scaled gate length devices would require very low equivalent oxide thickness (EOT) to maintain electrostatic control of the channel. This requires elimination of low-k layer at the interface of III–V and gate dielectric. Native oxide surface of III-V, for instance GaAs, consists of As oxidation states, As^{3+} and As^{5+}, corresponding to As_2O_3 and As_2O_5 [71], Ga^{3+} and Ga^{5+} as well as other oxide species such as Ga_2O [72, 73]. Having an all-in-situ process with III-V growth followed by high-k deposition can eliminate the formation of native oxides. But from manufacturing perspective, an ex-situ process flow has its incentive. In addition, atomic layer deposition technique (ALD) is already a well-developed process module for dielectric deposition in Si based industry.

The measured EOT includes the quantum mechanical thickness for III-V surface material as well, which is the distance between the maximum of the carrier wave-function and the semiconductor surface. For instance, by solving 1D Poisson and Schrödinger equation [74–77], the quantum mechanical thickness for $In_{0.53}Ga_{0.47}As$, when only lowest valley is occupied, is ~1.2 nm. For a better short channel control of scaled inversion channel devices it is imperative to incorporate a high-k dielectric as gate insulator. In addition, high-k dielectric is needed to improve subthreshold swing (SS). The ability of a MOSFET to switch on and off is described by the sub-threshold region. The value of subthreshold slope (SS) is dependent on competition between accumulation (oxide) capacitance (C_{ox}) and capacitance due to interface state density (C_{it}) [78]. Increasing C_{ox}, by using higher-k dielectric and scaling the device, can reduce the effect of C_{it} on SS.

Besides higher-k, it is critical that the Fermi level at the dielectric/III-V interface is weakly pinned or unpinned. Sensitivity of capacitance–voltage characteristics of the MOS device to the metal gate work function and low frequency dispersion [74, 79, 80] are also typically used as criteria for not strongly pinned Fermi level at the dielectric- III-V interface. Fermi-Level pinning (FLP), a term assigned as the reason why the barrier height of a GaAs Schottky diode is "pinned" at around 0.8–0.9 eV independent of the metal gate deposited, has since received extensive reviews [65]. FLP had also been tied to the key interfacial issues of the recent high-k CMOS development, such as the metal/high-k oxide interface and Si/high-k oxide interface [81]. FLP issues have now been eliminated by the replacement of the poly-Si gate with metal gates and also in large part due to the superior quality of Si/SiO_2 interface. In contrast, FLP by surface states, at the interface between III-V and the gate-dielectric due to the missing of the corresponding high quality native oxides, has imposed significant challenge to the development of surface-channel inversion-mode III-V MOSFETs.

The dielectric on III-V is being studied using various in-situ and ex-situ deposition methods. Literature reports are abundant with data from in-situ process, such as gallium-gadolinium oxide—GGO [82, 83], Si interlayer with e-beam HfO_2 [84] and ex-situ approaches, such as atomic layer deposition (ALD) dielectrics [71, 74, 85], arsenic capping [86–88], PVD Si, Ge interlayers [89, 90] etc. Typically GaAs and InGaAs layers grown on GaAs substrates or InGaAs (with varying In % and channel thickness) grown on InP substrates by MBE or MOCVD with appropriate buffer layers are used. Surfaces of group III-V semiconductors typically contains high density of traps resulting in Fermi level pinning and high surface recombination rates. It is typically accepted that these midgap traps are associated with surface oxidation (possibly with As-O bonds [91]), and therefore, are present at almost any interface with a metal or dielectric. This problem can be addressed by appropriate interface control using some sort of surface passivation. After more than three decades of research, a large number of passivation techniques were proposed and tested (Table 8.4). In 1990s and early 2000s, mainly due to efforts of researches from Bell Labs and later Motorola, $(GaGd)_2O_3$ gate oxide was positioned as the most promising material for III-V interface passivation, possibly due to formation of electrically inert bonds at the interface, such as Ga-O-Ga rather than As-O [91] ALD which is currently a mainstream technology for high-k gate stacks on Si is also very attractive for III-V's, as it may not need any special passivation [92, 93] (although sulfur or Si [94] passivation might be still beneficial). Effectiveness of a 1.5 nm thick amorphous Si interface passivation layer (IPL) for in-situ passivation of GaAs and InGaAs (with low In content on GaAs substrate) surface enabling good electrical properties and thermal stability with high-k oxide has also been recently reported [94]. Further improvements and mainly in-situ deposition of high-k oxide, resulted in virtually zero thickness of Si IPL [95, 96].

Table 8.4 Interface Passivation Technologies

Passivation Technology	References
In-situ high-k oxide on InGaAs	Ref. [48]
Amorphous ultrathin Si or Ge	Ref. [92, 95, 97, 98]
Atomic layer deposition of Al_2O_3, HfO_2	Ref. [93, 99–101]
E-beam or MBE of Ga_2O_3–Gd_2O_3	Ref. [102–106]
MBE–grown crystalline silicon	Ref. [107–109]
Atomic layer passivation with InP or GaP layers	Ref. [110, 111]
Sb-passivation of GaAs with MgO dielectric	Ref. [112]
Low-temperature-grown GaAs	Ref. [113]
Sulfur, Na_2S, $(NH_4)_2S$ passivation	Ref. [111, 114]
Hydrogen or Nitrogen plasma treatment	Ref. [115, 116]

Besides detailed physical characterization such as band offsets, interface structure and roughness, these reports also include electrical characterization using metal-oxide-semiconductor devices like capacitor and inversion or majority carrier long channel transistors with either conventional ion implantation or metal source/drain contacts. The learning from various published work is summarized in following sections.

8.5.1 Ex-Situ Dielectrics

From manufacturing perspective, an ex-situ dielectric deposition process is preferred. But poor native oxide of III-V substrates can be deleterious to the device performance. In reports of surface passivation behavior of $In_xGa_{1-x}As$ where high-k gate dielectrics such as Al_2O_3, HfO_2, ZrO_2 are grown by ALD [117–120], the challenge is how to maintain the surface of $In_xGa_{1-x}As$ clean enough in terms of minimizing formation of the interfacial layer. Ideally if subsequent ALD of high-k dielectrics can be done by in-situ process following MBE of III-V channel material, one can expect oxidations of components of III-V material can be minimized. However, for any practical applications of MOSFET, we need to assume some sort of "ex-situ" process before high-k ALD, which instigated a number of researches for ALD self-cleaning prior to growth of such high-k films for different precursors and ALD conditions.

Interestingly, several recent reports have revealed that ALD precursor chemistry can "self-clean" interfacial native oxides from III-V surfaces [121–123]. Also, the frequency dispersion and hysteresis on devices processed using different pre-cleans such as HF, HCl etc, prior to ALD dielectric deposition have not shown any appreciable dependence supporting the self-clean attribute of ALD process chemistry [123]. In addition, there are experimental evidence indicating the treatment of GaAs surfaces in sulfide forms Ga–S and As–S bonding which can resist further oxidation [124]. Inclusion of steps such as pre-clean and passivation in $(NH_4)_2S$ prior to ALD film deposition has been shown to improve interface characteristics [125]. These findings suggest that reduction in amount and type of oxides may be dependent on certain ALD precursor chemistries [126, 127].

ALD HfO_2 and ZrO_2 are promising candidates on III-V due to their higher dielectric constant value. The integration of such scaled high-k oxides can potentially help realize high I_{ON} and low I_{OFF} in MOSFETs. With these novel III-V materials in CMOS, it is essential to develop and orient various physical and electrical characterization techniques to probe and evaluate the interface and bulk characteristics effectively and correctly at the atomic level. Various analysis techniques are discussed in following sections. The experimental reports listed in these sections are mostly related to ex-situ dielectrics flow although it is emphasized that same techniques can also be used to evaluate devices with in-situ dielectrics.

8.5.1.1 ALD Self-Cleaning Mechanism: XPS

The Synchrotron Radiation Photoemission Spectroscopy (SRPES) has been intensively used to look into the changes in the interfacial layer behaviors due to its excellent in-depth resolution of compositional analysis [128, 129]. Figure 8.11 shows an example of Ga 3d, In 4d and Hf 4f and As 3d spectra from the interfacial layer which was observed after ALD of HfO_2 with TDEAH precursor on top of MBE grown $In_{0.53}Ga_{0.47}As$, where the native oxides were intentionally left in order to examine the transition of native oxides before and after ALD processes. It was followed by removing the HfO_2 leaving a couple of mono-layers and the interfacial layer left before SRPES measurement [129].

As compared to the initial surface of the $In_{0.53}Ga_{0.47}As$ shown in Fig. 8.12, the amount of native GaO_x, InO_x and AsO_x appear to be significantly reduced from the initial as-received substrates and surface elemental As-As bonding appears. It was reported that the thickness of the native oxide before the ALD growth of HfO_2 was estimated to be 2 nm and was reduced down to less than 1 nm after ALD HfO_2 depo-

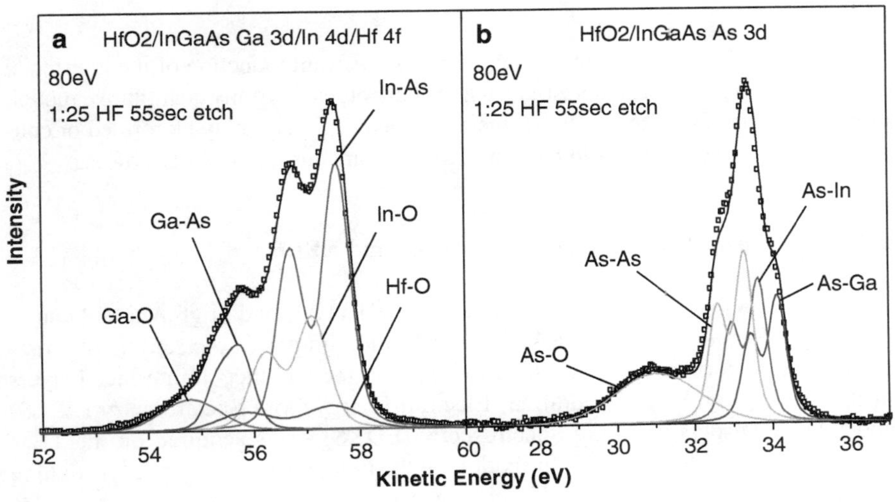

Fig. 8.11 Ga 3d/In 4d/Hf 4f spectrum at the interface of HfO_2/InGaAs after 55 sec etch-back

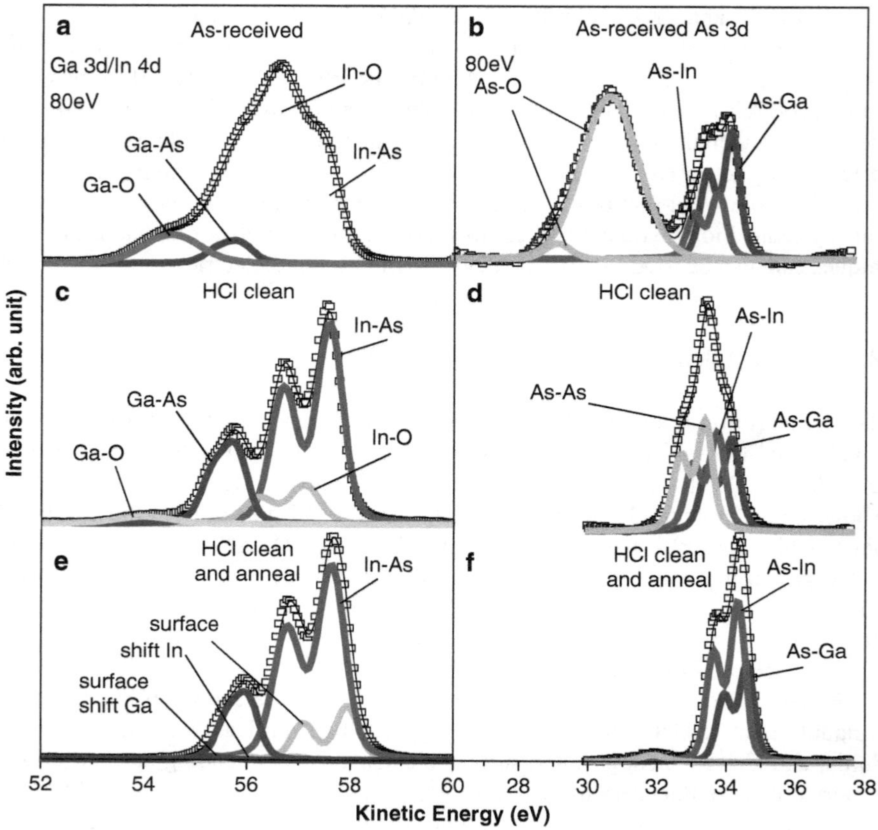

Fig. 8.12 Ga 3d/In 4d and As 3d spectrum of (**a**, **b**) as-received, (**c**, **d**) HCl clean, (**e**, **f**), HCl clean + annealed, InGaAs substrate

sition [119, 123]. The mechanism behind this significant reduction of the interfacial layer during ALD has been investigated now, but not reaching any quantitative model, though it is speculated that volatile interfacial layer products may be formed or conversion of interfacial oxides to hafnium oxide occurs during the ALD process.

8.5.1.2 Transmission Electron Microscopy and EELS

High-resolution (HRTEM), bright-field TEM (BFTEM) and High Angle Annular Dark-Field Scanning TEM (HAADF-STEM) micrographs of the sample in cross-section are powerful techniques that can highlight the presence of interfacial layers in the stacks, Fig. 8.13. In addition, Electron Energy Loss Spectrometry (EELS) and Energy Dispersive X-ray Spectroscopy (EDXS) scans acquired on the TEM samples can provide detailed information on the type of signals in each layer along with possible intermixing around interfacial regions.

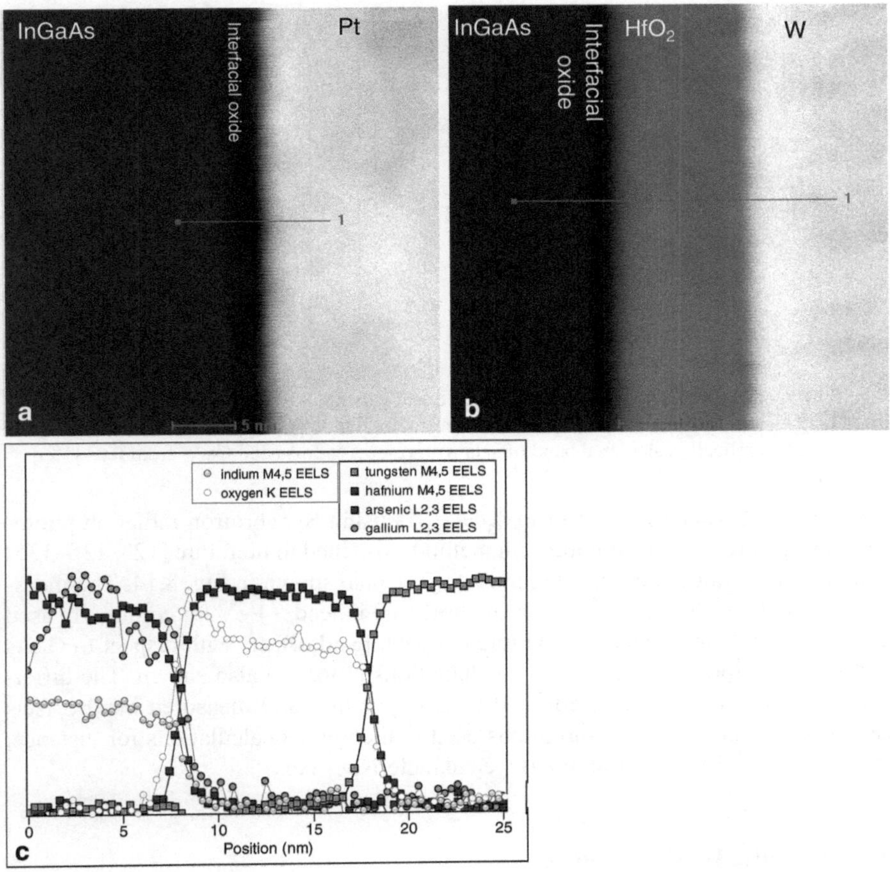

Fig. 8.13 High Angle Annular Dark-Field Scanning TEM (HAADF-STEM) micrograph of the sample in cross-section with corresponding Electron Energy Loss Spectrometry (EELS) scans acquired along the line from *left* to *right*. **a** 2 nm thick native oxide on $In_{0.53}Ga_{0.47}As$ surface after air exposure for several hours. **b** 10 nm HfO_2 on $In_{0.53}Ga_{0.47}As$ followed by 500 °C anneal showed about 1.2 nm thick interfacial oxide layer left after ALD self-cleaning and, **c** EELS spectra for **b**. The indium, arsenic and gallium signals can be clearly observed in the region corresponding to the InGaAs layer. In the region corresponding to the interfacial oxide layer a clear oxygen signal can be observed to overlap with the arsenic and gallium signals. The hafnium and oxygen signals can be clearly seen to overlap in the region corresponding to the HfO_2 layer. Finally, the tungsten signal can be observed in the region corresponding to the metal layer [123] (Reprinted with permission. Copyright 2008 American Institute of Physics)

8.5.1.3 Band Offsets

The dielectric should act as an insulator with low leakage current. Hence a requirement is that the valence and conduction band offsets of these dielectrics with respect to the semiconductor of interest be reasonably large, preferably > 1 eV. Since ALD chemistry can remove native oxides from InGaAs surfaces, it is possible to measure

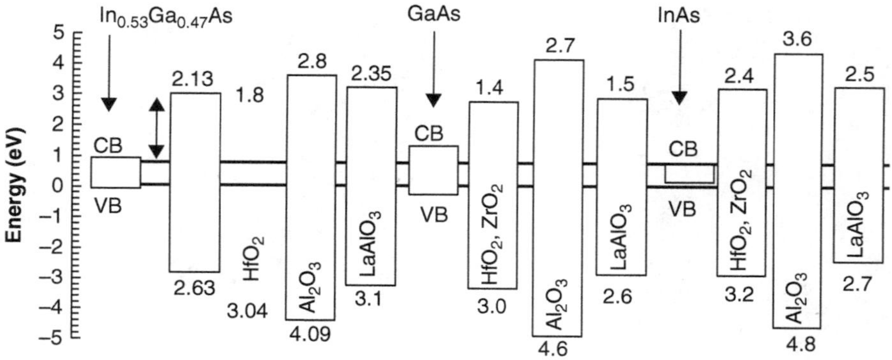

Fig. 8.14 Conduction and valence band offsets of various ALD dielectrics with respect to $In_{0.53}Ga_{0.47}As$ substrate as measured using synchrotron radiation photoelectron spectroscopy. Also shown are theoretically calculated band offsets with respect to InAs and GaAs from Ref. [136]

the band offsets with respect to $In_{0.53}Ga_{0.47}As$ using Synchrotron radiation photoelectron spectroscopy technique with method explained in literature [128, 130–135] and measured values are reported in Ref. [74], also shown in Fig. 8.14. The measured band offsets for both electrons and holes are indeed > 1 eV for commonly used dielectrics. In comparison to experimental data, band offsets with respect to GaAs and InAs as reported by theoretical calculations [136] are also shown. The differences in theoretical data determined by interpolation and measured band offsets can arise due to possible assumptions used in theoretical calculations for instance, crystallinity of the dielectric, interface characteristics etc.

8.5.1.4 Pulse IV Measurements

For detailed evaluation of gate stack issues, MOSCAPs and long channel nMOSFETs are good test vehicles, but not meant for performance comparisons. Based on several published results, ALD dielectrics appear promising for gate stack on surface channel devices with $In_xGa_{1-x}As$. It is noteworthy that these stacks appear thermally stable up to the dopant activation anneals (~700 °C). The turn on characteristics of the transistor, has been shown to be in accordance with $\phi_m - \phi_s$ again suggesting interface is not strongly pinned. The room temperature subthreshold slope (SS) is in the order of 94–120 mV/decade [74]. However, a detailed evaluation of interface states and charge trapping in these stacks is imperative to evaluate their true potential.

It is critical to evaluate the quality of the interfacial layer and charge trapping in high-k gate dielectric stacks as it may affect transistor performance and reliability. To determine surface state densities using transistors, only a few techniques such as deep level transient spectroscopy [137], 1/f noise [138], charge pumping [139, 140] etc., are available. One major issue in the introduction of high-k dielectrics to III-V MOSFETs is the charge-trapping phenomena. The charge trapping, which occurs while the gate voltage is ramping up during *I–V* measurements, may result in an uncontrolled threshold-voltage shift. Charge trapping/de-trapping greatly increases the complexity of characterizing high-k dielectrics. Pulse I-V [141] and charge pumping (CP) techniques [139, 140] can be used to understand source and impact of charge trapping.

Inversion charge can be trapped in bulk high-k and at interface in the presence of vertical E-field. It will be easier to trap in bulk if the interfacial layer is thinner. As charges are trapped in the gate dielectric, the threshold voltage of the transistor increases due to the built-in voltage in the gate capacitor; therefore, the drain current decreases. Pulse I-V technique can be used to investigate the presence of this issue in the gate stack [141]. Pulse and dc IV measurements are shown for $In_{0.53}Ga_{0.47}As$ MOSFET, Fig. 8.15.

To evaluate charge trapping, pulse-mode I-V measurements much like the case of high-k on Si [141], can be used for III-V based devices [74]. For InGaAs based surface channel devices, the single pulse measurement that by-passes the charge trapping issue shows a significant increase in I_{ON} and improvement in SS, Fig. 8.15, indicating that the actual mobility of carriers in the channel and subthreshold characteristics of surface channel MOSFETs are usually better than those measured in dc conditions. The transient measurement does not allow for de-trapping process, possibly leading to larger hysteresis. The mobility extracted from the surface channel MOSFETs shows an improvement after reducing the bulk charge contribution, Fig. 8.15. Note, this surface channel mobility is much lower than the buried-channel

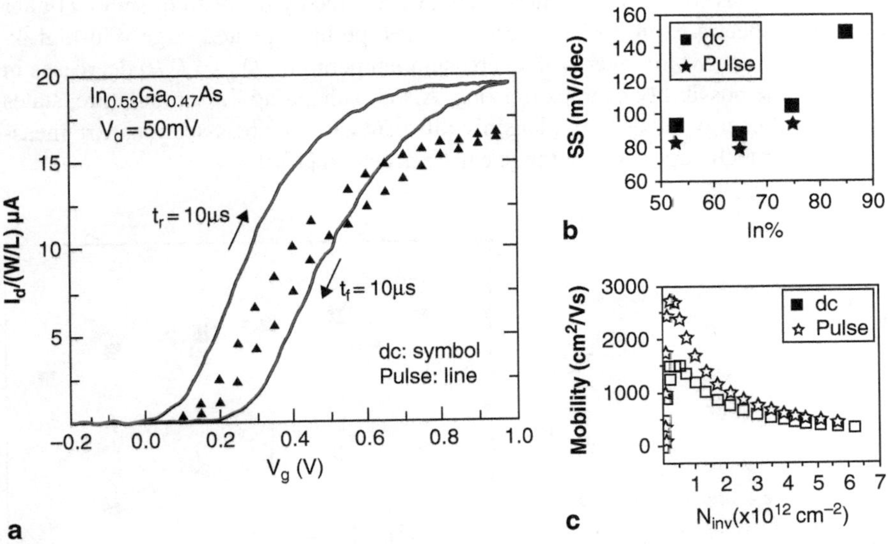

Fig. 8.15 **a** Pulse vs dc IV scan for $ZrO_2/In_{0.53}Ga_{0.47}As$ nMOSFET device. **b** Subthreshold slope measured from dc and pulse IV scans for various In % in InGaAs channel device. **c** Mobility as a function of N_{inv} as extracted from split CV using dc IV and pulse IV curves. Mobility enhancement after removal of bulk high-k charge contribution to device characteristics is seen

HEMT devices suggesting additional mechanisms (surface related) degrading mobility.

8.5.1.5 Charge Pumping Measurements

The charge pumping measurement can reliably quantitatively characterize MOS interfaces [139]. When performing CP measurements, the source and drain are either grounded or have a small reverse bias applied. Periodic pulses applied to the gate drive the channel region into inversion and accumulation. Recombination of trapped electrons and holes leads to a dc CP current as measured from the substrate. Keeping the pulse amplitude constant and sweeping the pulse base voltage from flatband to threshold voltage or vice versa, the CP current reaches its maximum at a certain pulse base voltage value. Area density of traps (N) within the detectable energy and spatial ranges for a given CP measurement condition can then be determined.

Figure 8.16 shows the charge trapping spectra for different frequencies on the same stack [74]. For SiO_2 gate dielectrics, most traps are located close to the Si/SiO_2 interface; therefore, N is typically considered to represent the interface traps. However, in general, traps in a dielectric may also be located away from the interface in the bulk of the dielectric film or the recombination can occur anywhere in the III-V layer structure. Therefore N deduced from these measurements has to be carefully analyzed before attributing it to just due to interface charge density. With decreasing measurement frequency, the probe depth increases. Both I_{cp} (Fig. 8.16) and V_{base} change as the frequency is decreased indicating that trap states are distributed spatially as well as in the energy level (measured by lower frequency). Higher frequency spectra exhibit high interface states, perhaps related largely to high-k/channel interface. Moreover with decreasing temperature, Q_{cp} (= I_{cp}/f) decreases in line with the possibility of traps freezing. Also by changing the temperature, states at different energy levels with plausibly different capture cross-sections, are measured and a much larger energy range can be accessed [130].

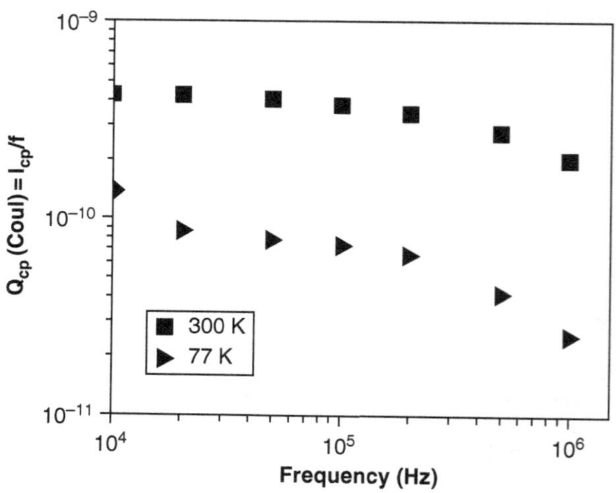

Fig. 8.16 The charge pumping spectra (I_{cp}/f) taken as a function of frequency and temperature on an n-channel $ZrO_2/In_{0.53}Ga_{0.47}As$ inversion carrier transistor [74] (Reprinted with permission. Copyright 2008 IEEE)

CP measurements have also been reported on Si interlayer $In_{0.53}Ga_{0.47}As$/high-k gate stacks [142] to estimate the density of interface states. The trap capture cross section can be determined to estimate the energy range probed by CP. The devices with and without a-Si interlayer showed $D_{it} = 6.1 \times 10^{13}$ cm^{-2} eV^{-1} and 1.6×10^{13} cm^{-2} eV^{-1}, respectively. The results supported measured SS trend in those devices and that the improved mobility observed for the samples with an a-Si interlayer [142] cannot be related to a reduction in D_{it} located near the intrinsic Fermi level.

8.5.1.6 Capacitance–Voltage and Conductance–Voltage Measurements

The charge density extracted from the high frequency charge pumping current value or SS values in a MOSFET can also be ascertained by capacitance and conductance measurements. Literature reports have indicated typical ALD MOS devices with D_{it} ~ high 10^{11}–10^{13} cm^{-2} range at dielectric/$In_{0.53}Ga_{0.47}As$ interface.

A few key issues, concerning the band bending behavior of III-V MOS systems impacted by FLP issues, have been compounding the overall understanding of III-V/dielectric interfaces. For example, with good capacitance (CV) or conductance (GV) measurements at room temperatures of a Si/SiO_2 MOS system we can safely assume that CV/GV at other temperatures (and hence at the other energies of the band gap) will also be well behaved. This is because the interfacial defect traps density D_{it} of a well-passivated Si/SiO_2 interface is mostly a bath-tub "U" like shape with $D_{it} \sim 10^{10}$ cm^{-2} eV^{-1} in most of the band gap [143–145]. In contrast, this assumption may not be applicable to III-Vs. In the case of GaAs, it is well known that it has one (or two large D_{it}) peak(s) in the midgap [146–148], as shown in Fig. 8.14. The midgap D_{it} peak(s) can not be revealed from the room temperature CV/GV measurements, which can only be observed if the samples are heated up to the required higher temperatures [148]. On the other hand, the high D_{it} peak of $In_{0.53}Ga_{0.47}As/Al_2O_3$ interface resides very close to the valence band, also shown in Fig. 8.17, can only be observed when the corresponding samples (p-type) are cooled down to very low temperatures [149]. The unusual D_{it} spectrum of III-Vs, not encountered in the Si/SiO_2 interface, obviously compounds the understanding of the FLP of the III-V MOS systems. One may be required to study the complete D_{it} spectrum before and after any specific passivation scheme introduced in order to fully understand the overall impact.

These D_{it} spectra, unfortunately, are very sensitive to many parameters. Using $In_xGa_{1-x}As$ as an example, the large D_{it} peak resides largely in the midgap region for In = 0% (GaAs/ALD-Al_2O_3) appears to be moved toward the valence band edge with high indium content (In = 53%, as in $In_{0.53}Ga_{0.47}As$/ALD-Al_2O_3), as shown in Fig. 8.17. In reality, this apparent relocation of D_{it} peak is attributed to the bandgap narrowing effect. Nonetheless, one still needs to modify the corresponding CV (or GV) measurements, as mentioned above (heating up the GaAs to 423 K and cooling down the InGaAs samples to 77 K), to comprehend this change [148–151].

Another example is about the common FLP removal criterion used for Si-MOS systems. It is generally believed that FLP removal efficiency can be measured from the flat-band voltage V_{fb} shift with respect to the metal gate work function change.

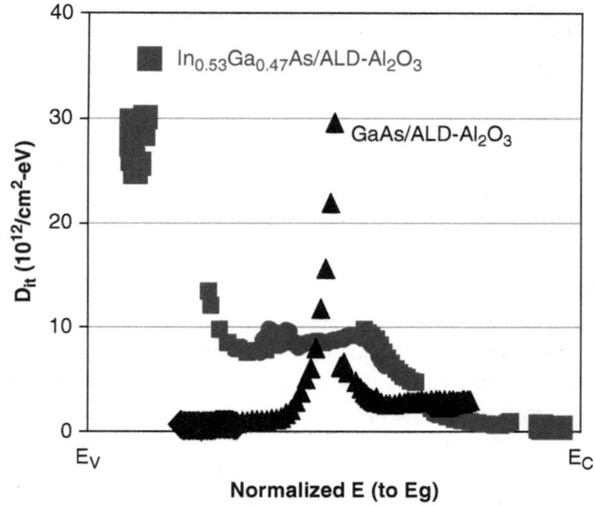

Fig. 8.17 Normalized D_{it} spectrums of GaAs/10 nm ALD-Al_2O_3/Pd capacitors (*solid triangles*) and $In_{0.53}Ga_{0.47}As$/10 nm ALD-Al_2O_3/Pd capacitors (*solid squares*), both systems have undergone exactly the same treatments including $(NH_4)_2S$ surface preclean before the atomic layer deposition of Al_2O_3 (in the same chamber using the same recipe) and a 300 °C, 30 min forming gas anneal (FGA). Note that the completed spectra were later constructed using conductance measurements carried out at temperatures ranging from 77 to 300 K for InGaAs samples and 300 to 423 K for GaAs samples [149]

This is because the shift of V_{fb} (surface potential $\psi_s = 0$) will translate into an exact shift of V_{th} ($\psi_s = \sim 2\psi_B$) for a well-passivated Si/SiO_2 interface ($D_{it} \sim 10^{10}$ cm^{-2} eV^{-1}). However, this practice, again, may not be directly applicable to those of the III-V MOS systems due to the large D_{it} in the band gap where ψ_s must travel through from V_{fb} ($\psi_s = 0$) to V_{th} ($\psi_s = \sim 2\psi_B$). The V_{fb} shift alone can no longer be representative for the overall band gap response. This suggests that the metal gate work function criterion mentioned above may need to be revised, such as including additional *direct* measurements of V_{th} shift to comprehend the D_{it} impact.

There is another complication from high D_{it}. The additional gate voltage ΔV_g required for a given surface potential displacement $\Delta\psi_s$ before reaching the inversion can be related as the *n*-factor (the reciprocal value of $d\psi_s/dV_g$). *n*-factor is close to its minimum and $d\psi_s/dV_g$ is close to its maximum (~unity) in the weak inversion region ($\psi_s = \psi_B$ to $\psi_s = 2\psi_B$) for a well-passivated Si-MOS systems [152]. However, these values can no longer maintain close to unity for the III-V interfaces. For example, $d\psi_s/dV_g$ measured by conductance measurements can be as low as 0.3 for an atomic hydrogen passivated lattice-matched p-$In_{0.53}Ga_{0.47}As$/10 nm ALD Al_2O_3/Pd capacitor [153]. Application of this $d\psi_s/dV_g$ (n-factor) for FLP evaluation, valid in the depletion and weak inversion region only (not in the accumulation or the strong inversion region), implies that FLP may have to be used in an *analogue* manner in describing the III-V interfaces, i.e., it can be 100% hard pinned or unpinned, or more generally as "weakly" or "moderately" pinned [153, 154]. For extremely high interfacial defect density D_{it}, surface potential ψ_s can no longer move (thus 100% pinned and $d\psi_s/dV_g = 0$), such is the case for GaAs with the very large midgap D_{it} peak preventing ψ_s from reaching the strong inversion. A weakly or moderately pinned ψ_s, on the other hand, implies that the response of ψ_s to V_g is weakened to some extent due to high D_{it}, under which some additional V_g is required to mitigate the D_{it} effect to keep the band bending and ψ_s moving to strong inversion, such as the case for the lattice-matched p-$In_{0.53}Ga_{0.47}As$/10 nm ALD-Al_2O_3/Pd capacitors in Fig. 8.17, and corresponding values of $d\psi_s/dV_g$ are between 0.3 and 0.5 in the upper part of the energy band. In any case, lower values of $d\psi_s/dV_g$ (higher values of n-factors) simply mean a worse FLP condition resulting in an undesirable higher V_{th} shift due to high D_{it}.

8.5.1.7 Split CV

Presence of traps in bulk high-k as well as interface with InGaAs is evident by pulse IV and CP techniques. Charges trapped in these levels can lead to Coulomb scattering of the charges in the inverted channel, as a result degrading mobility. Typically effective mobility of a MOSFET is generated using split CV technique where mobile inversion charges are determined by using gate to channel capacitance as a function of gate voltage [155]. But in low band gap and small effective mass materials like higher indium content InGaAs, traps can respond at even 1 MHz. As a result, the room temperature split CV curve can have contributions from interface state capacitance, hence overestimating Q_{inv} [74, 156].

A technique to remove the *C*it response in the split *C–V* measurement has been proposed [156]. Low-temperature split *C–V* is employed to freeze out the D_{it} response to the ac signal but still maintain the stretch-out caused by the same interface traps. Matching the experimental split *C–V* measurements with a simulated *C–V* at 77 K with the D_{it} distribution determined from any method enables the simulation of *C–V* curves at higher temperatures, providing an accurate measure of Q_{inv} and, hence, μ_{eff}. Although this correction provides a better insight into the intrinsic mobility and true potential of the device, the poor interfaces still need to be improved (by passivation efforts) to prevent threshold voltage variability and reliability issues in the transistor.

8.5.1.8 Leakage Current

The significant I_{OFF} contributors in electrically scaled InGaAs MOSFETs also require evaluation. The amount of activation of carriers in ion implanted source-drain junctions in InGaAs is restricted by material's lower thermal budget requirements. Figure 8.18 shows that both I_g and I_{bulk} contribute I_{OFF} [74]. However at lower temperature, the I_{OFF}, gate and junction leakage are significantly reduced. It is therefore evident that in the current status of the ion implanted source/drain region MOS devices, I_{OFF} may be defect-related.

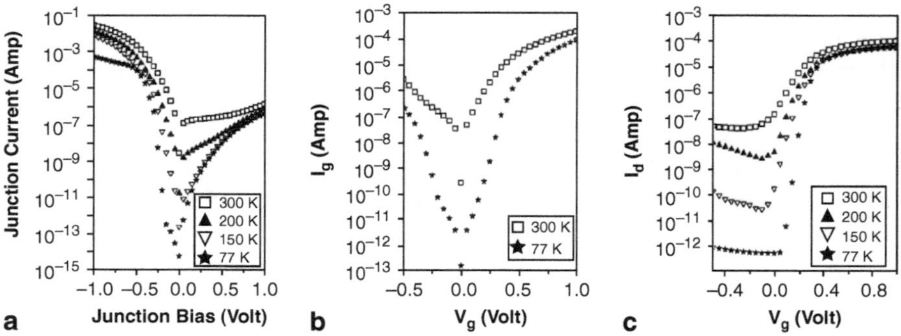

Fig. 8.18 **a** Junction leakage, **b** gate leakage and **c** off-state leakage current as functions of **a** junction bias or **b**, **c** gate bias and measurement temperature are shown for $ZrO_2/In_{0.53}Ga_{0.47}As$ n-channel MOSFET [74] (Reprinted with permission. Copyright 2008 IEEE)

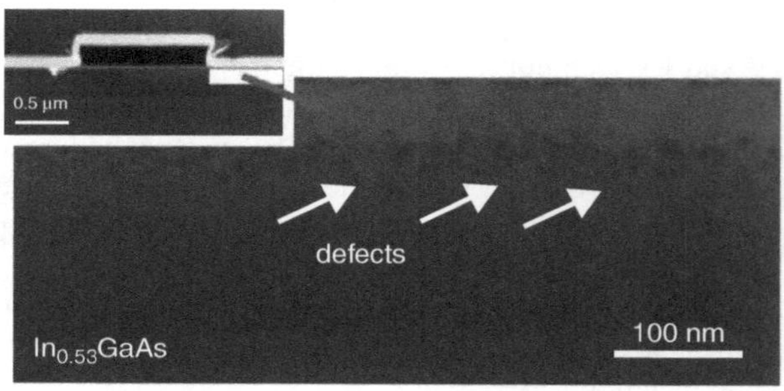

Fig. 8.19 High resolution transmission electron microscopy image (HRTEM) of the ion-implanted junction region post 700 °C RTA annealing in $In_{0.53}Ga_{0.47}As$. Clear signs of defects that have not been annealed out, are observed

HRTEM of the junction region clearly shows that the ion Implantation defects are not annealed out by the thermal treatment, typically in the range 600–750 °C, used for dopant activation, Fig. 8.19. These improvements at lower temperatures, Fig. 8.18, suggests the leakage current can be controlled to lower levels even with low band gap InGaAs channels and thereby possible to achieve a very high I_{ON}/I_{OFF}. It is therefore critical to evaluate novel passivation treatments for gate dielectric and annealing out the end of range defects if ion implanted source and drain regions are employed.

8.5.1.9 Passivation of III-V/Dielectric Interface

Through several detailed physical and electrical evaluation of various ALD dielectrics on InGaAs, the primary challenges associated with gate stacks on high electron mobility InGaAs channels can be delineated and addressed. Higher-k dielectrics appear as promising gate stack with acceptable leakage for low I_{OFF}, thermal stability allowing for integration into transistors with reasonable thermal budget. However, passivation of interface as well as bulk high-k traps is needed to realize higher mobility III-V nMOSFETs.

The issue of high density of states at the surface of group III-V semiconductors and their interfaces with gate dielectrics has been addressed using two major approaches, namely: (1) by maintaining appropriate control of the interfacial bonds with oxygen, and (2) by using interface passivation techniques. The first approach developed in 1990s to form electrically inert bonds at the interface, such as Ga-O-Ga rather than As-O [91], is still considered as a very promising technique for III-V interface passivation due to its in-situ nature enabling appropriate control of oxygen bonding. The second approach includes variety of techniques such as (1) pre-gate dielectric deposition treatment of the III-V surface with hydrogen or nitrogen plasma [115], (2) wet chemical surface passivation with sulfur [114], and/or (3) deposition of Si [91, 107, 157] and Ge [98] interface passivation layers (IPL). A few of these approaches are discussed next.

Hydrogen Passivation

Due to the highly reactive and amphoteric nature, hydrogen can play a significant role in altering the electrical properties of its host materials [158, 159]. The beneficial effect of molecular or atomic hydrogen anneal on the electrical performance of Si MOS systems, especially on reducing the interfacial defect density D_{it} (eV^{-1} cm^{-2}) of Si/SiO_2, is well documented [145, 160] and has been exercised as part of the standard process flow in modern Si-CMOS manufacturing. Hydrogen passivation effect on the bulk of III-V materials has also been studied extensively [161, 162]. Among them, Pearton [163] reported that the atomic hydrogen anneal is quite effective in passivating deep levels of GaAs: an E_c-0.36 eV electron trap and the EL2 complex associated with As_{Ga} antisites, and yet molecular hydrogen anneal is not.

Hydrogen passivation for improving D_{it} of III-V MIS systems has continued to be of interest. Callegari et al. [115] reported that D_{it} of GaAs/Ga_2O_3 interface can be reduced to ~10^{11} eV^{-1} cm^{-2} after the H_2 plasma clean of GaAs surface prior to the e-beam deposition of Ga_2O_3. Jauoad and Aimez [164] believed that during the post deposition anneal hydrogen incorporated from the Si_3N_4 deposition may move and repair the AlGaAs/Si_3N_4 interfacial damages. Li et al. [165] showed that hydrogen plasma treatment improved D_{it} of GaAs/Si_3N_4 interface to 9×10^{11} eV^{-1} cm^{-2} (at 0.57 eV above E_v). Passlack et al. reported that hydrogen anneal is very beneficial for improving GaAs/Ga_2O_3 interfacial quality [166]. Brammertz et al. [148, 149] also found similar benefit for GaAs/Al_2O_3 and $In_{0.53}Ga_{0.47}As$/Al_2O_3 interface.

Systematic study of D_{it} improvement from hydrogen passivation for InGaAs MIS systems is rather limited. Renaud et al. showed that D_{it} of $In_{0.53}Ga_{0.47}As$/~1.1 nm Ga(In)N_x/Si_3N_4 interface, and the corresponding electrical properties, can be improved using a low density hydrogen plasma [167]. Lin et al. and Brammertz et al. also used a hydrogen anneal, discussed in the subsequent part of this section, to improve $In_{0.53}Ga_{0.47}As$/Al_2O_3 interface [149, 168, 169].

Atomic hydrogen, more effective than molecular hydrogen in passivating bulk defects of GaAs and much less likely to harm interface like those from energetic hydrogen ions of plasma [170], is considered as a good candidate for passivating III-V/dielectric interfaces. Although difficult to be quantified exactly the amount produced, atomic hydrogen can be generated in many ways [171]. For example, atomic hydrogen can be generated from a remote plasma system with energetic ions filtered. Such systems are readily available from several equipment suppliers. Another simple technique, especially useful for a MIS system, is to have a platinum group metal (e.g., Pd or Pt) as the gate metal. Platinum group metal is renowned for its catalytic properties and can dissociate molecular hydrogen into atomic during a standard forming gas anneal for the MIS system [172]. This method, used extensively in the hydrogen sensor technology [173] and not directly applicable to actual CMOS manufacturing, is easy to apply, cost-effective, and can provide crucial interfacial passivation information in a timely manner. An example, applying this technique to the passivation of $In_{0.53}Ga_{0.47}As$/ALD-Al_2O_3 [149, 168, 169] and $In_{0.53}Ga_{0.47}As$/ALD-ZrO_2 interfaces, is described next.

The capacitors of Table 8.5, fabricated from latticed-matched (to InP) 1 μm 2×10^{17} cm^{-3} (n/p) $In_{0.53}Ga_{0.47}As$ substrates with either 10 nm ALD-Al_2O_3 or 10 nm ALD-ZrO_2 and Pd gate, are subjected to a typical 10% H_2/N_2 forming gas anneal for 30 min at 300 °C. It is worth noting that sample A and B had a $(NH_4)_2S$ solution pre-treatment and sample C had a NH_4OH pre-treatment right before the subsequent atomic layer deposition of the corresponding gate oxide.

The large frequency dispersion of p-$In_{0.53}Ga_{0.47}As/Al_2O_3$/Pd capacitor in the accumulation, shown in Fig. 8.20a, is very common to III-V/dielectric interfaces [65, 148]. The small *bump* observed at $V_g \sim 0.8$ V is a minority carrier (weak inversion) response to the high D_{it} in this region [148, 174]. In addition, true inversion response was not observed for this capacitor. This capacitor revealed the true inversion behavior, manifested as V_g-independent capacitance in this region due to the depletion layer reaching its maximum thickness [175, 176] (referring to the *horizontal* lines in the strong inversion region in right-hand-side of Fig. 8.20a), after the application of the hydrogen anneal. Hydrogen anneal, in this case, has somehow reduced the electron traps adequately enough to allow the previously pinned

Table 8.5 List of samples fabricated and the corresponding D_{it} values (cm^{-2} eV^{-1}, at the energy ~0.5 eV above E_v from conductance measurements) before and after the hydrogen anneal

	Sample description	D_{it} before (cm^{-2} eV^{-1})	D_{it} after (cm^{-2} eV^{-1})	Comments
A	1 μm 2E17 p-$In_{0.53}Ga_{0.47}As$/10 nm ALD-Al_2O_3/Pd capacitor	5E12	2E12	Hydrogen anneal only helped p-type in the inversion region
B	1 μm 2E17 n-$In_{0.53}Ga_{0.47}As$/10 nm ALD-Al_2O_3/Pd capacitor	1E13	4.5E12	Hydrogen anneal only helped n-type in the accumulation region
C	1 μm 2E17 p-$In_{0.53}Ga_{0.47}As$/10 nm ALD-ZrO_2/Pd capacitor	1.8E13	8E12	Hydrogen anneal benefit is similar to that of sample A
D	$In_{0.53}Ga_{0.47}As$	4E12	8.6E11	In-situ PH_3 passivation and MOCVD HfO_2 [184]
E	$In_{0.53}Ga_{0.47}As$	3E12	1.7E12	In-situ PH_3 passivation and MOCVD HfAlO [184]

Fig. 8.20 Comparison of CV (100 Hz to 1 MHz, varying logarithmically) of III-V capacitors from sample A (*top*), sample B (*middle*) and sample C (*bottom*) before (*on the left-hand-side*) and after the hydrogen anneal (*on the right-hand-side*), with the corresponding D_{it} values listed in Table 8.5

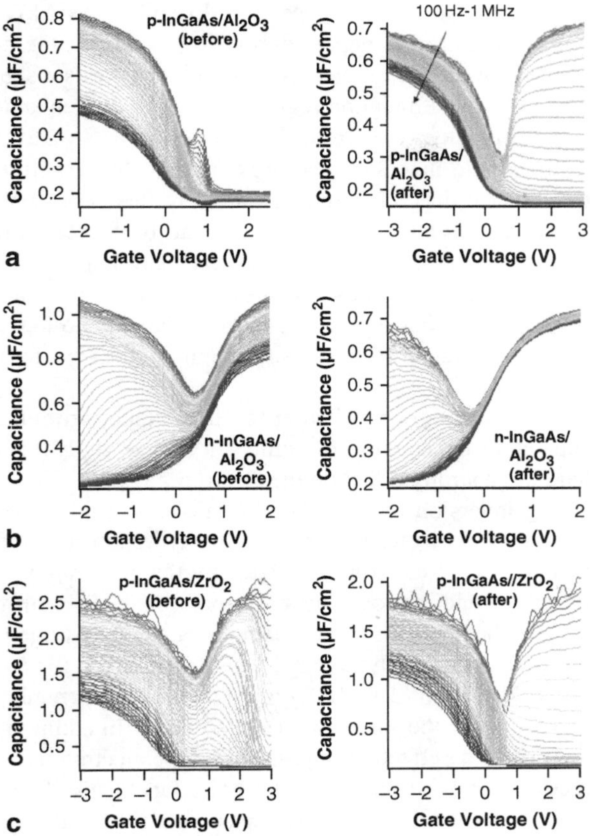

surface potential Ψ_S to continue its journey to reach strong inversion (by a distance greater than $2\Psi_B$, where Ψ_B is the bulk potential). It is worth noting that control experiments using a N_2 only anneal failed to obtain this improvement. Sample C, an alternative oxide (ZrO_2) with the same substrate, yielded a very similar behavior, as shown in Fig. 8.20c. However, hydrogen anneal applied here has only very limited success in reducing the dispersion in the accumulation region for p-$In_{0.53}Ga_{0.47}As$ capacitors with either Al_2O_3 or ZrO_2 oxide. The minor improvement observed in the accumulation region for p-type samples is due to the thermal effect, not hydrogen, as observed by the control N_2 anneal experiment.

It is worth noting that low-frequency CV like strong inversion behavior of p-type samples occurs even at relatively high frequencies (up to tens of kHz), which is attributed mainly to the high intrinsic carrier concentration of the substrate ($6 \times 10^{11} cm^{-3}$) [177]. The corresponding mechanism, such as generation-recombination and/or diffusion relating to the minority carrier behavior in this region, is very similar to the other substrate with high intrinsic carriers concentration n_i (e.g., Ge [178–180]) and is discussed further in details by Wang et al. [181].

In contrast, the response of n-$In_{0.53}Ga_{0.47}As/Al_2O_3/Pd$ capacitor to the same hydrogen anneal is very different from those of the p-type capacitors. Fig. 8.20b shows that the CV of an n-type capacitor reveals a significant reduction of dispersion in the accumulation region after the hydrogen anneal, yet with only negligible improvement in the strong inversion region. In addition, the inversion observed in the strong inversion region is believed to be fictitious since the corresponding capacitance depends strongly on V_g. In this case, surface potential Ψ_S (or Fermi-level) can not move from the upper part of the band gap to the E_v edge and beyond (a distance greater than $2\Psi_B$) and reach strong inversion due to the high D_{it} close to the E_v edge [149]. Thus, the strong inversion layer can not form and the depletion layer (and the corresponding capacitance) can no longer be shielded from the gate voltage change to remain constant, unlike the case for the p-type sample discussed previously.

The seemingly conflicting results obtained from these (n/p) type $In_{0.53}Ga_{0.47}As$ capacitors are actually complimentary. They imply that the hydrogen anneal applied here has a strong effect on the electron traps *only* since the dominant carriers in the strong inversion region of the p-type and in the accumulation of the n-type substrates are electrons. It seems that the hydrogen generated from the anneal somehow neutralizes the electron traps (but not the hole traps) at the $In_{0.53}Ga_{0.47}As$/dielectric interfaces and improves the corresponding CV performance.

The above results clearly demonstrate that hydrogen anneal applied in this study, with Pd as the catalytic metal gate, is beneficial for D_{it} reduction of $In_{0.53}Ga_{0.47}As$/ ALD-Al_2O_3 (or ALD-ZrO_2) interface in certain regions of the band gap (~0.5 eV above E_v, for the samples presented here). In addition, hydrogen from this anneal appears to be effective in passivating the electron traps of the corresponding interfaces, yet not so much for the hole traps. This observation may also imply that hydrogen anneal is not able to passivate D_{it} in other part of the band gap. In fact, D_{it} values of $In_{0.53}Ga_{0.47}As$/ALD-Al_2O_3 [149, 168, 169] and $In_{0.53}Ga_{0.47}As$/Ga(In)N/ Si_3N_4 interface [47] at regions close to the valence band are still undesirably trending high after the corresponding hydrogen treatment. Recently, D_{it} values of very high indium content $In_{0.75}Ga_{0.25}As$/ALD-Al_2O_3 interface reported by Xuan et al. are around ~10^{14} ranges at lower part of the band gap after the nitrogen anneal [47]. These results suggest that further advances in reducing the D_{it} across the whole band gap, using passivation schemes with hydrogen or with other novel methods, are still required before replacing silicon channel by high indium content InGaAs.

Recently, in-situ plasma PH_3 passivation of $In_{0.53}Ga_{0.47}As$ prior to dielectric deposition has also been exploited in III-V MOSFETs and showed reduction in mV/dec SS, increase in drain current and decrease in I_{OFF} compared to unpassivated control samples [182]. The room temperature interface trap layer densities (corresponding to the densities in the upper part of the band gap) in MOCVD high-k/passivated interface/InGaAs gate stack was reported to be 8.6×10^{11} cm^{-2} eV^{-1}, Table 8.5. In addition low frequency dispersion and hysteresis in C–V measurements were also observed.

Insertion of Interlayer

Scattered literature reports have included electrical and physical characteristics of various ALD dielectrics on InGaAs. Recently, the effect of insertion of various ALD interlayers (such as Al_2O_3, $HfAlO_x$, $LaAlO_x$, and $LaHfO_x$) between ALD HfO_2 and InGaAs has been demonstrated [183]. Having a thin Aluminum containing interlayer was seen to reduce the interface trap density, increase drive current, channel mobility and lowers SS, a consequence of reduced D_{it}, Fig. 8.21. Density Functional Theory calculation has recently also indicated that the a-Al_2O_3/$In_{0.5}Ga_{0.5}As$ interface has polar As-Al bonds and In/Ga-O bonds of opposite dipole direction, low lattice distortion, and no intermixing and may present as a better option for gate dielectric for III-V [184], in line with the experimental reports.

The use of Si IPL has been shown to improve the sub-threshold characteristics, drive current, and channel mobility of the surface channel $In_{0.53}Ga_{0.47}As$ n-MOSFETs with ALD Al_2O_3, PVD HfO_2 and MBD $LaAlO_3$ gate oxides, Fig. 8.22. One feasible approach involving Si passivation is based on a combination of wet chemical cleaning in sulfur containing solutions followed by ex-situ *a*-Si deposition using PVD or MBE tools [185, 186]. In this case high-k deposition can be done either

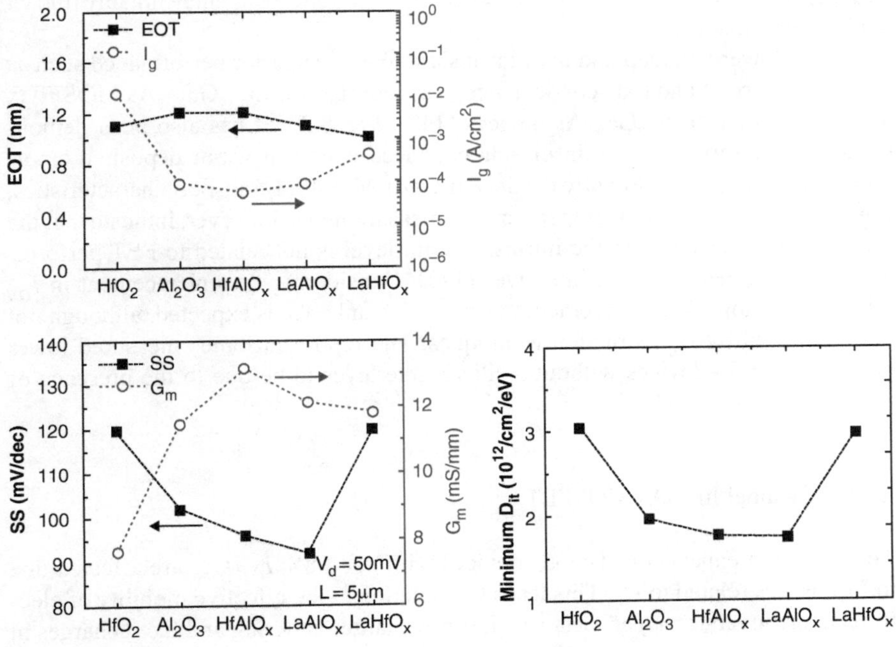

Fig. 8.21 EOT, gate leakage (at $V_g = 1$ V), SS, extrinsic transconductance and D_{it} numbers are compared for various interlayers in case of HfO_2/interlayer/$In_{0.53}Ga_{0.47}As$ surface channel MOSFETs [183] (Reprinted with permission. Copyright 2009 IEEE)

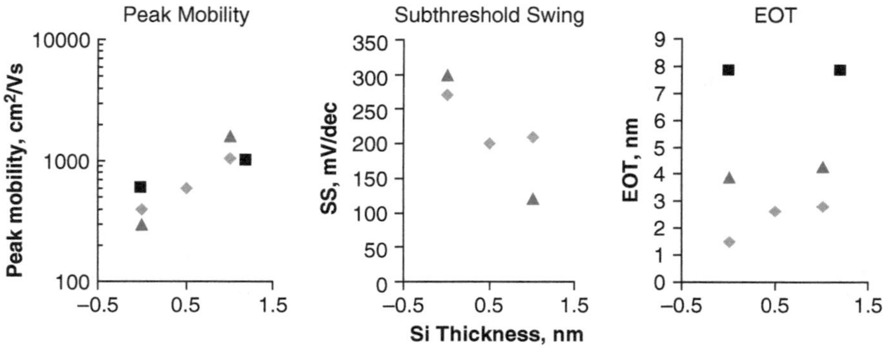

Fig. 8.22 Impact of Si IPL on channel mobility, subthreshold swing and EOT for n-MOSFETs with $In_{0.53}Ga_{0.47}As$ surface channel n-MOSFETs and various high-k dielectrics: *squares*—ALD Al_2O_3, *triangles*—PVD HfO_2, *diamonds*—MBD $LaAlO_3$

in-situ using PVD [185], or ex-situ using ALD [186] techniques. Si passivation layer can also be replaced with Ge, or bi-layer Si/Ge IPLs [187]. In contrast to GaAs, the $In_{0.53}Ga_{0.47}As$ interface with high-k dielectrics appears to provide not strongly pinned Fermi level [79]; however, $D_{it} > 10^{12}$ cm^{-2} eV^{-1} may notably impact device performance.

Due to lower bandgap and high intrinsic mobility, superior performance such as high drive current and transconductance, was reported for $In_{0.53}Ga_{0.47}As$ MOSFETs when compared to $In_{0.2}Ga_{0.2}As$ devices [142], Fig. 8.23. It has also been demonstrated that insertion of a thin plasma-enhanced chemical vapor deposited amorphous Si layer can significantly enhance InGaAs MOSFET transport characteristics, Fig. 8.23 [120]. The room temperature CP measurements, however, indicate that the interface trap density near the intrinsic Fermi level is not related to FET performance enhancement using Si interlayer [142]. Considering the enhancement in I_{ON} with Si insertion, D_{it} improvement towards the band edge is expected, although not yet confirmed by D_{it} extraction techniques. The report attributes measured lower peak mobility in devices without a silicon interlayer to be due to the presence of Coulomb Scattering [142].

Buried Channel InGaAs MOSFET

The inversion enhancement-mode devices with good SS, I_{ON}/I_{OFF} are affected due to FLP issues related to D_{it}. This leads to reduction in the effective mobility of electrons in the inversion layer. Besides interface charge scattering and fixed charges in oxide, are partly also intrinsically related to high-k. The high dielectric constant of insulators arises due to their large ionic polarizability (metal-oxygen bonds). These bonds are accompanied by soft optical phonons [188], interactions with which lead to remote phonon scattering [189].

The interface between channel and higher band-gap top barrier layer in HEMT structure, on the other hand, has much less FLP issue since the top layer is epitaxi-

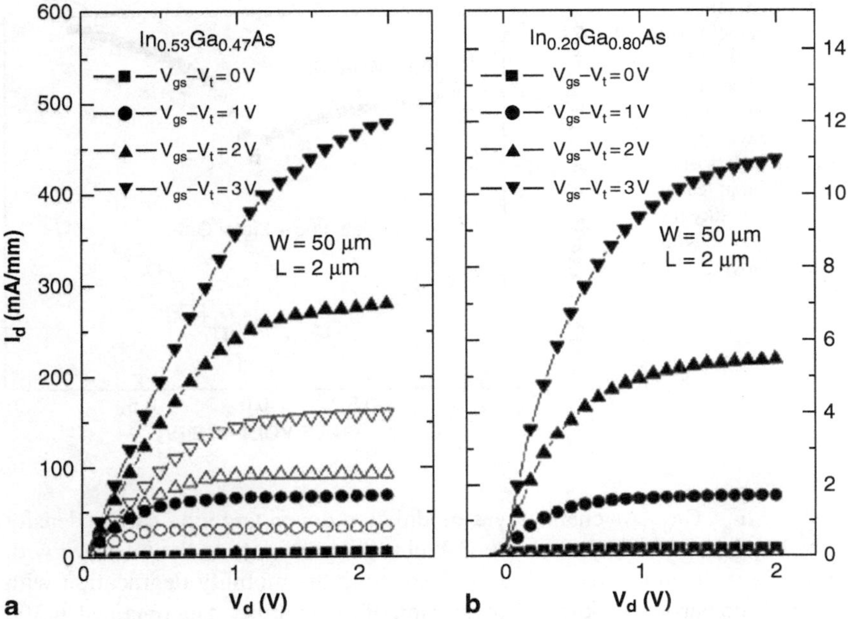

Fig. 8.23 I_d–V_d charactersitics of (**a**) $In_{0.53}Ga_{0.47}As$ and (**b**) $In_{0.20}Ga_{0.80}As$ n-MOSFETs with 2 μm channel length with (closed) and without (open) silicon interlayer as a function of $V_{gs}-V_t$. Note currents of $In_{0.20}Ga_{0.80}As$ devices without Si interlayer are so small that the data for devices without a-Si layer are not visible [142] (Reprinted with permission. Copyright 2008 American Institute of Physics)

ally grown on the channel. In addition, the higher barrier material with reasonable dielectric constant, itself acts as an almost negligible D_{it} insulator with the channel layer. InSb [26] and InGaAs [34] based Schottky FETs have been demonstrated. However, it was revealed that gate leakage current is high in these Schottky metal gate devices. The gate length independence of the gate leakage for the forward bias suggested that the gate leakage is dominated by edge leakage and can be reduced with an improved barrier recess technology. It has also been shown that the gate leakage can be suppressed by several orders of magnitudes by insertion of a gate dielectric, Fig. 8.24 [26, 37]. However, direct contact of high-k with III-V may create issues. Therefore, insertion of a barrier layer (semiconductor or dielectric) between high-k and channel can prevent effect of fixed gate oxide charges and oxide/top barrier interface charges on the mobility in the channel.

The recent study of scattering mechanisms in HfO_2 gated Hall structures [66] has shown that the $In_{0.53}Ga_{0.47}As/HfO_2$ interface is a significantly stronger source for remote scattering than $In_{0.53}Ga_{0.47}As/Si/HfO_2$ interface formed on the sample with *a*-Si passivation. The electron Hall mobility was measured in van der Pauw configuration samples at 77 K and 300 K in the magnetic field of 0.5–1 T. The structures with buried $In_{0.73}Ga_{0.27}As$ channel below the $In_{0.53}Ga_{0.47}As$ top barrier were grown by molecular beam epitaxy on semi-insulating (SI) Fe-doped InP(001) substrates with 0–10 nm thick HfO_2 in-situ deposited by e-beam evaporation. The

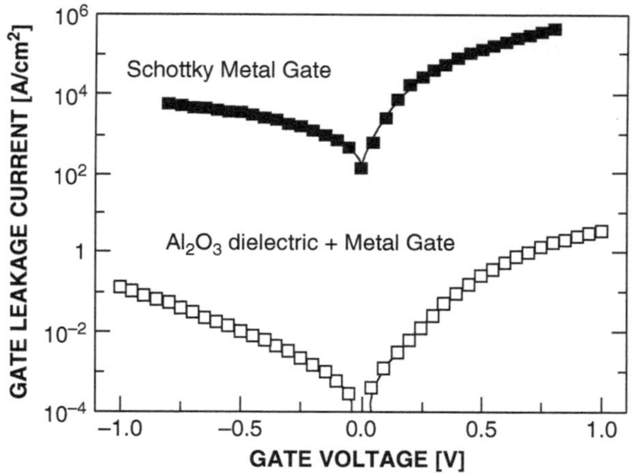

Fig. 8.24 Room temperature gate to channel leakage characteristics for ALD Al_2O_3 high-k dielectric and Al metal gate stack on AlInSb/InSb device layers showing four orders of magnitude reduction in leakage compared to Schottky metal gate [26] (Reprinted with permission. Copyright 2005 IEEE)

10 nm thick $In_{0.77}Ga_{0.23}As$ channel was modulation doped targeting carrier density of $1–2 \times 10^{12}$ cm^{-2}. The impact of the QW channel proximity to the interface with high-k oxide is demonstrated in Fig. 8.25 showing the mobility degradation with reduction of the barrier thickness. Comparison of the mobility data obtained at 300 and 77 K indicated [66] that in QW channels with thick 50 nm barrier phonon scattering was the major mechanism, whereas for thin barriers the mobility degradation was caused by remote Coulomb scattering of interface and/or oxide charges. To de-convolute contribution of the interface charges to mobility degradation, the HfO_2 thickness was varied while keeping the barrier thickness of 3 nm. The channel mobility decreased with HfO_2 thickness indicating the contribution of the oxide trapped charge or soft phonon scattering. The non-gated structure revealed reduced

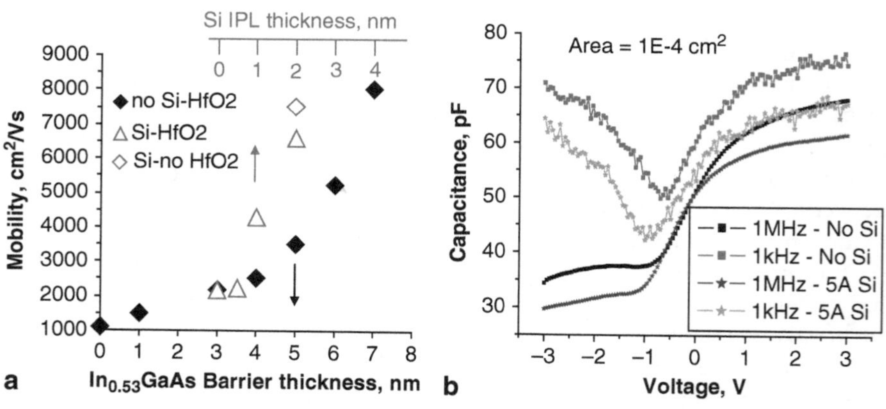

Fig. 8.25 **a** Mobility in Hall structures gated with HfO_2, *a*-Si and *a*-Si/HfO_2. **b** *C–V* characteristics of $In_{0.53}GaAs/HfO_2$ MOSCAPs with and without Si passivation

carrier concentration due to high D_{it}, but channel mobility was higher than on the HfO_2-gated samples.

Additional assessment of D_{it} contribution was obtained by comparing the structure with in-situ deposited amorphous Si and without Si passivation. Using *C–V* and conductance methods on the companion N-MOSCAPs the $In_{0.53}GaAs/HfO_2$ interface state density was estimated to be ~2×10^{12} cm^2 eV^{-1} for both samples with and without Si passivation (Fig. 8.25b). Increase of the *a*-Si layer thickness resulted in sharp increase of mobility (Fig. 8.25a): at the total distance of 5 nm from the interface the Si passivated QW structure exhibited higher mobility of 6200 cm^2/V s as compared with 3200 cm^2/V s on the sample without Si. The structure with 2 nm *a*-Si and without HfO_2 demonstrated the highest mobility of 7500 cm^2/V s. This result shows that the $In_{0.53}Ga_{0.47}As/HfO_2$ interface is a significantly stronger source for remote scattering than SiO_x/HfO_2 interface formed in the sample with *a*-Si passivation. On the other hand, the QW structures with HfO_2 directly deposited onto the InAlAs barrier, resulting in high D_{it} of ~5×10^{13} cm^{-2} eV^{-1}, revealed low mobility of ~900 cm^2/V s thus suggesting that scattering dominated by the interface states.

The device performance for buried $In_{0.7}Ga_{0.3}As$ and $In_{0.53}Ga_{0.47}As$ channel MOSFETs with a higher band-gap semiconductor top barrier layer (InP) and ALD Al_2O_3 gate dielectric has been found to provide higher transconductance, higher mobility (uncorrected value 4400 vs 2800 cm^2/V s) and lower SS than surface channel MOSFET counterparts [190]. Figure 8.26 shows I_d–V_g, G_m and SS comparisons between buried vs surface as well as $In_{0.53}Ga_{0.47}As$ vs $In_{0.7}Ga_{0.3}As$ channel devices [190].

Scaling of $In_{0.7}Ga_{0.3}As$ buried channel MOSFETs were also demonstrated down to 80 nm using $InAlAs/Si/Al_2O_3$ gate stack [37]. Good scaling behavior for on-state current, transconductance and virtual source velocity was observed although further improvements in short channel effect is expected with further decrease in EOT of the stack. Moreover, these devices have fairly large gate overlap region due to non-self-aligned integration scheme. However, a self-aligned process integration scheme will ultimately be required.

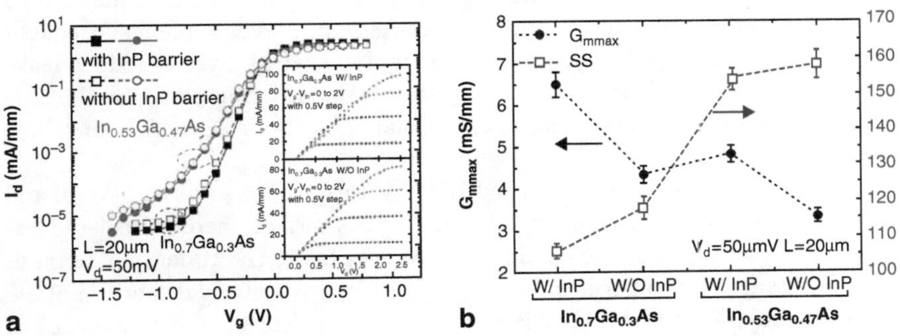

Fig. 8.26 **a** I_d–V_g curve (*inset* I_d–V_d curves with $V_d - V_t = 0$ to 2 V with 0.5 V step), **b** G_m and SS comparisons for $In_xGa_{1-x}As$ (x = 0.53, 0.7) 20 μm MOSFETs with and without InP barrier layer [190] (Reprinted with permission. Copyright 2009 American Institute of Physics)

8.5.2 *In-Situ Dielectrics*

8.5.2.1 Epitaxial Top Barrier (Dielectric) Layer

A highly scaled III-V FET will require very short channel lengths, fast carrier transport in the channel and careful device design. A PHEMT structure with a channel layer inserted between higher band gap barrier materials can provide such a platform. The barrier material itself has several advantages which include negligible D_{it} due to its in-situ epitaxial growth as well as prevention of scattering of carriers in the channel. Triple recess InAs PHEMTs with 10 nm channel thickness and 4 nm barrier layer (insulator) thickness has been shown to exhibit promising logic characteristics [34] and high f_T performance [19]. In addition, enhancement mode device (with positive V_t) has also been demonstrated [191]. The report however, points out that the gate leakage is still an issue and may still require improved barrier recess technology [191] or a still higher-k gate dielectric insertion.

8.5.2.2 MOSFET with GGO Dielectric

GaAs based MOSFET technology with buried $In_{0.3}Ga_{0.7}As$ quantum well channel layer has been much explored for possible future use in radio frequency applications at low voltage and for high efficiency, i.e., in wireless and mobile products. For those applications, enhancement mode high electron mobility MOSFET devices have been demonstrated using an oxide high-k gate dielectric stack developed using molecular beam epitaxy. In these MOSFETs, Ga_2O_3 film is typically deposited by effusive evaporation from a polycrystalline Ga_2O_3 source onto the pristine GaAs (001)–(2 × 4) surface in ultra-high vacuum. Through scanning tunneling microscopy and DFT calculations, it was shown that first monolayer of Ga_2O forms a charge balanced (2 × 2) surface order with a crystalline interface that is electronically unpinned. Further deposition proceeds via the formation of amorphous bulk Ga_2O_3 [111]. The template layer initially deposited on the surface of the III-V device is claimed to unpin the GaAs Fermi level while a subsequent $(Gd_xGa_{1-x})_2O_3$ bulk ternary layer forms the highly resistive layer to reduce leakage current through the dielectric stack [192, 193]. This gate stack on GaAs is reported to have room temperature interface state density of $\sim 2 \times 10^{11}$ cm^{-2} eV^{-1} [192].

On comparing CV characteristics and HRTEM image of ex-situ ALD Al_2O_3 and ALD Al_2O_3/in-situ GGO stack on $In_{0.15-0.2}Ga_{0.85-0.8}As$ surface, there exists large frequency dispersion in the single ex-situ film whereas the HRTEM does not point to any serious issues at the interface, Fig. 8.27. The larger frequency dispersion of the ALD-Al_2O_3/$In_{0.15}Ga_{0.85}As$ MOSCAP is caused by the existence of native oxides of In_2O_3, $In(OH)_3$, and Ga_2O_3 at the oxide/InGaAs interface, which was not observed at the GGO/$In_{0.2}Ga_{0.8}As$ interface [72, 83].

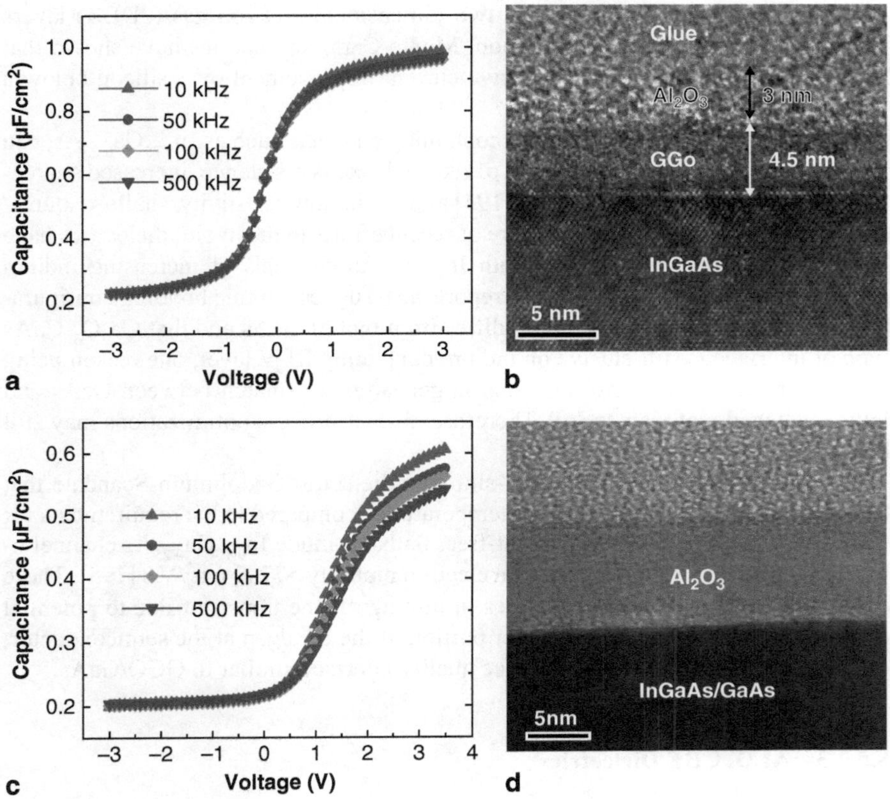

Fig. 8.27 **a** *C–V* characteristics and **b** cross-sectional HRTEM image for an Al_2O_3(3 nm)/GGO(8.5 nm)/$In_{0.2}Ga_{0.8}As$/GaAs MOS diode RTA to 850 °C in N_2 with the Al gate deposited afterwards. **c** *C–V* characteristics and **d** cross-sectional HRTEM image an ALD-Al_2O_3(8.5 nm)/$In_{0.15}Ga_{0.85}As$/GaAs MOS diode annealed at 500 °C in N_2 with the Au gate deposited afterwards [72] (Reprinted with permission. Copyright 1996 American Institute of Physics)

Similar device physics, in-situ deposition method, device design and test structures have also been recently employed to evaluate the potential in III-V CMOS arena. The potential of GaAs channel to provide necessary drive currents in the channel and access resistance appears to be limited. In order to boost mobility, indium is introduced to produce $In_xGa_{1-x}As$. Recently, a self-aligned process for fabricating inversion type n-channel metal–oxide–semiconductor field-effect-transistors (nMOSFETs) of strained $In_{0.2}Ga_{0.8}As$ on GaAs with $Ga_2O_3(Gd_2O_3)$ as high-k gate dielectric has been demonstrated [194]. The room temperature CV characteristics on $In_{0.2}Ga_{0.8}As$ MOSCAPs show low dispersion and better interface characteristics compared to ALD films. However, there is not much mobility advantage of low indium content InGaAs channels compared to GaAs and reducing contact resistance to shallow $In_{\leq 0.3}Ga_{\geq 0.7}As$ two dimensional electron gas (2DEG) layers may be challenging [195]. In addition, Monte Carlo simulations have shown that $In_{0.3}Ga_{0.7}As$ devices have limited drive current improvement over silicon below a gate length of 15 nm [196].

Alternatively, still higher indium containing channels such as $In_{0.75}Ga_{0.25}As$, can potentially address the requirements of reduced access resistance, increased source-injection velocity and drive current [195] along with low resistivity, shallow source/drain contacts [197]. As a result, more of recent efforts in the in-situ dielectrics have been refocused towards designs with $In_xGa_{1-x}As$ channels of increasing indium concentration (x > 0.3) [195]. These reports also suggest atomic bonding configurations at indium containing surfaces differ from that of GaAs and that Ga_2O_3-GaAs type of interface is still elusive on indium containing III-V layer, one reason being lower thermal budget constraints and larger lattice mismatch between GaAs and lattice matched materials to InP. Therefore, further process optimizations may still be needed.

As an alternative to GGO, an in-situ gate dielectric Gadolinium Scandate that can be deposited at relatively lower temperatures, compared to the requirements for GGO, has been evaluated in implant-free, flatband-mode $In_{0.75}Ga_{0.25}As$ channel n-MOSFETs and reported to provide an electron mobility >7700 $cm^2/V\,s$ [195]. These devices, however, showed difficulties in turning off the transistor due to potential donor-type traps located in the lower portion of the bandgap at the semiconductor/oxide interface, falling short of device quality interface similar to GGO/GaAs.

8.5.2.3 ALD, CBE Dielectrics

To date there has been no systematic comparison of MOS device electrical characteristics where the same type of dielectric is deposited in-situ and on air-exposed InGaAs wafer. Recently, other reports surfaced where in-situ dielectrics deposited by alternate techniques such as MOCVD, CBE, have also been investigated. C.-W. Cheng et al. [198] reported a room temperature D_{it}~2.5×10^{11} $cm^{-2}\,eV^{-1}$ for low thermal budget (≤ 400 °C) in-situ atomic layer deposition of Al_2O_3 on p-GaAs which is comparable to values reported for GGO/GaAs [192]. Chemical beam deposition (CBE) in ultrahigh vacuum on arsenic decapped InGaAs surface has also been explored [199]. The technique involved metal oxide precursors for which metal ion is bonded with oxygen, and no additional oxidant specie is needed. This has the potential of reducing the possibility of semiconductor surface oxidation. In addition, the high volatility of the precursor allows thermal evaporation at relatively low temperatures with high growth rates [199]. With this technique, a direct interface between TiO_2 and InGaAs was observed. However, a high diode leakage current was measured due to negligible conduction band offset between CBE deposited TiO_2 and $In_{0.53}Ga_{0.47}As$ [199]. On replacing the precursor to zirconium tert-butoxide (ZTB), where Zr is already bonded to oxygen, D_{it}~high10^{11} to low 10^{12} $eV^{-1}\,cm^{-2}$ around E_v + 0.5 eV was measured by the conductance method on in-situ deposited ZrO_2 on $In_{0.53}Ga_{0.47}As$ [200]. These reports are relatively new and require further detailed experimentation, for instance, effect of process temperature on interface degradation, amount of bulk trapping and controlled deposition.

8.5.2.4 Amorphous Si Interlayer

Besides above mentioned techniques and designs, another type of in-situ interlayer dielectric has been explored extensively. The method of passivation of GaAs with pseudomorphic Si layers in late 1980s and 1990s was designed to establish one-to-one registry between the semiconductor and the insulator and to prevent GaAs from oxidation. The use of an ultra-thin Si layer demonstrated significant improvement of $GaAs/SiO_2$ and $GaAs/Si_3N_4$ interfaces with the reduced interface state densities to low 10^{11} eV^{-1} cm^{-2} [201]. One of the problems associated with this method was that Si and GaAs required different processing temperatures resulting to an inevitable trade off between the crystallinity of Si and the stoichiometry of GaAs and poor reproducibility of their properties. Another problematic issue was related to formation of a parasitic surface quantum well due to strain induced band gap shrinkage [202]. In mid 2000s several research groups investigated GaAs based MOS devices gated with high-k oxides and containing amorphous Si IPL [95, 185, 186] and demonstrated significantly improved interface properties. In addition to strong reduction of interface state density resulting in un-pinning of Fermi level, the Si IPL also benefited to decrease of the gate leakage current, improvement of the thermal stability of the interface properties and in some cases to reduction of the gate oxide charges. The major disadvantage of the Si IPL is that Si is being fully or partially oxidized prior to or during high-k oxide deposition, and thus becomes a part of a gate stack contributing to EOT. Also, if a portion of *a*-Si adjusted to the III-V surface is not oxidized or bonded, the issue of Si in-diffusion at elevated temperatures may arise. The following part of this section discusses the benefits and challenges of Si IPL.

The Si IPL technique has several process modifications depending on method of high-k oxide deposition. In Ref. [95], the epitaxial MBE grown GaAs layer was in-situ encapsulated with a 1.5 nm thick amorphous Si layer followed by ex-situ deposition of HfO_2 gate oxide and TaN metal gate. The CV characteristics of MOS capacitors demonstrated low stretch-out and frequency dispersion, good scaling behavior of equivalent oxide thickness with EOT of 2.1 nm for 4.0 nm thick HfO_2 and leakage current of ~1 mA/cm^2 at V_{fb} + 1 V. The minimum thickness of the Si IPL of 1.5 nm was needed to prevent Fermi level pinning, the interfacial SiO_2 level being part of a high-k stack. Both the structure and chemistry of the interface between GaAs and Si IPL has been studied using XPS and TEM. The HRTEM confirmed that the Si IPL was nearly fully oxidized while the interface between the GaAs and *a*-Si remained atomically sharp without any sign of interfacial reaction even after 48 h exposure to air and post deposition anneal at 500 °C for 5 min. Non-oxidized Si at the interface and absence of As-O bonds in GaAs was correlated with unpinned Fermi level in MOS capacitors [95]. A significant drawback of the sputtered oxide is a large hysteresis of CV characteristics. It can be further argued that the charge capture in the oxide is responsible for the hysteresis, as the hysteresis is proportional to the oxide thickness and is reduced down to 0.25 V in the structure with a 4 nm—thick HfO_2 (inset Fig. 8.28). Another important feature of this gate stack is reproducibility of electrical results which is indicated by a linear scaling of EOT with HfO_2 thickness (Fig. 8.28). The slope of this line corresponds to the dielectric constant of 23 and ~1.4 nm offset is consistent with the thickness of oxidized IPL within the accuracy of measurements.

To further reduce the EOT the cluster MBE/High-k deposition tool was used enabling the full in-situ process of surface passivation and gate oxide deposition [50]. In this case, after the completion of MBE growth of the group III-V materials and ultra-thin (0.25 nm) amorphous Si IPL, a HfO_2 gate oxide was deposited in the same ultrahigh vacuum system using reactive electron beam evaporation at room temperature in an oxygen pressure of 3×10^{-6} Torr. In-situ deposition of the oxide prevented the exposure of the semiconductor structure with an ultra-thin IPL to atmosphere and thus reduced the native silicon oxide thickness due to environmental oxidation. Comparing the EOT for samples with ex-situ and in-situ grown HfO_2, (Fig. 8.26) the contribution of the interfacial passivation layer to EOT was found to be reduced by ~1.3 nm, approaching the value obtained on the Metal-Oxide-Metal structures without any interfacial layer. The dielectric constant of the e-beam deposited in-situ HfO_2 was found to be ~26 as measured on TaN based metal-oxide-metal (MOM) samples with scaled HfO_2 thickness (Fig. 8.28).

To increase process integration capabilities while maintaining low EOT the MBE process of in-situ passivation of III-V materials with *a*-Si IPL can be combined with various ex-situ processes of high-k oxide deposition using encapsulation with metallic arsenic, As_2. The feasibility of this approach was demonstrated in Ref. [197] by fabricating GaAs MOSCAPs with amorphous $LaAlO_3$ gate oxide layer grown by MBD, whereas the benefit of Si IPL was exhibited by comparison of Si passivated and non-passivated devices. It was shown that although, MOS capacitors without *a*-Si IPL demonstrated lower EOT than the Si-passivated samples, strong degradation of the gate leakage current occurred after anneal at 700 °C, whereas

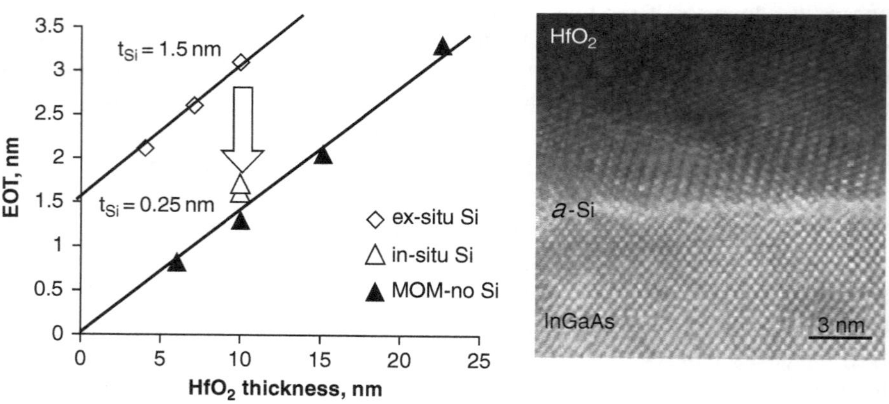

Fig. 8.28 Improvement of EOT scaling by using fully in-situ process of III-V growth, passivation and high-k deposition. The *a*-Si thickness can be reduced to 0.25 nm providing smooth atomically clean interface as shown by TEM

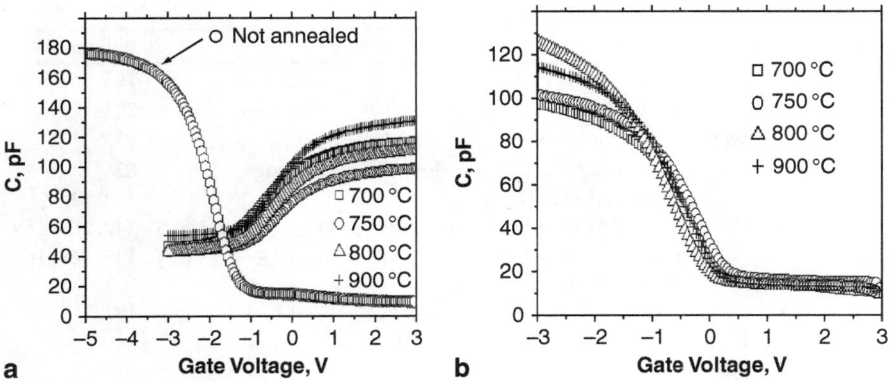

Fig. 8.29 MHz capacitance–voltage characteristics of p-doped GaAs MOS capacitors with (**a**) 1 nm and (**b**) 0.5 nm amorphous Si IPL. Inversion of conductivity due to Si diffusion into semiconductor is observed at annealing temperature > 650 °C on samples with thicker Si IPL. Change of accumulation capacitance is likely due to a densification

a-Si passivated GaAs exhibited not only excellent thermal stability of the electrical and structural properties up to 800 °C, but also significant improvement of both the high-k oxide and interface properties with annealing temperature demonstrating a reduction of EOT, hysteresis interface state density ($\leq 2 \times 10^{11}$ eV^{-1} cm^{-2}) [87].

Thermal stability of gate stack is mainly required by a relatively high processing temperature of MOSFETs with activation of ion-implanted impurities. As shown in [96], the stacks with thick Si IPL are less thermally stable than those with thin IPL against Si diffusion into semiconductor. Diffusion of Si into n-GaAs results in the increase of the donor concentration in the semiconductor. It was clearly observed via the rise of high-frequency capacitance in deep depletion when a sample was annealed above 650 °C. Another even more notable effect was inversion of the conductivity type from p-type to n-type after annealing as shown in Fig. 8.29. Remarkably, Si diffusion can be suppressed if the IPL thickness is reduced and its bonding status with oxygen and arsenic at the interface is controlled [96]. It is believed, that in the latter sample with 0.5 nm-thick IPL silicon was almost entirely oxidized, and therefore, bound to oxygen resulting in suppressed diffusion. In contrast, samples with excess IPL thickness containing non-oxidized Si exhibited Si diffusion at implant activation temperature.

8.6 Summary

The text discusses various dielectric deposition techniques on III-Vs and the impact of deposition, passivation, pre-cleans on electrical and physical characteristics of the gate stacks. The following Table 8.6 ends the chapter with a summary of reported D_{it} on various III-V MOS devices.

Table 8.6 Reported D_{it} numbers in the literature

Sample Details	Comments	D_{it} ($cm^{-2}\ eV^{-1}$)	Reference
GaAs	In-situ Ga_2O_3/GGO	~2×10^{11} at band edge	195
GaAs	In-situ Al_2O_3	~2×10^{11} at band edge	201
GaAs	In-situ Si/$LaAlO_3$	~2×10^{11} at band edge	87
$In_{0.2}Ga_{0.8}As$	In-situ Ga_2O_3/Gd_2O_3	low 10^{11} at band edge	82
$In_{0.15}Ga_{0.8}As$	Ex-situ ALD HfO_2	~10^{12}	119, 127
$In_{0.53}Ga_{0.47}As$	Ex-situ ALD Al_2O_3	5×10^{12} at 300 K (E_v + 0.5 eV)	149, 168, 169
$In_{0.53}Ga_{0.47}As$	Ex-situ ALD Al_2O_3; Hydrogen annealed	2×10^{12} at 300 K (E_v + 0.5 eV)	149, 168, 169
$In_{0.53}Ga_{0.47}As$	In-situ plasma PH_3surface passivation; MOCVD HfO_2	8.6×10^{11} at 300 K (E_v + 0.5 eV)	182
$In_{0.53}Ga_{0.47}As$ (2 × 4) Surface reconstruction	In-situ CBE 30 nm ZrO_2	7.8×10^{11} to low × 10^{12} at 300 K (E_v + 0.5 eV)	200
$In_{0.53}Ga_{0.47}As$	Ex-situ ALD ZrO_2	2 to 5×10^{12} at 300 K (E_v + 0.5 eV)	74
$In_{0.53}Ga_{0.47}As$	Ex-situ ALD Al_2O_3/HfO_2	2×10^{12} at 300 K (E_v + 0.5 eV)	183
GaAs	In-situ ALD Al_2O_3	2.5×10^{11} at band edge	198
$In_{0.53}Ga_{0.47}As$	In-situ *or* ex-situ *Si*	10^{12} to 10^{13} at 300 K	79, 142

Acknowledgments The authors acknowledge Mark Lundstrom for stimulating comments and reading the manuscript. The authors are grateful for financial support from Intel Corporation and MSD Focus Center Research Program.

References

1. M. Shur, *GaAs Devices and Circuits*. (Plenum Press, New York, 1987).
2. L. Pfeiffer and K. W. West, "The role of MBE in recent quantum Hall effect physics discoveries," *Physica E*, 20(1), 57–64 (2003).
3. V. Umansky, M. Heiblum, Y. Levinson, J. Smet, J. Nubler, and M. Dolev, "MBE growth of ultra-low disorder 2DEG with mobility exceeding 35×10^6 cm^2/Vs," *Journal of Crystal Growth*, 311(7), 1658–1661 (2009).
4. S. I. Long, "High speed digital circuit technology," in *Compound Semiconductor Electronics: the Age of Maturity*, M. Shur, Ed. (World Scientific Publishing Co., 1996), pp. 247–290.
5. T. Suemitsu and M. Tokumitsu, "InP HEMT technology for high-speed logic and communications," *IEICE Trans. on Fundamentals of Electronics, Communications and Computer Sciences*, E90-C(5), 917–922 (2007).
6. D. E. Grider, P. P. Ruden, J. C. Nohava, I. R. Mactaggart, J. J. Stronczer, and R. H. Tran, "0.7 micron gate length complementary $Al_{0.75}Ga_{0.25}As/In_{0.25}Ga_{0.75}As$/GaAs HIGFET technology for high speed/low power digital circuits," International Electron Devices Meeting (IEDM), *Technical Digest*, 331–334 (1992).
7. D. E. Fulkerson and S. Baier, "Comparison of GaAs-based heterostructure n-channel transistors to Si in complementary processes," *Solid State Electronics*, 43(1), 65–71 (1999).

8. A. Leuther, A. Thiede, K. Kohler, T. Jakobus, G. Weimann, K. H. Ploog, G. Trankle, and G. Weimann, "Complementary HFETs on GaAs with 0.2 Îμm gate length," Compound Semiconductors 1999, Proc. of 26th International Symposium on Compound Semiconductors (IOP Publishing, Bristol, UK, 2000), pp. 313–316.
9. T. Enoki, E. Sano, and T. Ishibashi, "Prospects of InP-based IC technologies for 100-Gbit/s-class lightwave communication systems," in *High-speed integrated circuit technology, towards 100 GHz logic*, M. J. W. Rodwell, Ed. (World Scientific, Singapore, 2001), pp. 137–158.
10. B. R. Nag, *Physics of Quantum Well Devices,* ch. 4, vol. 7. (Springer, New York, 2001).
11. W. Liu, *Fundamentals of III-V Devices: HBTs, MESFETs, and HFETs/HEMTs*. (John Wiley & Sons, New York, 1999).
12. J. M. Golio, *Microwave MESFETs And HEMTs*. (Artech House, London, 1991).
13. P. Mukhopadhyay, P. Das, S. Pathak, S. Kundu, E. Y. Chang, and D. Biswas, "A strategic review of recent progress in metamorphic quantum well based heterostructure electronic devices," Nanotechnology 2008, 8th IEEE Conference (NANO) (Piscataway, NJ, 2008), pp. 503–506.
14. S. Datta, G. Dewey, J. M. Fastenau, M. K. Hudait, D. Loubychev, W. K. Liu, M. Radosavljevic, W. Rachmady, and R. Chau, "Ultrahigh-speed 0.5 V supply voltage quantum-well transistors on silicon substrate," *IEEE Electron Device Letters*, 28(8), 685–687 (2007).
15. T. Ashley, L. Buckle, S. Datta, M. T. Emeny, D. G. Hayes, K. P. Hilton, R. Jefferies, T. Martin, T. J. Phillips, D. J. Wallis, P. J. Wilding, and R. Chau, "Heterogeneous InSb quantum well transistors on silicon for ultra-high speed, low power logic applications," *Electronics Letters*, 43(14), 777–779 (2007).
16. Y. Jin-Delorme, A. De Lustrac, P. Crozat, K. Yazbek, R. Adde, G. Vernet, Y. Yin, B. Etienne, and H. Launois, "Electric parameter evolutions against gatelength and bias in ultrashort gate AlGaAs/GaAs HEMTs," *Electronics Letters*, 29(7), 642, 643 (1993).
17. I. G. Thayne, G. U. Jensen, M. C. Holland, Y. Chen, W. Li, A. Paulsen, J. H. Davies, S. P. Beaumont, and P. K. Bhattacharya, "Comparison of 80–200 nm gate length $Al_{0.25}GaAs/GaAs/(GaAs:AlAs)$, $Al_{0.3}GaAs/In_{0.15}GaAs/GaAs$, and $In_{0.52}AlAs/In_{0.65}GaAs/InP$ HEMTs," *IEEE Transactions on Electron Devices*, 42(12), 2047–2055 (1995).
18. Y. Yamashita, A. Endoh, K. Shinohara, M. Higashiwaki, K. Hikosaka, T. Mimura, S. Hiyamizu, and T. Matsui, "Ultra-short 25-nm-gate lattice matched InAlAs/InGaAs HEMTs within the range of 400 GHz cutoff frequency," *IEEE Electron Device Letters*, 22, 367–370 (2001).
19. D.-H. Kim and J. A. del Alamo, "30-nm InAs pseudomorphic HEMTs on an InP substrate with a current-gain cutoff frequency of 628 GHz," *IEEE Electron Device Letters*, 29(8), 830–833 (2008).
20. S.-J. Yeon, M. Park, J. Choi, and K. Seo, "610 GHz $InAlAs/In_{0.75}GaAs$ metamorphic HEMTs with an ultra-short 15-nm-gate," International Electron Devices Meeting, *Technical Digest*, 613–616 (2007).
21. T. Enoki, K. Arai, and Y. Ishii, "Delay Time Analysis for 0.4- to 5-μm-Gate InAlAs-InGaAs HEMT's," *IEEE Electron Device Letters*, 11(11), 502–504 (1990).
22. Y. Yamashita, A. Endoh, K. Shinohara, K. Hikosaka, T. Matsui, S. Hiyamizu, and T. Mimura, "Pseudomorphic $In_{0.52}Al_{0.48}As/In_{0.7}Ga_{0.3}As$ HEMTs with an ultrahigh f_T of 562 GHz," *IEEE Electron Device Letters*, 23(10), 573–575 (2002).
23. S. Lee, B. Jagannathan, S. Narasimha, A. Chou, N. Zamdmer, J. Johnson, R. Williams, L. Wagner, J. Kim, J.-O. Plouchart, J. Pekarik, S. Springer, and G. Freeman, "Record RF performance of 45-nm SOI CMOS technology," International Electron Devices Meeting, *Technical Digest*, 255–258 (2007).
24. A. Khakifirooz and D. A. Antoniadis, "MOSFET performance scaling. Part I. Historical trends," *IEEE Transactions on Electron Devices*, 55(6), 1391–1400 (2008).
25. R. Chau, S. Datta, and A. Majumdar, "Opportunities and challenges of III-V nanoelectronics for future high-speed, low-power logic applications," IEEE Compound Semiconductor Integrated Circuit Symposium, 2005, *Technical Digest*, 17–20 (2005).

26. S. Datta, T. Ashley, J. Brask, L. Buckle, M. Doczy, M. Emeny, D. Hayes, K. Hilton, R. Jefferies, T. Martin, T. J. Phillips, D. Wallis, P. Wilding, and R. Chau, "85 nm gate length enhancement and depletion mode InSb quantum well transistors for ultra high speed and very low power digital logic applications," International Electron Devices Meeting, *Technical Digest*, 783–786 (2005).
27. ITRS, "International Technology Roadmap for Semiconductors, 2007 Edition," 2007.
28. A. Khakifirooz and D. A. Antoniadis, "MOSFET performance scaling-Part II: future directions," *IEEE Transactions on Electron Devices*, 55(6), 1401–1408 (2008).
29. G. G. Shahidi, D. A. Antoniadis, and H. I. Smith, "Electron velocity overshoot at room and liquid nitrogen temperatures in silicon inversion layers," *IEEE Electron Device Letters*, 9(2), 94–96 (1988).
30. M. Lundstrom, "Elementary scattering theory of the Si MOSFET," *IEEE Electron Device Letters*, 18(7), 361–363 (1997).
31. A. Lochtefeld, I. J. Djomehri, G. Samudra, and D. A. Antoniadis, "New insights into carrier transport in n-MOSFETs," *IBM Journal of Research and Development*, 46(2), 347–357 (2002).
32. G. Dewey, M. K. Hudait, K. Lee, R. Pillarisetty, W. Rachmady, M. Radosavljevic, T. Rakshit, and R. Chau, "Carrier transport in high-mobility III-V quantum-well transistors and performance impact for high-speed low-power logic applications," *IEEE Electron Device Letters*, 29(10), 1094–1097 (2008).
33. D.-H. Kim and J. A. del Alamo, "Scaling behavior of $In_{0.7}Ga_{0.3}As$ HEMTs for logic," International Electron Devices Meeting, *Technical Digest*, 1–4 (2006).
34. D.-H. Kim and J. A. del Alamo, "Logic performance of 40 nm InAs HEMTs," International Electron Devices Meeting, *Technical Digest*, 629–632 (2007).
35. D.-H. Kim and J. A. del Alamo, "30 nm E-mode InAs PHEMTs for THz and future logic applications," International Electron Devices Meeting, *Technical Digest*, 1–4 (2008).
36. H. C. Lin, T. Yang, H. Sharifi, S. K. Kim, Y. Xuan, T. Shen, S. Mohammadi, and P. D. Ye, "Enhancement-mode GaAs metal-oxide-semiconductor high-electron-mobility transistors with atomic layer deposited Al_2O_3 as gate dielectric," *Applied Physics Letter*, 91(21), 212101-1-3 (2007).
37. Y. Sun, E. W. Kiewra, J. P. de Souza, J. J. Bucchignano, K. E. Fogel, D. K. Sadana, and G. G. Shahidi, "Scaling of $In_{0.7}Ga_{0.3}As$ buried-channel MOSFETs," International Electron Devices Meeting, *Technical Digest*, 67–70 (2008).
38. K. C. Saraswat, "High mobility channel materials for future CMOS," *2007 International Symposium on VLSI Technology Systems and Applications (VLSI-TSA)* (IEEE, Piscataway, NJ, USA, 2007), pp. 116–119.
39. N. Harada, S. Kuroda, T. Katakami, K. Hikosaka, T. Mimura, and M. Abe, "Pt-based gate enhancement-mode InAlAs/InGaAs HEMTs for large-scale integration," Indium Phosphide and Related Materials, Proc. of 3rd International Conference (IEEE, New York, 1991), pp. 377–380.
40. A. Mahajan, M. Arafa, P. Fay, C. Caneau, and I. Adesida, "160 GHz enhancement-mode InAlAs/InGaAs/InP high electron mobility transistor," Device Research Conference, *Technical Digest*, 132, 133 (1996).
41. D. Moran, E. Boyd, H. McLelland, K. Elgaid, Y. Chen, D. S. Macintyre, S. Thoms, C. R. Stanley, and I. G. Thayne, "Novel technologies for the realisation of GaAs pHEMTs with 120 nm self-aligned and nanoimprinted T-gates," *Microelectronic Engineering*, 67, 68, 769–774 (2003).
42. A. Cetronio, F. Giannini, and G. Leuzzi, "Self-aligned gate technology for analogue and digital GaAs integrated circuits," Gallium Arsenide applications Symposium. 2000, Paris, 2–6 October (2000).
43. T.-W. Kim, D.-H. Kim, S.-H. Shin, S.-J. Jo, J. H. Jang, and J.-I. Song, "Characteristics of 0.2μm depletion and quasi-enhancement mode self-aligned gate capless p-HEMTs," *Electronics Letters*, 42(20), 1178–1180 (2006).

44. N. Waldron, D.-H. Kim, and J. A. del Alamo, "90 nm self-aligned enhancement-mode InGaAs HEMT for logic applications," International Electron Devices Meeting, *Technical Digest*, 633–636 (2007).
45. I. Watanabe, A. Endoh, T. Mimura, and T. Matsui, "35-nm-gate $In_{0.7}Ga_{0.3}As/In_{0.52}Al_{0.48}As$ HEMT with 520-GHz f_T," *International Conference on Indium Phosphide and Related Materials*, 28–31 (2007).
46. R. J. W. Hill, D. A. J. Moran, X. Li, H. Zhou, D. Macintyre, S. Thoms, A. Asenov, P. Zurcher, K. Rajagopalan, J. Abrokwah, R. Droopad, M. Passlack, and I. G. Thayne, "Enhancement-mode GaAs MOSFETs with an $In_{0.3}Ga_{0.7}As$ channel, a mobility of over 5000 cm^2/V s, and transconductance of over 475 μmS/μm," *IEEE Electron Device Letters*, 28(12), 1080–1082 (2007).
47. Y. Xuan, Y. Q. Wu, and P. D. Ye, "High-performance inversion-type enhancement-mode InGaAs MOSFET with maximum drain current exceeding 1 A/mm," *IEEE Electron Device Letters*, 29(4), 294–296 (2008).
48. S. Koveshnikov, R. Kambhampati, V. Tokranov, M. Yakimov, D. Schlom, M. Warusawithana, C. Adamo, W. Tsai, and S. Oktyabrsky, "Thermal Stability of Electrical and Structural Properties of GaAs based MOS Capacitors with Amorphous $LaAlO_3$ Gate Oxide," *Applied Physics Letter*, 93(1), 012903-1-3 (2008).
49. T. D. Lin, P. Chang, H. C. Chiu, Y. C. Chang, C. A. Lin, W. H. Chang, Y. J. Lee, Y. H. Chang, M. L. Huang, J. Kwo, and M. Hong, "Nano-electronics of high-k dielectrics on InGaAs for key technologies beyond Si CMOS," Indium Phosphide and Related Materials, Proceedings of International Conference. (IEEE, Piscataway, NJ, 2009), pp. 94–99.
50. S. Oktyabrsky, V. Tokranov, M. Yakimov, R. Moore, S. Koveshnikov, W. Tsai, F. Zhu, and J. C. Lee, "High-k gate stack on GaAs and InGaAs using in situ passivation with amorphous silicon," *Materials Science & Engineering B*, 135(3), 272–276 (2006).
51. I. Ok, H. Kim, M. Zhang, F. Zhu, S. Park, J. Yum, S. Koveshnikov, W. Tsai, V. Tokranov, M. Yakimov, S. Oktyabrsky, and J. C. Lee, "Metal gate HfO2 metal-oxide-semiconductor structures on InGaAs substrate with varying Si interface passivation layer and postdeposition anneal condition," *Journal of Vacuum Science & Technology B*, 25(4), 1491–1494 (2007).
52. Y. Xuan, Y. Q. Wu, H. C. Lin, T. Shen, and P. D. Ye, "Submicrometer inversion-type enhancement-mode InGaAs MOSFET with atomic-layer-deposited as gate dielectric," *IEEE Electron Device Letters*, 28(11), 935–938 (2007).
53. C. P. Chen, T. D. Lin, Y. J. Lee, Y. C. Chang, M. Hong, and J. Kwo, "Self-aligned inversion n-channel $In_{0.2}Ga_{0.8}As$/GaAs metal-oxide-semiconductor field-effect-transistors with TiN gate and $Ga_2O_3(Gd_2O_3)$ dielectric," *Solid State Electronics*, 52(10), 1615–1618 (2008).
54. M. Passlack, M. Hong, and J. P. Mannaerts, "Quasistatic and high frequency capacitance–voltage characterization of Ga/sub 2/O/sub 3/-GaAs structures fabricated by in situ molecular beam epitaxy," *Applied Physics Letter*, 68(8), 1099–1101 (1996).
55. K. Hiruma, M. Yazawa, H. Matsumoto, O. Kagaya, M. Miyazaki, and Y. Umemoto, "Selective growth of ultra-low resistance GaAs/InGaAs for high performance InGaAs FETs," *Journal of Crystal Growth*, 124(1), 255–259 (1992).
56. C. Liao, D. Cheng, C. Cheng, K. Y. Cheng, M. Feng, T. H. Chiang, J. Kwo, and M. Hong, "Inversion-channel enhancement-mode GaAs MOSFETs with regrown source and drain contacts," *Journal of Crystal Growth*, 311(7), 1958–1961 (2009).
57. U. Singisetti, M. A. Wistey, G. J. Burek, A. K. Baraskar, J. Cagnon, B. Thibeault, A. C. Gossard, S. Stemmer, M. J. W. Rodwell, E. Kim, B. Shin, P. C. McIntyre, and Y.-J. Lee, "Enhancement mode $In_{0.53}Ga_{0.47}As$ MOSFET with self-aligned epitaxial source/drain regrowth," 2009 IEEE International Conference on Indium Phosphide & Related Materials (IEEE, Piscataway, 2009), pp. 120–123.
58. J. B. Boos, B. R. Bennett, N. A. Papanicolaou, M. G. Ancona, J. G. Champlain, R. Bass, and B. V. Shanabrook, "High mobility p-channel HFETs using strained Sb-based materials," *Electronics Letters*, 43(15), 834, 835 (2007).

59. M. K. Hudait and R. Chau, "Integrating III-V on silicon for future nanoelectronics," 2008 IEEE Compound Semiconductor Integrated Circuits Symposium (IEEE, Piscataway, NJ, 2008), pp. 1, 2.
60. C. L. Dohrman, K. Chilukuri, D. M. Isaacson, M. L. Lee, and E. A. Fitzgerald, "Fabrication of silicon on lattice-engineered substrate (SOLES) as a platform for monolithic integration of CMOS and optoelectronic devices," 2006 International SiGe Technology and Device Meeting (IEEE, Piscataway, 2006), pp. 44, 45.
61. Z. Yu, R. Droopad, D. Jordan, J. Curless, Y. Liang, C. Overgaard, H. Li, A. Talin, T. Eschrich, B. Craigo, K. Eisenbeiser, R. Emrick, J. Finder, X. Hu, Y. Wei, J. Edwards, Jr., D. Convey, K. Moore, D. Marshall, and J. Ramdani, "GaAs-based heterostructures on silicon," 2002 GaAs MANTECH Conference. *Technical Digest*, 276–279 (2002).
62. F. Letertre, "Formation of III-V semiconductor engineered substrates using smart Cut layer transfer technology," *AIP Conference Proceedings*, 1068, 185–196 (2008).
63. J. W. Lee, Y. W. Ahn, J. H. Song, B. G. Cho, and I. H. Ahn, "AlGaAs/InGaAs PHEMT with multiple quantum wire gates," *Microelectronics Journal*, 36(3), 389–391 (2005).
64. S. A. Fortuna and X. Li, "GaAs MESFET with a high-mobility self-assembled planar nanowire channel," *IEEE Electron Device Letters*, 30(6), 593–595 (2009).
65. H. Hasegawa and M. Akazawa, "Surface passivation technology for III-V semiconductor nanoelectronics," *Applied Surface Science*, 255(3), 628–632 (2008).
66. S. Oktyabrsky, P. Nagaiah, V. Tokranov, S. Koveshnikov, M. Yakimov, R. Kambhampati, R. Moore, and W. Tsai, "Electron scattering in buried InGaAs MOSFET channel with HfO_2 gate oxide," *MRS Proceedings*, 1155, C02-3 (2009).
67. T. Matsuoka, E. Kobayashi, K. Taniguchi, C. Hamaguchi, and S. Sasa, "Temperature dependence of electron mobility in InGaAs/InAlAs heterostructures," *Japan Journal of Applied Physics Pt. 1*, 29(10), 2017–2025 (1990).
68. M. A. Negara, K. Cherkaoui, P. Majhi, C. D. Young, W. Tsai, D. Bauza, G. Ghibaudo, and P. K. Hurley, "The influence of HfO_2 film thickness on the interface state density and low field mobility of n channel HfO_2/TiN gate MOSFETs," *Microelectronic Engineering*, 84(9–10), 1874–1877 (2007).
69. K. Maitra, M. M. Frank, V. Narayanan, V. Misra, and E. A. Cartier, "Impact of metal gates on remote phonon scattering in titanium nitride/hafnium dioxide n-channel metal-oxide-semiconductor field effect transistors-low temperature electron mobility study," *Journal of Applied Physics*, 102(11), 114507–1141 (2007).
70. S. Barraud, L. Thevenod, M. Casse, O. Bonno, and M. Mouis, "Modeling of remote Coulomb scattering limited mobility in MOSFET with HfO_2/SiO_2 gate stacks," *Microelectronic Engineering*, 84(9, 10), 2404–2407 (2007).
71. P. D. Ye, G. D. Wilk, B. Yang, J. Kwo, S. N. G. Chu, S. Nakahara, H.-J. L. Gossmann, J. P. Mannaerts, M. Hong, K. K. Ng, and J. Bude, "GaAs metal-oxide-semiconductor field-effect transistor with nanometer-thin dielectric grown by atomic layer deposition," *Applied Physics Letter*, 83(1), 180 (2003).
72. M. L. Huang, Y. C. Chang, C. H. Chang, Y. J. Lee, P. Chang, J. Kwo, T. B. Wu, and M. Hong, "Surface passivation of III-V compound semiconductors using atomic-layer-deposition-grown Al_2O_3," *Applied Physics Letter*, 87, 252104 (2005).
73. H. Sugahara, M. Oshima, H. Oigawa, H. Shigekawa, and Y. Nannichi, "Synchrotron radiation photoemission analysis for $(NH_4)_2S_x$-treated GaAs" *Journal Applied Physics*, 69 (4349) (1991).
74. N. Goel, D. Heh, S. Koveshnikov, I. Ok, S. Oktyabrsky, V. Tokranov, R. Khambapati, M. Yakimov, Y. Sun, P. Pianetta, C. K. Gaspe, M. B. Santos, J. Lee, S. Datta, P. Majhi, and W. Tsai, "Addressing The Gate Stack Challenge For High Mobility $In_xGa_{1-x}As$ Channels For NFETs," *IEDM Technical Digest*, 363 (2008).
75. K. Majumdar, Private Communications, Unpublished (2008).
76. S. Uno, S. H. Abebe, and E. Cumberbatch, "Analytical formulae of quantum-mechanical electron density in inversion layer in planar MOSFETs," *IWCM* p. 25 (2006).
77. C. A. Richter et al., "Differences between quantum-mechanical capacitance–voltage simulators," *Simulation of Semiconductor Processes and Devices: Sispad* p. 340 (2001).

78. S. M. Sze, "Physics of Semiconductor Devices 2nd" p. 447 (1985).
79. S. Koveshnikov, N. Goel, P. Majhi, H. Wen, M. B. Santos, S. Oktyabrsky, V. Tokranov, R. Kambhampati, R. Moore, F. Zhu, J. Lee, and W. Tsai, "$In_{0.53}Ga_{0.47}As$ based metal oxide semiconductor capacitors with atomic layer deposition ZrO_2 gate oxide demonstrating low gate leakage current and equivalent oxide thickness less than 1 nm," *Applied Physics Letter*, 92, p. 222904 (2008).
80. W. Tsai, N. Goel, S. Koveshnikov, P. Majhi, and W.-E Wang, "Challenges of integration of high-k dielectric with III-V materials," *Microelectronic Engineering*, 86, 1540 (2009).
81. R. Chau et al., "Devices: Opportunities and Challenges for Future High-Performance and Low-Power Computational Applications" in: Proceedings of Technical Papers, IEEE VLSI-TSA International Symposium on VLSI Technology, Hsinchu, Taiwan, April 2005, pp. 13–16.
82. K. H. Shiu, T. H. Chiang, P. Chang, L. T. Tung, M. Hong, J. Kwo, and W. Tsai, "1 nm equivalent oxide thickness in $Ga_2O_3(Gd_2O_3)/In_{0.2}Ga_{0.8}As$ metal-oxide-semiconductor capacitors," *Applied Physics Letter*, 92, 172904 (2008).
83. M. Passlack, M. Hong, and J. P. Mannaerts, "Quasistatic and high frequency capacitance–voltage characterization of Ga_2O_3–GaAs structures fabricated by *in situ* molecular beam epitaxy," *Applied Physics Letter*, 68, 1099 (1996).
84. S. Oktyabrsky, M. Yakimov, V. Tokranov, R. Kambhampati, H. Bakhru, S. Koveshnikov, W. Tsai, F. Zhu, and J. Lee, "Challenges and progress in III-V MOSFETs for CMOS circuits," *International Journal of High Speed Electronics and Systems* 18, 761–772 (2008).
85. K. Y. Lee, Y. J. Lee, P. Chang, M. L. Huang, Y. C. Chang, M. Hong, and J. Kwo, "Achieving 1 nm capacitive effective thickness in atomic layer deposited HfO_2 on $In_{0.53}Ga_{0.47}As$," *Applied Physics Letter*, 92, 252908 (2008).
86. N. Goel, P. Majhi, W. Tsai, M. Warusawithana, D. G. Schlom, M. B. Santos, J. S. Harris, and Y. Nishi, "High-indium-content InGaAs metal-oxide-semiconductor capacitor with amorphous $LaAlO_3$ gate dielectric," *Applied Physics Letter*, 91, 093509 (2007).
87. S. Koveshnikov, R. Kambhampati, V. Tokranov, M. Yakimov, D. Schlom, M. Warusawithana, C. Adamo, W. Tsai, and S. Oktyabrsky, "Thermal Stability of Electrical and Structural Properties of GaAs based MOS Capacitors with Amorphous $LaAlO_3$ Gate Oxide," *Applied Physics Letter*, 93(1), 012903-1-3 (2008).
88. P. C. McIntyre, Y. Oshima, E. Kim, and K. C. Saraswat, "Interface studies of ALD-grown metal oxide insulators on Ge and III–V semiconductors," *Microelectronic Engineering*, 86, 1536–1539 (2009).
89. InJo Ok, H. Kim, M. Zhang, F. Zhu, S. Park, J. Yum, H. Zhao, Domingo Garcia, Prashant Majhi, N. Goel, W. Tsai, C. K. Gaspe, M. B. Santos, and Jack C. Lee, "Self-aligned *n*-channel metal-oxide-semiconductor field effect transistor on high-indium-content $In_{0.53}Ga_{0.47}As$ and InP using physical vapor deposition HfO_2 and silicon interface passivation layer," *Applied Physics Letter*, 92, 202903, (2008).
90. Hyoung-Sub Kim, Injo Ok, Feng Zhu, M. Zhang, S. Park, J. Yum, H. Zhao, Prashant Majhi, Domingo I. Garcia-Gutierrez, Niti Goel, W. Tsai, C. K. Gaspe, M. B. Santos, and Jack C. Lee, "High mobility HfO_2-based $In_{0.53}Ga_{0.47}As$ *n*-channel metal-oxide-semiconductor field effect transistors using a germanium interfacial passivation layer," *Applied Physics Letter*, 93, 132902 (2008).
91. M. J. Hale, S. I. Yi, J. Z. Sexton, A. C. Kummel, and M. Passlack, "Scanning tunneling microscopy and spectroscopy of gallium oxide deposition and oxidation on GaAs(001)-c(2×8)/(2×4)," *Journal of Chemical Physics*, 119, 6719–6728 (2003).
92. D. Shahrjerdi, M. M. Oye, A. L. Holmes, Jr., and S. K. Banerjee, "Unpinned metal gate/high-k GaAs capacitors: fabrication and characterization," *Applied Physics Letter*, 89, 43501–431 (2006).
93. Y. Xuan, H.-C. Lin, and P. D. Ye, "Simplified surface preparation for GaAs passivation using atomic layer-deposited high-kappa dielectrics," *IEEE Transactions on Electron Devices*, 54, 1811–1817, 2007.
94. S. Oktyabrsky, S. Koveshnikov, V. Tokranov, M. Yakimov, R. Kambhampati, H. Bakhru, F. Zhu, J. Lee, and W. Tsai, "InGaAs and GaAs/InGaAs channel enhancement mode n-MOS-

FETs with HfO_2 gate oxide and a-Si interface passivation layer," *Digest, Device Research Conference*, p. 203 (2007).
95. S. Koveshnikov, W. Tsai, I. Ok, J. Lee, V. Torkanov, M. Yakimov, and S. Oktyabrsky, "Metal-Oxide-Semiconductor Capacitors on GaAs with High-k Gate Oxide and Amorphous Silicon Interface Passivation Layer," *Applied Physics Letter*, 88, 022106-8, 2006.
96. R. Kambhampati, S. Koveshnikov, V. Tokranov, M. Yakimov, R. Moore, W. Tsai, and S. Oktyabrsky, "In-Situ Deposition of High-k Gate Stack on InGaAs and GaAs for Metal-Oxide-Semiconductor Devices with Low Equivalent Oxide Thickness," *ECS Transactions*, 11, 431–439, 2007.
97. S. Oktyabrsky, V. Tokranov, M. Yakimov, R. Moore, S. Koveshnikov, W. Tsai, F. Zhu, and J. C. Lee, "High-k gate stack on GaAs and InGaAs using *in-situ* passivation with amorphous silicon," *Materials Science & Engineering B (Solid-State Materials for Advanced Technology)*, 135, 272–276 (2006).
98. H.-S. Kim, I. Ok, M. Zhang, T. Lee, F. Zhu, L. Yu, J. C. Lee, S. Koveshnikov, W. Tsai, V. Tokranov, M. Yakimov, and S. Oktyabrsky, "Depletion-mode GaAs metal-oxide-semiconductor field-effect transistor with HfO_2 dielectric and germanium interfacial passivation layer," *Applied Physics Letters*, 89, 222904–2221 (2006).
99. P. D. Ye, G. D. Wilk, B. Yang, J. Kwo, S. N. G. Chu, S. Nakahara, H. J. L. Gossmann, J. P. Mannaerts, M. Hong, K. K. Ng, and J. Bude, "GaAs metal-oxide-semiconductor field-effect transistor with nanometer-thin dielectric grown by atomic layer deposition," *Applied Physics Letters*, 83, 180–182 (2003).
100. P. D. Ye, G. D. Wilk, B. Yang, J. Kwo, H.-J. L. Gossmann, M. Hong, K. K. Ng, and J. Bude, "Depletion-mode InGaAs metal-oxide-semiconductor field-effect transistor with oxide gate dielectric grown by atomic-layer deposition," *Applied Physics Letters*, 84, 434–436 (2004).
101. D. Shahrjerdi, E. Tutuc, and S. K. Banerjee, "Impact of surface chemical treatment on capacitance–voltage characteristics of GaAs metal-oxide-semiconductor capacitors with Al_2O_3 gate dielectric," *Applied Physics Letter*, 91, 063501–063503 (2007).
102. M. Passlack, M. Hong, and J. P. Mannaerts, "Quasistatic and high frequency capacitance–voltage characterization of Ga_2/O_3-GaAs structures fabricated by *in-situ* molecular beam epitaxy," *Applied Physics Letters*, vol. 68, pp. 1099–1101 (1996).
103. F. Ren, M. Hong, W. S. Hobson, J. M. Kuo, J. R. Lothian, J. P. Mannaerts, J. Kwo, S. N. G. Chu, Y. K. Chen, and A. Y. Cho, "Demonstration of enhancement-mode p- and n-channel GaAs MOSFETs with $Ga_2O_3/(Gd_2/O_3)$ as gate oxide," *Solid-State Electronics*, pp. 1751–1753 (1997).
104. M. Hong, F. Ren, W. S. Hobson, J. M. Kuo, J. Kwo, J. P. Mannaerts, J. R. Lothian, M. A. Marcus, C. T. Liu, A. M. Sergent, T. S. Lay, Y. K. Chen, M. Melloch, and M. A. Reed, "Growth of $Ga_2O_3/(Gd_2O_3)$ using molecular beam epitaxy technique-key to first demonstration of GaAs MOSFETs," in *Compound Semiconductors 1997. Proceedings of the IEEE Twenty-Fourth International Symposium on Compound Semiconductors*, M. Melloch and M. A. Reed, Eds. (IEEE, USA New York, 1998), pp. 319–324.
105. M. Passlack, J. K. Abrokwah, R. Droopad, Z. Yu, C. Overgaard, S. I. Yi, M. Hale, J. Sexton, and A. C. Kummel, "Self-aligned GaAs p-channel enhancement mode MOS heterostructure field-effect transistor," *IEEE Electron Device Letters*, 23, pp. 508–510 (2002).
106. R. J. W. Hill, D. A. J. Moran, X. Li, H. Zhou, D. Macintyre, S. Thoms, R. Droopad, M. Passlack, and I. G. Thayne, "180 nm metal gate, high-k dielectric, implant-free III-V MOSFETs with transconductance of over 425 μS/μm," *Electronics Letters*, 43, pp. 543–545 (2007).
107. S. Tiwari, S. L. Wright, and J. Batey, "Unpinned GaAs MOS capacitors and transistors," *IEEE Electron Device Letters*, 9, pp. 488–490 (1988).
108. Z. Wang, M. E. Lin, D. Biswas, B. Mazhari, N. Teraguchi, Z. Fan, X. Gui, and H. Morkoc, "Si/sub 3/N/sub 4//Si/n-GaAs capacitor with minimum interface density in the 10/sup 10/ eV/sup -1/ cm/sup -2/range," *Applied Physics Letters*, 62, pp. 2977–2979 (1993).

109. H. Hasegawa, "MBE growth and applications of silicon interface control layers," presented at Thin Solid Films Third International Workshop on Molecular Beam Epitaxy-Growth Physics and Technology (MBE-GPT), Switzerland Warsaw, Poland (2000).
110. Y. Wada and K. Wada, "GaAs surface passivation by deposition of an ultrathin InP-related layer," *Applied Physics Letters*, 63, 379 (1993).
111. Y. Wada and K. Wada, "Nearly ideal characteristics of GaAs metal-insulator-semiconductor diodes by atomic layer passivation," *Journal of Vacuum Science & Technology B (Microelectronics and Nanometer Structures)*, 12, pp. 3084–3089 (1994).
112. E. J. Tarsa, X. H. Wu, J. P. Ibbetson, J. S. Speck, and J. J. Zinck, "Growth of epitaxial MgO films on Sb-passivated (001)GaAs: Properties of the MgO/GaAs interface," *Applied Physics Letters*, 66, pp. 3588–3590 (1995).
113. S. Lodha, D. B. Janes, and N.-P. Chen, "Unpinned interface Fermi-level in Schottky contacts to n-GaAs capped with low-temperature-grown GaAs; experiments and modeling using defect state distributions," *Journal of Applied Physics*, 93, pp. 2772–2779 (2003).
114. B. J. Skromme, C. J. Sandroff, E. Yablonovitch, and T. Gmitter, "Effects of passivating ionic films on the photoluminescence properties of GaAs," *Applied Physics Letters*, 51, pp. 2022–2024 (1987).
115. A. Callegari, P. D. Hoh, D. A. Buchanan, and D. Lacey, "Unpinned gallium oxide/GaAs interface by hydrogen and nitrogen surface plasma treatment," *Applied Physics Letters*, 54, pp. 332–334 (1989).
116. A. Paccagnella and A. Callegari, "GaAs surface plasma treatments for Schottky contacts," *Solid-State Electronics*, 34, pp. 1409–1414 (1991).
117. P. D. Ye, G. D. Wilk, B. Yang, J. Kwo, H.-J. L. Gossmann, M. Hong, K. K. Ng, and J. Bude, "Depletion-mode InGaAs metal-oxide-semiconductor field-effect transistor with oxide gate dielectric grown by atomic-layer deposition," *Applied Physics Letters*, 84, p. 434 (2004).
118. M. K. Hudait, G. Dewey, S. Datta, J. M. Fastenau, J. Kavalieros, W. K. Liu, D. Lubyshev, R. Pillarisetty, W. Rachmady, M. Radosavljevic, T. Rakshit, and R. Chau, "Heterogeneous integration of enhancement mode $In_{0.7}Ga_{0.3}As$ quantum well transistor on silicon substrate using thin (<2 mm) composite buffer architecture for high-speed and low-voltage (0.5 V) logic applications," Technical Digest – International Electron Devices Meet. 625 (2007).
119. C.-H. Chang, Y.-K. Chiou, Y.-C. Chang, K.-Y. Lee, T.-D. Lin, T.-B. Wu, and M. Hong, "Interfacial self-cleaning in atomic layer deposition of HfO_2 gate dielectric on $In_{0.15}Ga_{0.85}As$," *Applied Physics Letters* 89, p. 242911 (2006).
120. M. M. Frank, G. D. Wilk, D. Starodub, T. Gustafsson, E. Garfunkel, Y. J. Chabal, J. Grazul, and D. A. Muller, "HfO_2 and Al_2O_3 gate dielectrics on GaAs grown by atomic layer deposition," *Applied Physics Letters*, 86, p. 152904 (2005).
121. G. Hollinger, R. Skheyta-Kabbani, and M. Gendry, "Oxides on GaAs and InAs surfaces: An x-ray-photoelectron-spectroscopy study of reference compounds and thin oxide layers," *Physics Review B*, 49 (11159) (1994).
122. C. L. Hinkle, A. M. Sonnet, E. M. Vogel, S. McDonnell, G. J. Hughes, M. Milojevic, B. Lee, F. S. Aguirre-Tostado, K. J. Choi, H. C. Kim, J. Kim, and R. M. Wallace, "GaAs interfacial self-cleaning by atomic layer deposition," *Applied Physics Letters*, 92, p. 071901 (2008).
123. M. Kobayashi, P. T. Chen, Y. Sun, N. Goel, P. Majhi, M. Garner, W. Tsai, P. Pianetta, and Y. Nishi, "Synchrotron radiation photoemission spectroscopic study of band offsets and interface self-cleaning by atomic layer deposited HfO_2 on $In_{0.53}Ga_{0.47}As$ and $In_{0.52}Al_{0.48}As$," *Applied Physics Letters*, 93, p. 182103 (2008).
124. Y. Sun, P. Pianetta, P.-T. Chen, M. Kobayashi, Y. Nishi, N. Goel, M. Garner, and W. Tsai, "Arsenic-dominated chemistry in the acid cleaning of InGaAs and InAlAs surfaces," *Applied Physics Letters*, 93, p. 194103 (2008).
125. M. Passlack, Z. Yu, R. Droopad, J. K. Abrokwah, D. Braddock, S.-I. Yi, M. Hale, J. Sexton, A. C. Kummel, "Gallium oxide on gallium arsenide: Atomic structure, materials, and devices," III-V semiconductor heterostructures: physics and devices, edited by W. Z. Cai, Research Signpost, pp. 327–355 (2003).

126. H. Sugahara, M. Oshima, H. Oigawa, H. Shigekawa, Y. Nannichi, “Synchrotron radiation photoemission analysis for $(NH_4)_2S_x$-treated GaAs,” *Journal of Applied Physics*, 69, p. 4349 (1991).
127. N. Goel, P. Majhi, C. O. Chui, W. Tsai, D. Choi and J. S. Harris, “InGaAs metal-oxide-semiconductor capacitors with HfO_2 gate dielectric grown by atomic-layer deposition,” *Applied Physics Letters*, 89, p. 163517 (2006).
128. N. Goel, W. Tsai, C. M. Garner, Y. Sun, P. Pianetta, M. Warusawithana, D. G. Schlom, H. Wen, C. Gaspe, J. C. Keay, M. B. Santos, L. V. Goncharova, E. Garfunkel, and T. Gustafsson, “Band offsets between amorphous $LaAlO_3$ and $In_{0.53}Ga_{0.47}As$,” *Applied Physics Letters*, 91, p. 113515 (2007).
129. P. T. Chen, Y. Sun, E. Kim, P. C. McIntyre, W. Tsai, M. Garner, P. Pianetta, Y. Nishi, and C. O. Chui, “HfO_2 gate dielectric on $(NH_4)_2S$ passivated (100) GaAs grown by atomic layer deposition,” *Journal of Applied Physics*, 103, 034106 (2008).
130. E. A. Kraut, R. W. Grant, J. R. Waldrop, and S. P. Kowalczyk, “Precise determination of the valence-band edge in x-ray photoemission spectra: application to measurement of semiconductor interface potentials,” *Physics Review Letter*, 44, 1620 (1980).
131. E. A. Kraut, R. W. Grant, J. R. Waldrop, and S. P. Kowalczyk, “Semiconductor core-level to valence-band maximum binding-energy differences: Precise determination by x-ray photoelectron spectroscopy,” *Physics Review* B, 28, 1965 (1983).
132. Z. Liu, Y. Sun, F. Machuca, P. Pianetta, W. E. Spicer, and R. F. W. Pease, “Optimization and characterization of III–V surface cleaning,” *Journal of Vacuum Science Technology* B, 21, 1953 (2003).
133. S. A. Chambers, Y. Liang, Z. Yu, R. Droopad, J. Ramdani, and K. Eisenbeiser, “Band discontinuities at epitaxial $SrTiO_3$/Si(001) heterojunctions,” *Applied Physics Letters*, 77, p. 1662 (2000).
134. S. A. Chambers, Y. Liang, Z. Yu, R. Droopad, and J. Ramdani, “Band offset and structure of SrTiO3 ÕSi.001. heterojunctions,” *Journal Vacuum Science Technology*, A 19, 934 (2001).
135. Miyazaki, “Photoemission study of energy-band alignments and gap-state density distributions for high-k gate dielectrics,” *Journal Vacuum Science Technology*, B 19, 2212 (2001); S. A. Chambers, T. Droubay, T. C. Kaspar, and M. Gutowski, *ibid*, 22, 2205 (2004).
136. J. Robertson, and B. Falabretti, “Band offsets of high K gate oxides on high mobility semiconductors,” *Materials Science Engineering*, 135, 267 (2006).
137. K. L. Wang, and A. O. Evwaraye, “Determination of interface and bulk trap states of IGFET’s using deep-level transient spectroscopy,” *JAP*, 47, 4574 (1976).
138. F. M. Klassen, “Characterization of low 1/f noise in MOS transistors,” *IEEE Transactions on Electron Devices*, ED-18, 887 (1971).
139. G. Groeseneken, H. E. Maes, N. Beltran, and R. F. de Keersmaecker, A reliable approach to charge pumping measurements in MOS transistors, *IEEE Transactions on Electron Devices*, ED-31, 42 (1984).
140. G. V den Bosch, G. V. Groeseneken, P. Heremans, and H. E. Maes, “Spectroscopic charge pumping: A new procedure for measuring interface trap distributions on MOS transistors,” *IEEE Transactions on Electron Devices*, 38, 1820 (1991).
141. D. Heh, et. al, “A novel bias temperature instability characterization methodology for high-k nMOSFETs,” IEEE Elect. Development Letter, 27, pp. 849–851, 2006; “Experimental evidence of the fast and slow charge trapping/detrapping processes in high-k dielectrics subjected to PBTI stress,” 29, 180–182 (2008).
142. M. Sonnet, C. L. Hinkle, M. N. Jivani, R. A. Chapman, G. P. Pollack, R. M. Wallace, and E. M. Vogel, “Performance enhancement of n-channel inversion type InGaAs metal oxide semiconductor field effect transistor using *ex-situ* deposited thin film amorphous silicon layer,” *Applied Physics Letter*, 93, 122109 (2008).
143. M. H. White and J. R. Cricchi, “Characterization of thin-oxide MNOS transistors,” *IEEE TED*, 19 (12), 1280–1288 (1972).

144. R. Castagne and A. Vapaille, "Description of the SiO_2—Si interface properties by means of very low frequency MOS capacitance measurements," *Surface Sceince*, 28, 157–193 (1971).
145. M. Shulz, "Interface states at the SiO_2-Si interface," *Surface Science*, 132, 422–455 (1983).
146. W. E. Spicer, T. Kendelewicz, N. Newman, K. K. Chin, and I. Lindau, "The mechanisms of Schottky barrier pinning in III-V semiconductors: Criteria developed from microscopic (atomic level) and macroscopic experiments," *Surface Science*, 168, 240–259 (1986).
147. W. E. Spicer, Z. Liliental-Weber, E. Weber, N. Newman, T. Kendelewicz, R. Cao, C. Mc-Cants, P. Mahowald, K. Miyano, and I. Lindau, "The advanced unified defect model for Schottky barrier formation," *Journal of Vacuum Science Technology* B, 6 (4), 1245–1251 (1988).
148. G. Brammertz, H.-C. Lin, K. Martens, D. Mercier, S. Sioncke, A. Delabie, W.-E Wang, M. Caymax, M. Meuris and M. Heyns, "Capacitance–voltage characterization of GaAs-Al_2O_3 interfaces," *Applied Physics Letter*, 93, 183504-1-3 (2008).
149. G. Brammertz, H. C. Lin, K. Martens, A. Alian, C. Merckling, J. Penaud, D. Kohen, W.-E Wang, S. Sioncke, A. Delabie, M. Meuris, M. Caymax, and M. Heyns, "Electrical properties of III-V/oxide interfaces," *Electrical Chemical Society spring meetings*, pp. 375–383 (2009).
150. H.-C. Lin, W.-E Wang, G. Brammertz, M. Meuris, and M. Heyns, "Inversion behavior on n and p type ALD—$Al_2O_3/In_{0.53}Ga_{0.47}As$ MOS capacitors," *IEEE Semiconductor Interface Specialists Conference*, San Diego, CA, December (2008).
151. G. Brammertz, K. Martens, S. Sioncke, A. Delabie, M. Caymax, M. Meuris, and M. Heyns, "Characteristic trapping lifetime and capacitance–voltage measurements of GaAs metal-oxide-semiconductor structures," *Applied Physics Letter*, 91, 133510 (2007).
152. Y. Tsividis, in *Operation and Modeling of the MOS Transistor*, McGraw-Hill, New York (2003).
153. H.-C. Lin, G. Brammertz, K. Martens, G. de Valicourt, L. Negre, W.-E Wang, W. Tsai, M. Meuris, and M. Heyns, "Application of Fermi Level Efficiency Method to the analysis of high interface trap density oxide-semiconductor interfaces," *Applied Physics Letter*, 94, 153508-1-3 (2009).
154. K. Martens, C. O. Chui, G. Brammertz, B. De Jaeger, D. Kuzum, M. Meuris, M. M. Heyns, T. Krishnamohan, K. Saraswat, H. E. Maes, and G. Groeseneken, "On the correct extraction of interface trap density of MOS devices with high-mobility semiconductor," *IEEE Transaction on Electron Devices*, 55(2), 547–549 (2008).
155. D. K. Schroder, Semiconductor Material and device characterization, Wiley-Interscience publication, Second Edition, p. 541 (1998).
156. C. L. Hinkle, A. M. Sonnet, R. A. Chapman, E. M. Vogel, Extraction of the Effective Mobility of $In_{0.53}Ga_{0.47}As$ MOSFETs, *IEEE Electron Device Letters*, 30, pp. 316–318 (2009).
157. A. Callegari, D. K. Sadana, D. A. Buchanan, A. Paccagnella, E. D. Marshall, M. A. Tischler, and M. Norcott, "Properties of SiO_2/Si/GaAs structures formed by solid phase epitaxy of amorphous Si on GaAs," *Applied Physics Letter*, 58, 2540 (1991).
158. C. G. Van de Walle and J. Neugebauer, "Universal alignment of hydrogen levels in semiconductors, insulators, and solutions," *Nature*, 423, 626 (2003).
159. S. J. Pearton, J. W. Corbett, and T. S. Shi, "Hydrogen in crystalline semiconductors," *Applied Physics* A, 43, 153 (1987).
160. Y. C. Cheng, "Electronic states at the silicon-silicon dioxide interfaces," *Progress in Surface Science*, 8, 181–218 (1977).
161. M. Chuskey and E. E. Haller, "Hydrogen on III-V and II-V semiconductors," in *Hydrogen in Semiconductor II*, vol. 61, edited by N. H. Nickel, Academic Press, NY, pp. 273–440 (1999).
162. L. Pavesi, in *Properties of Aluminium Gallium Arsenide*, edited by S. Adachi, IET (1993).
163. S. J. Pearton, "Hydrogen passivation of a bulk donor defect (E_C-0.36 eV) in GaAs," *Journal of Applied Physics*, 53, pp. 4509–4511 (1982).

164. A. Jaouad and V. Aimez, "Passivation of air-exposed AlGaAs using low frequency plsma-enhance chemical vapour deposition of silicon nitride," *Applied Physics Letter,* 89, 092125-1-3 (2001).
165. W. P. Li, X. M. Wang, Y. X. Liu, S. I. Shim, and T. P. Ma, "Demonstration of unpinned GaAs surface and surface inversion with gate dielectric made of Si_3N_4," *Applied Physics Letter*, 90, 193503-1-3 (2001).
166. M. Passlack, R. Droopad, K. Rajagopalan, J. Abrokwah, and P. Zurcher, "High mobility III-V MOSFET technology," *CS MANTECH conference*, May 14–17, Austin, Texas, pp. 235–238 (2007).
167. M. Renaud, P. Boher, J. Barrier, J. Scheider, and J. P. Chane, "Multiple plasma treatments of $In_{0.53}Ga_{0.47}As$ surface for MIS devices application," Proceedings of ESSDERC, 17, 129–133 (1987).
168. H.-C. Lin, W.-E Wang, G. Brammertz, M. Meuris, and M. Heyns, "Inversion behavior on n and p type ALD—$Al_2O_3/In_{0.53}Ga_{0.47}As$ MOS capacitors," *IEEE Semiconductor Interface Specialists Conference*, San Diego, CA, December (2008).
169. H.-C. Lin, G. Brammertz , K. Martens, G. de Valicourt, L. Negre, W.-E Wang, W. Tsai, M. Meuris, and M. Heyns, "Application of Fermi Level Efficiency Method to the analysis of high interface trap density oxide-semiconductor interfaces," *Applied Physics Letter*, 94, 153508 (2009).
170. A. Y. Polyakov, A. V. Pakhomov, M. V. Tishkin, and E. M. Omeljanovsky, "Hydrogen passivation: Relevance to semi-insulating III-V materials," *6th Conference on Semi-insulating III-V Materials*, Toronto, Canada, pp. 247–256 (1990).
171. C. H. Seager, "Hydrogenation methods," in Hydrogen in semiconductor, *semiconductors and semimetals*, 34, edited by J. I. Pankove and N. M. Johnson, Academic Press, NY, pp. 17–31 (1991).
172. I. Lundström, "Hydrogen sensitive MOS-structures, Part 1: principles and applications," *Sensors and Actuators*, 1, 403–426 (1981).
173. Ryzhikov, F. Robaut, M. Labeau, and A. Gaskov, "New gas sensitive MIS structure Pt/Al_2O_3 (M=Pt/Ru)/Si with a granular dielectric layer," *Sensors and Actuators* B, 133, 613–616 (2008).
174. K. Martens, C. O. Chui, G. Brammertz, B. De Jaeger, D. Kuzum, M. Meuris, M. M. Heyns, T. Krishnamohan, K. Saraswat, H. E. Maes, and G. Groeseneken, "On the correct extraction of interface trap density of MOS devices with high-mobility semiconductor," *IEEE Transaction on Electron Devices*, 55(2), 547–549 (2008).
175. E. H. Nicollian, and J. R. Brews, in *MOS (Metal Oxide Semiconductor) Physics and Technology*, Wiley, NY, 1981.
176. Y. Taur and T. K. Ning, *Fundamentals of modern VLSI devices*, Cambridge University Press, New York, 1998.
177. http://www.ioffe.rssi.ru/SVA/NSM/
178. P. Batude, X. Garros, L. Clavelier, C. Le Royer, J. M. Hartmann, V. Loup, P. Besson, L. Vandroux, Y. Capidelli, S. Deleonibus, and F. Boulanger,"Insights on fundamental mechanisms impacting Ge metal oxide semiconductor capacitors with high-k/metal gate stacks," *Journal of Applied Physics*, 102, 034514-1-8 (2007).
179. Y. Fukuda, Y. Otani, Y. Itayama, and T. Ono, "Electrical analyses of germanium MIS structure and spectroscopic measurement of the interface trap density in an insulator/germanium interface at room temperature," *IEEE Transaction on Electron Devices*, 54, 2878–2883 (2007).
180. C.-C. Cheng, C.-H. Chien, G.-L. Luo, Y.-T. Ling, R.-D. Chang, C.-C. Kei, C.-N. Hsiao, J.-C. Liu, and C.-Y. Chang, "Effect of minority-carrier response behaviour on Ge MOS capacitor characteristics: experimental measurements and theoretical simulations," *IEEE Transaction on Electron Devices*, 56, 1118–1127 (2009).
181. W.-E Wang, H.-C. Lin, G. Brammertz, J. Miatrd, A. Delabie, S. Sioncke, C. Merckling, J. Penaud, R. Yang, E. Simoen, W. Tsai, M. Caymax, M. Meuris, and M. Heyns, "Minority carrier response of hydrogen annealed $In_{0.53}Ga_{0.47}As/Al_2O_3/Pd$ capacitors at room temperature,"

Applied Physics Letter, to be submitted; W.-E Wang, H.-C. Lin, G. Brammertz, A. Delabie, E. Simoen, M. Caymax, M. Meuris, and M. Heyns, "Catalytic-forming gas anneal on III-V/Ge MOS systems," *Invited Paper for MRS Fall Meeting* (2009, submitted).

182. J. Lin, S. Lee, H.-J. Oh, W. Yang, G. Q. Lo, D. L. Kwong, and D. Z. Chi, "Inversion-mode self-aligned $In_{0.53}Ga_{0.47}As$ N-channel metal-oxide-semiconductor field-effect transistor with HfAlO gate dielectric and TaN metal gate," *IEEE Electron Device Letter*, 29, 977 (2008).
183. H. Zhao, Y. Chen, J. Yum, Y. Wang, and J. C. Lee, "HfO_2-based $In_{0.53}Ga_{0.47}As$ MOSFETs (EOT~10Å) using various interfacial dielectric layers," DRC Device Conference June 2009.
184. E. Chagarov, A. C. Kummel, "Density-Functional Theory Molecular Dynamics Simulations of a-Al_2O_3/Ge(100)(2×1), a-Al_2O_3/$In_{0.5}Ga_{0.5}As$, a-Al_2O_3/$In_{0.5}A_{l0.5}As$/$In_{0.5}Ga_{0.5}As$," AVS meeting 2008.
185. I. Ok, H.-S. Kim, M. Zhang, C.-Y. Kang, S. J. Rhee, C. Choi, S. A. Krishnan, T. Lee, F. Zhu, G. Thareja, and J. C. Lee, "Metal gate-HfO_2 MOS structures on GaAs substrate with and without Si interlayer," *IEEE Electron Device Letter*, 27, 145 (2006).
186. S. J. Koester, E. W. Kiewra, Yanning Sun, D. A. Neumayer, J. A. Ott, M. Copel, D. K. Sadana, D. J. Webb, J. Fompeyrine, J.-P. Locquet, C. Marchiori, M. Sousa, and R. Germann, "Evidence of electron and hole inversion in GaAs metal-oxide-semiconductor capacitors with HfO_2 gate dielectrics and α-Si/SiO_2 interlayers," *Applied Physics Letter*, 89, 042104 (2006).
187. H.-S. Kim, I. Ok, M. Zhang, T. Lee, F. Zhu, L. Yu, and J. C. Lee, "Metal gate-HfO_2 metal-oxide-semiconductor capacitors on *n*-GaAs substrate with silicon/germanium interfacial passivation layers," *Applied Physics Letter*, 89, 222903-1-3 (2006).
188. M. V. Fischetti, D. A. Neumayer and E. A. Cartier, "Effective electron mobility in Si inversion layers in metal–oxide–semiconductor systems with a high-k insulator: The role of remote phonon scattering," *Journal Applied Physics*, 90, 4587 (2001).
189. M. V. Fischetti, S. E. Laux, "Monte Carlo study of electron transport in silicon inversion layers," *Physics Review* B, 48, 2244 (1993).
190. H. Zhao, Y.-T. Chen, J. H. Yum, Y. Wang, N. Goel, and J. C. Lee, "High performance $In_{0.7}Ga_{0.3}As$ metal-oxide-semiconductor transistors with mobility >4400 cm^2/V s using InP barrier layer," *Applied Physics Letters*, 94, 193502 (2009).
191. D. H. Kim et al., "30 nm E-mode InAs PHEMTs for THz and future logic applications," D. H. Kim et al., IEDM *Technical Digest*, 719 (2008).
192. R. Droopad, K. Rajagopalan, J. Abrokwah, P. Zurcher, and M. Passlack, "Compound semiconductor MOSFETs," *Microelectronic Engineering*, 84, 2138–2141 (2007).
193. F. Ren, M. Hong, J. M. Kuo, W. S. Hobson, J. R. Lothian, H. S. Tsai, J. Lin, J. P. Mannaerts, J. Kwo, S. N. G. Chu, Y. K. Chen, and A. Y. Cho, "III-V compound semiconductor MOSFETs using $Ga_2O_3(Gd_2O_3)$ as gate dielectric," 19th Annual Technical Digest, Gallium Arsenide Integrated Circuit (GaAs IC) Symposium, 1997.
194. C. P. Chen, T. D. Lin, Y. J. Lee, Y. C. Chang, M. Hong, and J. Kwo, "Self aligned inversion n-channel $In_{0.2}Ga_{0.8}As$/GaAs metal-oxide-semiconductor field effect transistors with TiN gate and $Ga_2O_3(Gd_2O_3)$ dielectric," *Solid State Electronics*, 52, 1615–1618 (2008).
195. R. J. W. Hill, R. Droopad, D. A. J. Moran, X. Li, H. Zhou, D. Macintyre, S. Thoms, O. Ignatova, A. Asenov, K. Rajagopalan, P. Fejes, I. G. Thayne, and M. Passlack, "1 μm gate length, $In_{0.75}Ga_{0.25}As$ channel, thin body n-MOSFET on InP substrate with transconductance of 737 mS/mm," *Electronics Letter*, 44 (2008).
196. K. Kalna, et al.: 'Monte Carlo simulations of InGaAs nano-MOSFETs', *Microelectron. Engineering*, 84(9, 10), 2150–2153 (2007).
197. J. D. Zimmerman and A. C. Gossard, S. R. Bank, "Low resistance, nonalloyed Ohmic contacts to InGaAs," *Applied Physics Letter*, 91, 192114 (2007).
198. C.-W. Cheng, E. A. Fitzgerald, "*In situ* metal-organic chemical vapor deposition atomic-layer deposition of aluminum oxide on GaAs using trimethyaluminum and isopropanol precursors," *Applied Physics Letter*, 93, 031902 (2008).

199. Y. Hwang, M. A. Wistey, J. Cagnon, R. Engel-Herbert, and S. Stemmer, "Metal-oxide-semiconductor capacitors with erbium oxide dielectrics on $In_{0.53}Ga_{0.47}As$ channels," *Applied Physics Letter*, 94, 122907 (2009).
200. S. Stemmer, Y. Hwang, R. Engel-Herbert, J. Cagnon, and N. Goel, "Atomic scale control of high-k/InGaAs interfaces and correlation with interface trap densities," MRS Fall meeting (2009).
201. M. Tao, A. E. Botchkarev, D. Park, J. Reed, S. J. Chey, J. E. Van Nostrand, D. G. Cahill, and H. Morkoc, "Improved Si_3N_4/Si/GaAs metal-insulator-semiconductor interfaces by *in situ* anneal of the as-deposited Si," *Journal of Applied Physics*, 77 4113 (1995).
202. Z. Chen, S. N. Mohammad, D.-G. Park, D. M. Diatezua, H. Morkoc, and Y. C. Chang, "Band structure and confined energy levels of the Si_3N_4/Si/GaAs system," *Journal of Applied Physics Letter*, 82, 275 (1997).

Chapter 9
InGaAs, Ge, and GaN Metal-Oxide-Semiconductor Devices with High-k Dielectrics for Science and Technology Beyond Si CMOS

M. Hong, J. Kwo, T. D. Lin, M. L. Huang, W. C. Lee and P. Chang

Abstract An overview is given on advances of science and devices of InGaAs, Ge, and GaN MOS capacitors and inversion-channel and depletion-mode MOSFET's, with emphasis on the results using ultra-high vacuum (UHV) deposited $Ga_2O_3(Gd_2O_3)$ [GGO] and atomic layer deposited (ALD) oxides as high-k dielectrics. Very importantly, no interfacial layers are employed in these MOS devices. Low interfacial densities of states (D_{it}) as well as low electrical leakage currents have been obtained. Moreover, thermodynamic stability was attained with GGO/InGaAs, as the heterostructures were rapid thermal annealed to 800–900 °C. The oxide remains amorphous and the interface retains its atomic smoothness and sharpness. The oxide scalability with capacitance equivalent thickness (CET) of ≤1 nm has been achieved in GGO and ALD-HfO_2 on InGaAs. Interfacial chemical properties and band parameters of valence band offsets, conduction band offsets, and oxide band gaps in the high-k's/InGaAs are determined using x-ray photoelectron spectroscopy and electrical leakage transport. Inversion-channel and/or depletion mode InGaAs, GaN, and Ge MOSFET's were fabricated; in particular, a self-aligned inversion-channel $In_{0.53}Ga_{0.47}As$ MOSFET, made of Al_2O_3 (2 nm)/GGO (5 nm) gate oxide and TiN metal gate at 1 µm gate length, has reached a ***world-record*** drain current and transconductance.

9.1 Introduction

Electrons "move" much faster in III-V compound semiconductors of GaAs and InGaAs than in Si, a phenomenon already known since late 1950s. The electron mobility of $In_xGa_{1-x}As$ (x = 0 to 1) is 6–18 times of that of Si. The high electron mobility is an important parameter for building high speed devices. In SiO_2/Si, the dimensional scaling, in gate length and in gate dielectric thickness, during the past

M. Hong (✉)
Department of Materials Science and Engineering, National Tsing Hua University,
Hsinchu 30013, Taiwan
e-mail: mhong@mx.nthu.edu.tw

S. Oktyabrsky, P. D. Ye (eds.), *Fundamentals of III-V Semiconductor MOSFETs*,
DOI 10.1007/978-1-4419-1547-4_9, © Springer Science+Business Media LLC 2010

40 years, has simultaneously provided high-density, low-cost, and high performance integrated circuits (ICs). Further scaling in the transistors, however, does not guarantee device performance advantages. New materials and novel device architectures have to be employed for achieving the required performance; for example, HfO_2-based high-k dielectrics + metal gates are now replacing the long-standing, reliable SiO_2 + poly-silicon. The new breakthrough technology has already been in production for the 45 nm node (and will be for the 32 nm node) complementary metal-oxide-semiconductor (CMOS) circuits.

Looking ahead beyond the 22–16 nm node ICs, the consensus is that the combination of high-k dielectrics with high mobility channels of the III-V's (InGaAs) and Ge will have to be integrated onto Si; this trend of "hybrid chip" may occur in 2015–2020. The realization of these new MOS field-effect-transistors (MOSFETs) has posted unprecedented challenges for material scientists, processing engineers, condensed matter and device physicists. Moreover, direct band-gap and band-gap engineering, not available in the systems based on Si and Ge, provide novel designs, including integrating MOS, photonic and high power devices into highly-performed optoelectronic circuits.

One of the key challenges in compound semiconductor and Ge device technology was to find thermodynamically stable insulators on InGaAs, Ge, and GaN that provide a low density of interfacial states (D_{it}). The most intensively studied and widely used compound semiconductor is GaAs. In early days, thermal, anodic (including plasma), and photochemical oxidation of GaAs surface produced highly resistive films, but could not provide the oxide-GaAs interfaces with a low D_{it}. Approaches of using various dry, wet, and photochemical surface treatments prior to the deposition of insulating films produced limited success, since major sources of interfacial states such as non-stoichiometry, structural defects, and surface contamination still exist [1].

Ultra-high-vacuum (UHV) deposited amorphous mixed oxide $Ga_2O_3(Gd_2O_3)$ [GGO] [2, 3] and single crystal Gd_2O_3 [4] on GaAs have firstly unpinned the Fermi level in the oxide/compound semiconductors, and opened up a new era for the III-V MOS devices. Very low D_{it} of $<10^{11}\,cm^{-2}\,eV^{-1}$ and low leakage currents of $10^{-8}\,A/cm^2$ were demonstrated. In addition to the UHV-GGO and Gd_2O_3, another approach that has achieved a low D_{it} and a low electrical leakage current density without invoking interfacial layers is the atomic layer deposited (ALD) Al_2O_3 [5–7] and HfO_2 [8]. Good electrical properties were also reported in MOS devices employing Si [9, 10] and Ge [11] as interfacial passivation layers between GaAs and gate dielectrics.

In 1990s, the UHV-GGO on GaAs and InGaAs have enabled the demonstration of the first non-self-aligned inversion-channel GaAs and $In_{0.53}Ga_{0.47}As$ MOSFETs [12–14]. The device performances of the inversion-channel GaAs MOSFETs were further improved in 1999 [15]. Recently, with ALD high-k oxides as the gate dielectrics, inversion channel InGaAs (with In content ≥ 53%) MOSFETs without any interfacial layers have given excellent device performances [16–18]. With Si (Silane) interfacial passivation layers, GaAs MOSFETs with ALD and MOCVD high-k oxides as the gate dielectrics have also shown good drain currents [10, 19]. In the same time, the MOS-capacitors (MOSCAPs) of GGO-$In_{0.2}Ga_{0.8}As$ with in-situ Al_2O_3 cap have shown remarkably excellent *C–V* characteristics and thermodynamic stability at high temperatures [20, 21]. With the Al_2O_3/GGO as the gate dielectric, self-aligned inversion-channel $In_{0.53}Ga_{0.47}As$ MOSFET's have exhibited very high drain currents and transconductance [22]. These InGaAs MOSFETs have showed inversion-channel characteristics, similar to those exhibited in the traditional SiO_2/Si. On the other hand, non-inversion-channel enhancement-mode (E-mode) III-V MOSFETs have been demonstrated in the configurations of implant-free [23] or buried channel type [24].

In this chapter, we reviewed mainly the recent advances on our research results using high-k's of GGO and ALD oxides on GaAs, InGaAs, Ge, and GaN, which will benefit the future development of III-V MOSFETs. The interfacial chemical properties and the band parameters in the oxides/III-V hetero-structures are studied using x-ray photoelectron spectroscopy (XPS) and electrical transport measurements of Fowler–Nordheim (F–N) tunneling. The mechanism of III-V surface passivation with correlation between the chemical and electrical characteristics is discussed. The band parameters are very useful in the further understanding and modeling for the III-V MOS devices. Device performances of inversion-channel and/or depletion-mode (D-mode) InGaAs, Ge, and GaN MOSFETs are reported. Various E-mode InGaAs MOSFETs are discussed, and their device performances are compared. Moreover, the interface trap densities and the degree of freedom of Fermi-level movement in high-k/InGaAs interface were studied using quasi-static capacitance–voltage (QSCV), charge pumping, and capacitance–voltage/conductance–voltage (*C–V*/*G–V*) characteristics. No interfacial layers were used in all of our devices, very distinctly different from most other approaches, which emphasize the necessity or inevitability of using interfacial layers.

9.2 Material Growth, Device Fabrication, and Measurement

9.2.1 Preparation of Samples

9.2.1.1 InGaAs

$In_xGa_{1-x}As$ (x = 0, 0.15, and 0.25) and $In_{0.53}Ga_{0.47}As$ epi-layers were molecular beam epitaxy (MBE)-grown on 2-inch GaAs and InP substrates, respectively, in a multi-chamber UHV system [2, 3]. Afterwards, the Al_2O_3/GGO dual-layer gate dielectrics were in-situ e-beam evaporated onto the $In_xGa_{1-x}As$ surfaces [20, 21]. Also, ALD-Al_2O_3 and ALD-HfO_2 films were ex-situ deposited on top of the $In_xGa_{1-x}As$ layer [6, 8]. The Al_2O_3 films were grown at a wafer temperature of 300 °C and a chamber pressure of 1 Torr using alternating pulses of $Al(CH_3)_3$(TMA) and H_2O as precursors; the ALD-HfO_2 films were grown at a wafer temperature of 200 °C, using tetrakis (ethylmethylamino) hafnium, $Hf(NCH_3C_2H_5)_4$ and H_2O as the precursors. High-purity Ar (99.999%) was used as diluted and purged gas. Deposition rates of 0.084 nm of 0.104 nm/cycle were measured for the ALD-Al_2O_3 and -HfO_2 films, respectively, using spectroscopic ellipsometer, x-ray reflectivity, and high resolution transmission electron microscopy (HR-TEM). After deposition of the gate dielectrics, metal electrodes were formed by e-beam/thermal deposition or sputtering for MOSCAPs. The samples for fabricating MOSFETs were covered by TiN 0.16–0.2 μm thick by reactive-sputtering from a pure Ti target in Ar/N_2 RF-plasma.

9.2.1.2 Ge

The 2-inch n-type Ge (100) wafers with a resistivity of 0.07–0.3 Ω cm (Sb-doped) were dipped in diluted HF (2%) and rinsed in de-ionized water, followed by UHV annealing at ~450 °C to remove the residual native oxides. A detailed description of the oxide growth has been given previously [25, 26]. After growth of nm-thick GGO, GGO/Ge samples were in-situ transferred to the analysis chamber for XPS studies.

9.2.1.3 GaN

The GaN epitaxial layers were grown on *c*-plane sapphire by metal-organic chemical-vapor-deposition (MOCVD), consisting of a well of a *p*-GaN epi-layer 0.3 μm thick with Mg doping of $2 \times 10^{17}\,cm^{-3}$ grown on an undoped GaN 1.5 μm thick. Gd_2O_3 layers 20 nm thick grown by molecular beam epitaxy (MBE) [27] were deposited on the samples to protect GaN surface during the high-temperature dopant-activation annealing. The n^+ source/drain contact regions were achieved by Si implantation with multiple-energy conditions ($120\,keV/1 \times 10^{15}\,cm^{-2}$; $80\,keV/6 \times 10^{14}\,cm^{-2}$; $40\,keV/4 \times 10^{14}\,cm^{-2}$). This implant scheme was designed to produce a uniform Si profile to a depth of ~0.1 μm. Rapid thermal annealed (RTA) to 1100 °C for 5 min under He ambient was used for implant activation. Afterwards, the Gd_2O_3 layer was removed by HCl etching. ALD-Al_2O_3 was then used as a gate dielectric [6].

9.2.2 *Fabrication and Measurements of Devices*

9.2.2.1 InGaAs

After TiN-sputtering, the definition and formation of the metal-gates was carried out using inductively coupled plasma reactive ion etching (ICP-RIE). The dry-etched metal gates and photo-resist defined the S/D regions, onto which Si+ ions were implanted through the gate dielectrics. Activation of the implanted ions was accomplished with RTA in He or N_2 ambience at 650–700 °C. After wet removal of oxides at the S/D contact region, the ohmic contacts were subsequently formed by e-beam/thermal evaporation of Pd/Ge/Ti/Pt metal stack and a lift-off process, which was followed by annealing at 400 °C for 10 s in a N_2 ambience to achieve a low contact resistance [22, 28]. D-mode MOSFETs were fabricated in a similar manner, however, without ion implantation [29].

9.2.2.2 Ge

The fabrication procedure of MOS capacitors (MOSCAPs) and self-aligned Al_2O_3 (~2 nm, protecting GGO)/GGO (~5.5 nm)/Ge pMOSFETs, and their characterizations were similar to what has been given previously [30].

9.2.2.3 GaN

After using dilute buffered oxide etch (BOE) solution to remove Al_2O_3 in source/drain regions, Ti/Al/Pt/Au (30 nm/120 nm/80 nm/120 nm) was deposited as the ohmic metal, followed by annealing at 750 °C for 30 s under Helium ambient using RTA. Finally, Pt/Au (30 nm/120 nm) was used as the gate metal. The fabricated MOSFETs have gate width of 100 μm and gate lengths varying from 1 to 16 μm. The cross-sectional schematic of the fabricated device structure is shown in Ref. [31]. The Al_2O_3/GaN interface remains atomically smooth after high temperature 750 °C annealing under Helium ambient, with the oxide thickness and the roughness determined from x-ray reflectivity (XRR) to be 12 and 0.5 nm (not shown), respectively. A cross-sectional HR-TEM image (not shown) also showed an abrupt transition from amorphous Al_2O_3 to crystalline GaN even after high temperature annealing, in agreement with the XRR data.

9.3 Devices

9.3.1 *InGaAs MOS Devices*

9.3.1.1 InGaAs Enhancement-Mode Devices

Enhancement-mode (E-mode) devices in III-V InGaAs compound semiconductors include inversion-channel (similar to SiO_2/Si) and non-inversion-channel configurations. The latter includes implant-free and buried-channel type devices, similar to high electron mobility transistors (HEMTs), with the channel being far away from the gate. In general, the state-of-the-art HEMT-like device structures are very complicate, and moreover, they need highly precise epi-layer growth (in the order of less than 1 nm). In comparison, the device structures in inversion-channel are simpler. Among the inversion-channel devices, the self-aligned approach is preferred over its counterpart of the non-self-aligned one, considering device scaling and simplicity of fabrication.

Previously, the inversion-channel InGaAs MOSFETs, however, were shied away, due to the unavailability of high-k's/InGaAs interfaces with low D_{it}'s. Furthermore, the difficulty in the past in the attainment of high-temperature thermodynamic stability of high-k's/InGaAs interfaces has prevented the fabrication of the self-aligned inversion-channel InGaAs MOSFETs.

Now, the breakthrough results on achieving the above critical material properties of low D_{it}'s and high-temperature thermodynamic stability have been made. As a consequence, excellent device performances have been achieved in self-aligned and non-self-aligned inversion-channel $In_{0.53}Ga_{0.47}As$ MOSFET's, as will be discussed in Sect. "Inversion-channel devices".

Inversion-Channel Devices

Due to the interaction between $Ga_2O_3(Gd_2O_3)$ (in fact, it was the hydro-oxides) and GaAs (InGaAs) during the high-temperature anneal (at least >750 °C) for effective activating ion implants in the source and drain of the MOSFET, the devices in 1990s at Bell Labs were processed by ion implantation on GaAs (InGaAs) wafers, followed by activation annealing, and then gate-oxide deposition [12, 13]. However, the GaAs (InGaAs) surface became very rough as observed using RHEED, despite extra efforts to protect the surface. The interfacial smoothness (with a roughness ≤1 Å) is very important in this type of devices since carriers are confined to the top of GaAs (InGaAs), which is within 50–100 Å next to the high-k-GaAs (InGaAs) interface. Some device fabrication started with deposition of oxides (e.g. SiO_2, and GGO), which were etched off after implant activation anneal. The native oxides (of GaAs or InGaAs) on the device wafers were then thermally desorbed with annealing in the solid state GaAs-based MBE chamber under As overpressure to prevent the loss of Ga (In) and to preserve the surface smoothness. Consequently, in-situ GGO deposition took place in a separate oxide chamber. These devices were, of course, non-self-aligned. The earlier inversion-channel GGO/$In_{0.53}Ga_{0.47}As$ MOSFETs exhibited an I_d of 375 μA/μm (1 μm gate length (L_g)) and a transconductance (g_m) of 190 μS/μm (0.75 μm L_g) [14].

Adopting the approach above, but with an ALD-Al_2O_3 protecting layer during activation, Xuan et al. have fabricated non-self-aligned inversion-channel ALD-Al_2O_3 (8 nm)/$In_{0.53}Ga_{0.47}As$ MOSFETs (0.5 μm L_g); the devices gave an I_d of 430 μA/μm, and a g_m of 160 μS/μm [16]. Also, an ALD-Al_2O_3(10 nm)/$In_{0.65}Ga_{0.35}As$ MOSFET (0.4 μm L_g) showed an I_d of 1.05 mA/μm, and a g_m of 350 μS/μm [17]. The protecting ALD-Al_2O_3 was etched off after activation annealing, and a new ALD-Al_2O_3 was then deposited as the gate dielectric.

Nonetheless, the non-self-aligned process is impractical due to the complexity of mask alignment as well as the unavoidable parasitic resistance.

Recently, self-aligned inversion-channel $In_{0.53}Ga_{0.47}As$ MOSFETs of 1 μm gate length by employing a UHV-Al_2O_3 (2 nm)/GGO (5 nm) and TiN metal gate have been successfully fabricated [22]. In-situ deposited Al_2O_3 was used to protect GGO from absorbing moisture during fabrication of devices, and TiN was chosen due to its suitable work function and thermal stability. The TiN/Al_2O_3/GGO/$In_{0.53}Ga_{0.47}As$ MOSFET has demonstrated *world record* performance of a maximum drain current of 1.05 mA/μm, a transconductance (g_m) of 714 μS/μm, and a peak mobility of 1300 $cm^2/V\,s$, the highest ever reported for III-V InGaAs inversion-channel devices of 1 μm gate-length (Figs. 9.1 and 9.2). Notice that the drain current demonstrated by this 1 μm device is comparable to the Si NMOSFET in 90 nm node, which has a physical gate length of 45 nm.

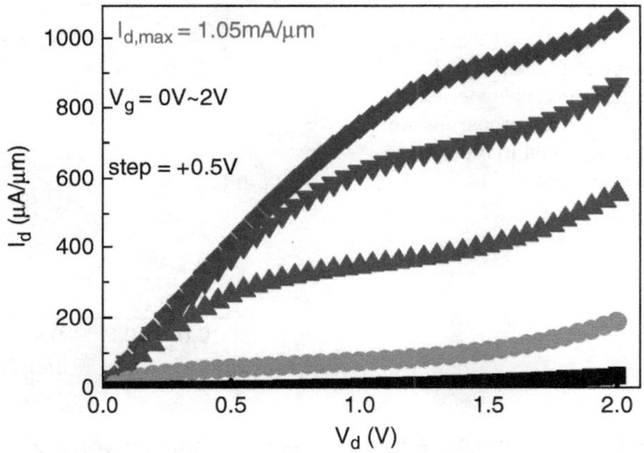

Fig. 9.1 Output characteristics I_d vs V_d of an 1 × 10 μm inversion-channel $In_{0.53}Ga_{0.47}As$ MOSFET. A maximum I_d of 1.05 mA/μm is measured at $V_g = 2$ V and $V_d = 2$ V [62] (Reprinted with permission. Copyright 2009 Materials Research Society)

Significantly, for the self-aligned inversion-channel $TiN/Al_2O_3/GGO/In_{0.53}Ga_{0.47}As$ MOSFET, the device performance of drain currents and transconductance scales with the gate length, as shown in Fig. 9.3. Since the I_d and the g_m are proportional to the inverse gate length, L_g^{-1}, a maximum I_d of ~2 mA/μm and a maximum g_m of about 1.4 mS/μm are expected for a device with 0.5 μm gate-length, assuming that all the other parameters remain unchanged. Even higher performance can be anticipated by scaling down the gate-length to less than 0.1 μm; however, parasitic resistance and short channel effect may start to play a role.

More recently, an 1 μm-gate-length self-aligned inversion-channel $In_{0.53}Ga_{0.47}As$ MOSFET with ALD-Al_2O_3 (10 nm) as the gate dielectric [18] has exhibited a maximum I_d of 288 μA/μm and a peak g_m of 93 μS/μm.

Capacitance equivalent thicknesses (CET) of ≤1 nm in UHV-GGO with in-situ Al_2O_3 cap [21] and ALD-HfO_2 on InGaAs [32] have been achieved, which allow further scaling and improvement of the inversion-channel InGaAs MOSFETs. CET is simply the thickness that is derived directly from the relationship of $CET(V) = (K_{SiO_2})*(\varepsilon_0)*(Area)/C(V)$, where ε_0 is the permittivity of free space, and

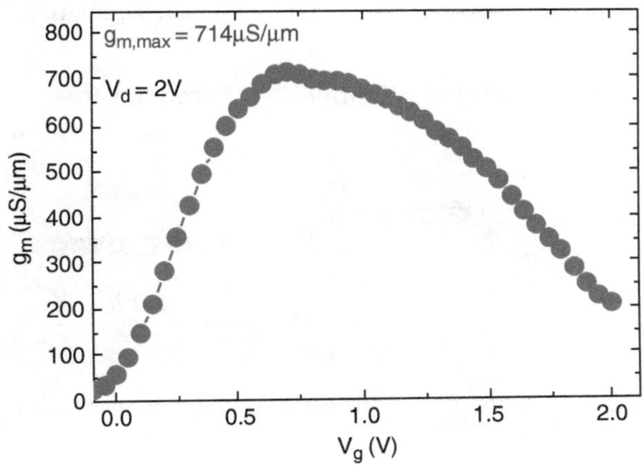

Fig. 9.2 The transfer characteristics and g_m curve of the same device showing a maximum g_m of 714 μS/μm, measured at $V_g = 0.7$ V and $V_d = 2$ V [62] (Reprinted with permission. Copyright 2009 Materials Research Society)

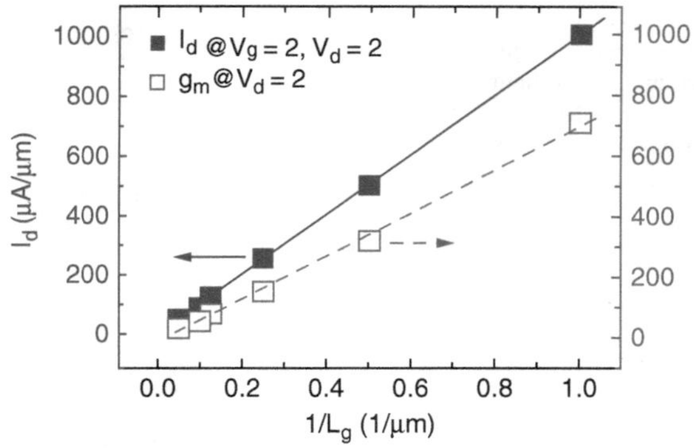

Fig. 9.3 Drain current at $V_d = 2\,V$, $V_g = 2\,V$ and peak transconductance at $V_d = 2\,V$ versus inverse gate-length (L_g^{-1}). Both properties are proportional to the inverse gate length

C(V) is the measured capacitance at a biasing voltage V. After quantum-mechanical corrections, the equivalent oxide thickness (EOT) is even lower than CET.

Implant-Free and Buried-Channel Type Devices

Figure 9.4 shows schematic diagrams of inversion-channel, implant-free and buried-channel type devices. Notice that the later two structures are conceptually identical, and both are derivatives of a HEMT structure.

A 1 μm-gate-length implant-free $In_{0.3}Ga_{0.7}As$ channel MOSFET using a 10 nm GaGdO gate dielectric exhibited a maximum I_d of 407 μA/μm and a peak g_m of 477 μS/μm [23]; a 0.26 μm-gate-length ALD-Al_2O_3 (7 nm)/$In_{0.7}Ga_{0.3}As$ buried—channel MOSFET exhibited a maximum I_d of ~177 μA/μm and a peak g_m of 157 μS/μm [24].

Comparisons

Maximum I_d and peak g_m of representative E-mode (inversion-channel or non inversion-channel) InGaAs MOSFETs are summarized in Fig. 9.5a, b, respectively.

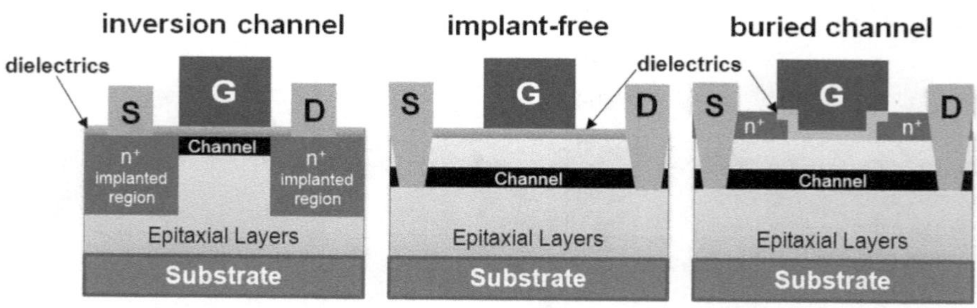

Fig. 9.4 Schematic cross-sections of inversion-channel, implant-free, and buried-channel type NMOSFETs

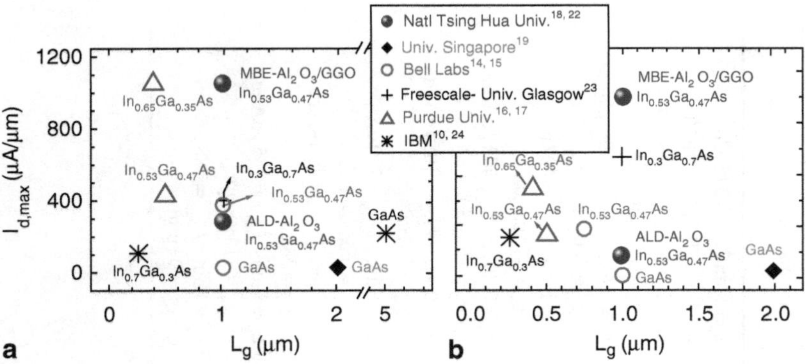

Fig. 9.5 Summary of **a** maximum drain currents and **b** peak transconductances of representative work on III-V enhancement-mode nMOSFETs reported in the last decade [62] (Reprinted with permission. Copyright 2009 Materials Research Society)

The details on the device configurations, Indium contents, gate dielectrics, and the device performances are summarized in Table 9.1. The drain current achieved in the self-aligned 1 μm-gate-length TiN/Al_2O_3/GGO/$In_{0.53}Ga_{0.47}As$ channel in this work is as high as that of the 0.4 μm-gate-length non-self-aligned ALD-Al_2O_3 MOSFETs with a higher In-content channel of $In_{0.65}Ga_{0.35}As$ [17]; the value is the highest ever reported among all types of E-mode III-V MOSFETs, fabricated by self-aligned and non-self-aligned processes.

Moreover, the peak g_m using the UHV in-situ Al_2O_3/GGO is also the highest. A clean GGO/InGaAs interface, free of In_2O_3, $In(OH)_3$, and Ga_2O_3, may attribute to the exceedingly high device performance [33]. These undesired oxides inevitably exist at the ALD-Al_2O_3/InGaAs interface.

In general, MOSFETs using InGaAs channel with high Indium contents have shown much better device performances than those using GaAs or InGaAs with low Indium contents channel. Good device results were also reported in the work using UHV-deposited GGO [15], ALD-Al_2O_3 with Si IPL [10], and MOCVD HfAlO with silane passivation [19].

GaAs and $In_{0.2}Ga_{0.8}As$ MOSFETs with the advanced Al_2O_3/GGO are now being fabricated for a direct comparison between MOSFETs using channels with different In-content, since our previous work using GGO only (without in-situ Al_2O_3 cap) has not exhibited the expected device performance, due to a degraded oxide/III-V interface [15, 28].

Comparing the inversion-channel and non-inversion-channel E-mode III-V InGaAs MOSFETs, the inversion-channel devices demonstrated in our work [22] and in Xuan et al. [16, 17] have shown device performances better than those of implant-free [23], or buried channel-type [24] of E-mode (non inversion-channel) III-V devices. The reason may be attributed to a relatively thick EOT of the structures of implant-free or buried channel-type devices, which is the obstacle for the further scaling of the non-inversion-channel E-mode III-V MOSFETs. Taking

Table 9.1 Summary of representative E-mode nMOSFETs reported in the last decade

No.	Year Published [Ref]	Research group	Channel type/materials	Gate dielectric (passivation)	Device characteristics
(1)	1998 [14]	Bell Labs Ren, et al.	Inversion channel/$In_{0.53}Ga_{0.47}As$	UHV-GGO[a]	$L_g = 1\ \mu m$ $I_{dmax} = 375\ \mu A/\mu m$ $L_g = 0.75\ \mu m$ $g_m = 190\ \mu S/\mu m$
(2)	1999 [15]	Bell Labs Wang, et al.	Inversion channel/GaAs	UHV-GGO[a]	$L_g = 1\ \mu m$ $I_{dmax} = 30\ \mu A/\mu m$ $g_m = 4\ \mu S/\mu m$
(3)	2007 [16]	Purdue Univ. Xuan, et al.	Inversion channel/$In_{0.53}Ga_{0.47}As$	ALD-Al_2O_3	$L_g = 0.5\ \mu m$ $I_{dmax} = 430\ \mu A/\mu m$ $g_m = 160\ \mu S/\mu m$
(4)	2007 [23]	Freescale/U. Glasgow Hill, et al.	Implant-free/$In_{0.3}Ga_{0.7}As$	UHV-Ga_2O_3/GaGdO[a]	$L_g = 1\ \mu m$ $I_{dmax} = 407\ \mu A/\mu m$ $g_m = 477\ \mu S/\mu m$
(5)	2007 [24]	IBM Sun, et al.	Buried channel/$In_{0.7}Ga_{0.3}As$	ALD-Al_2O_3	$L_g = 0.26\ \mu m$ $I_{dmax} = 117\ \mu A/\mu m$ $g_m = 157\ \mu S/\mu m$
(6)	2008 [17]	Purdue Univ. Xuan, et al.	Inversion channel/$In_{0.65}Ga_{0.35}As$	ALD-Al_2O_3	$L_g = 0.4\ \mu m$ $I_{dmax} = 1050\ \mu A/\mu m$ $g_m = 350\ \mu S/\mu m$
(7)	2008 [10]	IBM de Souza, et al.	Inversion channel/GaAs	ALD-Al_2O_3 (a-Si)	$L_g = 5\ \mu m$ $I_{dmax} = 220\ \mu A/\mu m$ g_m = unknown
(8)	2008 [19]	Univ. Singapore Chin, et al.	Inversion channel/GaAs	MOCVD-HfAlO (SiH_4)	$L_g = 2\ \mu m$ $I_{dmax} = 46\ \mu A/\mu m$ $g_m = 28\ \mu S/\mu m$
(9)	2008 [22]	Natl Tsin-Hua Univ. Lin, et al.	Inversion channel/$In_{0.53}Ga_{0.47}As$	UHV-Al_2O_3/GGO[a]	$L_g = 1\ \mu m$ $I_{dmax} = 1050\ \mu A/\mu m$ $g_m = 714\ \mu S/\mu m$
(10)	2009 [18]	Natl Tsin-Hua Univ. Chiu, et al.	Inversion channel/$In_{0.53}Ga_{0.47}As$	ALD-Al_2O_3	$L_g = 1\ \mu m$ $I_{dmax} = 288\ \mu A/\mu m$ $g_m = 93\ \mu S/\mu m$

[a] E-beam evaporated.

the advantages of using a structure in the derivatives of high mobility devices, the implant-free [23] III-V MOSFETs, nevertheless, showed a good g_m.

9.3.1.2 InGaAs Depletion-Mode Devices

D-mode Al_2O_3 (2 nm)/GGO (6 nm)/$In_{0.2}Ga_{0.8}As$/GaAs MOSFETs with 12 μm gate-lengths have been fabricated by two different processes (denoted as process A and B), with process steps listed in Table 9.2. Both process A and B demonstrated excellent performances, as shown in Figs. 9.6 and 9.7. The drain current–voltage (I_d–V_d) curve in Fig. 9.6 exhibits a high saturation drain current ($I_{d,sat}$) of 135 μA/μm (at V_g = 2 V and V_d = 5 V) with process A and a higher one of 175 μA/μm (at V_g = 2 V and V_d = 5 V) with process B. The data reveal that the Al_2O_3 capping layer has effectively protected GGO in both processing approaches. Moreover, better device performances exhibited by D-mode MOSFET fabricated with process B may be due to a better interface between the gate dielectric and the channel. This may be attributed to the fact that the gate dielectric/$In_{0.2}Ga_{0.8}As$ interface in process B has been kept away from chemicals/water during the processing steps.

When normalized to a gate length of 0.8 μm with the consideration of both the source-gate and drain-gate distances, the maximum drain current density is ~460 μA/μm for the device underwent process B. The performance is comparable to that of a conventional D-mode device configuration reported by Wang et al. in a GaAs MOSFET with GGO gate dielectric 38 nm thick [34], Tsai et al. in an $In_{0.2}Ga_{0.8}As$ MOSFET with GGO 54 nm thick [35], and Ye et al. in an $In_{0.2}Ga_{0.8}As$ MOSFET with ALD-Al_2O_3 16 nm thick [5].

Our data, however, compare very favorably with those of ring-gate D-mode MOSFET's with PVD HfO_2 gate dielectrics and Si and Ge as interfacial layers. [11,

Table 9.2 Process flow of $In_{0.2}Ga_{0.8}As$/GaAs ring-gate D-mode nMOSFET

Process	A metal-gate-last	B metal-gate-first
Process flow	1. deposition of Al_2O_3/GGO on n-$In_{0.2}Ga_{0.8}As$/n-GaAs in UHV system 2. post-deposition-annealing 3. S/D contact region patterning, wet-etching of oxide, contact metal deposition 4. contact metal lift-off & ohmic alloying 5. gate region patterning & Ti/Au deposition 6. gate metal lift-off	1. deposition of Al_2O_3/GGO on n-$In_{0.2}Ga_{0.8}As$/n-GaAs in UHV system 2. post-deposition-annealing 3. TiN sputtering 4. gate region patterning & dry-etching 5. S/D contact region patterning, wet-etching of oxide, contact metal deposition 6. contact metal lift-off & ohmic alloying

Reprinted with permission from Ref. [29]. Copyright 2009 Elsevier

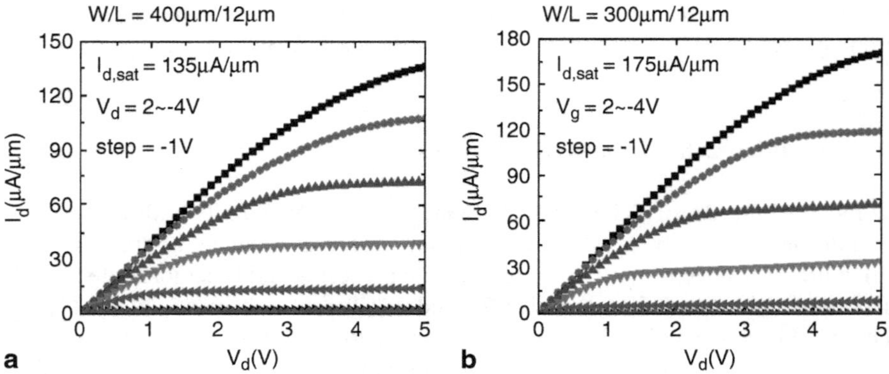

Fig. 9.6 I_d–V_d curves of the ring-gate D-mode Al_2O_3/GGO/$In_{0.2}Ga_{0.8}As$/GaAs nMOSFETs fabricated with **a** "metal-gate-last" process and **b** "metal-gate-first" process [29] (Reprinted with permission. Copyright 2009 Elsevier)

36] Note that in the D-mode devices, drain currents are not simply linearly scaled with gate length. The S–G and D–G distances have to be taken into consideration. Therefore, the extrapolation of drain current with gate length reported in Refs. [11, 36] is inadequate.

A similar phenomenon is also observed from the transfer characteristics in Fig. 9.7. The device undergone process B exhibits a maximum extrinsic transconductance of 48 μS/μm, which is higher than 38 μS/μm obtained from the device undergone processes A. For process B, a better interface between the gate dielectric and the channel is implied, and the extrinsic transconductance of 130 μS/μm (if normalized to 0.8 μm gate length with the consideration of the S–G and D–G distances) is also comparable with other works.

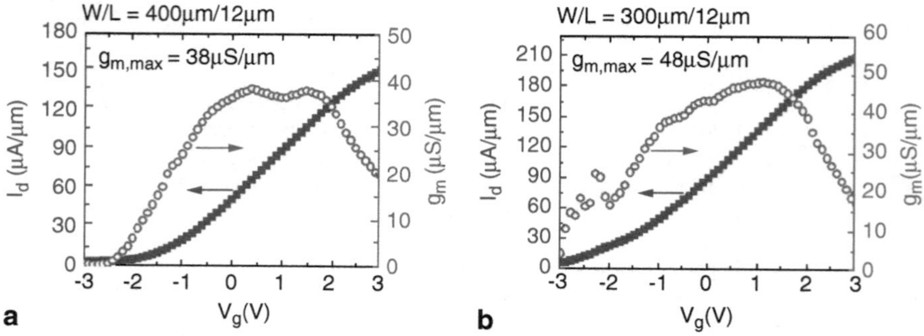

Fig. 9.7 g_m–V_g and I_d–V_g characteristics of the ring-gate D-mode Al_2O_3/GGO/$In_{0.2}Ga_{0.8}As$/GaAs nMOSFETs fabricated with **a** "metal-gate-last" process **b** "metal-gate-first" process [29] (Reprinted with permission. Copyright 2009 Elsevier)

9.3.2 Ge Inversion Channel MOSFET's

After a moderate anneal (450 °C, 20 min) for boron dopant activation, the GGO in the pMOSFET remains amorphous, and the GGO/Ge interface remains abrupt, as shown in a cross-sectional HR-TEM (Fig. 9.8a). The dc output characteristics of the pMOSFET (Fig. 9.8b illustrating its schematic cross-sectional view) with 1 μm gate length (L_g) and 50 μm gate width (W_g) are shown in Fig. 9.8c, d for drain current

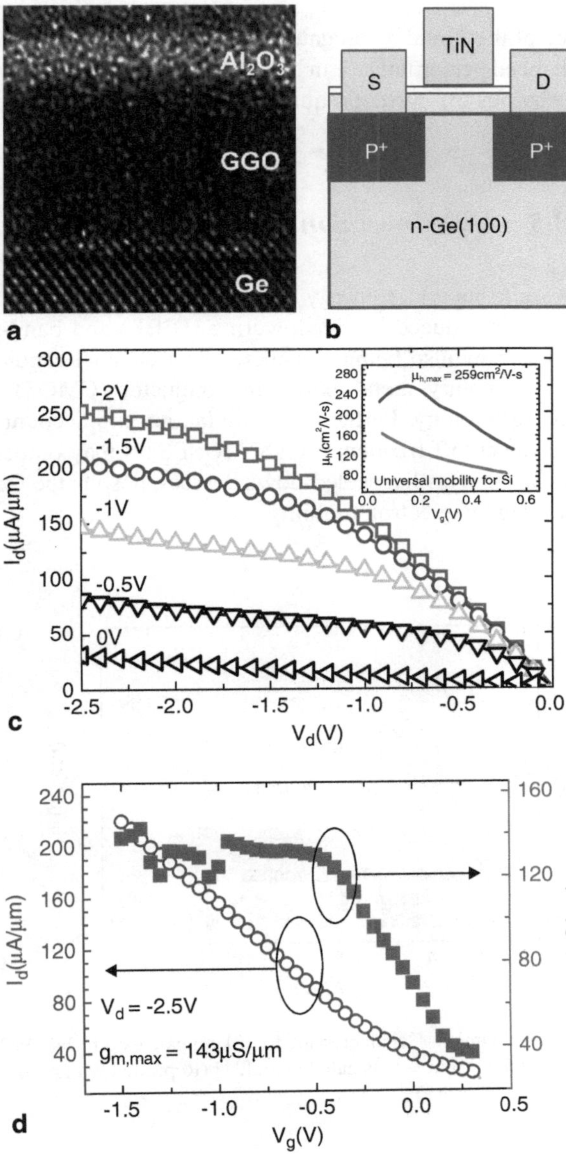

Fig. 9.8 **a** HR-TEM micrograph of Al_2O_3/GGO/Ge (100) hetero-structures after a 450 °C-20 min anneal; **b** Schematic cross-section of the Ge pMOSFET with TiN metal gate and UHV-deposited Al_2O_3/GGO bi-layer dielectrics; **c** Output characteristics I_d vs V_d of the Ge pMOSFET of 50 μm gate width, and 1 μm gate length. A maximum drain current I_d of 252 μA/μm was measured at $V_g = -2$ V and $V_d = -2.5$ V; the *inset* shows the mobility graph; **d** The transfer characteristics and transconductance g_m curve of the Ge pMOSFET. A peak g_m of 143 μS/μm was measured at $V_g = -1.4$ V and $V_d = -2.5$ V. [41] (Reprinted with permission. Copyright 2009 American Institute of Physics)

density (I_d) vs drain voltage (V_d), and transconductance (g_m) and I_d vs gate voltage (V_g), respectively. A high saturation I_d of 252 mA/mm at a V_g of −2 V, and a maximum g_m of 143 mS/mm are achieved. Furthermore, a peak hole mobility extracted from the linear regime drain current [37] reaches 259 cm^2/V s (inset of Fig. 9.8c), which is roughly 1.5 times greater than the universal hole mobility for Si [38]. The drain current characteristics of our pMOSFET ($I_d \times L_g$ ~188 μA at $V_g - V_{th} = -1.5$ V with $L_g = 1$ μm) are comparable to the previous MOSFET data using high-k/GeO_xN_y ($I_d \times L_g$ ~195 μA at $V_g - V_{th} = -1.5$ V with $L_g = 10$ μm) [39] and high-k/Si stacks ($I_d \times L_g$ ~130 μA at $V_g - V_{th} = -1.5$ V with $L_g = 0.125$ μm) [40]. Systematic dependence of the I_d and g_m on gate lengths varying from 20 to 1 μm indicates that a much enhanced performance can be expected with further device down-scaling. However, an obvious off-current requires further optimization of junction formation.

9.3.3 GaN Inversion Channel MOSFET's

Owing to its wider energy band gap (3.4 eV) which alleviates the adverse affects like drain-induced barrier lowering (DIBL) and band-to-band tunneling (BTBT), GaN is now also being considered as a channel candidate for the next generation complementary metal-oxide-semiconductor (CMOS) devices beyond the 22 nm node technology. Furthermore, by taking into account of the short channel effect with the cutoff frequency, GaN MOSFET's may outperform its counterparts of Si and GaAs in further scaled-down devices, despite the fact that GaN offers no special advantage in electron mobility.

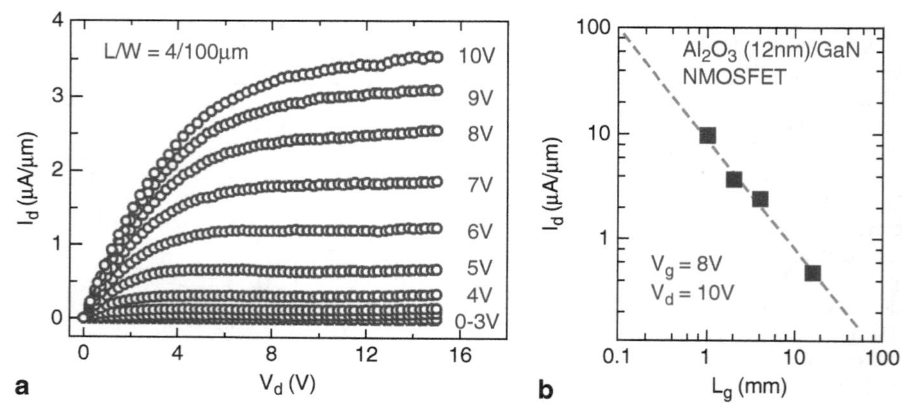

Fig. 9.9 **a** Drain *I-V* characteristic for a 4 μm gate length GaN MOSFET; **b** the scaling characteristic of drain current versus gate length [31] (Reprinted with permission. Copyright 2008 American Institute of Physics)

Well-behaved drain *I–V* characteristics of a GaN MOSFET are shown in Fig. 9.9a with a clean pinch-off. For a device of a 4 μm gate length and a 100 μm gate width, the maximum drain current is 3.5 μA/μm at a gate voltage V_g of 10 V, and a drain voltage V_d of 15 V. The maximum drain current was improved to ~10 μA/μm in an 1 μm gate-length device, measured at a gate voltage of 8 V and a drain voltage of 10 V. Our devices also showed normal drain *I–V* characteristics and transfer characteristics as expected for a typical inversion-channel MOSFET. The measured drain current is scaled with gate length, with the scaling dependence displayed in Fig. 9.9b. The overall performances of these devices are markedly improved over the previously reported results of the inversion-channel GaN MOSFETs based on MgO and SiN_x high-k dielectrics [42, 43].

The transfer characteristics (L/W = 4/100 μm) in Fig. 9.10 were obtained with V_g sweeping from 0 to 10 V, and V_d set at 0.1 V. A high I_{on}/I_{off} ratio of 2.5×10^5 is achieved with a very low off-state leakage current of 4×10^{-13} A/μm, as determined with the drain current exhibited in a logarithmic scale on the left-side y-axis. The low gate leakage and the low n+/p junction leakage current are attributed to the good device characteristics. The device is normally off with a threshold voltage of 2.8 V using a linearly scaled drain current shown on the right-side y-axis, and a sub-threshold slope of 290 mV/dec. The large threshold voltage is mainly due to a large potential difference between intrinsic Fermi energy and Fermi level (which is caused by the wide band gap property of GaN), and a high work function of Pt as the gate metal. Choosing gate metals with lower work functions and decreasing acceptor doping concentration in the *p* epi-layer well are expected to reduce the threshold voltage. Peak intrinsic transconductance and mobility are at least ~4.0 μS/μm and ~10 cm²/V s, respectively, by taking the consideration of the measured high sheet resistance (6×10^4 Ω/square) and contact resistance (3×10^{-3} Ω cm²) in the device. The transconductance and mobility will be higher when a self aligned process is

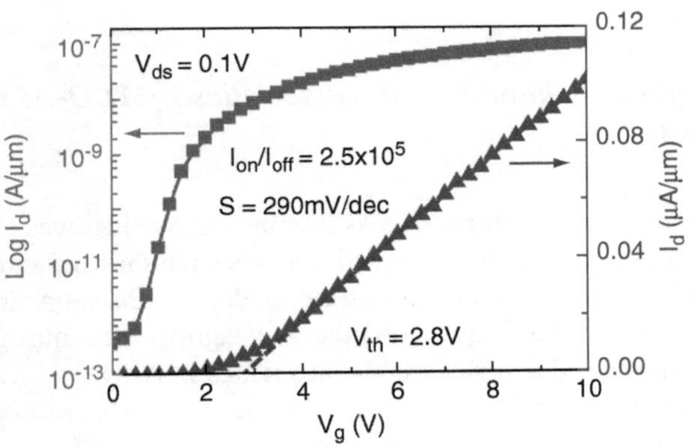

Fig. 9.10 Transfer characteristics for a 4 μm gate length GaN MOSFET at a drain voltage V_d of 0.1 V [31] (Reprinted with permission. Copyright 2008 American Institute of Physics)

implemented to further reduce the sheet resistance and a better-quality GaN substrate is employed to increase the electron mobility.

9.4 Interfacial Chemical Properties

The chemical composition in the bulk of the oxide thin film at the oxide/semiconductor interfaces with a thickness no more than 20 Å can best be studied using x-ray photoelectron spectroscopy (XPS).

9.4.1 Interfacial Chemical Characteristics in UHV-GGO on GaAs

For the materials properties of the oxide, Kwo et al. established that Gd_2O_3 is essential in the oxide mixture for effective passivation of GaAs [25, 44]. Furthermore, the initial growth of GGO on GaAs was found to be of a single crystal and contains only pure Gd_2O_3 [45]. Reflection high-energy electron diffraction and x-ray diffraction (XRD) studies show that the thin oxide film is epitaxially grown on GaAs with the surface normal (110) and in-plane axis [001] parallel to (100) and [011] of GaAs, respectively, and has a structure isomorphic to Mn_2O_3. Studies using high-resolution transmission electron microscopy on the oxide–GaAs interface indicate some atomic registry between the oxide and GaAs during the initial growth. The chemical composition of the oxide film was determined by XPS to be unequivocally pure Gd_2O_3 [45]. There is no interfacial layer such as Ga_2O_3, (In_2O_3), and $Ga(OH)_3$ ($In(OH)_3$) using the in-situ deposition, in contrast to the ex-situ approach using ALD.

9.4.2 Interfacial Chemical Characteristics in ALD-Al_2O_3 on InGaAs

A sharp transition from crystalline InGaAs to amorphous oxides was observed using HR-TEM. As-grown and oxygen-annealed samples have shown a similar morphology in terms of the amorphous nature of the oxide, and the roughness of surface and interface. The interface shows good thermal stability after nitrogen annealing at 500 °C. The interfacial and surface characteristics in $Al_2O_3/In_{0.15}Ga_{0.85}As/GaAs$ [6] was studied using high-resolution x-ray photoelectron spectroscopy (HR-XPS) using synchrotron radiation, with a sample of native oxides/$In_{0.15}Ga_{0.85}As/GaAs$ served as a reference (Fig. 9.11). The XPS results showed that

1. the native arsenic oxide on the air-exposed InGaAs is As_2O_3;
2. the arsenic oxide on top of Al_2O_3 after ALD deposition is As_2O_5;

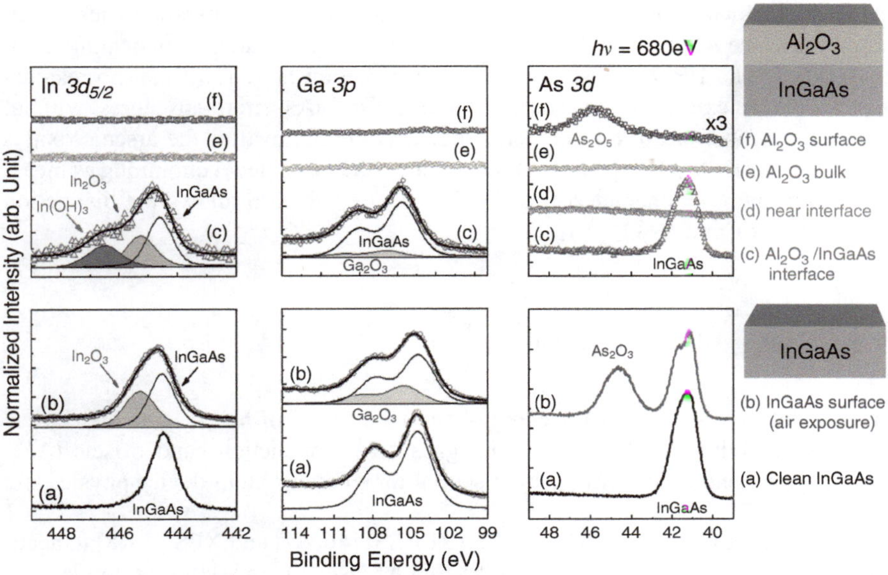

Fig. 9.11 In *3d*, As *3d* core, and Ga *3p* level spectra recorded from two samples. Al_2O_3/$In_{0.15}Ga_{0.85}As$/GaAs (*top*) and native oxide/$In_{0.15}Ga_{0.85}As$/GaAs (*bottom*); **a** clean $In_{0.15}Ga_{0.85}As$; **b** at the surface of air-exposed $In_{0.15}Ga_{0.85}As$; **c** at the interface of Al_2O_3/$In_{0.15}Ga_{0.85}As$; **d** near the oxide/$In_{0.15}Ga_{0.85}As$ interface; **e** immediately below the Al_2O_3 surface in the bulk of Al_2O_3; and **f** at the surface of Al_2O_3 [33] (Reprinted with permission. Copyright 2007 The Japan Society of Applied Physics)

3. there is no detectable arsenic oxides within the Al_2O_3 and at the Al_2O_3/InGaAs interface;
4. In_2O_3 and Ga_2O_3, residues of the native oxide, existed at the Al_2O_3/InGaAs interface.

During the ALD process, As_2O_3, the native oxide on InGaAs surface, first interacted with $Al(CH_3)_3$ and was transformed to become arsenic and Al_2O_3. [Note that As_2O_3 was reduced by $Al(CH_3)_3$. However, it is not known that the arsenic form is As, As_2, or As_4.] The arsenic then reacted with H_2O to become As_2O_5, which then reacts again with the next incoming pulse of $Al(CH_3)_3$ to become As and Al_2O_3. As_2O_5 has a lower Gibbs free energy than As_2O_3. This reaction has repeated itself during the process. Most of the arsenic oxides evaporated since the melting point of As_2O_5 is low around 280 °C with a small amount of residual As_2O_5 left on the skin-depth top of Al_2O_3.

The removal of arsenic oxides from the Al_2O_3/InGaAs hetero-structures ensures the Fermi level unpinning, which was observed in the *C–V* measurements. In_2O_3 and Ga_2O_3 of the original native oxides were left at the Al_2O_3/InGaAs interface due to their higher thermal stability. We have achieved good electrical properties in

ALD-Al_2O_3/InGaAs hetero-structures without any surface preparation such as HF dip prior to the Al_2O_3 deposition or post thermal treatment such as annealing in O_2 at ~600 °C. The HR-XPS using synchrotron radiation has shown that no residual arsenic oxides exist at the interface of the oxide/InGaAs hetero-structures, without post deposition annealing in O_2. This indicates the removal of the arsenic oxides during the ALD-Al_2O_3 deposition, thus ensuring the Fermi level unpinning as measured from the *C–V* characteristics. Moreover, the HR-XPS allows the distinction between As_2O_3 and As_2O_5.

9.5 Energy-Band Parameters

In addition to low D_{it}, and J's, energy-band parameters of high-k gate dielectrics/InGaAs, including oxide energy-band gaps (E_g), conduction-band offsets (ΔE_c) and valence-band offsets (ΔE_v), are essential for studying MOS device physics and design.

We have combined reflection-EELS (REELS) [46, 47] and XPS [7, 48] to accurately determine E_g, ΔE_c and ΔE_v of ultra thin ALD-Al_2O_3 and -HfO_2 on $In_xGa_{1-x}As$ (x = 0, 0.15, 0.25, and 0.53). The advantages of REELS over the commonly used electron energy loss spectroscopy (EELS) [7] in acquiring E_g values are given elsewhere [47]. The energy-band parameters of UHV-GGO on GaAs were published earlier [49]. The band offsets of the studied high-k's on InGaAs are adequate for reliable gate dielectric operation. The ΔE_c values obtained from the HR-XPS and REELS analyses are in good agreement with those estimated from the electrical measurement of Fowler-Nordheim tunneling [8, 47, 49].

E_v values at the oxide-InGaAs interface have been determined using HR-XPS by measuring the differences of binding energy between core-level (CL) and valence band maximum (VBM) of the oxides and InGaAs with the method proposed by Kraut et al. [48] from the following equation:

$$\Delta E_V = \left(E_{CL}^{InGaAs} - E_{VBM}^{InGaAs}\right) - \left(E_{CL}^{oxide} - E_{VBM}^{oxide}\right) + \Delta E_{CL}, \tag{9.1}$$

where E_{CL} and E_{VBM} are the binding energy of core-level and VBM, and $\left(E_{CL}^{InGaAs} - E_{VBM}^{InGaAs}\right)$ and $\left(E_{CL}^{oxide} - E_{VBM}^{oxide}\right)$ are the binding energy differences of core-level to VBM for InGaAs and oxides, respectively. The last term ΔE_{CL} is the energy separation between the core-level of oxides and that of semiconductors across the oxides/InGaAs interface.

The experiments on each of the following steps were carefully and thoroughly carried out: (i) acquisition of the VBM and As *3d* core level spectra of the sputter-cleaned $In_xGa_{1-x}As$ (x = 0, 0.15, 0.25, and 0.53) surface prior to oxide deposition for $\left(E_{CL}^{InGaAs} - E_{VBM}^{InGaAs}\right)$. The VBM values were determined from the intersection of a line fit to the leading edge of the valence band spectrum with the background level, (ii) attainment of the Al *2p* and Hf *4f* core level spectra along with the corresponding VBM of the Al_2O_3 and HfO_2 films after the deposition of oxides for $\left(E_{CL}^{oxide} - E_{VBM}^{oxide}\right)$, and (iii) determination of the energy separation ΔE_{CL} across the $Al_2O_3/In_xGa_{1-x}As$ and $HfO_2/In_xGa_{1-x}As$ interface. Note that some oxides with thickness >3 nm were thinned down by sputtering, while no sputtering was needed for thinner oxides.

Figure 9.12a, b illustrate the As *3d*, Al *2p*, Hf *4f* core level along with the VBM of ALD-Al_2O_3 and -HfO_2 on GaAs with incident photon energies of 680 eV. The same analytical methods were also carried out to investigate the ΔE_v of Al_2O_3 and HfO_2

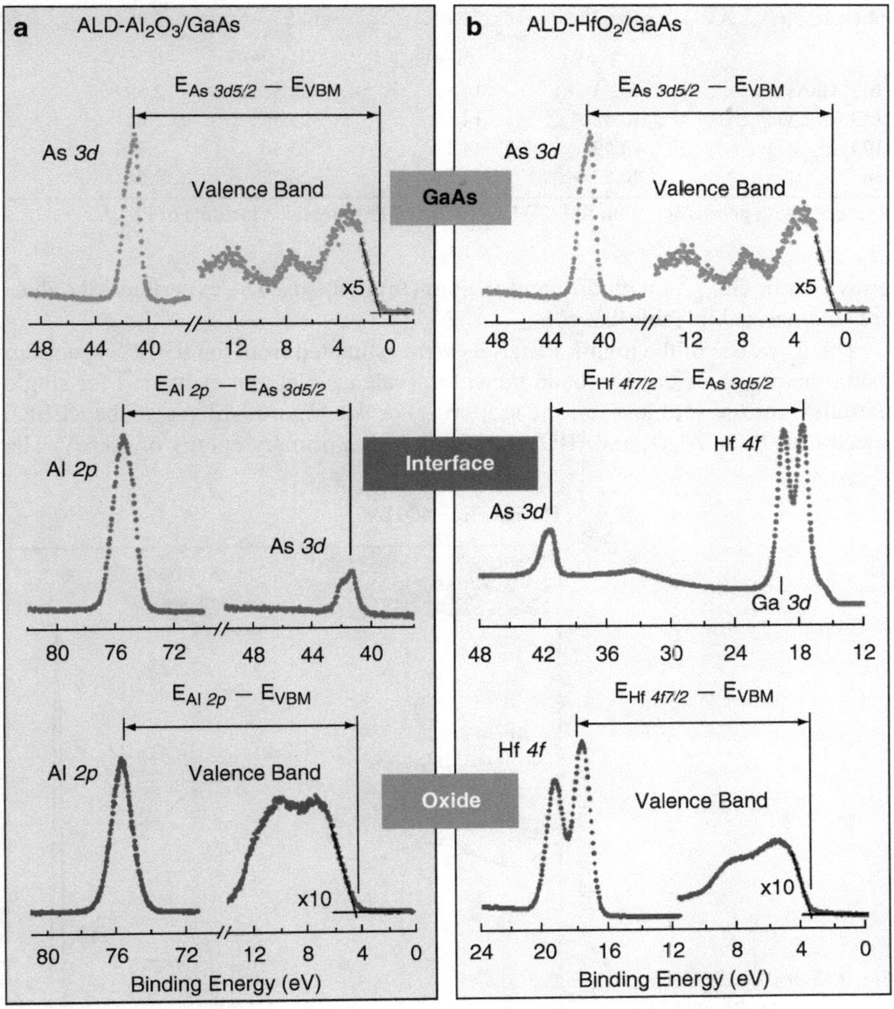

Fig. 9.12 XPS spectra of **a** As *3d* core level and valence band of GaAs film, Al *2p* and As *3d* core levels at the ALD-Al_2O_3/GaAs interface and Al *2p* core level and valence band of Al_2O_3 film **b** As *3d* core level and valence band of GaAs film, Hf *4f* and As *3d* core levels at the ALD-HfO_2/GaAs interface and Hf *4f* core level and valence band of HfO_2 film [47] (Reprinted with permission. Copyright 2009 American Institute of Physics)

Table 9.3 Parameters of core-level and valence band maximum energy difference of (i) $E_{As3d5/2}^{InGsAs} - E_{VBM}^{InGaAs}$, (ii) $E_{Al2p}^{Al_2O_3} - E_{VBM}^{Al_2O_3}$, and (iii) $E_{Al2p}^{Al_2O_3} - E_{As3d5/2}^{InGaAs}$, for the ALD-$Al_2O_3$; (iv) $E_{As3d5/2}^{InGsAs} - E_{VBM}^{InGaAs}$, (v) $E_{Hf4f7/2}^{HfO_2} - E_{VBM}^{HfO_2}$, and (vi) $E_{Hf4f7/2}^{HfO_2} - E_{As3d5/2}^{InGaAs}$ for -HfO_2 grown on $In_xGa_{1-x}As$ along with the corresponding estimated values of E_V

	(i) (eV)	(ii) (eV)	(iii) (eV)	E_V (eV)
Al_2O_3/GaAs	40.36	71.14	34.45	3.67
$Al_2O_3/In_{0.15}Ga_{0.85}As$	40.42	71.14	34.50	3.78
$Al_2O_3/In_{0.25}Ga_{0.75}As$	40.49	71.14	34.51	3.86
$Al_2O_3/In_{0.53}Ga_{0.47}As$	40.57	71.14	34.53	3.96
	(iv) (eV)	(v) (eV)	(vi) (eV)	E_V (eV)
HfO_2/GaAs	40.36	14.19	−23.58	2.59
$HfO_2/In_{0.15}Ga_{0.85}As$	40.42	14.19	−23.56	2.67
$HfO_2/In_{0.25}Ga_{0.75}As$	40.49	14.19	−23.56	2.74
$HfO_2/In_{0.53}Ga_{0.47}As$	40.57	14.19	−23.52	2.86

grown on $In_xGa_{1-x}As$ with different indium (In) contents. The experimental values are summarized in Table 9.3.

The E_g values of the high-k materials were estimated from the REELS spectrum using the onsets of band-to-band transition (valence-electron excitation for single particles) for the zero loss (elastic scattering) peaks. Figure 9.13 shows the REELS spectra for SiO_2, Al_2O_3 and HfO_2 thin films with a primary energy of 400 eV. The

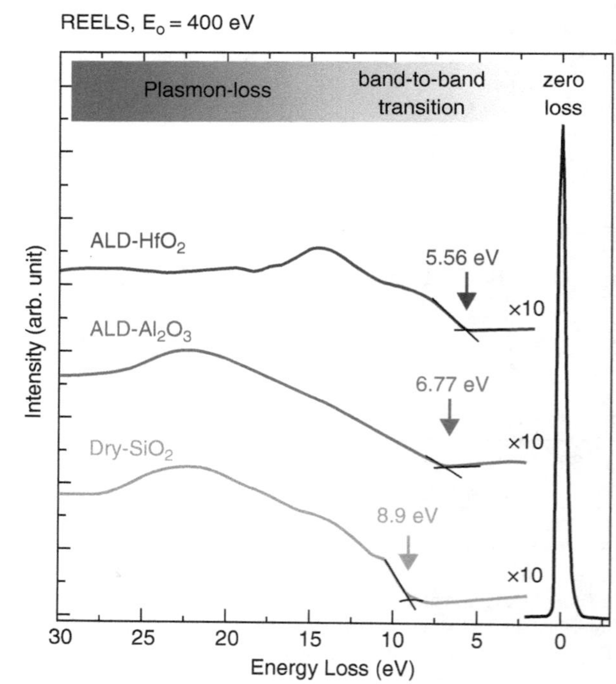

Fig. 9.13 Reflection electron energy loss spectra for ALD-HfO_2, ALD-Al_2O_3 and SiO_2 thin films at the primary energy of 400 eV [47] (Reprinted with permission. Copyright 2009 American Institute of Physics)

elastic peaks were normalized to the same area. The E_g value of SiO_2 appears at ~8.9 ± 0.05 eV, which is in good agreement with the commonly obtained value of 8.8–9.0 eV, demonstrating the validity of our REELS measurement.

The bulk plasmon peaks were located at ~22.5, ~22.1 and ~14.5 eV for SiO_2, Al_2O_3, and HfO_2. Experimental data of E_g values were 6.77 = 0.05 and 5.56 ± 0.05 eV for Al_2O_3 and HfO_2, respectively. These results are consistent with those from REELS analyses reported on Al_2O_3 and HfO_2 [46]. In addition, the impurity content in the films, such as residual carbon and nitrogen, was less than 1 at% by the XPS analyses (not shown).

Combining the experimental data of ΔE_v and the oxide E_g with the band gap values of $In_xGa_{1-x}As$ (x = 0, 0.15, 0.25, and 0.53), the conduction band offsets (ΔE_c), important energy band parameters for MOS devices, can be simply derived from the following equation:

$$\Delta E_c = E_g(oxide) - E_g(InGaAs) - \Delta E_v \tag{9.2}$$

The energy band parameters of ALD-oxides/InGaAs hetero-structures are summarized in Fig. 9.14. Note that the ΔE_c and ΔE_v of ALD-Al_2O_3 and $HfO_2/In_xGa_{1-x}As$ increase as x varying from 0 to 0.53. The lattice parameters of the strained $In_xGa_{1-x}As$ layers have been obtained through thorough high-resolution x-ray diffraction (HR-XRD) measurements using synchrotron radiation. The value of x was

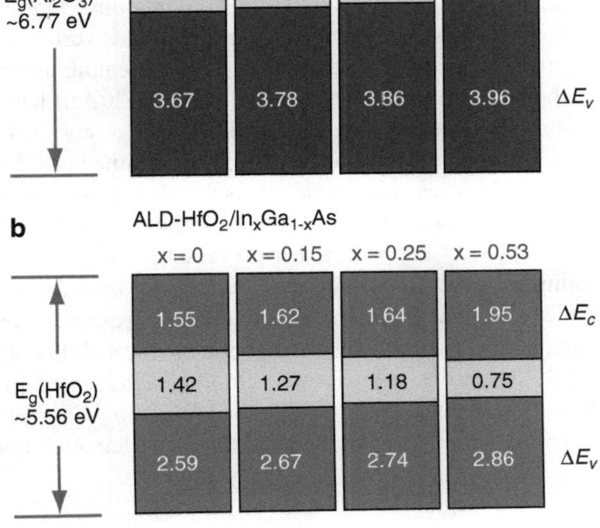

Fig. 9.14 Energy-band parameters for **a** $Al_2O_3/In_xGa_{1-x}As$ and **b** $HfO_2/In_xGa_{1-x}As$ hetero-structures acquired using HR-XPS and REELS [47] (Reprinted with permission. Copyright 2009 American Institute of Physics)

then determined, with an inaccuracy of 3–5%. The samples after XRD were then used to calibrate the In/Ga ratio studied using in-situ XPS analyses. The E_g values for the strained $In_xGa_{1-x}As/GaAs$ layers were achieved by using optical measurement in previous reports (Ref. [50]). Therefore, the E_g values of $In_xGa_{1-x}As$ were obtained since the x values were known. The E_g values of 1.42 eV for unstrained GaAs, 1.27 eV and 1.18 eV for strained-$In_{0.15}Ga_{0.85}As$ and -$In_{0.25}Ga_{0.75}As$ on GaAs and 0.75 eV for unstrained $In_{0.53}Ga_{0.47}As$ on InP were used in our calculation. The derived ΔE_c values are consistent with those determined using Fowler–Nordheim (F–N) tunneling analysis [7, 8].

Moreover, the band parameters presented here were obtained from a wide range of oxide thickness varying from 3 to 8 nm, with the same/similar energy-band values regardless of the oxide thickness. These, thus, have proved the uniqueness of using HR-XPS and REELS for acquiring the band parameters of ultra thin oxide films on semiconductors. In contrast, it is difficult, if not impossible, in utilizing electrical transport analysis, such as Fowler–Nordheim tunneling, to derive ΔE_C for the oxide thickness less than 4 nm, due to the direct tunneling effect.

9.6 Thickness Scalability of $Ga_2O_3(Gd_2O_3)$ on InGaAs with Low D_{it}, Low Leakage Currents, and High-Temperature Thermodynamic Stability

Besides unpinning surface Fermi level of the III-V's, the first step for realizing the inversion-channel devices, scaling high-k oxides to nanometer range has been one main focus of recent high-k research on high mobility channel materials. Among all the dielectrics passivating GaAs (InGaAs) [3–5, 8, 14, 15, 20, 21, 32, 51, 52], GGO was found to give the lowest interfacial densities of states [51]. Nevertheless, GGO thickness was in the range of 25–55 nm for the studies of MOS capacitors and MOSFET's in the 1990's. [3] GGO, containing rare earth oxide of Gd_2O_3, tends to absorb moisture during air exposure, thus resulting in degradation of electrical properties and device performance. Annealing under UHV or nitrogen ambient of the air-exposed oxides (with thickness >20 nm) has been effective in expelling absorbed moisture. Low D_{it} and low leakage currents were restored and smooth oxide-semiconductor interfaces were maintained [28, 53]. However, further scaling of GGO thickness was hindered, as the air-exposed degraded oxide/III-V interfaces may not be recovered with UHV annealing for thinner GGO.

In-situ Al_2O_3 cap, effective in preventing GGO from absorbing moisture, thus passivating the GGO/III-V interfaces [20, 21], has been deposited on $GGO/In_{0.2}Ga_{0.8}As$ to achieve a 1 nm CET, along with a low leakage current density, a high dielectric constant, a low D_{it}, and excellent thermodynamic stability at temperatures above 800 °C. These properties are essential for the self-aligned processed inversion-channel $Al_2O_3/GGO/InGaAs$ MOSFETs. The $GGO/In_{0.2}Ga_{0.8}As$ (InGaAs) hetero-structures were fabricated with the oxide thickness systematically reduced from 33 to 4.5 nm.

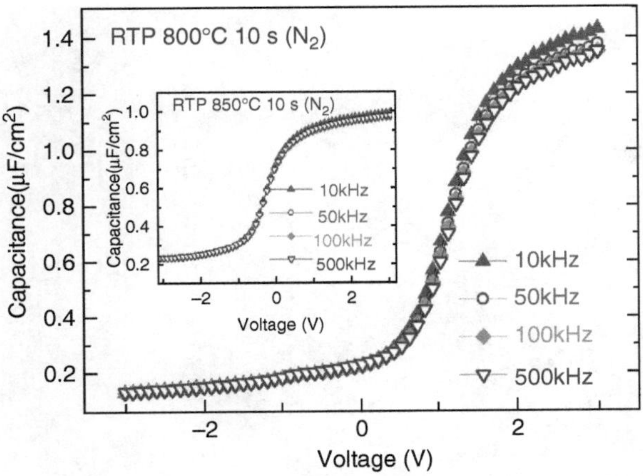

Fig. 9.15 *C–V* characteristics for an Al_2O_3 (3 nm)/GGO (4.5 nm)/$In_{0.2}Ga_{0.8}As$/GaAs MOS diode RTA to 800 °C with the Au gate deposited afterwards. The *inset* shows *C–V* characteristics of another 8.5 nm sample, which was RTA to 850 °C with the Al gate [62] (Reprinted with permission. Copyright 2009 Materials Research Society)

MOS capacitors with the dual-dielectric layer on InGaAs have withstood rapid-thermal-anneals (RTA) to 800–900 °C. Excellent *C–V* characteristics in terms of small flat-band voltage shift, a small frequency dispersion of measured capacitances at accumulation, D_{it}'s in the low $10^{11}\,cm^{-2}\,eV^{-1}$, the κ values of GGO remaining ~14–16, were achieved for all GGO film thicknesses (Fig. 9.15). The CET is linearly scaled with the GGO physical film thickness as shown in Fig. 9.16. Figure 9.17 is a HR-TEM on an Al_2O_3/GGO/InGaAs being RTA to 800 °C.

GGO remains amorphous along with the atomically sharp GGO/InGaAs interface, critical for achieving excellent device performances. These remarkable proper-

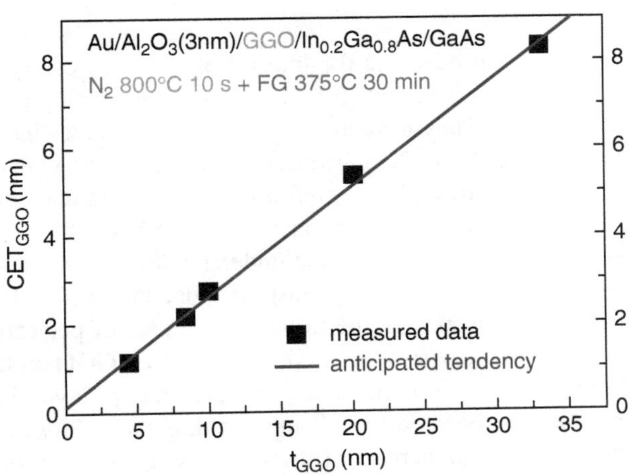

Fig. 9.16 CET of GGO with various physical thicknesses [62] (Reprinted with permission. Copyright 2009 Materials Research Society)

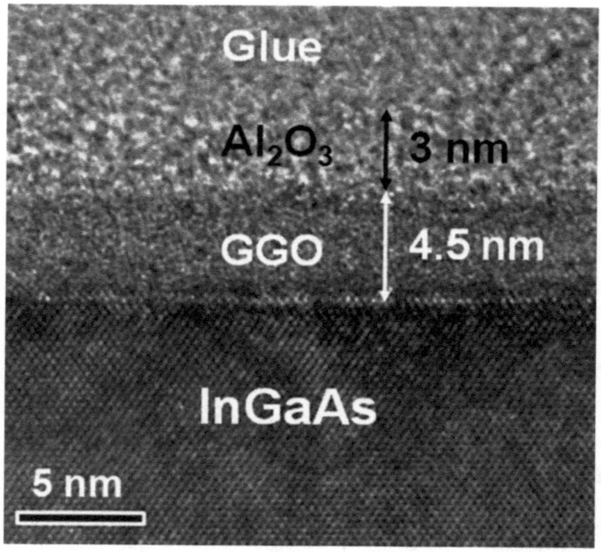

Fig. 9.17 Cross-sectional HR-TEM image of the 4.5 nm sample after 800 °C RTA [21] (Reprinted with permission. Copyright 2008 American Institute of Physics)

ties have aided our high-performance self-aligned inversion-channel $In_{0.53}Ga_{0.47}As$ MOSFET and have greatly enhanced the prospect for realistic device applications.

9.7 Interface Trap Densities and Efficiency of Fermi-Level Movement

Interface trap densities (or interfacial densities of states (D_{it}'s) and efficiency of Fermi-level movement in SiO_2/Si were thoroughly studied and well understood, since 1960s. They have also been intensively investigated in the high-k dielectrics and metal gates on Si in the last decades, however, with a less degree of comprehension. The studies are essential for successful implementation and mass production of high-k's and metal gates on 45 nm node Si CMOS, and soon on 32 (28) nanometre node. Comparing with the understanding of D_{it}'s and efficiency of Fermi-level movement in SiO_2/Si, our grasp of those in the high-k's/InGaAs is in its infancy. The main reason was due to the less degree of perfection of the high-k's/InGaAs interface in the past. With in-situ UHV-Al_2O_3/GGO on InGaAs and ex-situ short air exposure between ALD-Al_2O_3 and InGaAs epi-layer growth, the high-k's/InGaAs interface has been perfected, approaching that in SiO_2/Si. Now, we are able to study the interface trap densities (or interfacial densities of states (D_{it}'s) and efficiency of Fermi-level movement in the high-k's/InGaAs with conventional quasi-static CV

(QSCV) measurement and charge pumping, and conductance method under various temperatures, with meaningful experimental results.

9.7.1 QSCV Measurements

The *C–V* curves of an $In_{0.53}Ga_{0.47}As$ MOSCAP with 10 nm ALD-Al_2O_3 with frequencies varying from 1 kHz to 1 MHz are shown in Fig. 9.18a, displaying clear accumulation, depletion, and inversion with negligible frequency dispersion at the accumulation. Quasi-static *C–V* was measured [54], Fig. 9.18b, with a gate bias sweep of 1 mV/sec from −3 to +3 V at room temperature in dark, to minimize the effects of bias sweeping rate.

The Ψ–V relationship (where Ψ is the surface potential) obtained using Berglund's integral indicates the Fermi-level movement across the entire band gap (0.74 eV) of $In_{0.53}Ga_{0.47}As$, with a high efficiency of 63% near the mid-gap compared to the value obtained in an ideal Ψ–V curve with $D_{it} = 0$.

According to the *C–V* characteristics displaying negligible frequency dispersion at accumulation and nearly ideal inversion behavior, a D_{it} ~2–3 × 10^{12} cm^{-2} eV^{-1} at

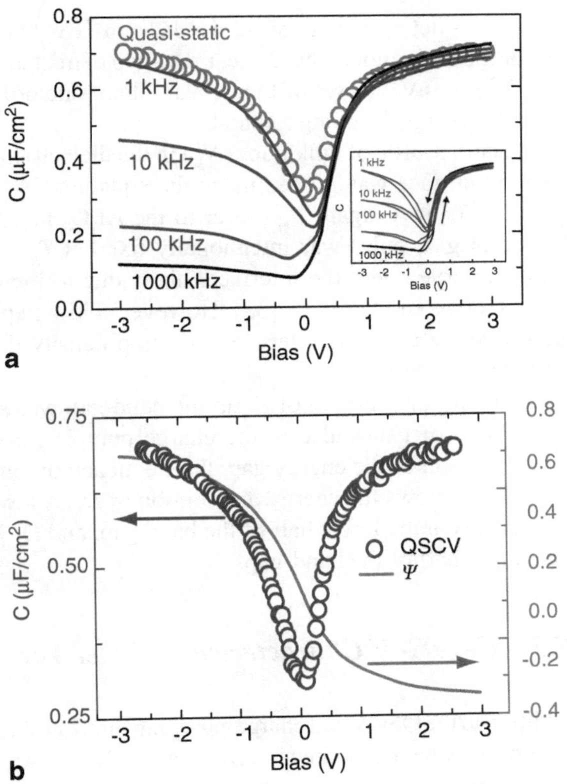

Fig. 9.18 **a** *C–V* characteristics of a TiN/Al_2O_3 (10 nm)/*n*-$In_{0.53}Ga_{0.47}As$ MOSCAP subjected to 650 °C RTA for 15 sec in He. **b** Ψ–V relationship extracted from qusai-static curve; the definition of Fermi-level movement efficiency (%) = 100% × (measured Ψ–V slope) ÷ (ideal Ψ–V slope) [54] (Reprinted with permission. Copyright 2008 American Institute of Physics)

mid-gap extracted by the conductance method at room temperature appears to be unreasonably high. The overestimate in D_{it} may be caused with the contribution of weak inversion in $In_{0.53}Ga_{0.47}As$ [55]. A more accurate determination of D_{it} was achieved using the charge pumping method [54] described in the following:

9.7.2 Charge Pumping Method

Under various frequencies, the curves of I_{cp} versus gate bias were obtained with sweeping the gate base level from −1 to +1 V and a gate pulse of a constant amplitude (1 V), as shown in the Fig. 9.19a. The rise/fall time was fixed at 100 ns. For SiO_2/Si, charges pumped per cycle, $Q_{cp} = I_{cp}$/frequency, slightly depend on frequency. However, for high-k's on semiconductors, Q_{cp} arises with decreasing frequency because low frequency provides sufficient time for electrons to tunnel into the oxide traps, known as bulk traps (N_{bt}) [56]. Thus, an I_{cp} at 1 MHz was used here to calculate mean D_{it} according to the following equation [57]:

$$I_{cp} = 2q\overline{D}_{it}fA_GkT\left[\ln\left(v_{th}n_i\sqrt{\sigma_n\sigma_p}\right) + \ln\left(\frac{|V_{TH} - V_{FB}|}{|V_h - V_b|}\sqrt{T_RT_F}\right)\right] \quad (9.3)$$

The $\overline{D}_{it}$ was determined to be ~2.5×10^{11} cm^{-2} eV^{-1} near mid-gap (±0.2 eV). However, for the same diode, the D_{it} near mid-gap extracted by conductance method was ~3×10^{12} cm^{-2} eV^{-1} (inset of Fig. 9.19a), about one order of magnitude higher than that by the charge pumping method.

The depth profile of bulk traps (N_{bt}) in the dielectric layer as a function a distance from the interface was probed using the equations in Ref. [58], and is depicted in Fig. 9.19b. To investigate N_{bt} closer to the $Al_2O_3/In_{0.53}Ga_{0.47}As$ interface, the rise/fall time of gate pulse was intentionally fixed at $T_R = T_F = 10$ ns. The increase of N_{bt} was observed near the interface, likely due to the existence of residual native oxides, such as In_2O_3, Ga_2O_3 [33]. However, a low trap density ~7×10^{18} cm^{-3} was observed away from the interface. Our trap density data is lower than the earlier reported value [59].

The energy distribution of D_{it} in the band-gap was studied with modulating the rise/fall time of gate pulse in the charge pumping measurement [54]. To scan a broader fraction of the energy gap, the frequency of gate pulse was held at 50 kHz. Figure 9.19c shows the energy distribution of D_{it}. A low D_{it} of 2–4 × 10^{11} cm^{-2} eV^{-1} was obtained in the lower half of the band gap, and D_{it} increases to ~10^{12} cm^{-2} eV^{-1} in the upper half of the band-gap.

9.7.3 *C–V/G–V* Characteristics Under Various Temperatures

In addition to the QSCV and charge-pumping on ALD-$Al_2O_3/In_{0.53}Ga_{0.47}As$, *C–V/G–V* measurements have also been carried out for UHV-Al_2O_3/GGO/$In_{0.2}Ga_{0.8}As$ under various temperatures [60]. Note that the D_{it} value measured at room temperature

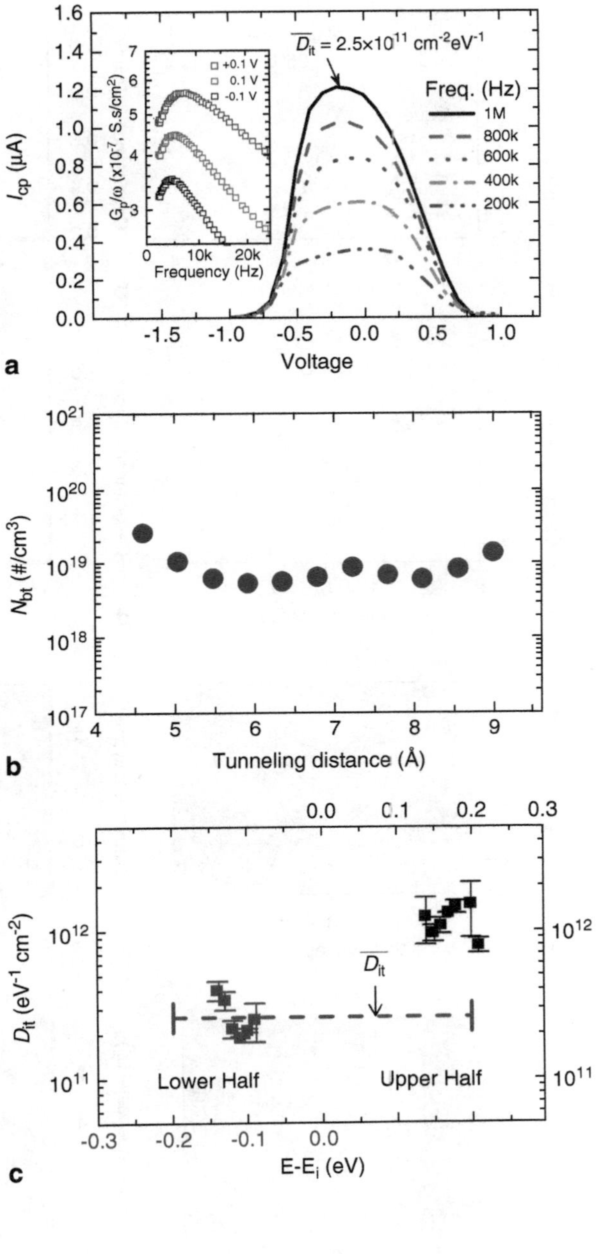

Fig. 9.19 **a** Charge pumping current (I_{cp}) as function of gate pulse with frequency dependence. The *inset* shows G_p/ω versus frequency under different voltages measured by the conductance method. **b** Profile of the volume densities of bulk traps (N_{bt}) in ALD-Al_2O_3 as a function of distance from the $In_{0.53}Ga_{0.47}As$ surface. **c** Energy dependence of interfacial densities of states (D_{it}) in ALD-Al_2O_3/$In_{0.53}Ga_{0.47}As$ [54] (Reprinted with permission. Copyright 2008 American Institute of Physics)

using conductance measurement may not cover the important mid-gap region. For MOSCAP with n-$In_{0.2}Ga_{0.8}As$ (Fig. 9.20b), for example, very small frequency dispersion at accumulation at room and low temperature indicates low D_{it} for most of the upper bandgap.

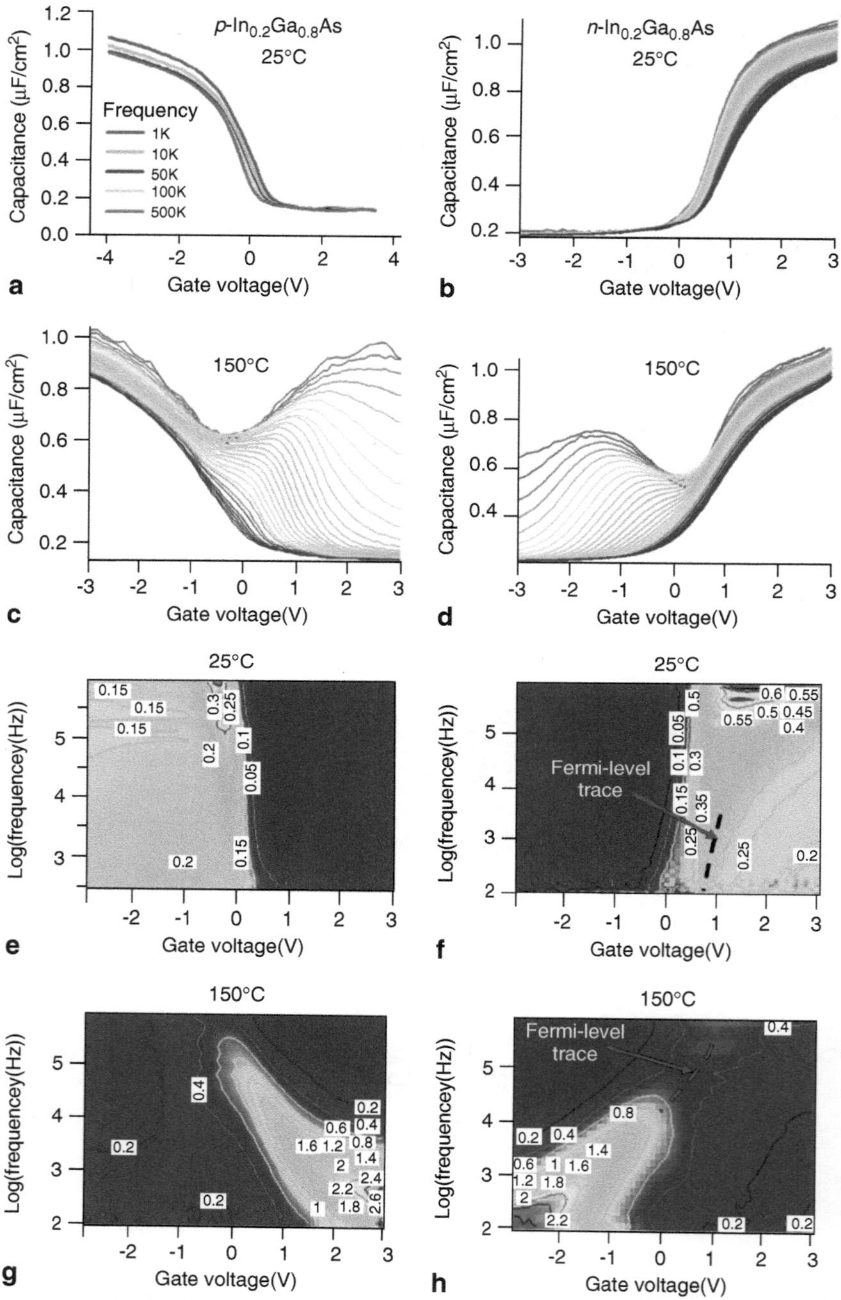

Fig. 9.20 **a–d** *C–V* and **e–h** *G–V* characteristics of Au/Al_2O_3 (3 nm)/GGO (8.5 nm)/*n*-$In_{0.2}Ga_{0.8}As$/n-GaAs (*left column*) and Au/Pt/Al_2O_3 (3 nm)/GGO (8.5 nm)/p-$In_{0.2}Ga_{0.8}As$/p-GaAs (*right column*) MOSCAPs measured at 25 °C and 150 °C

The relatively high substrate temperature (150 °C) can shift and extend the measurement window, and allow the mid-band region be probed properly with the measurement frequency of 100 ~ 1 M Hz [61]. Figure 9.20a–d show *C–V* characteristics of UHV—Al_2O_3/GGO on *n*-type and *p*-type $In_{0.2}Ga_{0.8}As$ measured at 25 °C and 150 °C.

Differentiating between the conventional methods to extract D_{it} values, the *G–V* measurements were applied to study the interface trap distribution within the bandgap. Figure 9.20e–h are *G–V* spectroscopy maps of UHV-Al_2O_3/GGO on *n*-type and *p*-type $In_{0.2}Ga_{0.8}As$ measured at 25 °C and 150 °C. From the *G–V* results, unobstructed Fermi-level movement was observed in the upper and lower 1/3 of the bandgap. In these regions, steep Fermi-level traces with respect to gate bias reveal high modulation efficiency, low electrical friction at the GGO/$In_{0.2}Ga_{0.8}As$ interface and low D_{it}'s.

High temperature *C–V*'s at 150 °C reveal the minority carrier traps near mid-gap region for both n- and p-type samples. A similar conclusion can be drawn from the corresponding conductance maps. The decreasing slopes of Fermi-level movement with respect to bias also confirm the trapping of the carriers and an increase in friction against gate modulation near the mid-bandgap. It appears that higher inversion bias is required to move the Fermi-level beyond this region to achieve strong inversion. High temperature *C–V* and *G–V* results show signs of high interface traps around midgap, as revealed by the *C–V* bumps in the inversion region of Fig. 9.20c, d, as well as the increasing friction against gate modulation near the same region on Fig. 9.20g, h.

9.8 Conclusion

The quest for the InGaAs MOSFETs since early 1960s has been answered with the discovery of UHV-$Ga_2O_3(Gd_2O_3)$ on GaAs in 1990s, resulting in the first inversion-channel III-V (GaAs and InGaAs) MOSFETs with good device performance. In the last few years since 2003, tremendous progress and very important discoveries have been made in tailoring the growth in an atomic scale and in probing the interfaces of ultra-thin (nm thick) high-k dielectric on the III-V's. Record-high drain currents and transconductance in self-aligned inversion-channel Al_2O_3/GGO/$In_{0.53}Ga_{0.47}As$ MOSFET's have been achieved; the outperformance over the other E-mode MOSFETs may be attributed to the superior oxide/semiconductor interface, free of native oxides encountered inevitably in the ex-situ ALD approach. Depletion-mode InGaAs MOSFET's were also fabricated, with excellent device properties, suitable for power applications. Inversion-channel GaN MOSFET's with high k dielectrics were demonstrated, for the first time. Furthermore, inversion-channel GGO/Ge MOSFET's exhibited the device characteristics, in parallel with the top-performed Ge MOSFET's using other approaches.

No interfacial layers were employed in our inversion-channel InGaAs, Ge, and GaN MOSFET's (self-aligned or non-self-aligned). Interfacial layers are needed and/or inevitably formed using other deposition methods; they, nevertheless, increase the thickness budget of the gate dielectrics, the D_{it}'s, but may hamper the high-temperature annealing of the hetero-structures of the high-k's/InGaAs, Ge, and GaN, due to the less thermodynamic stability.

Many critical material issues for high-performance devices have been solved: low D_{it}s' of $10^{11}\,cm^{-2}\,eV^{-1}$, low leakage current densities of 10^{-8}–10^{-9} A/cm^2 at $V_{fb} \pm 1$ V, high temperature (800–900 °C) thermal stability of GGO/InGaAs, scalability of high-k's to 1 nm, and comprehensive mapping of energy band parameters of high-k's on InGaAs. Moreover, the high-k's/InGaAs, Ge, and GaN interfaces have been perfected using in-situ GGO and ex-situ ALD oxides with short air exposure. As a consequence, quasi-static CV and charge pumping, conventional methods for probing SiO_2/Si, are now being employed to study the new high-k/high mobility channel with meaningful properties, along with temperature-variation CV/GV characteristics.

Acknowledgments The authors wish to thank National Science Council, Ministry of Education, Taiwan, Republic of China, Intel, AOARD (U.S. Air Force), TSMC, and IBM for supporting this work. The contributions from Y. J. Lee, Y. H. Chang, Y. D. Wu, T. H. Chiang, L. K. Chu, C. A. Lin, W. H. Chang, H. C. Chiu, R. L. Chu, and Y. C. Chang are greatly appreciated.

References

1. Hong M, Liu CT, Reese H, Kwo J (1999): Semiconductor-insulator interfaces. In: Webster JG (ed), Encyclopedia of electrical and electronics engineering, John Wiley & Sons, New York, pp 87–100.
2. Hong M, Passlack M, Mannaerts JP, Kwo J, Chu SNG, Moriya N, Hou SY, Fratello VJ (1996): Low interface state density oxide-GaAs structures fabricated by in situ molecular beam epitaxy. J. Vac. Sci. Technol. B 14:2297–2300.
3. Hong M, Mannaerts JP, Bower JE, Kwo J, Passlack M, Hwang W-Y, Tu LW (1997): Novel $Ga_2O_3(Gd_2O_3)$ passivation techniques to produce low Dit oxide-GaAs interfaces. J. Cryst. Growth 175,176:422–427.
4. Hong M, Kwo J, Kortan AR, Mannaerts JP, Sergent AM (1999): Epitaxial cubic gadolinium oxide as a dielectric for gallium arsenide passivation. Science 283:1897–1900.
5. Ye PD, Wilk GD, Yang B, Kwo J, Gossmann H-JL, Hong M, Ng KK, Bude J (2004): Depletion-mode InGaAs metal-oxide-semiconductor field-effect transistor with oxide gate dielectric grown by atomic-layer deposition. Appl. Phys. Lett. 84:434–436.
6. Huang ML, Chang YC, Chang CH, Lee YJ, Chang P, Kwo J, Wu TB, Hong M (2005): Surface passivation of III-V compound semiconductors using atomic-layer-deposition-grown Al_2O_3. Appl. Phys. Lett. 87:252104.
7. Huang ML, Chang YC, Chang CH, Lin TD, Kwo J, Wu TB, Hong M (2006): Energy-band parameters of atomic-layer-deposition Al_2O_3/InGaAs heterostructure. Appl. Phys. Lett. 89:012903.
8. Chang YC, Huang ML, Lee KY, Lee YJ, Lin TD, Hong M, Kwo J, Lay TS, Liao CC, Cheng KY (2008): Atomic-layer-deposited HfO_2 on $In_{0.53}Ga_{0.47}As$: Passivation and energy-band parameters. Appl. Phys. Lett. 92:072901.
9. Koveshnikov S, Tsai W, Ok I, Lee JC, Torkanov V, Yakimov M, Oktyabrsky S (2006): Metal-oxide-semiconductor capacitors on GaAs with high-k gate oxide and amorphous silicon interface passivation layer. Appl. Phys. Lett. 88:022106.

10. de Souza JP, Kiewra E, Sun Y, Callegari A, Sadana DK, Shahidi G, Webb DJ, Fompeyrine J, Germann R, Rossel C, Marchiori C (2008): Inversion mode n-channel GaAs field effect transistor with high-k/metal gate. Appl. Phys. Lett. 92:153508.
11. Kim H-S, Ok I, Zhang M, Lee T, Zhu F, Yu L, Lee JC, Koveshnikov S, Tsai W, Tokranov V, Yakimov M, Oktyabrsky S (2006): Depletion-mode GaAs metal-oxide-semiconductor field-effect transistor with HfO_2 dielectric and germanium interfacial passivation layer. Appl. Phys. Lett. 89:222904.
12. Ren F, Hong MW, Hobson WS, Kuo JM, Lothian JR, Mannaerts JP, Kwo J, Chen YK, Cho AY (1996): Enhancement-mode p-channel GaAs MOSFETs on semi-insulating substrates, *IEEE International Electron Devices Meeting*, San Francisco, CA, pp 943–945.
13. Ren F, Hong M, Hobson WS, Kuo JM, Lothian JR, Mannaerts JP, Kwo J, Chu SNG, Chen YK, Cho AY (1997): Demonstration of enhancement-mode p- and n-channel GaAs MOSFETs with $Ga_2O_3(Gd_2O_3)$ as gate oxide. Solid-State Electron. 41:1751–1753.
14. Ren F, Kuo JM, Hong M, Hobson WS, Lothian JR, Lin J, Tsai HS, Mannaerts JP, Kwo J, Chu SNG, Chen YK, Cho AY (1998): $Ga_2O_3(Gd_2O_3)$/InGaAs enhancement-mode n-channel MOSFETs. IEEE Electron Device Lett. 19:309–311.
15. Wang YC, Hong M, Kuo JM, Mannaerts JP, Kwo J, Tsai HS, Krajewski JJ, Weiner JS, Chen YK, Cho AY (1999): Advances in GaAs MOSFETs using $Ga_2O_3(Gd_2O_3)$ as gate oxide. In: Hasegawa H, Lu ZH, Pearton SJ (ed), *Mat. Res. Soc. Symp. Proc.*, pp 219–225.
16. Xuan Y, Wu YQ, Shen T, Yang T, Ye PD (2007): High performance submicron inversion-type enhancement-mode InGaAs MOSFETs with ALD Al_2O_3, HfO_2, and HfAlO as gate dielectrics, *IEEE International Electron Devices Meeting*, Washington, DC, pp 637–640.
17. Xuan Y, Wu YQ, Ye PD (2008): High-performance inversion-type enhancement-mode InGaAs MOSFET with maximum drain current exceeding 1 A/mm. IEEE Electron Device Lett. 29:294–296.
18. Chiu HC, Lin TD, Chang YH, Chang P, Kwo J, Lin YS, Hsu SSH, Tsai W, Hong M (2009): DC and RF Characteristics of Self-aligned Inversion Channel $In_{0.53}Ga_{0.47}As$ N-MOSFET with ALD-Al_2O_3 as a Gate Dielectric, International Symposium on VLSI Technology, Systems and Applications, (VLSI-TSA), Hsinchu, Taiwan, pp 141–142.
19. Chin HC, Zhu M, Samudra GS, Yeo YC (2008): n-channel GaAs MOSFET with TaN/HfAlO gate stack formed using in situ vacuum anneal and silane passivation. J. Electrochem. Soc. 155:H464–H468.
20. Shiu KH, Chiang CH, Lee YJ, Lee WC, Chang P, Tung LT, Hong M, Kwo J, Tsai W (2008): Oxide scalability in $Al_2O_3/Ga_2O_3(Gd_2O_3)/In_{0.20}Ga_{0.80}As$/GaAs heterostructures. J. Vac. Sci. Technol. B 26:1132–1135.
21. Shiu KH, Chiang TH, Chang P, Tung LT, Hong M, Kwo J, Tsai W (2008): 1 nm equivalent oxide thickness in $Ga_2O_3(Gd_2O_3)/In_{0.2}Ga_{0.8}As$ metal-oxide-semiconductor capacitors. Appl. Phys. Lett. 92:172904.
22. Lin TD, Chiu HC, Chang P, Tung LT, Chen CP, Hong M, Kwo J, Tsai W, Wang YC (2008): High-performance self-aligned inversion-channel $In_{0.53}Ga_{0.47}As$ metal-oxide-semiconductor field-effect-transistor with $Al_2O_3/Ga_2O_3(Gd_2O_3)$ as gate dielectrics. Appl. Phys. Lett. 93:033516.
23. Hill RJW, Moran DAJ, Li X, Zhou H, Macintyre D, Thoms S, Asenov A, Zurcher P, Rajagopalan K, Abrokwah J, Droopad R, Passlack M, Thayne LG (2007): Enhancement-mode GaAs MOSFETs with an $In_{0.3}Ga_{0.7}As$ channel, a mobility of over 5000 cm^2/V s, and transconductance of over 475 μS/μm. IEEE Electron Device Lett. 28:1080–1082.
24. Sun Y, Kiewra EW, Koester SJ, Ruiz N, Callegari A, Fogel KE, Sadana DK, Fompeyrine J, Webb DJ, Locquet JP, Sousa M, Germann R, Shiu KT, Forrest SR (2007): Enhancement-mode buried-channel $In_{0.7}Ga_{0.3}As/In_{0.52}Al_{0.48}As$ MOSFETs with high-k gate dielectrics. IEEE Electron Device Lett. 28:473–475.
25. Kwo J, Murphy DW, Hong M, Opila RL, Mannaerts JP, Sergent AM, Masaitis RL (1999): Passivation of GaAs using. $(Ga_2O_3)_{1-x}(Gd_2O_3)x$, $0 <= x <= 1.0$ films. Appl. Phys. Lett. 75:1116–1118.
26. Lee CH, Lin TD, Tung LT, Huang ML, Hong M, Kwo J (2008): Molecular beam epitaxy grown $Ga_2O_3(Gd_2O_3)$ high kappa dielectrics for germanium passivation—x-ray photoelectron spectroscopy and electrical characteristics. J. Vac. Sci. Technol. B 26:1128–1131.

27. Hong M, Kwo J, Chu SNG, Mannaerts JP, Kortan AR, Ng HM, Cho AY, Anselm KA, Lee CM, Chyi JI (2002): Single-crystal GaN/Gd_2O_3/GaN heterostructure. J. Vac. Sci. Technol. B 20:1274–1277.
28. Chen CP, Lee YJ, Chang YC, Yang ZK, Hong M, Kwo J, Lee HY, Lay TS (2006): Structural and electrical characteristics of $Ga_2O_3(Gd_2O_3)$/GaAs under high temperature annealing. J. Appl. Phys. 100:104502.
29. Lin CA, Lin TD, Chiang TH, Chiu HC, Chang P, Hong M, Kwo J (2009): Depletion-mode $In_{0.2}Ga_{0.8}As$/GaAs MOSFET with molecular beam epitaxy grown $Al_2O_3/Ga_2O_3(Gd_2O_3)$ as gate dielectrics. J. Cryst. Growth 311:1954–1957.
30. Chu LK, Lee WC, Huang ML, Chang YH, Tung LT, Chang CC, Lee YJ, Kwo J, Hong M (2009): Metal-oxide-semiconductor devices with molecular beam epitaxy-grown Y_2O_3 on Ge. J. Cryst. Growth 311:2195–2198.
31. Chang YC, Chang WH, Chiu HC, Tung LT, Lee CH, Shiu KH, Hong M, Kwo J, Hong JM, Tsai CC (2008): Inversion-channel GaN metal-oxide-semiconductor field-effect transistor with atomic-layer-deposited Al_2O_3 as gate dielectric. Appl. Phy. Lett. 93:053504.
32. Lee KY, Lee YJ, Chang P, Huang ML, Chang YC, Hong M, Kwo J (2008): Achieving 1 nm capacitive effective thickness in atomic layer deposited HfO_2 on $In_{0.53}Ga_{0.47}As$. Appl. Phys. Lett. 92:252908.
33. Hong MW, Kwo JR, Tsai PC, Chang YC, Huang ML, Chen CP, Lin TD (2007): III-V metal-oxide-semiconductor field-effect transistors with high kappa dielectrics. Jpn. J. Appl. Phys. 46:3167–3180.
34. Wang YC, Hong M, Kuo JM, Mannaerts JP, Kwo J, Tsai HS, Krajewski JJ, Chen YK, Cho AY (1999): Demonstration of submicron depletion-mode GaAs MOSFETs with negligible drain current drift and hysteresis. IEEE Electron Device Lett. 20:457–459.
35. Tsai PJ, Chu LK, Chen YW, Chiu YN, Yang HP, Chang P, Kwo J, Chi J, Hong M (2007): Depletion-mode GaAs-based MOSFET with $Ga_2O_3(Gd_2O_3)$ as a gate dielectric. J. Cryst. Growth 301, 302:1013–1016.
36. Zhu F, Koveshnikov S, Ok I, Kim HS, Zhang M, Lee T, Thareja G, Yu L, Lee JC, Tsai W, Tokranov V, Yakimov M, Oktyabrsky S (2007): Depletion-mode GaAs metal-oxide-semiconductor field-effect transistor with amorphous silicon interface passivation layer and thin HfO_2 gate oxide. Appl. Phys. Lett. 91:043507.
37. Currie MT, Leitz CW, Langdo TA, Taraschi G, Fitzgerald EA, Antoniadis DA (2001): Carrier mobilities and process stability of strained Si n- and p-MOSFETs on SiGe virtual substrates. J. Vac. Sci. Technol. B 19:2268–2279.
38. Kai C, Wann HC, Ko PK, Chenming H (1996): The impact of device scaling and power supply change on CMOS gate performance. IEEE Electron Device Lett. 17:202–204.
39. Ritenour A, Khakifirooz A, Antoniadis DA, Lei RZ, Tsai W, Dimoulas A, Mavrou G, Panayiotatos Y (2006): Subnanometer-equivalent-oxide-thickness germanium p-metal-oxide-semiconductor field effect transistors fabricated using molecular-beam-deposited high-k/metal gate stack. Appl. Phy. Lett. 88:132107.
40. Nicholas G, De Jaeger B, Brunco DP, Zimmerman P, Eneman G, Martens K, Meuris M, Heyns MM (2007): High-performance deep submicron Ge pMOSFETs with halo implants. IEEE Trans. Electron Devices 54:2503–2511.
41. Chu LK, Lin TD, Huang ML, Chu RL, Chang CC, Kwo J, Hong M (2009): Ga_2O_3 (Gd_2O_3) on Ge without interfacial layers—energy band parameters and metal oxide semiconductor devices. Appl. Phys. Lett. 94:202108.
42. Irokawa Y, Nakano Y, Ishiko M, Kachi T, Kim J, Ren F, Gila BP, Onstine AH, Abernathy CR, Pearton SJ, Pan CC, Chen GT, Chyi JI (2004): MgO/p-GaN enhancement mode metal-oxide semiconductor field-effect transistors. Appl. Phys. Lett. 84:2919–2921.
43. Saripalli YN, Pei L, Biggerstaff T, Ramachandran S, Duscher GJ, Johnson MAL, Zeng C, Dandu K, Jin Y, Barlage DW (2007): Transmission electron microscopy studies of regrown GaN Ohmic contacts on patterned substrates for metal oxide semiconductor field effect transistor applications. Appl. Phys. Lett. 90:204106.

44. Kwo J, Hong M, Mannaerts JP, Wu YD, Lee QY, Yang B, Gustafsson T (2004): Fundamental study and oxide reliability of the MBE grown $Ga_{2-x}Gd_xO_3$ dielectrics for compound semiconductor MOSFETs. In: D Kumar RKS, Ramesh R, Guha S, Koinuma H (ed), *Mat. Res. Soc. Symp. Proc.*, Warrendale, PA: Materials Research Society, pp. E1.12.1–E1.12.6.
45. Hong M, Lu ZH, Kwo J, Kortan AR, Mannaerts JP, Krajewski JJ, Hsieh KC, Chou LJ, Cheng KY (2000): Initial growth of $Ga_2O_3(Gd_2O_3)$ on GaAs: Key to the attainment of a low interfacial density of states. Appl. Phys. Lett. 76:312–314.
46. Jin H, Oh SK, Kang HJ, Tougaard S (2006): Electronic properties of ultrathin HfO_2, Al_2O_3, and Hf-Al-O dielectric films on Si(100) studied by quantitative analysis of reflection electron energy loss spectra. J. Appl. Phys. 100:083713.
47. Huang ML, Chang YC, Chang YH, Lin TD, Kwo J, Hong M (2009): Energy-band parameters of atomic layer deposited Al_2O_3 and HfO_2 on $In_xGa_{1-x}As$. Appl. Phys. Lett. 94:052106.
48. Kraut EA, Grant RW, Waldrop JR, Kowalczyk SP (1980): Precise determination of the valence-band edge in x-ray photoemission spectra: Application to measurement of semiconductor interface potentials. Phys. Rev. Lett. 44:1620–1623.
49. Lay TS, Hong M, Kwo J, Mannaerts JP, Hung WH, Huang DJ (2001): Energy-band parameters at the GaAs—and GaN-$Ga_2O_3(Gd_2O_3)$ interfaces. Solid-State Electron. 45:1679–1682.
50. Singh J (1993): In: Weiss BL (ed), Properties of lattice-matched and strained Indium Gallium Arsenide. INSPECS, London, p 61.
51. Kwo J, Hong M, Busch B, Muller DA, Chabal YJ, Kortan AR, Mannaerts JP, Yang B, Ye P, Gossmann H, Sergent AM, Ng KK, Bude J, Schulte WH, Garfunkel E, Gustafsson T (2003): Advances in high kappa gate dielectrics for Si and III-V semiconductors. J. Cryst. Growth 251:645–650.
52. Ye PD, Wilk GD, Yang B, Kwo J, Chu SNG, Nakahara S, Gossmann HJL, Mannaerts JP, Hong M, Ng KK, Bude J (2003): GaAs metal-oxide-semiconductor field-effect transistor with nanometer-thin dielectric grown by atomic layer deposition. Appl. Phys. Lett. 83:180–182.
53. Huang YL, Chang P, Yang ZK, Lee YJ, Lee HY, Liu HJ, Kwo J, Mannaerts JP, Hong M (2005): Thermodynamic stability of $Ga_2O_3(Gd_2O_3)$/GaAs interface. Appl. Phys. Lett. 86:191905.
54. Chiu HC, Tung LT, Chang YH, Lee YJ, Chang CC, Kwo J, Hong M (2008): Achieving a low interfacial density of states in atomic layer deposited Al_2O_3 on $In_{0.53}Ga_{0.47}As$. Appl. Phys. Lett. 93:202903.
55. Martens K, Chi On C, Brammertz G, De Jaeger B, Kuzum D, Meuris M, Heyns M, Krishnamohan T, Saraswat K, Maes HE, Groeseneken G (2008): On the correct extraction of interface trap density of MOS devices with high-mobility semiconductor substrates. IEEE Trans. Electron Devices 55:547–556.
56. Martens K, Kaczer B, Maes H, Groeseneken G (2008): Electrical interface characterization of non-si based MOSFETs: Challenges and solutions, *International SiGe Technology and Device Meeting*, pp 82–83.
57. Groeseneken G, Maes HE, Beltran N, De Keersmaecker RF (1984): A reliable approach to charge-pumping measurements in MOS transistors. IEEE Trans. Electron Devices 31: 42–53.
58. Bauza D, Maneglia Y (1997): In-depth exploration of Si-SiO_2 interface traps in MOS transistors using the charge pumping technique. IEEE Trans. Electron Devices 44:2262–2266.
59. Jakschik S, Avellan A, Schroeder U, Bartha JW (2004): Influence of Al_2O_3 dielectrics on the trap-depth profiles in MOS devices investigated by the charge-pumping method. IEEE Trans. Electron Devices 51:2252–2255.
60. Chiang TH, Lee WC, Lin TD, Lin D, Shiu KH, Kwo J, Wang WE, Tsai W, Hong M (2008): Approaching fermi level unpinning in oxide-$In_{0.2}Ga_{0.8}As$, *IEEE International Electron Devices Meeting*, San Francisco, CA, pp 375–378.
61. Brammertz G, Martens K, Sioncke S, Delabie A, Caymax M, Meuris M, Heyns M (2007): Characteristic trapping lifetime and capacitance-voltage measurements of GaAs metal-oxide-semiconductor structures. Appl. Phys. Lett. 91:133510.
62. Hong M, Kwo J, Lin TD, Huang ML (2009): InGaAs metal-oxide-semiconductor devices with Ga_2O_3 (Gd_2O_3) high k dielectrics for science and technology beyond Si CMOS. MRS Bull. 34:514–521.

Chapter 10
Sub-100 nm Gate III-V MOSFET for Digital Applications

K. Y. (Norman) Cheng, Milton Feng, Donald Cheng and Chichih Liao

Abstract The performance of nano-CMOS digital ICs is characterized in terms of key figure of merits, which provides a benchmark for III-V compound semiconductor MOSFETs. The selection of III-V channel material is discussed based on the transport properties, i.e., the intrinsic electron mobility and hole mobility, of different III-V alloys. To improve the hole mobility for III-V p-MOSFETs, the effects of strain-induced hole mobility enhancement are reviewed. Critical process issues for self-aligned III-V MOSFET are thermal stability of the oxide-semiconductor interface and source/drain doping limitations. The advantages of self-aligned GaAs enhancement-mode MOSFETs using regrown source and drain regions are demonstrated. Finally, the state-of-the-art device performance of sub-100 nm gate III-V FETs is compared with Si MOSFETs.

10.1 Introduction

Silicon based CMOS technology is the basis for the majority of digital integrated circuits (DICs) in today's semiconductor industry. As CMOS continues to scale in order to sustain performance increases every year, energy efficiency becomes increasingly important. Scaling power supply voltage and reducing leakage currents are the common methods to curtail the rising power consumption in ICs. In addition, this performance increase also relies in part on channel carrier velocity improvement through innovative channel designs such as incorporating strains into the channel. However, in view of the material physical properties, the performance increase of Si-based ICs is unlikely to continue in the next decade. Therefore, non-Si electronic materials have been explored for future IC applications. Among several emerging nanoscale devices, III-V compound semiconductor based FETs

---

K. Y. (Norman) Cheng (✉)
Department of Electrical and Computer Engineering, University of Illinois, Urbana-Champaign, Urbana, IL 61801, USA
e-mail: kycheng@illinois.edu

S. Oktyabrsky, P. D. Ye (eds.), *Fundamentals of III-V Semiconductor MOSFETs*,
DOI 10.1007/978-1-4419-1547-4_10,

are the most attractive devices due to their high electron mobility, low supply voltage, and demonstrated manufacturability. In fact, III-V material based high performance transport devices such as high electron mobility transistors (HEMTs) and heterojunction bipolar transistors (HBTs) have been widely used in communication products for many years. But, due to the lack of robust high-quality native oxides, III-V based CMOS was demonstrated only a decade ago [1]. Since then, magnificent progress has been made in the development of III-V MOSFETs for future high-speed and low-power applications. To gauge these advancements toward future high-speed and low-power digital applications, it is necessary to benchmark them against the state-of-the-art Si MOSFET data using a set of adequate device metrics, as introduced in Sect. 10.2. In Sect. 10.3, the selection of high carrier mobility III-V materials is reviewed. The enhancement of hole mobility in p-type III-V materials by introducing strains is discussed. Section 10.4 contains a discussion of alternatives and challenges of self-aligned sub-100 nm III-V MOSFET fabrication techniques. Finally, in Sect. 10.5, state-of-the-art sub-100 nm III-V FETs are benchmarked against Si CMOS and followed by a brief conclusion and future outlooks.

10.2 MOSFET Figures of Merit for Digital Applications

For low-power and high-performance CMOS digital ICs, their performances can be characterized with several figures of merit. These metrics that successfully characterize current digital CMOS IC technology can provide critical parameters related to the feasibility of emerging III-V-based devices and are discussed in more detail below.

10.2.1 Intrinsic Delay Time

The MOSFET intrinsic switching delay time, $\tau = C_g V_{DD}/I$, is one of the key figures of merit that determines the maximum switching speed of the MOSFET [2]. This metric can be interpreted with the transistor circuit shown in Fig. 10.1 where the

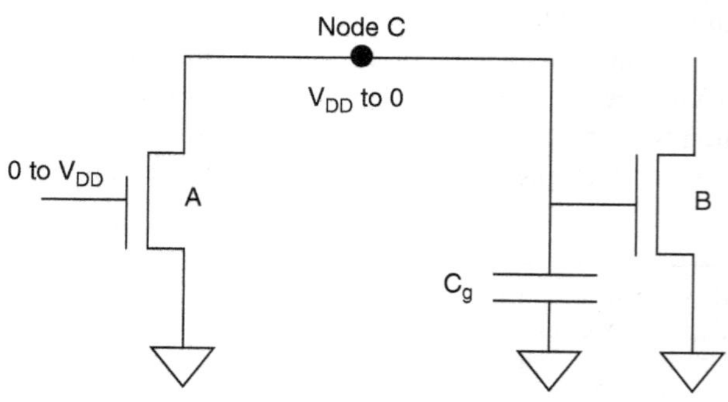

Fig. 10.1 *CV/I* of MOSFETs is used to evaluate their digital application performance. The circuit shows that only V_{DD}, L_g, and drain current are needed to estimate *CV/I*

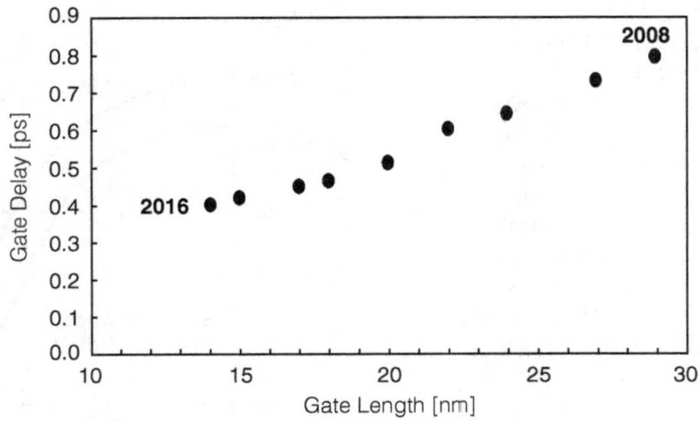

Fig. 10.2 ITRS projected gate delay trend. The gate node will be scaled down to 15 nm with a 0.4 ps gate delay in 2016

transistors A and B are identical. The intrinsic gate delay of the device is defined as the time required for the voltage at node C to swing between zero and the supply voltage V_{DD}, and is characterized with three device parameters C_g, V_{DD}, and I. The gate capacitance (C_g) of transistor B is defined as gate area times the gate-oxide capacitance per unit area. Physical gate length is used to calculate C_g, which includes both channel capacitance and overlap capacitance. The current I in the equation is the saturated drive current ($I_{D,sat}$) of the transistor. The measured intrinsic MOSFET delay τ determines the maximum intrinsic switching frequency $1/\tau$. The improvement of $1/\tau$ in each subsequent technology node is a critical advance toward increasing the chip's clock speed. Each new generation of the nanometer scale MOSFET reduces the gate length, increases $I_{D,sat}$, decreases C_g and reduces V_{DD} to provide 30–40% improvement in gate delay and achieve a better energy-delay product. The trend of the intrinsic gate delay for each CMOS IC technology node from 2008 to 2016 is displayed in Fig. 10.2, as projected by the International Technology Roadmap for Semiconductors (ITRS).

10.2.2 Subthreshold Swing (SS)

The source/drain subthreshold leakage current is the second key parameter that determines whether the chip's static power dissipation is within tolerable limits. In an ideal case, the current decreases to zero as soon as the gate bias is below threshold. In reality, drain conduction below threshold does exist and is known as subthreshold conduction. This subthreshold leakage current is defined as the drain current per micrometer of device width with the drain bias held at V_{DD} and gate, source, and substrate grounded (0 V). It dominates the total off-state leakage current, I_{OFF}, which is the sum of the NMOS subthreshold, gate, and junction leakage currents. This subthreshold conduction behavior can be described best by the subthreshold swing (SS), which measures the dependence of the drain current I_{DS} on gate bias voltage V_G. The subthreshold characteristics shown in Fig. 10.3 include the important relation shown below [3]:

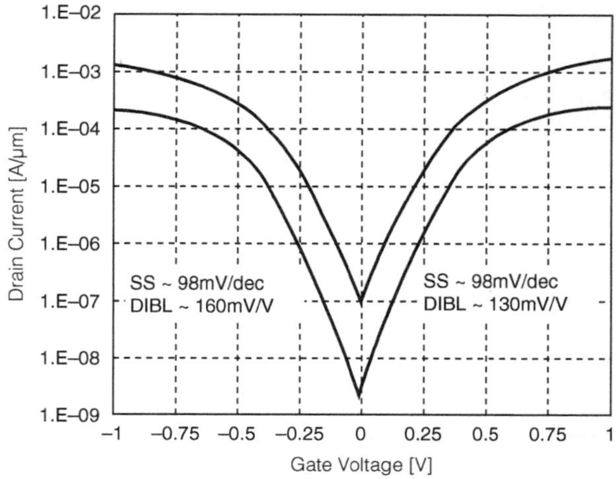

Fig. 10.3 Subthreshold curves of 32 nm-node MOSFETs [3]. The *upper* and *lower* curves have $V_{DS} = 1.5$ V and 50 mV, respectively. The n-channel device has $DIBL = 130$ mV/V and $SS = 98$ mV/decade. The p-channel has $DIBL = 160$ mV/V and $SS = 98$ mV/decade

$$SS = \frac{dV_G}{d\left(\log_{10} I_{DS}\right)} = 2.3\frac{kT}{q}\left(1 + \frac{C_{dm}}{C_{ox}}\right) \tag{10.1}$$

where *k, T,* and *q* are the Boltzmann constant, temperature (K), and electronic charge, respectively. C_{dm} is the channel depletion layer capacitance and C_{ox} is the gate-oxide capacitance. A small subthreshold swing (corresponding to lower mV/decade) is more desirable for achieving fast switching time between the ON and OFF states. Because C_{dm} includes the interface state capacitance, *SS* is large (lower switching speed) if the semiconductor-oxide interface has many fast interface states. Therefore, *SS* reflects the dielectric/semiconductor interface quality of a MOSFET.

10.2.3 Drain Induced Barrier Lowering (DIBL)

DIBL is another critical parameter for MOSFET characterization. In short channel devices, the source-drain potential has a strong effect on the band bending over a significant portion of the device. Hence, the threshold voltages and the subthreshold currents of short channel devices vary with the drain bias. This effect is referred to as *DIBL* and illustrated in Fig. 10.4. The *DIBL* adds to the barrier lowering due to gate bias and enhances the electron injection from the source into the channel, leading to a noticeable increase of the channel current. Figure 10.3 illustrates the drain currents under $V_{DS} = 50$ mV and 1.5 V, respectively. When a high drain voltage is applied to a short-channel device, it lowers the barrier height, resulting in decreased threshold voltage V_{TH}. The threshold voltage roll-off due to *DIBL* consequently creates a trade-off between device performance and power dissipation. Lower threshold voltage can result in higher drive current because the drain current is proportional to the difference between gate voltage and threshold voltage; hence, fast switching

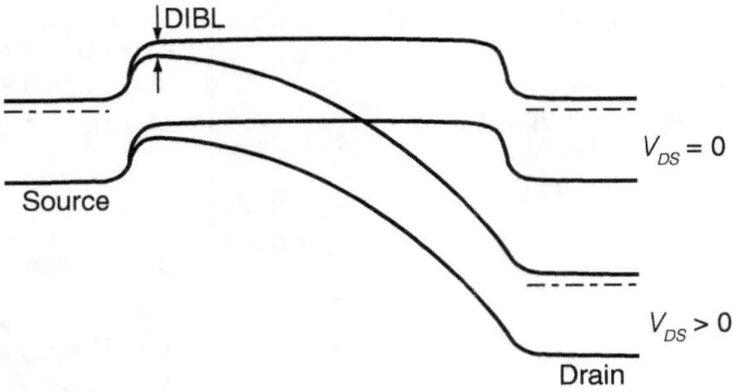

Fig. 10.4 Energy band diagrams along the channel of a MOSFET illustrating the *DIBL* and *SS*

speed can be realized. Nonetheless, the subthreshold (off-state) leakage current sets a lower limit for threshold voltage. Actually, the threshold voltage limitation sets a boundary on the supply voltage V_{DD}. The speed/power dilemma is thus evident: Reducing V_{DD} is essential to lower the device's active power dissipation and improves the switching speed. The threshold voltage also needs to be scaled down simultaneously to achieve high enough current drive. However, the effect of speed improvement is counteracted by the current drive reduction of the transistor.

The adverse effects caused by higher off-state leakage current stemming from the reduced threshold voltage in short channel devices are collectively called the short channel effect. To ameliorate the short channel effect, the gate oxide thickness can be reduced, but this will increase the transverse electric field and result in higher gate tunneling leakage current. When scaling the gate dielectric of Si CMOS ICs down to the sub-nanometer scale, the gate leakage becomes a critical issue. This gate leakage degrades the high input impedance of the MOS transistor and lowers the circuit performance severely. Figure 10.5 shows the improvement of gate leakage by using a high-k gate dielectric plus metal gate stack in the 45 nm production mode [4]. It can be seen that for the same equivalent oxide thickness (EOT), the high-k dielectric exhibits more than four times of gate leakage improvement over SiO_2. The EOT is defined as the thickness that SiO_2 would have for a given capacitance value.

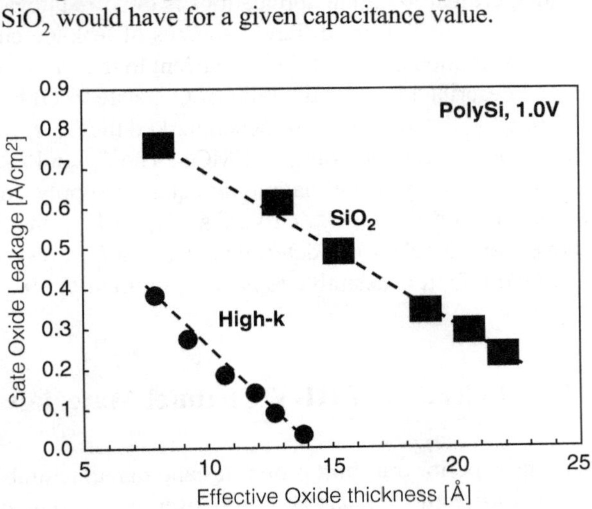

Fig. 10.5 Comparison of gate leakage between SiO_2 and high-k gate material [4]. The high-k dielectric has more than four times of gate leakage improvement over SiO_2

Fig. 10.6 Significant subthreshold leakage current is observed when V_{DG} is high due to gate induced drain leakage current (GIDL) [5]

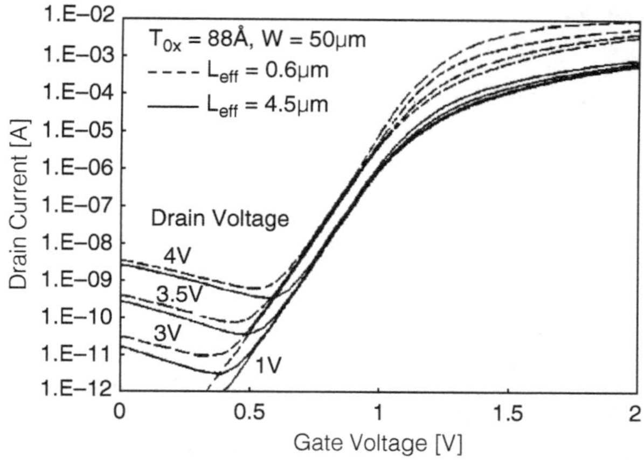

Another major leakage source in nano-scale CMOS devices is the gate-induced drain leakage (GIDL). Figure 10.6 shows that the leakage current near zero gate bias increases with increasing drain voltage [5]. We can explain this phenomenon as being due to the band-to-band tunneling (BTBT) current in the gate-to-drain overlap region: As the drain bias is raised for a fixed gate bias, a narrow depletion region forms in the heavily doped drain region overlapping with the gate. If the induced band bending is more than the band gap across a depletion region, electron-hole pairs will be created. Electrons can then tunnel into the drain, contributing to the drain leakage current, GIDL.

10.2.4 I_{ON}/I_{OFF} Ratio

Finally, critical to digital applications is the off-state power dissipation which contains components from different sources of leakage currents as mentioned above. The contribution of the OFF state current to the speed/power tradeoff of MOSFETs can be described best by the ratio of ON state to OFF state current (I_{ON}/I_{OFF} ratio) at a fixed V_{DD}. Chau et al. [6] benchmarked the I_{ON}/I_{OFF} ratio for a number of nanoelectronic devices including Si CMOS. The I_{ON} and I_{OFF} are determined by setting a gate bias swing of the magnitude equal to supply voltage V_{DD} around threshold voltage V_{TH} on the I_{DS}–V_G curves. At $V_{DS} = V_{DD}$, the position of V_G swing at 2/3 above and 1/3 below V_{TH} determines I_{ON} and I_{OFF}, respectively. For sub-100 nm III-V MODFETs, it is desirable to have a minimum $I_{ON}/I_{OFF} > 1000$ at a $V_{DD} = 0.5$ V.

10.3 Selection of III-V Channel Materials

With its superior transport properties and manufacturability, III-V compound semiconductor-based FET has been proposed as a promising device option for future high-speed, low-power digital logic applications in addition to its demonstrated

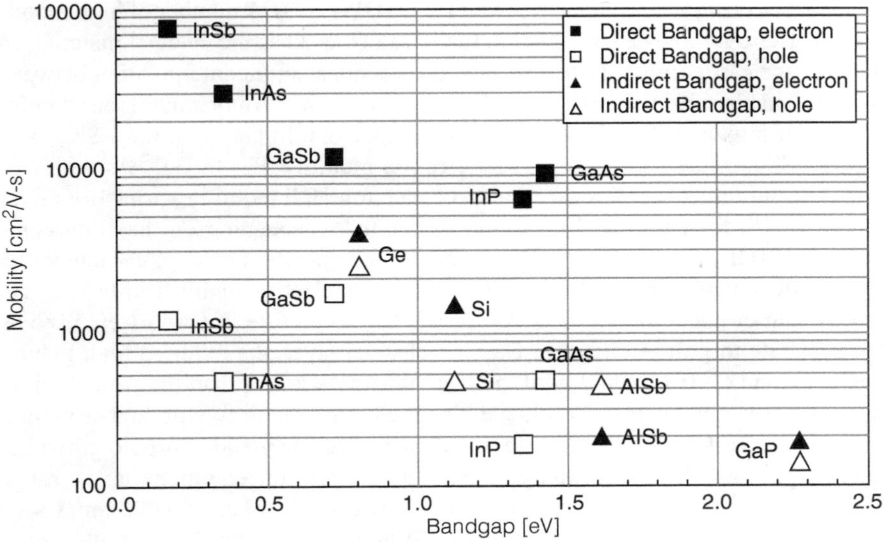

Fig. 10.7 Carrier mobility as a function of minimum energy band gap for elemental and III-V compound semiconductors

applications in communication and optoelectronic products. High electron mobility and conductivity in III-Vs give rise to high transistor drive current and low gate delay, which are very important for high-speed logic applications. Since the underlying physics of carrier mobility is based on scattering, one would expect the mobility to be a function of energy band structure and inversely proportional to the effective mass. By considering the effect of band interaction, n-type semiconductors with a small, direct energy gap show proportionally high electron mobility. This trend of increasing electron mobility with decreasing energy gap in direct band gap semiconductors is clearly seen in Fig. 10.7, where elemental and III-V semiconductors are compared [7]. Apparently, InSb and InAs possess the highest electron mobility among all semiconductors. Nevertheless, the small energy gap semiconductors are more susceptible to thermal generation of excess carriers, creating leakage current. Specifically, it increases the subthreshold leakage current and band-to-band tunneling current in a MOSFET. Therefore, the selection of III-V compounds for the channel material needs to take the size of energy band gap into consideration to avoid a high leakage current or a low I_{ON}/I_{OFF} ratio.

By mixing two binaries with a common element, one can form a ternary compound that can also be used as the channel material. Due to the added degree of freedom, ternary alloys, with a wide selection of composition or lattice constant, can be used to form heterojunction devices. This allows the design of III-V FETs using either a surface channel structure (similar to Si MOSFET) or a buried channel structure. The buried channel structure further relaxes the lattice-match constraints in selecting channel material since a very thin channel layer can be used in a pseudomorphic structure. The very thin conduction channel forms a quantum well and can be doped either locally or remotely as in HEMTs. Thus the buried channel FETs are collectively termed as quantum well FETs (QWFETs). A number of high mobility ternaries over a wide composition range can be used as the channel material. In general, the carrier mobility of a ternary alloy is not a simple interpolation between the two end binaries. It is complicated by alloy scattering. Among numerous combinations of high mobility III-V binary semiconductors, $In_xGa_{1-x}As$, $InAs_xSb_{1-x}$ and $In_xGa_{1-x}Sb$ are those with the highest electron mobility. For $In_xGa_{1-x}As$ with carrier concentrations near $5 \times 10^{16}\,cm^{-3}$, the electron Hall mobility varies from over $6{,}000\,cm^2/V\,s$ for GaAs to $20{,}000\,cm^2/V\,s$ for InAs depending on the alloy composition [8]. It shows a minimum at an indium composition of $x \sim 20\%$ due to the strong effect of alloying scattering (Fig. 10.8a). The electron mobility of $In_xGa_{1-x}As$ is less than or equal to that of GaAs for $x \leq 0.4$. Therefore, if GaAs were used as the substrate to grow strained $In_xGa_{1-x}As$ channel layer, the required high indium composition ($x \geq 0.4$) would limit the layer thickness to less than the critical thickness of a few nanometers. To achieve higher electron mobility with higher indium composition in $In_xGa_{1-x}As$, the use of InP substrates or pseudomorphic growth is necessary. For $InAs_xSb_{1-x}$ ternary alloys with carrier concentrations in the range 10^{16}–$10^{17}\,cm^{-3}$, the electron mobility varies, between 20,000 and $60{,}000\,cm^2/V\,s$, as a function of composition [8] (Fig. 10.8b). It is found that the effect of alloy scattering is not large in the whole range of composition. The electron mobility as a function of composition has a convex shape with a maximum near $x \sim 0.2$. The rise

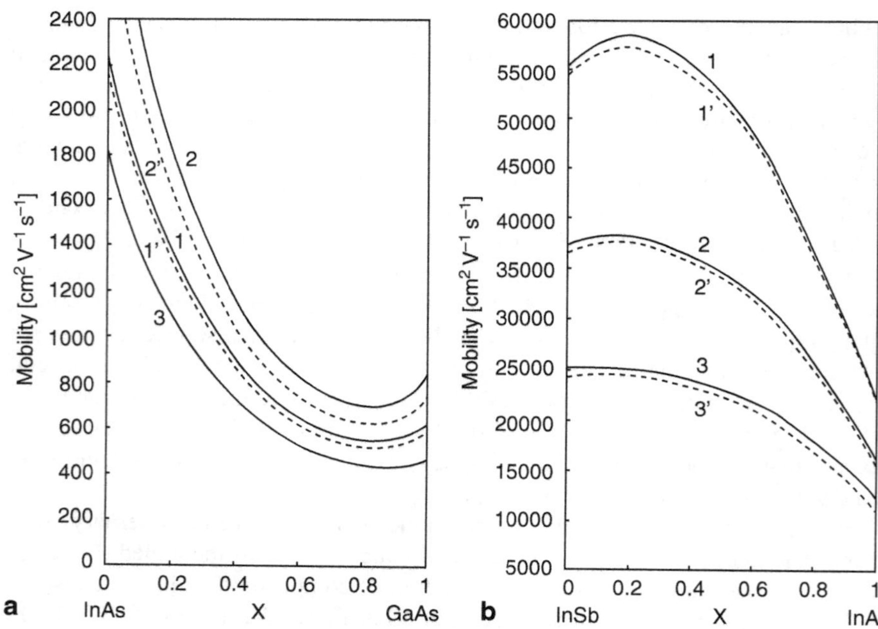

Fig. 10.8 Electron hall mobility in **a** $Ga_xIn_{1-x}As$ and **b** $InAs_xSb_{1-x}$ as a function of composition x [8]. For $Ga_xIn_{1-x}As$, the doping concentrations in curve 1, 2, and 3 are 4×10^{16}, 3×10^{15}, and $2.3 \times 10^{17}\,cm^{-3}$, respectively. For $InAs_xSb_{1-x}$, the doping concentrations in curve 1, 2, and 3 are 5×10^{16}, 7.5×10^{16}, and $2.5 \times 10^{17}\,cm^{-3}$, respectively. The *dashed lines* represent electron drift mobility

in mobility is due to the fact that the effective mass decreases at first with increase in x. The electron mobility of the ternary compound $In_xGa_{1-x}Sb$ doped with 10^{15}–$10^{16}\,cm^{-3}$ donors shows no sign of alloy scattering and varies near linearly from $8{,}000\,cm^2/V\,s$ for GaSb to $70{,}000\,cm^2/V\,s$ for InSb with the composition [9, 10]. Compared with $InAs_xSb_{1-x}$, which covers a similar lattice constant span, $In_xGa_{1-x}Sb$ is more advantageous due to its larger energy gap and higher hole mobility.

For p-type semiconductors, the hole mobility is derived by taking into account the degenerated light and heavy hole bands at the zone center and characterized by two effective masses. This unique feature leads to the possibility of interband scattering between light-hole band and heavy-hole band. These scattering events are different from the intervalley scattering seen in n-type semiconductors with indirect energy gap. As a result, shown in Fig. 10.7, the trend of measured hole mobility as a function of the minimum energy gap is less clear for these common semiconductors. In general, due to the greater effective mass of holes as compared to electrons, the hole mobility is, for example, in the case of Sb compounds, many orders of magnitude smaller than the electron mobility. Although many III-V compound semiconductors have much higher electron mobility than silicon, they are not superior in hole mobility. In fact, Ge has the highest hole mobility among all semiconductors discussed here. The huge difference between electron and hole mobility would make the design of CMOS a great challenge. Therefore, how to enhance the hole mobility in III-V semiconductors in order to balance p- and n-channel FETs becomes an important topic to explore.

In one approach, it has been proved that the incorporation of strain in Si or SiGe MOSFET is effective in improving the hole mobility [11]. The enhancement factor can be as large as 1.5× or even higher over unstrained silicon, depending on the strain induced by the composition change around the channel. The main effect of strain on energy band structure is to increase the warping of valence band curves, especially heavy-hole, and split open heavy-hole and light-hole bands, as shown in Fig. 10.9 [12]. In either biaxial or uniaxial strained case, the effective mass around the energy band maximum decreases because the top-most valence band curve becomes light-hole-like. The curvature of the top-most valence band curve increases and effective hole mass decreases; therefore, hole mobility increases. Besides the mobility enhancement by effective mass reduction, under high uniaxial or biaxial stress, the reduction of interband scattering could also contribute to the enhancement of hole mobility [13]. The same phenomenon of warping and splitting of valence band surface is also expected in strained III-V semiconductors [14]. In the biaxial strained case, the curvature of the topmost valence band would be larger for compressive strain than in tensile strain, which means compressive strain can enhance in-plane hole mobility more effectively. For the uniaxial strained case, the warping and splitting of the valence bands depend on material parameters. Hole mobility enhancement is more prominent in semiconductors with larger difference between heavy-hole and light-hole masses such as GaAs and Ge. They show a higher mobility enhancement factor than Si under the same uniaxial stress, especially when the stress is larger than 2 GPa [13]. Therefore, the hole mobility of compound semiconductors might be enhanced with strain, which can be introduced by

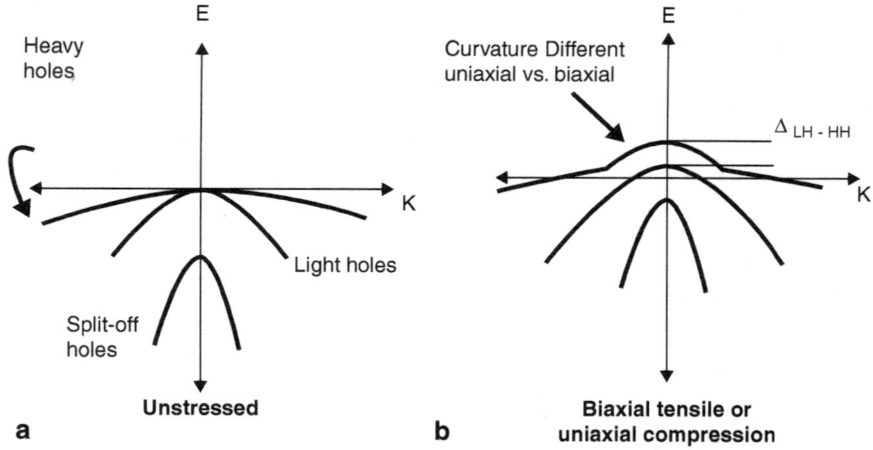

Fig. 10.9 Simplified valence band structure for longitudinal in-plane direction **a** unstrained and **b** strained-Si [12]

simply inserting pseudomorphic layer designs into the device structure. Recently, researchers at Intel have demonstrated both n- and p-channel sub-100 nm QWFETs with a modulation-doped biaxial-strained pseudomorphic InSb channel grown on $Al_xIn_{1-x}Sb$ buffer layers [15, 16]. The channel mobility of 30,000 and 1230 cm²/V s, and the maximum cut-off frequency, f_T, of 305 and 140 GHz have been demonstrated for n- and p-channel FETs, respectively. The hole mobility of the p-channel InSb QWFET shows a ~3× improvement over an unstrained but otherwise identical QWFET. In addition, by incorporating lattice-mismatched regrown source and drain contact regions in sub-100 nm devices, it is possible to further enhance in-plane hole mobility by uniaxial strain along the channel.

10.4 Self-Aligned III-V MOSFET Structures

In general, III-V compound semiconductors have much higher electron mobility and lower conduction band effective mass than Si. These attributes lead to a higher intrinsic speed, i.e., lower gate delay, and lower energy-delay product, than Si MOSFETs at a given transistor gate length L_g. However, it is observed that the intrinsic gate delay tends to "saturate" with reducing transistor gate length to below ~200 nm (Fig. 10.10) [17]. This saturation is mainly caused by parasitic source and drain resistances R_s and R_d, and output resistance R_{ds} on unity-gain cutoff frequency f_T. As shown in Fig. 10.11, most current III-V FET technologies use a re-aligned gate design where the drain-to-source separation (L_{DS}) is much larger (≥10×) than the physical gate length L_g. The parasitic resistances degrade the cutoff frequency dramatically despite the intrinsic high electron mobility. For example, the cutoff frequency of a strained InAs HEMT with a 30 nm gate length (with $L_{DS} \sim 5\ L_g$)

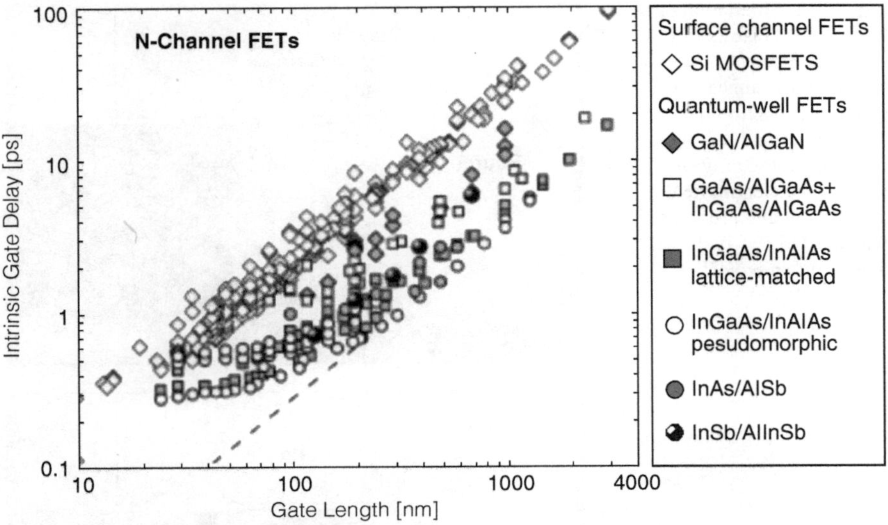

Fig. 10.10 Intrinsic gate delay of various channel materials saturates with scaled gate length. The *dashed line* indicates the projected *CV/I* in the absence of source and drain parasitic resistances [17]

improves only 1.5× over its self-aligned Si NMOS counterpart, even though the electron mobility in bulk InAs is at least 100× higher than that in Si. Therefore, a self-aligned III-V channel MOSFET process is the viable approach to tackle this issue. Reducing the gate-to-source and gate-to-drain distances using a self-aligned technology should reduce R_s and R_d, while reducing the gate-to-channel separation should improve short-channel effects, hence increasing R_{ds}. It is then expected that the performance improvement trend will continue with L_g scaling as shown by the dashed line in Fig. 10.10.

For scaling Si MOSFETs down to the sub-nanometer scale, ion-implantation and rapid thermal annealing (RTA) are required to generate a diffusionless source-

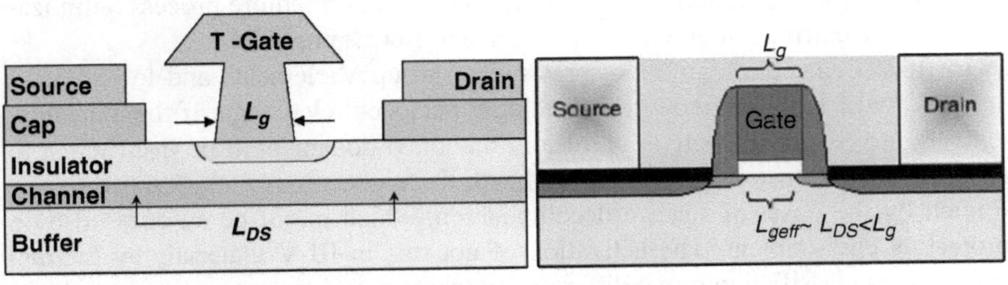

Fig. 10.11 Structure comparison of realigned versus self-aligned gate III-V MOSFETs. In the realigned gate structure (*left*), L_g is much shorter than the source-to-drain distance, L_{DS}, leading to large source and drain parasitic resistances. For the self-aligned gate structure (*right*), the effective gate length, L_{geff}, is equal to or less than the physical gate length, L_g

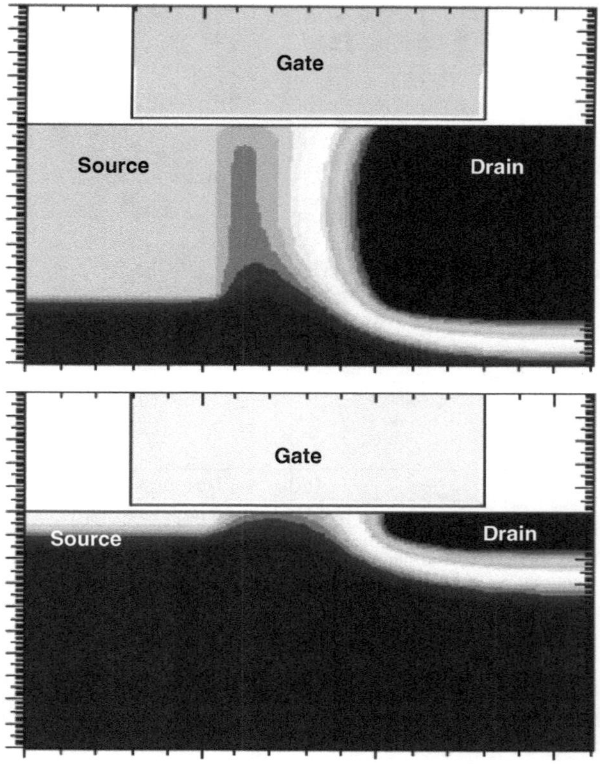

Fig. 10.12 Simulation of electric field distribution along the channel shows that improper annealing makes the junction too deep for vertical scaling to match lateral scaling (*top panel*). The electric field distribution of a properly scaled MOSFET is shown in the *bottom panel* [18]

drain profile (toward vertical scaling) so that a source-drain extension (SDE) can be formed to suppress the short channel effect (Fig. 10.12). High dopant activation is necessary to enhance the on-state current. These demands are fulfilled through a high-temperature and short-duration annealing process using a millisecond laser to raise the surface temperature close to the Si melting point [18]. The success of this process relies on the unique material properties of Si such as its low vapor pressure, high impurity solid solubility, and robust dielectric/semiconductor interface. The source/drain (S/D) junction profile and dopant activation require process optimization to make extrinsic source/drain resistance as low as possible.

In III-V FETs, the high vapor pressure of group V elements and low thermal stability make high temperature annealing a persistent challenge. If the annealing temperature is low, the activation of ion implanted dopants will be insufficient to produce S/D extensions. On the other hand, the highest annealing temperature is limited by the onset of surface decomposition, which might be suppressed by a protective environment. The activation of dopants in III-V materials by thermal annealing is classified into two different approaches: cap-less annealing in a chamber with group V overpressure and annealing with a protective dielectric cap. Both ways have been demonstrated and utilized in the production of III-V MESFET ICs. Nevertheless, due to the solid solubility limitation, the maximum activated impurity concentration, e.g., in GaAs, can only reach mid 10^{18} cm^{-3}, which makes the formation of deep source-drain extension regions by direct ion implantation questionable. Furthermore, due to the multiple-elements nature of compound semiconductors, the thermal instability at the gate-oxide and III-V semiconductor interface poses another critical issue. Therefore, scaling self-aligned gate III-V MOSFETs requires more research in addition to the ion implantation approach.

Regrowth of source/drain regions in III-V MOSFETs, similar to the regrowth of highly doped SiGe source/drain in Si CMOS, is an alternative approach to address the thermal stability and doping limitation issues encountered by the ion implantation process. Using the gate metal as a mask on an n-channel III-V MOSFET to remove the gate oxide and the p-type material in source/drain regions, a self-aligned highly doped n-type material can be regrown in the source and drain regions by a low temperature epitaxy process using, e.g., molecular beam epitaxy (MBE) [19, 20]. The highest temperature the sample encountered over the whole MBE process can be confined to the in situ surface cleaning temperature, which is 580 °C for GaAs, much lower than the post-ion-implantation annealing temperature. Therefore, the dielectric/semiconductor interface integrity can be easily maintained. Further, the regrowth approach has another major advantage over the ion-implantation method in that it can form lateral heterojunctions between the source/drain and the channel regions. The consequence of this ability is tremendous. First, hot carriers from the source region with a larger energy band gap than that of the channel material will enhance the carrier initial injection velocity and, thus, the speed of the device. This is especially advantageous in sub-100 nm devices since the carrier transient time is too short to gain a great amount of energy from the field. Second, materials with a high impurity solid solubility can be incorporated as the source/drain material to increase the doping density and lower the contact resistances R_s and R_d. Both the channel current and cutoff frequency of the device can be improved. Finally, when a lattice-mismatched material is used as the regrown source/drain, either compressive or tensile stress can be introduced to enhance the carrier mobility and, thus, the current conduction capability as well as cutoff frequency. Of course, misfit dislocations could be incorporated into the device structure when using lattice-mismatched heterojunctions. However, since the regrown materials are confined to the neutral part of the junction, the misfit dislocations will not add device leakage current.

In the source-drain regrowth approach, the main challenge is to generate growth selectivity between metal and semiconductor surfaces, addressed by optimized surface cleaning prior to regrowth and adjustment of growth parameters such as substrate temperature, V/III flux/flow ratio, and growth rate. Different epitaxial growth methods, including MBE-related and metal-organic chemical vapor deposition (MOCVD) techniques, should be investigated to achieve the required growth selectivity.

Currently, the development of self-aligned III-V MOSFETs using the regrowth approach is still in its infancy. For example, only recently, initial device results on regrown GaAs enhancement-mode MOSFETs have been reported by researchers at the University of Illinois at Urbana-Champaign collaborating with the National Tsing-Hua University in Taiwan [19, 20]. Figures 10.13a–d show the surface pro-

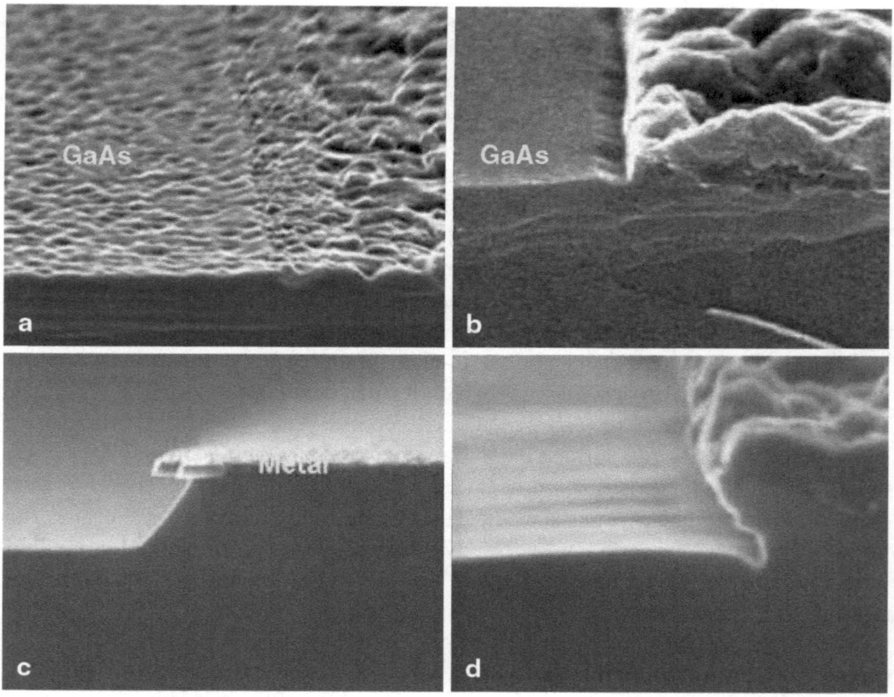

Fig. 10.13 **a** Regrown source/drain GaAs MOSFET sample without proper surface cleaning before epitaxy. **b** Smooth regrown GaAs surface. **c** Patterned GaAs with undercut (before regrowth). **d** Side profile of the regrown sample

files of regrown GaAs on the metal gate surface using solid-source MBE. A smooth surface profile is achieved after the regrowth by adjusting the growth temperature and V/III flux ratio, and using a migration enhancement growth mode. However, the growth selectivity is not ideal and calls for further investigations. Nevertheless, encouraging initial device results have been observed. Figure 10.14 shows I-V curves for the III-V n-type enhancement-mode MOSFET with Ga_2O_3 gate oxide layer and a $4 \times 64\,\mu m^2$ gate size. The gate voltage is swept from 0 to 5 V in 0.2 V increments, and the drain voltage is swept to 8 V. The maximum source-drain current density is 74 mA/mm. The shift in threshold voltage is attributed to the fixed charge in the oxide layer. Due to the shift of threshold voltage, the transconductance is enhanced from 40 to 80 mS/mm. This transconductance is at least an order of magnitude higher than all reported values of E-mode GaAs MOSFETs, indicating the advantage of using a low regrowth temperature to better preserve the dielectric/semiconductor interface. However, in view of the turn-on voltage shifting with different gate bias, which is evidence of gate-drain/gate-source leakage, the regrowth process needs to be optimized, or an alternative epitaxy technique should be selected, to achieve selective growth between gate and source/drain trenches. Recently, Barlage et al. showed that MOCVD regrown source/drain in GaN material system is applicable [21]. However, the growing temperature (>800 °C) is too high for III-V MOSFETs.

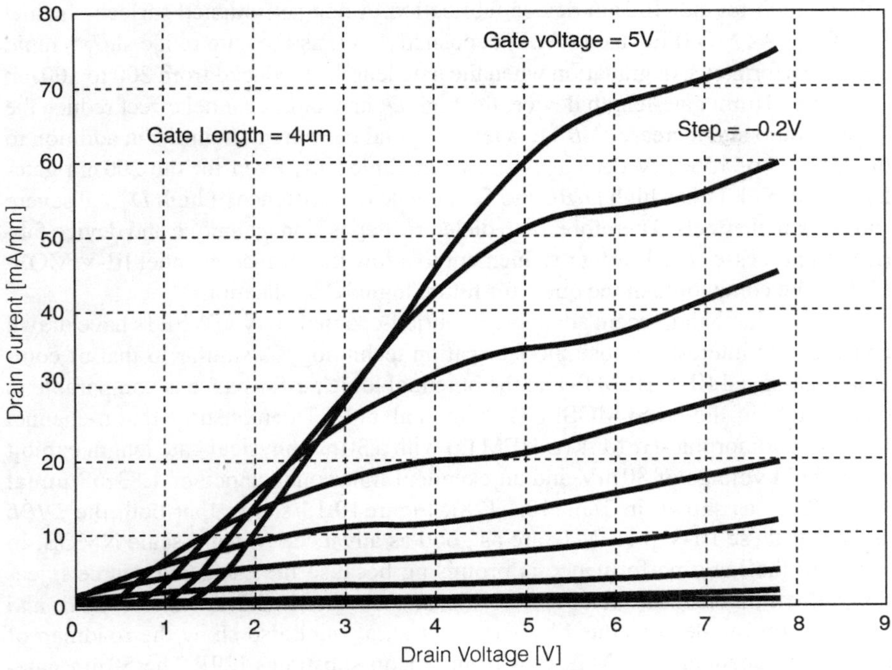

Fig. 10.14 The I_{DS}–V_{DS} characteristics of a 4 × 64 µm self-aligned regrowth S/D n-channel enhancement-mode GaAs MOSFET measured at different gate voltages between 0 and 5 V with steps of 0.2 V [20]

10.5 Benchmark of III-V FET with Si CMOS

A high-quality gate dielectric stack with very low interface trap density (D_{it}) is essential for the realization of high-performance III-V MOSFETs for advanced CMOS logic applications. Although a great deal of progress has been made, the dielectric/III-V material interface properties are still not fully understood at present and are under intense investigations. It is also recognized that the selection of channel material plays an important role, as important as the dielectric/semiconductor interface quality, for achieving high performance III-V MOSFETs [22]. For example, the performance of surface-channel inversion-mode $In_xGa_{1-x}As$ MOSFETs improves with the increase of indium composition in channel material. Currently, low interface trap density interfaces have been achieved using either an in situ ultra-high vacuum deposition approach where the electron-gun evaporated $Al_2O_3/Ga_2O_3(Gd_2O_3)$ was used as the gate dielectric [23], or using an ex situ atomic layer deposition (ALD) technique to deposit Al_2O_3 on an $In_xGa_{1-x}As$ surface [24]. The surface inversion-channel enhancement-mode $In_xGa_{1-x}As$ MOSFETs ($L_g \geq 0.4$ µm) with gate dielectric formed by either method showed a large drain current over 1 A/mm and a high intrinsic transconductance of over 550 mS/mm, indicating the potential of this technology. However, the interface trap density achieved in these materials (low 10^{12} cm^{-2}) is still too high for sub-100 nm device fabrication. A non-self-aligned surface-channel $In_{0.75}Ga_{0.25}As$ MOSFET using ALD deposited Al_2O_3 as the gate oxide shows rapid device performance degradation when the gate length is reduced from 200 to 100 nm [25]. In a 110 nm gate-length device, the high D_{it} and short channel effect reduce the I_{ON}/I_{OFF} ratio and increase *DIBL* to a level beyond measurement limits in addition to the high *SS* of 1500 mV/decade as shown in Table 10.1. Even for the 200 nm gate-length MOSFETs, the high *DIBL* and *SS* indicate the existence of high D_{it} and severe short channel effects. Therefore, both dielectric deposition technique and device fabrication processes need further refinement to allow the surface-channel III-V MOSFETs to be compatible in the quest for future logic IC applications.

On the other hand, recent advances in buried-channel III-V QWFETs have drawn considerable interest because the fabrication technology is similar to that of common high-speed III-V HEMTs and the potential logic performance is comparable to that of state-of-the-art Si MOSFETs. Kim et al. at MIT demonstrate that n-channel InAs pseudomorphic-HEMTs (p-HEMTs) with a 30 nm physical gate length exhibit a threshold voltage of 80 mV and an excellent transconductance of 1.83 mS/μm at $V_{DS} = 0.5$ V as shown in Table 10.1 [28]. Figure 10.15 shows that both the *DIBL* and *SS* of these III-V p-HEMTs are as good as advanced Si nano-scale NMOS. In addition, the logic performance is promising because the extracted source injection velocity is about a factor of two higher than Si-NMOS. Advanced work and excellent results demonstrated by Hudait et al. at Intel also show the roadmap of $In_xGa_{1-x}As$ channel p-HEMTs grown on silicon substrates [29]. The 80 nm gate-length $In_{0.7}Ga_{0.3}As$ p-HEMTs grown on Si with a 1.3 μm buffer layer achieve *DIBL* and *SS* similar to the sub-100 nm scale Si-NMOS with a similar device dimension. The enhancement-mode InGaAs p-HEMTs on Si exhibit >10× reduction in DC power dissipation for the same speed performance or >2× gain in speed performance for the same power.

In Table 10.1, we compare recent research results of sub-100 nm III-V MOSFETs and QWFETs to Si NMOS. With a similar physical gate length, most HEMTs demonstrate very promising *DIBL* and subthreshold swing (*SS*). Under low power operation (low V_{DD}), III-V HEMTs exhibit acceptable gate leakage current and reduction of short channel effect. However, the I_{ON}/I_{OFF} ratio of HEMTs is much lower than that of Si MOSFETs. This drawback is due to the limited barrier height of the Schottky gate. This large gate leakage current contributes to the total I_{OFF} and is undesired for the low power dissipation requirement.

Sun et al. at IBM proposed an insulated-gate buried-channel QWFET approach with amorphous silicon as the gate insulator to suppress the gate leakage [31]. These devices have shown six orders of magnitude lower gate leakage current than Schottky-gate devices. However, the subthreshold swing is much higher (worse) even though the gate-length is larger than other published results on III-V HEMTs [27–30]. This can be explained in terms of the aspect ratio (or scaling parameters) [32]. Because of the extra gate-insulator thickness, which increases the distance between gate and channel, the gate loses its effectiveness in modulating the channel current and, as a result, the *SS* increases. This argument can also apply to other HEMT structures where enhancement-mode devices show better *DIBL* and *SS*

Table 10.1 Logic performance comparison chart of Si and III-V FETs

	Si CMOS Intel 2005 [26]	E-mode InGaAs MOSFET Purdue 2009 [25]		InGaAs HEMT MIT 2007 [27]	E-mode InAs HEMT MIT 2008 [28]	InGaAs QWFET on Si Intel 2007 [29]	InSb QWFET on Si Intel 2007 [30]	InGaAs insulate gate QWFET IBM 2009 [31]
L_g (nm)	65	200	110	90	30	80	85	160
V_{DD} (V)	1.2	0.8	0.8	0.5	0.5	0.5	0.5	1
SS (mV/decade)	90	154	1500	70	73	90	94	340
DIBL (mV/V)	80	137	N/A* (>956)	55	85	90	82	N/A
I_{ON}/I_{OFF}	>10000	2700	N/A*(<60)	2000	N/A	2150	370	N/A

* These items are either too large or too small to be determined. The number quoted in the parenthesis is estimated from devices with larger L_g

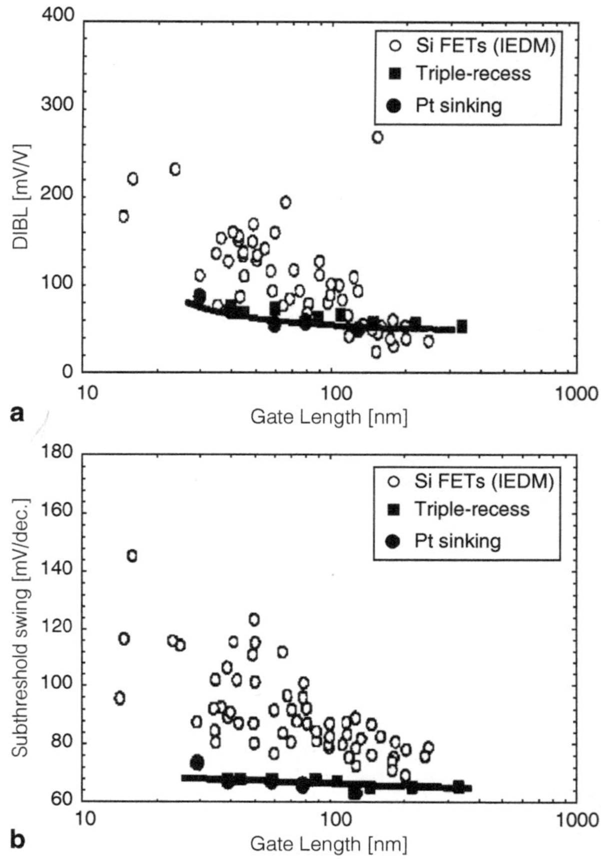

Fig. 10.15 **a** *DIBL* and **b** *SS* of E-mode and D-mode InAs HEMTs against L_g, compared to Si NMOS [28]

than depletion mode devices [27, 30]. This is due to the fact that enhancement-mode devices are fabricated with a deeper gate recess, where the barrier thickness is reduced, than that of depletion-mode devices. Datta at Intel also incorporated high-k dielectrics in the InSb p-HEMT process [30]. The gate leakage current of the insulated-gate InSb QWFET shows four orders of magnitude reduction compared to Schottky metal gate devices. However, more effort is needed to overcome the gate-leakage (power dissipation) problem due to the relatively high interface states so that III-V QWFETs can become more viable for logic applications in the future.

10.6 Outlook and Conclusions

Several emerging nanotechnologies addressing the need for the end of the Si CMOS roadmap after 10 nm gate length are under intense investigation worldwide. The efforts in the US have been led by a consortium of university researchers managed by the Semiconductor Research Corporation (SRC) with the support of the Defense Advanced Research Projects Agency (DARPA) and the semiconductor industry. The emerging technologies such as carbon nanotubes, graphene, and spintronics have stirred up considerable scientific interest, but will unlikely be incorporated into the future DICs because they lack the scaling and integration capabilities and infrastructure necessary to viably contend for cost effective IC manufacturing. The only promising solution is III-V based FETs, in forms of MOSFETs and/or QWFETs, due to the extensive materials, device processing, heterostructure, band gap engineering and integration technologies acquired in the past 45 years. However, despite the established infrastructures and the widespread usage of III-V devices in wireless communications and optoelectronics, some fundamental problems with scaling of low-power III-V CMOS remain to be resolved, specifically in the development of high quality and robust dielectric/semiconductor interface with very low interface trap density. The promising recent results of III-V QWFETs developed at various institutions and laboratories provide a roadmap toward future high-speed low-power logic DICs with better performance than Si CMOS of the same gate length. Other important parameters that need to be improved are the enhancement of I_{ON}/I_{OFF} ratio by developing high-k dielectrics on the III-V surface with a low interface trap density, and heterogeneous integration of III-V FETs on Si substrates. Finally, we need to investigate strained p-channel III-V QWFETs to achieve ultra-low-power III-V CMOS logic in order to reach large-scale integration. All these tasks may take another 5–10 years of considerable efforts. Once accomplished, III-V-based CMOS technology will make a big impact on society comparable to or exceeding that of today's silicon DICs.

Acknowledgements This work is supported in part by the MARCO MSD Focus Center, one of five research centers funded under the Focus Center Research Program, a Semiconductor Research Corporation program.

References

1. M. Hong, J. N. Baillargeon, J. Kwo, J. P. Mannaerts, and A. Y. Cho, "First demonstration of GaAs CMOS," *Proceedings of 2000 IEEE International Symposium on Compound Semiconductors*, 2000, p. 345.
2. M. Bohr, "MOS transistors: Scaling and performance trends," *Semiconductor International*, June 1995, pp. 75–80.
3. S. Natarajan et al., "A 32 nm logic technology featuring 2nd-generation high-k + metal-gate transistors, enhanced channel strain and SRAM cell size in a 291 Mb array," *IEDM Tech. Dig.*, 2008, pp. 1–3.
4. B. Doyle et al., "Transistor elements for 30 nm physical gate length and beyond," *Intel Technol. J.*, **6**, 42 (2002).
5. J. Chen, T. Y. Chan, I. C. Chen, P. K. Ko, and C. Hu, "Subbreakdown drain leakage current in MOSFET," *IEEE Electron Device Lett.*, **8**, 515 (1987).
6. R. Chau et al., "Benchmarking nanotechnology for high-performance and low-power logic transistor applications," *IEEE Trans. Nanotechnol.*, **4**, 153 (2005).
7. Sadao Adachi, *Properties of group-IV, III-V and II-VI semiconductors*, (John Wiley & Sons, 2005), p. 322.

8. D. Chattopadhyay, S. K. Sutradhar, and B. R. Nag, "Electron transport in direct-gap III-V ternary alloys," *J. Phys. C*, **14**, 891 (1981).
9. H. Miki, K. Segawa, M. Otsubo, K. Shirahata, and K. Fujibayashi, "Growth of $In_xGa_{1-x}Sb$ by liquid phase epitaxy," in *GaAs and Related Compounds* (Inst. Phys. Conf. Ser. no. 24, Inst. of Phys., London and Bristol, 1975), pp. 16–21.
10. M. Kawashima and S. Kataoka, "Electron velocity-field characteristics of $Ga_xIn_{1-x}Sb$ measured by a microwave heating technique," *Jpn. J. Appl. Phys.*, **18**, 1311 (1979).
11. S. E. Thompson et al., "A 90nm logic technology featuring strained-silicon," *IEEE Trans. Electron Devices*, **51**, 1790 (2004).
12. M. L. Lee and E. A. Fitzgerald, "Hole mobility enhancements in nanometer-scale strained-silicon structures grown on Ge-rich relaxed $Si_{1-x}Ge_x$," *J. Appl. Phys.*, **94**, 2590 (2003).
13. Y. Sun, S. E. Thompson, and T. Nishida, "Physics of strain effects in semiconductors and metal-oxide-semiconductor field-effect transistors," *J. Appl. Phys.*, **101**, 104503 (2007).
14. S. L. Chuang, "Efficient band-structure calculations of strained quantum wells," *Phys. Rev. B*, **43**, 9649 (1991).
15. S. Datta et al., "85nm gate length enhancement and depletion mode InSb quantum well transistors for ultra high speed and very low power digital logic applications," *IEDM Tech. Dig.*, 2005, pp. 763–766.
16. M. Radosavljevic et al., "High-performance 40nm gate length InSb p-channel compressively strained quantum well field effect transistors for low-power (V_{CC} = 0.5V) logic applications," *IEDM Tech. Dig.*, 2008, pp. 1–4.
17. R. Chau, S. Datta, and A. Majumdar, "Opportunities and challenges of III-V nanoelectronics for future high-speed, low-power logic applications," *IEEE Compound Semiconductor Integrated Circuit Symposium*, 2005, pp. 17–20.
18. H. W. Kennel et al., "Kinetics of shallow junction activation: Physical mechanism," *Proceedings International Conference on Advanced Thermal Processing*, 2006, pp. 85–91.
19. D. Cheng, C. Liao, K. Y. Cheng, and M. Feng, "Process development and characteristics of nano III-V MOSFET," *International Conference on Compound Semiconductor Manufacturing Technology (CS MANTECH) Technology Digest*, 2008, pp. 339–342.
20. C. Liao, D. Cheng, C. Cheng, K. Y. Cheng, M. Feng, T. H. Chiang, J. Kwo, and M. Hong, "Inversion-channel enhancement-mode GaAs MOSFETs with regrown source and drain contacts," *J. Crystal Growth*, **311**, 1958 (2009).
21. D. Barlage et al., "Low temperature selected area regrowth of ohmic contacts for III-N FETs," *Mater. Res. Soc. Symp.*, **892**, (2006).
22. P. D. Ye, "Main determinants for III-V metal-oxide-semiconductor field-effect-transistors," *J. Vac. Sci. Technol. A*, **26**, 697 (2008).
23. T. D. Lin, H. C. Chiu, P. Chang, L. T. Tung, C. P. Chen, M. Hong, J. Kwo, W. Tsai, and Y. C. Wang, "High-performance self-aligned inversion-channel $In_{0.53}Ga_{0.47}As$ metal-oxide-semiconductor field-effect-transistor with $Al_2O_3/Ga_2O_3(Gd_2O_3)$ as gate dielectrics," *Appl. Phys. Lett.*, **93**, 033516 (2008).
24. Y. Xuan, Y. Q. Wu, and P. D. Ye, "High-performance inversion-type enhancement-mode InGaAs MOSFET with maximum drain current exceeding 1A/mm," *IEEE Electron Device Lett.*, **29**, 294 (2008).
25. Y. Q. Wu, W. K. Wang, O. Koybasi, D. N. Zakharov, E. A. Stach, S. Nakahara, J. C. M. Hwang, and P. D. Ye, "0.8V supply voltage deep-submicrometer inversion-mode $In_{0.75}Ga_{0.25}As$ MOSFET," *IEEE Electron Device Lett.*, **30**, 700 (2009).
26. C. H. Jan et al., "A 65nm ultra low power logic platform technology using uni-axial strained silicon transistors," *IEDM Tech. Dig.*, 2007, pp. 60–63.
27. N. Waldron, D. Kim, and J. A. Alamo, "90nm self-aligned enhancement-mode InGaAs HEMT for logic applications," *IEDM Tech. Dig.*, 2007, pp. 633–636.
28. D. Kim and J. A. Alamo, "30nm E-mode InAs pHEMTs for THz and future logic applications," *IEDM Tech. Dig.*, 2008, pp. 830–833.

29. M. Hudait et al., "Heterogeneous integration of enhancement mode $In_{0.7}Ga_{0.3}As$ quantum well transistor on silicon substrate using thin composite buffer architecture for high-speed and low-voltage (0.5 V) logic applications," *IEDM Tech. Dig.*, 2007, pp. 625–628.
30. S. Datta, "III-V field-effect transistors for low power digital logic applications," *Microelectronic Eng.*, **84**, 2133 (2007).
31. Y. Sun et al., "High-performance $In_{0.7}Ga_{0.3}As$-channel MOSFETs with high-k gate dielectrics and α-Si passivation," *IEEE Electron Device Lett.*, **30**, 5 (2009).
32. R. Yan, A. Ourmazd, and K, Lee, "Scaling the Si MOSFET: From bulk to SOI to bulk," *IEEE Trans Electron Dev.*, **39**, 1704 (1992).

Chapter 11
Electrical and Material Characteristics of Hafnium Oxide with Silicon Interface Passivation on III-V Substrate for Future Scaled CMOS Technology

Injo Ok and Jack C. Lee

Abstract Electrical properties of Hf-based high-k dielectric materials on compound semiconductor such as GaAs, InGaAs and InP are studied based on understanding of physics, reliability, and process development in MOSCAP and MOSFET characteristics. Our research focus on several approaches to develop and understanding the electrical characteristics of TaN/HfO_2/GaAs MOS capacitors with Si interface passivation layer (IPL) under various post-deposition anneal (PDA) condition and various physical vapor deposition (PVD) Si deposition temperature/time will be described. The insertion of the ultra thin Si layer is designed to establish one-to-one registry between the semiconductor and the insulator and also, in the case of SiO_2, prevents the GaAs from oxidation. Depletion mode transistor and self-aligned n- and p-channel enhancement GaAs MOSFETs using HfO_2 and silicon IPL with the results of plausible characteristics will be presented. We will also present the investigation $In_{0.2}Ga_{0.8}As$, high-indium-content $In_{0.53}Ga_{0.47}As$, and InP MOSCAP and MOSFET using the same oxide of HfO_2 as gate insulator with Si IPL. We present the electrical characteristics of TaN/HfO_2/InGaAs (and InP) MOS capacitors with Si IPL under various PDA (post-deposition anneal) condition and various Si deposition time. Also, we will review the effects of hydrogen incorporation for $In_{0.2}Ga_{0.8}As$ MOSCAP using H_2 annealing.

11.1 Introduction

CMOS (complementary metal oxide semiconductor) devices have been scaled down. For the last four decades, the scaling down of physical thickness of SiO_2 gate dielectrics has improved the speed of output drive current by shrinking of transistor area in front-end-process of integrated circuits. A higher number of transistors on chip resulting in faster speed and lower cost can be allowable by

J. C. Lee (✉)
Cockrell School of Engineering, The University of Texas at Austin, Austin, Texas 78758, USA
e-mail: jacklee@uts.cc.utexas.edu

S. Oktyabrsky, P. D. Ye (eds.), *Fundamentals of III-V Semiconductor MOSFETs*,
DOI 10.1007/978-1-4419-1547-4_11, © Springer Science+Business Media LLC 2010

the scaling down and these fruitful achievements have been mainly made by the thinning thickness of one key component—gate dielectric—at Si based MOSFET (metal-oxide-semiconductor field effect transistor) devices. So far, SiO_2 (silicon dioxide) gate dielectric having the excellent material and electrical properties such as good interface (i.e., D_{it} ~$2 \times 10^{10}\,eV^{-1}\,cm^{-2}$), low gate leakage current, higher dielectric breakdown immunity (≥10 MV/cm) and excellent thermal stability at typical Si processing temperature has been commonly used as the leading gate oxide material.

The next generation Si based MOSFETs will require more aggressive gate oxide scaling to meet the required specifications. Since high-k dielectrics provide the same capacitance with a thicker film, the leakage current reduction, therefore, less the standby power consumption is one of the huge advantages. Also, it is easier to fabricate during the process because the control of film thickness is still not in the critical range compared to the same leakage current characteristic of SiO_2 film. HfO_2 based gate dielectric is considered as the most promising candidate among materials being studied since it shows good characteristics with conventional Si technology and good device performance has been reported. However, it has still many problems like insufficient thermals stability on silicon such as low crystallization temperature, low k interfacial regrowth, charge trapping and so on. The integration of hafnium based high-k dielectric into CMOS technology is also limited by major issues such as degraded channel mobility and charge trapping. One approach to overcome these obstacles is using alternative substrate materials such as SiGe, GaAs, InGaAs, and InP to improve channel mobility. However, the key challenge for III-V-based metal-oxide-semiconductor field effect transistor (MOSFET) is the lack of high-quality, thermodynamically stable insulators that passivate the interface states and prevents Fermi level pinning at III-V-gate dielectric interface.

Electrical properties of Hf-based high-k dielectric materials on compound semiconductor such as GaAs, InGaAs and InP in this study are based on understanding of physics, reliability, and process development in MOSCAP and MOSFET characteristics.

In this chapter, we will discuss several approaches to develop and understanding the electrical characteristics of TaN/HfO_2/GaAs MOS capacitors with Si interface passivation layer (IPL) under various post-deposition anneal (PDA) condition and various physical vapor deposition (PVD) Si deposition temperature/time will be described. The insertion of the ultra thin Si layer is designed to establish one-to-one registry between the semiconductor and the insulator and also, in the case of SiO_2, prevents the GaAs from oxidation. Depletion mode transistor and self-aligned n- and p-channel enhancement GaAs MOSFETs using HfO_2 and silicon IPL with the results of plausible characteristics will be presented.

We will also present the investigation $In_{0.2}Ga_{0.8}As$, high-indium-content $In_{0.53}Ga_{0.47}As$, and InP MOSCAP and MOSFET using the same oxide of HfO_2 as gate insulator with Si IPL. We present the electrical characteristics of TaN/HfO_2/InGaAs (and InP) MOS capacitors with Si IPL under various PDA (post-deposition anneal) condition and various Si deposition time. Also, we will review the effects of hydrogen incorporation for $In_{0.2}Ga_{0.8}As$ MOSCAP using H_2 annealing.

11.2 MOSCAPs and MOSFETs on GaAs with Si, SiGe Interface Passivation Layer (IPL)

11.2.1 Process Development: Si IPL on GaAs

Several approaches to develop and understand the electrical characteristics of TaN/HfO_2/GaAs MOS capacitors with Si IPL under various PDA condition and various PVD Si deposition temperature/time have been attempted. Depletion mode transistor and self-aligned n- and p-channel GaAs MOSFETs using HfO_2 and silicon IPL will be presented. Also, a new approach of high-k era on GaAs, which results in significantly improvement in electrical and material aspects for future CMOS technology, will be demonstrated.

11.2.1.1 Test Structures

MOS capacitors were fabricated on n-type GaAs (100) wafer doped with Si. The surface oxides were removed with an HCl clean followed by $(NH_4)_2S$ dip, resulting in a clean S-passivated GaAs surface. Then Si IPL was deposited by RF sputtering of Si at 500 °C. PVD HfO_2 films were deposited using the modulation technique [1], followed by PDA at 600 °C in N_2 (O_2 5%) ambient (Fig. 11.1). The thickness of Si IPL layer varied from ~8 Å (1 min deposition time) to ~24 Å (3 min); and the physical thickness of HfO_2 layer was ~100 Å. PVD TaN (220 nm) was used as gate electrode. After low-resistance ohmic contact was formed by using AuGe/Ni/Au alloy on the backside, the samples were annealed at 450 °C.

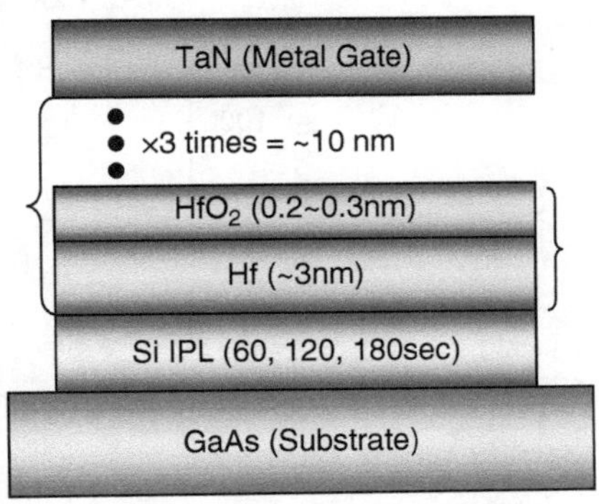

Fig. 11.1 Layer structure of the Si-capped sample with Hf of three times modulation deposition

11.2.1.2 MOSCAP Characteristics on n-type GaAs

Typical capacitance-voltage characteristics of TaN/HfO_2/GaAs as a function of frequency are shown in Fig. 11.2. The samples consist of thin Si IPL layer thickness (with Si deposition time of 60 s). It can be observed that well-behaved C–V characteristics have been obtained.

The frequency dispersion on accumulation capacitance is an important issue for high-k dielectrics on III-V and the magnitude of frequency dispersion has been correlated with the interface state density.

Figure 11.3 summarizes the frequency dispersion characteristics percent (definition of $(C_{1KHz} - C_{1MHz})/C_{1MHz} \times 100$ [%] at 2.5 V) and ΔmV (definition of $|V_{1MHz} - V_{1KHz}|$ at V_{fb}) defined in Fig. 11.4) versus PDA time as a function of Si IPL deposition condition. With Si capping layer, low frequency dispersion (<5%, <100 mV) can be obtained. In contrast, without Si capping layer, frequency dispersion was around 25%. In general, shorter Si deposition time (1 or 2 min) and longer PDA time resulted

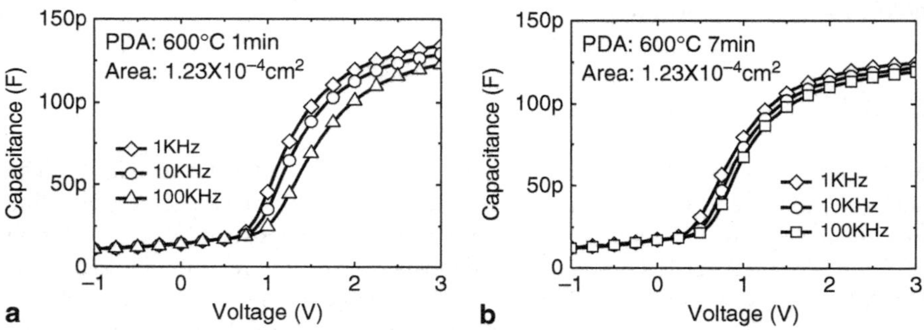

Fig. 11.2 Typical C–V characteristics of TaN/HfO_2/GaAs as a function of frequency for 60 s Si IPL deposition time and varying PDA time: **a** 1 min and **b** 7 min

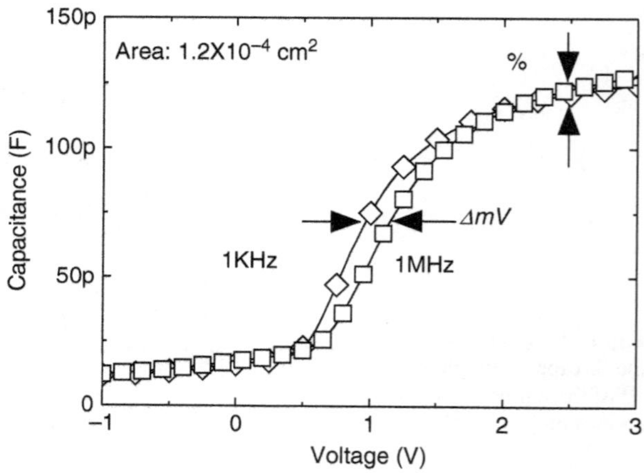

Fig. 11.3 Frequency dispersion percent (definition of $(C_{1KHz} - C_{1MHz})/C_{1MHz} \times 100$ [%] at 2.5 V) and ΔmV (definition of $|V_{1MHz} - V_{1KHz}|$ at V_{fb})

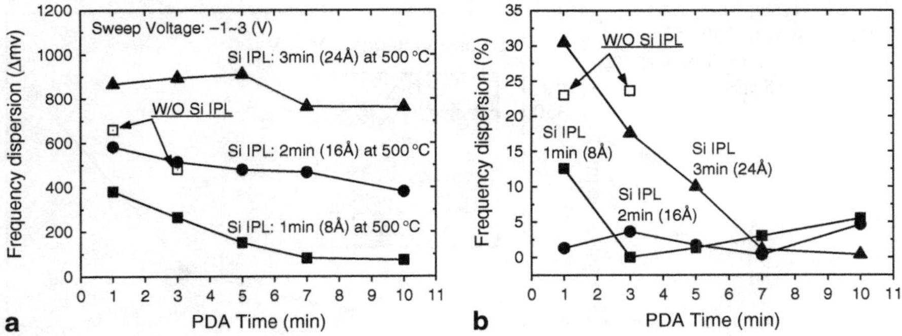

Fig. 11.4 Frequency dispersion versus PDA time as a function of Si IPL deposition condition: **a** voltage difference (mV) at flat band voltage; and **b** capacitance difference (%) between 1 MHz and 1 KHz

in reduced frequency dispersion, relative to shorter PDA. The results suggest that there is negligible Fermi level pinning at the surface.

High-k dielectrics are known to exhibit hysteresis. Figure 11.5 shows the effects of Si IPL thickness on hysteresis behavior of MOSCAPs with different PDA condition obtained from bidirectional high frequency CV measurements at 1 MHz (sweep rate 500 mV/s with sweep range from −1 to 3 V). Hysteresis was measured in constant voltage regime between third and fourth sweep at each V_{fb} (Fig. 11.5a). It has been found that the hysteresis was reduced significantly using the Si IPL (Fig. 11.5b).

In general, longer Si deposition time led to lower hysteresis. However, longer Si deposition time also resulted in more severe stretched-out C–V curve. As we decreased the HfO_2 thickness we had smaller hysteresis (Fig. 11.6) and increased PDA condition we have a little hysteresis improvement. So most hysteresis is coming from HfO_2 itself and second reason is from reasonable PDA condition. And also interface quality affects the hysteresis from comparing with and without Si IPL hysteresis result. Figure 11.7a shows the leakage current density at $V_g - V_{fb} = 1$ as

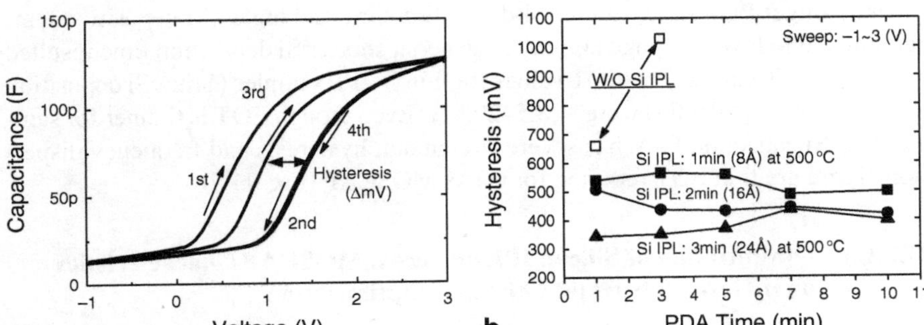

Fig. 11.5 **a** Definition of hysteresis. **b** Hysteresis versus PDA time as a function of Si IPL deposition condition

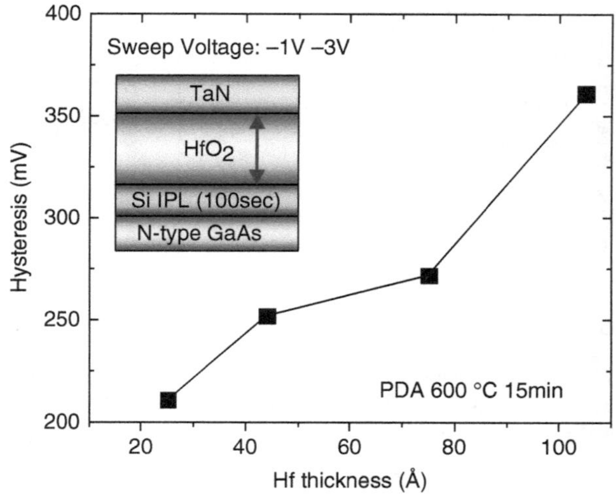

Fig. 11.6 Hysteresis versus Hf thickness with 100 s Si IPL deposition condition and PDA 600 °C 15 min

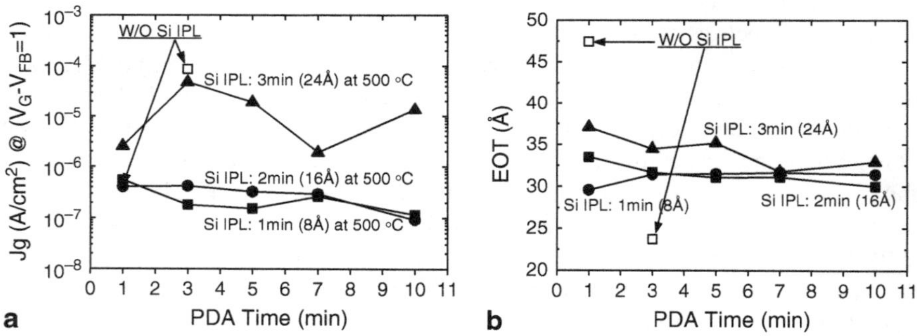

Fig. 11.7 EOT versus PDA time as a function of Si IPL deposition condition with different Si IPL thickness and deposition temperature

a function of PDA time. Using the Si IPL significantly reduced leakage current. In general, longer PDA time, shorter Si-deposition time and higher Si deposition temperature led to lower leakage current. In general, shorter Si deposition time resulted in thinner EOT values. The EOT values for thin Si IPL samples (1 min Si deposition w/PDA time of 3–10 min) were ~28.5–29.5 Å. Even though EOT is thinner for sample W/O Si capacitor, C–V has severe stretch out, hysteresis and frequency dispersion. There are thin EOT reported for GaAs MOSCAP (Fig. 11.7b).

11.2.1.3 Optimization of Silicon IPL on GaAs. MOSCAP Characteristics on n-GaAs Substrate: Thickness Optimization

TaN/HfO_2/GaAs MOS capacitors on n-type GaAs (100) wafer have been fabricated. Various deposition time (thicknesses) of Si IPL deposited at 400 °C under various PDA (post-deposition anneal) condition have been used.

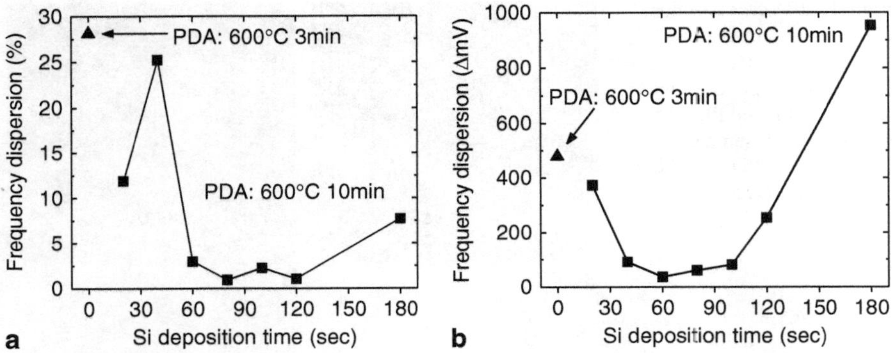

Fig. 11.8 Frequency dispersion versus Si IPL deposition condition as a function of PDA time: **a** voltage difference (mV) at flat band voltage; and **b** capacitance difference (%) between 1 MHz and 1 KHz

Figure 11.8 summarizes the frequency dispersion characteristics versus Si IPL deposition condition as a function of PDA time. With appropriate thickness of Si capping layer, low frequency dispersion (<50 mV and <5%) can be obtained. In contrast, without Si capping layer (0 Si deposition time) frequency dispersion was around ~300 mV and 30%. With PDA of 10 min, 60 s was minimum thickness to unpin the Fermi level (Fig. 11.8a, b). In general, Si deposition time of 60–100 s and longer PDA time resulted in reduced frequency dispersion. Frequency dispersion is more depended on Si deposition time than on PDA condition. However samples with thicker than Si 60 s deposition time show worse frequency dispersion and D_{it} compare to Si 60 s deposition (Figs. 11.8 and 11.9). D_{it} was measured by using conductance method with parameter fitting at different applied voltage for each GaAs samples.

Figure 11.10a shows transmission electron microscopy (TEM) on the MOSCAP with 10 nm HfO_2 with PDA of 600 °C 3 min. The interface between HfO_2 and GaAs surface was degraded severely after 600 °C 3 min PDA for sample without Si IPL.

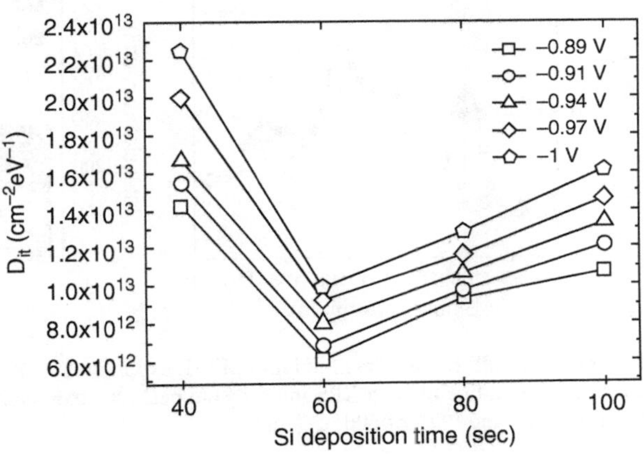

Fig. 11.9 D_{it} comparisons with Si IPL of 40, 60, 80, and 100 s with PDA 600 °C 10 min

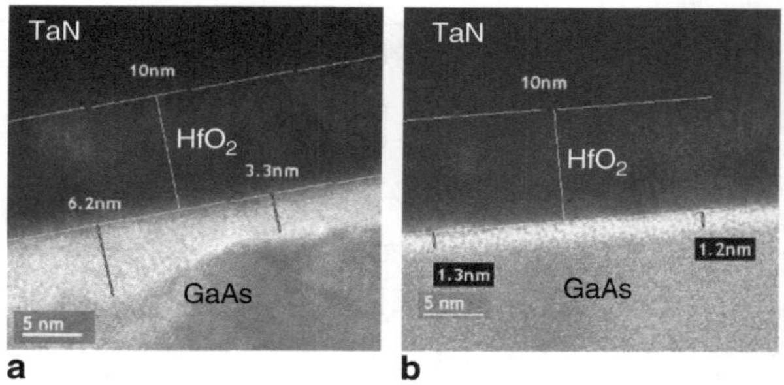

Fig. 11.10 A cross-sectional high resolution transmission electron microscopy (HRTEM) image for sample **a** W/O Si IPL and PDA of 600 °C 3 min **b** sample Si IPL (60 s) and PDA of 600 °C 10 min in $N_2(O_2$ 5%)

However, Fig. 11.10b shows Si IPL protected the interface between HfO_2 and GaAs surface without degradation after 600 °C 10 min PDA.

Figure 11.11a show X-ray photoelectron spectroscopy (XPS) that with 1 min Si deposition time (minimum Si IPL thickness), Si was partially oxidized, and after 7 min PDA in N_2 (O_2 5%), the Si IPL was oxidized to SiO_2. XPS showed that with increasing Si deposition time, the remained Si measured from reference (minimum line of each XPS intensity) to peak value of Si spectra on GaAs increased while the formation of SiO_2 saturated at a range. Small amount of Si with 60 s Si deposition time shows that almost all Si IPL transformed to SiO_2 (Fig. 11.11b). The interfacial quality is affected by remained Si IPL and thinner remained Si IPL show better interface quality after PDA.

Note that even though the 20 s Si resulted in sharp C–V curve, it also exhibited large frequency dispersion, large hysteresis and non-saturated C–V.

It has been found that the hysteresis was reduced significantly using the Si IPL (Fig. 11.12). In general, longer Si deposition time (thicker Si IPL) led to lower hysteresis (ΔmV). Hysteresis is believed to be due to charge trapping/detrapping

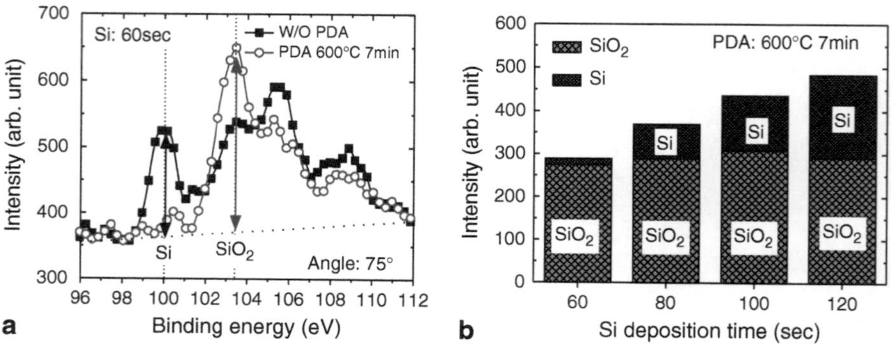

Fig. 11.11 **a** XPS (Si spectra) for 1 min Si IPL as function of PDA time. **b** Si spectra from reference (minimum line of each XPS intensity) to peak value on GaAs with Si deposition time of 60, 80, 100, 120 s and PDA of 600 °C 7 min

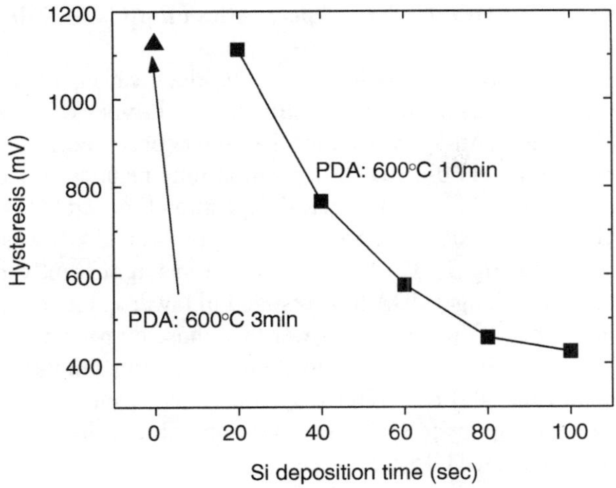

Fig. 11.12 Hysteresis versus Si IPL deposition condition

between high-k layer and the substrate as the voltage is swept back and forth. Thus, a thicker SiO_2 IPL layer could lead to larger distance between HfO_2 and GaAs substrate and thereby lower hysteresis. However, Si deposition time of >100 s resulted in more severe stretched-out C–V curve and have high frequency dispersion as discussed above. Figure 11.13a shows the leakage current density at $V_g - V_{fb} = 1$ as a function of Si deposition time. Using the Si IPL significantly reduced leakage current. In general, longer Si-deposition time at longer PDA time led to lower leakage current. Leakage current was found to be more depended on PDA condition than Si deposition time. In general, Si deposition time of 60–100 s resulted in thinner EOT values (Fig. 11.13b). Note that even though EOT is thinner for sample 20 s Si cap layer, C–V has severe stretch out, larger hysteresis and frequency dispersion.

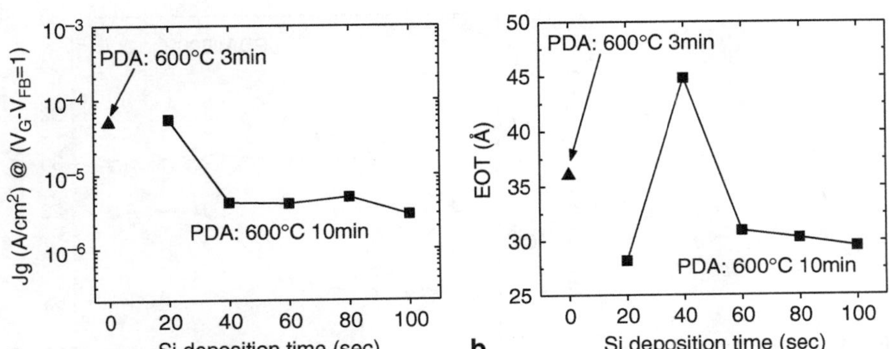

Fig. 11.13 **a** Leakage current density and **b** EOT versus Si IPL deposition time

11.2.1.4 MOSCAP Characteristics on p-GaAs Substrate

In this section, we will present the electrical characteristics of MOSCAPs on p-type GaAs using the same structure as n-type MOSCAPs. In general, frequency dispersion (ΔmV) was larger than those observed in n-type substrate capacitors in 60–100 s and 60–120 s Si deposition time resulted in reduced frequency dispersion (<5%) (Fig. 11.14a). Longer Si deposition time led to lower hysteresis (Fig. 11.14b). Generally, hysteresis observed in p-type MOSCAPs was lower than those on n-type substrate (Fig. 11.12). Leakage current was significantly reduced using the Si cap technique and longer PDA time resulted in lower J_l. Generally, leakage current observed in p-type MOSCAPs was lower than those on n-type substrate (Fig. 11.13a).

In general, longer Si deposition time resulted in thinner EOT values. Si IPL layer differences did not affect EOT thickness. Even though EOT is thinner for sample 120 s Si cap layer, C–V has severe stretch out, leakage current, and frequency dispersion (Fig. 11.15).

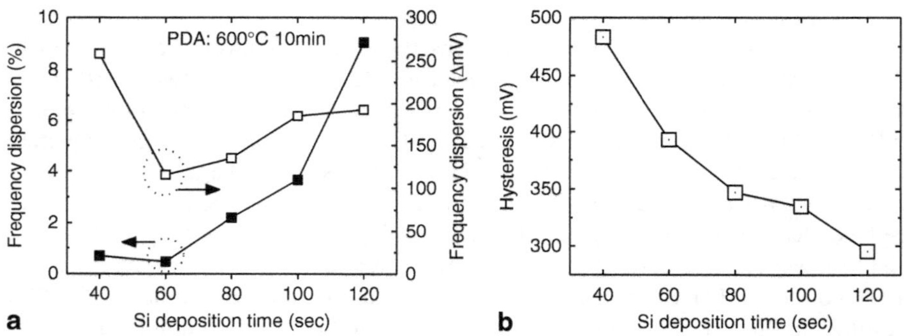

Fig. 11.14 **a** Frequency dispersion versus Si IPL deposition condition as a function of PDA time: capacitance difference (%) between 1 MHz and 1 KHz and voltage difference (mV) at flat band voltage. **b** Hysteresis versus Si IPL deposition condition

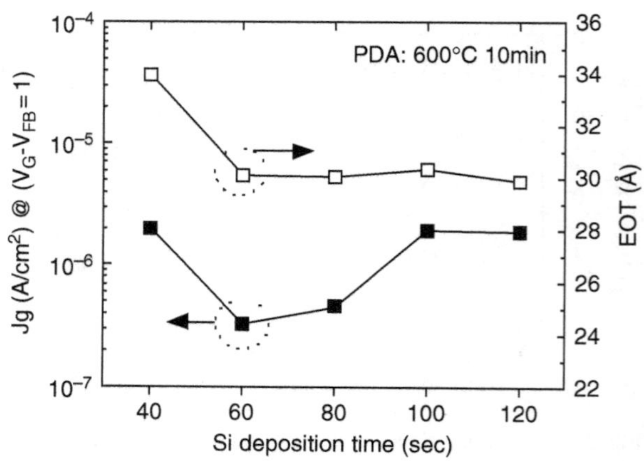

Fig. 11.15 Leakage current density and EOT versus Si IPL deposition condition

11.2.1.5 MOSCAP Characteristics on n-GaAs Substrate: PDA Optimization

The optimum thickness of the Si (with SiO_2) layer preventing Fermi level pinning at GaAs–HfO_2 interface was found to be ~1.2 nm (Si deposition time 60 s) after PDA of 600 °C 10 min for capacitor structure. In this chapter, we studied the influence of the PDA on the electrical and material characteristics of TaN/HfO_2/GaAs GaAs n-type Metal-Oxide-Semiconductor Capacitors (NMOSCAP). Among the different PDA time and temperature but with a fixed thickness of Si (60 s deposition times), PDA time of 7 min shows the smallest frequency dispersion (Fig. 11.16). This implies the lowest interface trap densities.

Figure 11.17a shows transmission electron microscopy (TEM) on the MOSCAP with 10 nm HfO_2 without PDA. Electron energy loss spectroscopy (EELS) shows that Si was partially oxidized without PDA simply due to exposure to air after Hf deposition process (Fig. 11.17b).

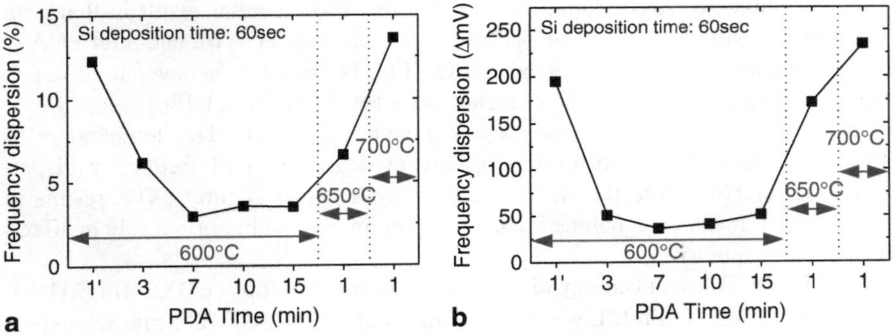

Fig. 11.16 **a** Frequency dispersion of % and **b** Frequency dispersion of ΔmV versus different PDA condition with Si deposition time of 60 s (minimum thickness to unpin the Fermi level)

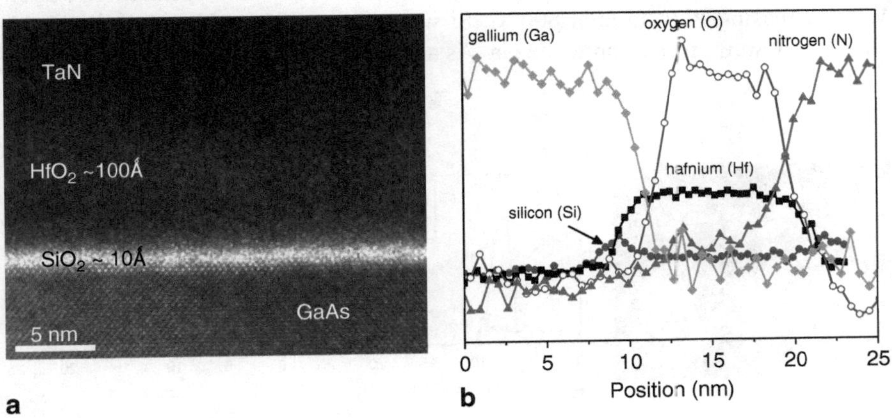

Fig. 11.17 **a** HRTEM for sample with 1 min Si IPL as deposited with **b** Electron energy loss spectroscopy (EELS) analysis

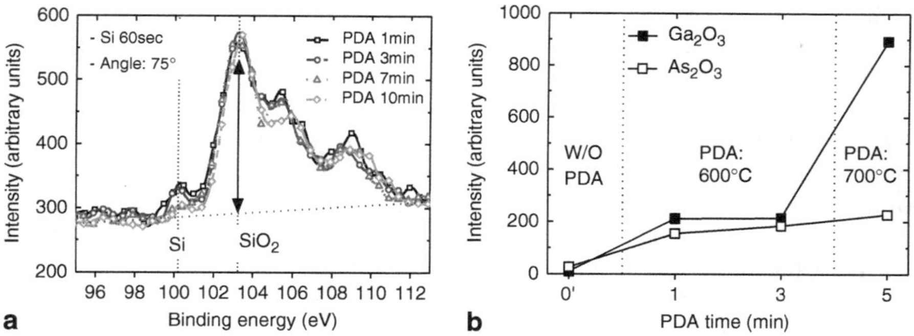

Fig. 11.18 **a** XPS (Si spectra) for 1 min Si IPL as function of PDA time. **b** XPS spectra from reference to peak value for Ga_2O_3 and As_2O_3 as function of PDA time

X-ray photoelectron spectroscopy (XPS) also show similar result in that with 1 min Si deposition time, Si was partially oxidized (Fig. 11.17b), and after PDA in $N_2(O_2$ 5%), the Si IPL was oxidized to SiO_2 (Fig. 11.18a). It is believed to be due to the increasing As_2O_3 and Ga_2O_3 contents (XPS results in Fig. 11.18b).

Note that the Ga_2O_3 signal increased significantly if PDA temperature is above 650 °C with the corresponding severe degradation of frequency dispersion (Fig. 11.16); while the As_2O_3 signal increased just slightly (XPS results in Fig. 11.18b). It can be concluded that Ga_2O_3 might play an important role in affecting interface quality.

TEM with EELS and energy dispersive X-ray spectroscopy (EDXS) (Fig. 11.19) show that Si with 1 min IPL was fully oxidized after PDA 700 °C 5 min while protecting GaAs surface and Ga_2O_3 and As_2O_3 increasing on surface resulted in C–V degradation (Fig. 2.16). IPL thickness also changed. Figure 11.20 show XPS with 0–7 min PDA for 120 s Si deposition time on (100) GaAs wafer without HfO_2 layer. With increasing PDA time at 600 °C, Si was continuously oxidized to SiO_2 with Ga_2O_3 and As_2O_3 formation while Ga–As and As–As bond decrease.

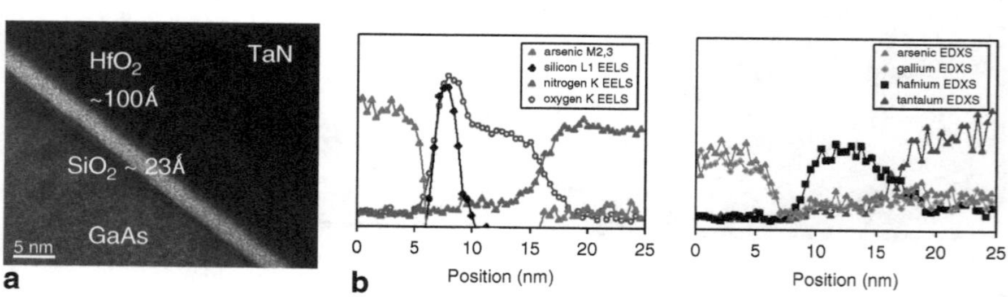

Fig. 11.19 **a** HRTEM image for sample 1 min Si IPL and PDA of 700 °C 5 min in $N_2(O_2$ 5%) with **b** EELS and energy dispersive X-ray spectroscopy (EDXS) analysis

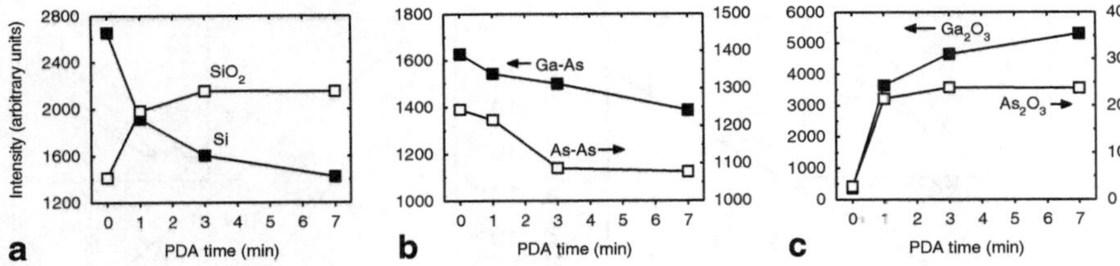

Fig. 11.20 **a** XPS spectra from reference to peak value for **a** Si and SiO_2 **b** Ga–As and As–As and **c** Ga_2O_3 and As_2O_3 with Si 120 s IPL on GaAs as function of PDA time

11.2.2 Demonstration of Depletion Mode Transistor on GaAs: Materials and Electrical Analysis

Demonstration of depletion mode MOSFET using optimum capacitance condition with material and electrical characteristics of TaN/HfO_2/GaAs MOS capacitors using Si IPL under various PDA (post-deposition anneal) condition and various Si deposition time will be presented. Electrical characteristics with thin EOT (~1.6 nm), low frequency dispersion (<5%) and high maximum mobility (568 cm^2/V s) have been obtained.

11.2.2.1 MOSCAP Characteristics on GaAs Substrate

Flat band voltage shift with different metal gate also provided good evidence for unpinned Fermi level (Fig. 11.21). Metal gate was fabricated by using lift-off process for both n- and p-type GaAs substrate. Using Si passivation layer (60 s), flatband voltage shifts according to metal electrode on n- and p-type GaAs which suggests good interface quality and unpinned Fermi level.

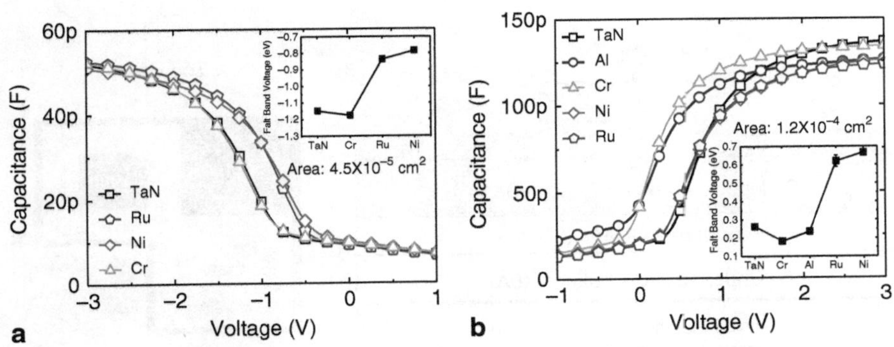

Fig. 11.21 Flatband voltage shift with different metal gates on **a** p-type and **b** n-type GaAs substrate with flatband voltage trend with different metal gate (*inset figure*)

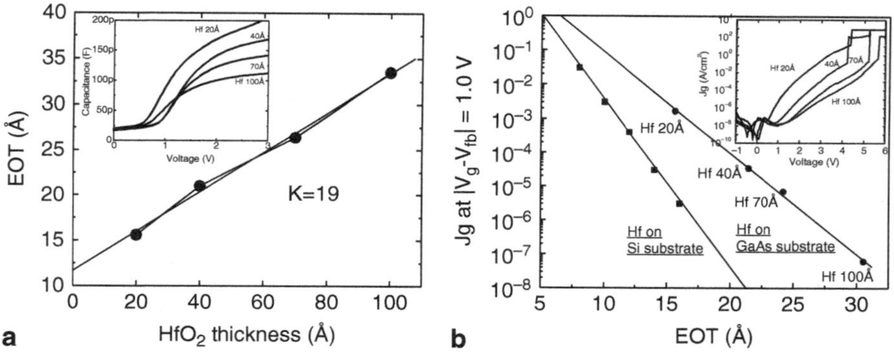

Fig. 11.22 **a** EOT versus physical thickness of HfO_2. **b** J_g versus EOT of HfO_2 with PVD Si passivation

Figure 11.22a demonstrates the linear dependence of the equivalent oxide thickness (EOT) on the physical thickness of HfO_2 with PVD Si 60 s passivation with PDA 7 min. The k-value of the gate HfO_2 was found to be k = 19.5 (with k SiO_2 (native oxide) = 3.9) [2]. In general, varying Si deposition time (>60 s) with same Hf thickness still resulted in similar EOT values (Fig. 11.15). Thin amorphous Si IPL between HfO_2 and GaAs substrate did not significantly contribute EOT change after PDA. EOT was mostly related to HfO_2 thickness. With increasing PDA time, negligible increase in SiO_2 thickness was observed (Fig. 11.18). Ga_2O_3, As_2O_3 and Si possibly behave as defects. Thus Ga_2O_3, As_2O_3 and SiO_2 contributed to minimal change to EOT. Leakage current was reduced (to ~10^{-5} A/cm^2) by using the Si IPL for 4.0 nm thick high-k dielectric in n-type GaAs wafer (Fig. 11.22b).

11.2.2.2 Demonstration of Depletion Mode Transistor

Depletion mode MOSFET was fabricated by using the optimum thickness and PDA condition (Si deposition time of 60 s and 7 min PDA) of the Si passivation layer. The

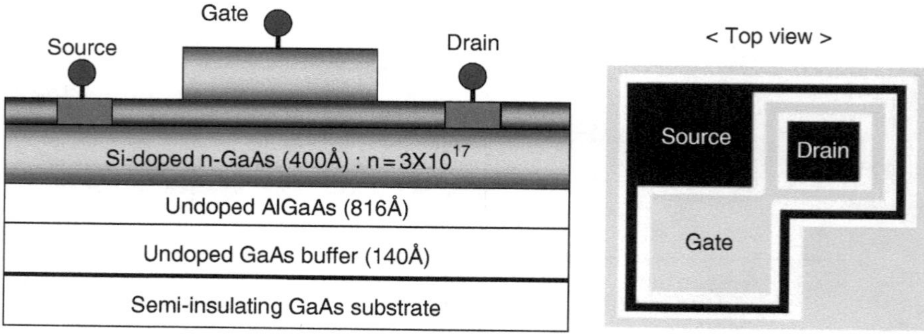

Fig. 11.23 Schematic cross section and top view of depletion mode GaAs MOSFET with Si Passivation

schematic cross section and top view (ring-type) of depletion mode GaAs MOSFET with Si passivation is shown in Fig. 11.23. The thickness of molecular beam epitaxy (MBE) grown n-type GaAs epi-layer layer on semi-insulating substrate was 40 nm with Si dopant concentration in the epi-layer of $3 \times 10^{17}\,cm^{-3}$.

The I_d–V_g and I_d–V_d characteristics are shown in Fig. 11.24. Depletion mode MOSFET with PVD Si passivation layer shows V_t of −1.41 V, excellent off-current I_{off} of 69 pA, transconductance G_{m_max} of 123 μA/V and swing S of 113 mV/decade with width 400 μm and length 5 μm at drain voltage (V_d) = 50 mV. Figure 11.24b shows the output characteristics of MOSFET with I_{d_max} of 2.58 mA at $V_g - V_{th}$ = 1 V and high series resistance was observed. The effective electron mobility (μ_{eff}) was evaluated for the same device from the I_d–V_g plot (measured at a drain voltage of 50 mV) and its corresponding split C–V for the n-type GaAs. Electron mobility for GaAs MOSFET with PVD Si passivation was estimated to be 568 cm^2/V s (Fig. 11.25).

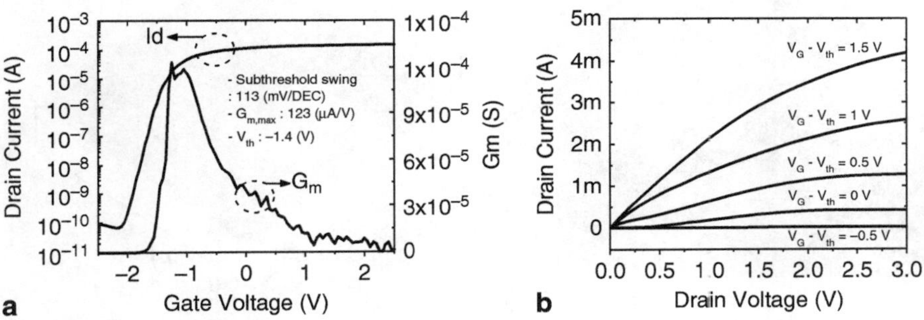

Fig. 11.24 **a** I_d–V_g and Gm of depletion mode GaAs MOSFET. **b** Mobility of depletion mode GaAs MOSFET with Si passivation (W 400 μm × L 5 μm)

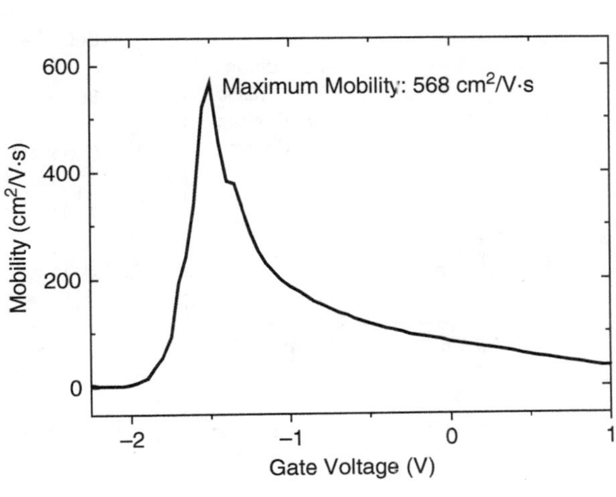

Fig. 11.25 Mobility of depletion mode GaAs MOSFET with Si passivation (W 400 μm × L 5 μm)

11.2.3 Demonstration of Enhancement Mode Transistor

Using Si IPL the electrical characteristics of TaN/HfO_2/GaAs both p- and n-MOSFET made on GaAs substrates with high temperature PMA on undoped GaAs will be presented.

11.2.3.1 MOSCAP Characteristics on GaAs Substrate with PMA

The Si IPL on undoped GaAs exhibited a small roughness of 0.086 nm (see Fig. 11.26 for atomic force microscopy result).

After PMA at 900 °C, frequency dispersion of 60–80 s IPL devices were still in excellent range (~5%, $\Delta V < 100\,mV$) (Fig. 11.27).

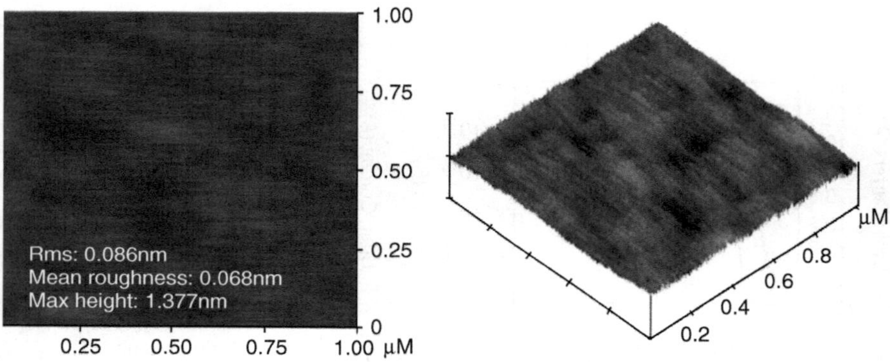

Fig. 11.26 AFM analysis: surface roughness (RMS: 0.086 nm) after Si IPL deposition (3 min) at 400 °C

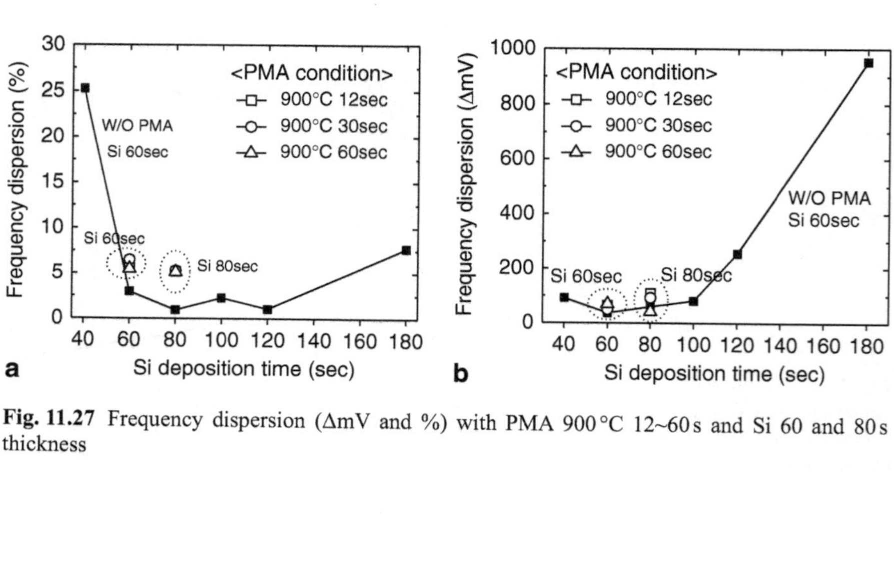

Fig. 11.27 Frequency dispersion (ΔmV and %) with PMA 900 °C 12~60 s and Si 60 and 80 s thickness

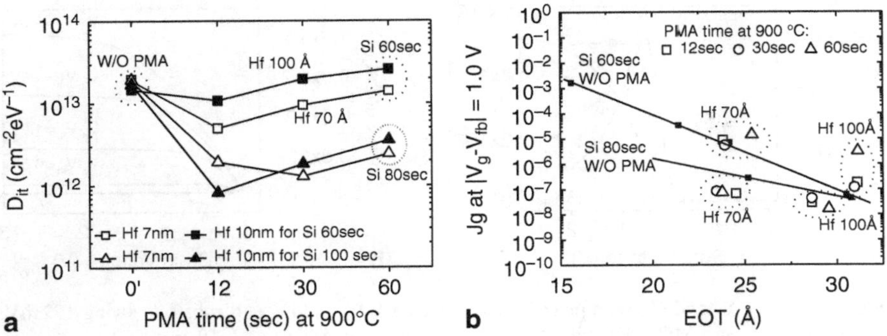

Fig. 11.28 **a** D_{it} with PMA 900 °C 12–60 s with Si 60 s and 80 s deposition time using conductance method. **b** J_g versus EOT of HfO_2 70 Å and 100 Å with PVD Si passivation 60 s and 80 s thickness after PMA

D_{it} measured by conductance method was improved for thicker (80 s Si deposition time) IPL samples (Fig. 11.28a) compare to 60 s Si deposition time. In Fig. 11.28b, the leakage current values as function of EOT are compared. The leakage currents of PMA samples were also found to improve with slightly thicker IPL.

11.2.3.2 MOSFET Characteristics on Undoped GaAs Substrate

n-MOSFETs on undoped GaAs substrate using the Si IPL (80 s) and 900 °C PMA were fabricated. The schematic cross section and top view of n-MOSFET with Si passivation is shown in Fig. 11.29. Source and drain were doped with Si ion implantation of dose 1×10^{14} and energy 25 KeV.

n-channel MOSFET with PVD Si passivation layer shows V_{th} of 0.24 V, excellent on–off ratio of ~10^7, transconductance G_{m_max} of 516.7 μA/V and swing S of 121.81 mV/decade with width 800 μm and length 10 μm at drain voltage (Vd = 50 mV); along with excellent I_d–V_d characteristics that I_{d_max} = 12.2 mA at $V_g - V_{th}$ = 2.0 V (Fig. 11.30).

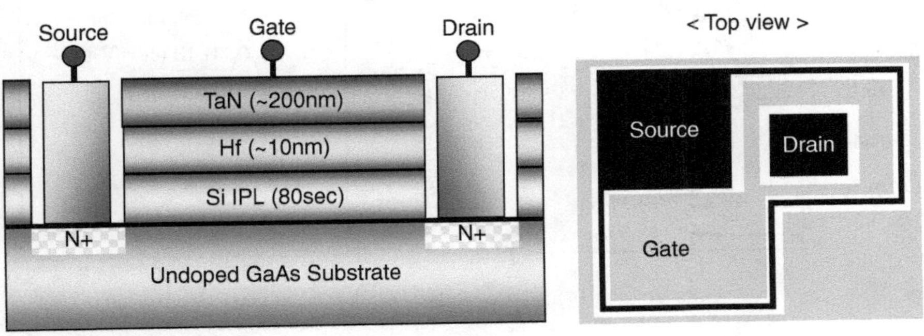

Fig. 11.29 Schematic cross section and top view of n-MOSFET with Si passivation 80 s

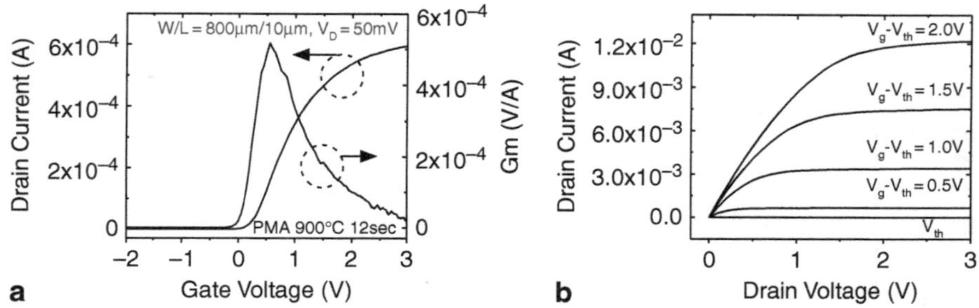

Fig. 11.30 **a** I_d–V_g and G_m with peak G_m 516.7 (μA/V), V_{th} 0.24 V, and substrate swing 122 mV/decade. **b** I_d–V_d of n-MOSFET with Si (80 s) passivation (W 800 μm × L 10 μm)

Inversion C–V characteristic between gate and source combined with drain was measured from 1 MHz to 1 KHz. Inversion C–V curves for frequency from 1 KHz to 1 MHz were very close each other (Fig. 11.31a). The hysteresis was reduced to 288 mV after PMA using the Si IPL (inset in Fig. 11.31a). Using inversion capacitance (Fig. 11.31a) and I_d–V_g, high peak mobility (1213 cm²/V s) has been obtained for n-MOSFET on undoped GaAs (Fig. 11.31b).

Identical layout for p-MOSFET with varying PMA temperature 750–900 °C on undoped GaAs was used with schematic cross section shown in Fig. 11.32. Source and drain were doped with Zn ion implantation of dose 2×10^{14} and energy 60 KeV.

p-channel MOSFET with PVD Si passivation layer shows V_{th} of −0.03 V, off current of ~−8 × 10^{-10} A, transconductance G_{m_max} of 6.2 μA/V and swing S of 178 mV/decade with width 800 μm and length 10 μm at drain voltage (V_d = 50 mV); along with excellent I_d–V_d characteristics that I_{d_max} = 156 μA at V_g = −2.0 V (Fig. 11.33). Frequency dispersion with inversion capacitance between 1 MHz and 100 KHz was found to be large for p-MOSFET (Fig. 11.34a). Using the inversion capacitance and I_d–V_g, excellent peak mobility (191 cm²/V s) has been obtained for p-MOSFET on undoped GaAs substrate (Fig. 11.34b).

With increasing PMA temperature, 1 MHz inversion capacitance increased along with improved frequency dispersion, higher peak G_m and mobility (Fig. 11.35).

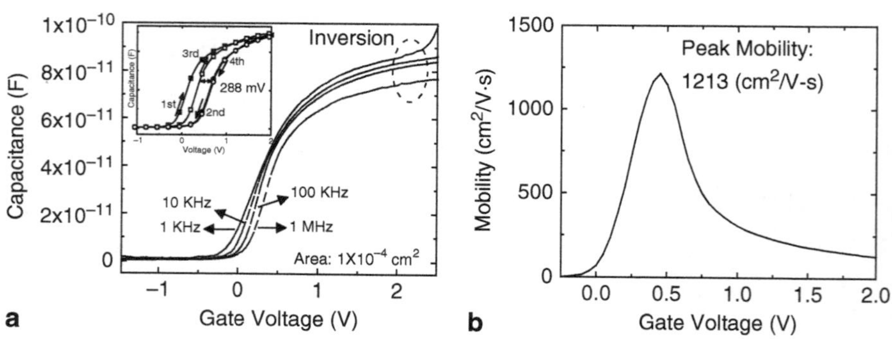

Fig. 11.31 **a** Frequency dispersion in inversion area between gate and source combined with drain of n-MOSFET with Si 80 s and PMA 900 °C (*inset* figure show hysteresis 288 mV). **b** Mobility with Si passivation (W 800 μm × L 10 μm)

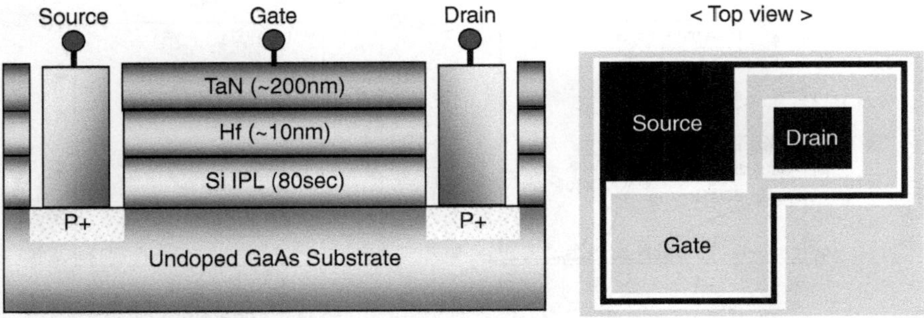

Fig. 11.32 Schematic cross section and top view of p-MOSFET on undoped GaAs substrate with Si passivation 80 s

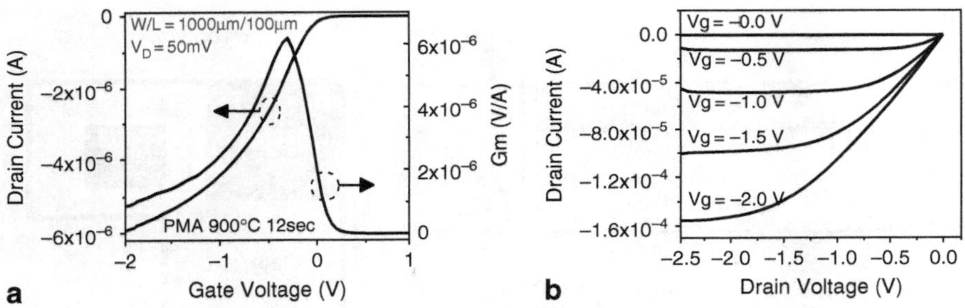

Fig. 11.33 **a** p-MOSFET I_d–V_g and G_m with peak G_m 6.2 (μA/V), V_{th} −0.03 V, and substrate swing 178 mV/decade. **b** I_d–V_d of p-MOSFET with Si (80 s) passivation (W 1000 μm × L 10 μm)

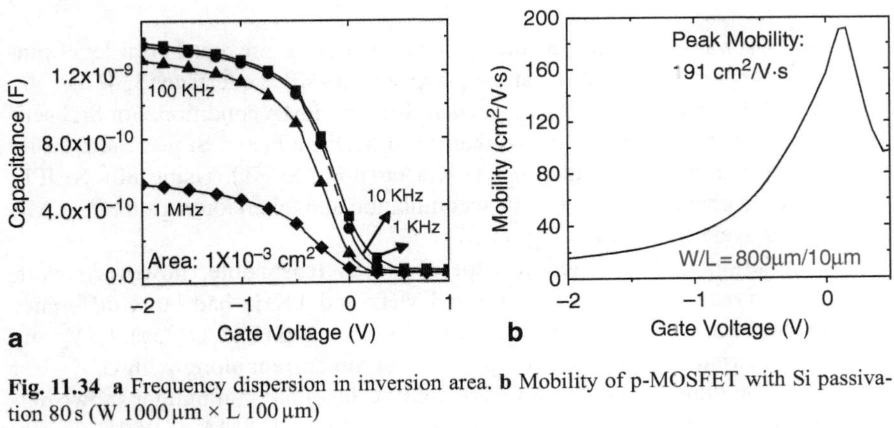

Fig. 11.34 **a** Frequency dispersion in inversion area. **b** Mobility of p-MOSFET with Si passivation 80 s (W 1000 μm × L 100 μm)

11.2.3.3 n-MOSFET Characteristics on p-GaAs Substrate

n-MOSFETs with Si IPL of 80 s were fabricated employing the same gate stack using a ring-FET on p-type GaAs substrate. The schematic cross section and top view of n-MOSFET with Si passivation is shown in Fig. 11.36. Si IPL layer with

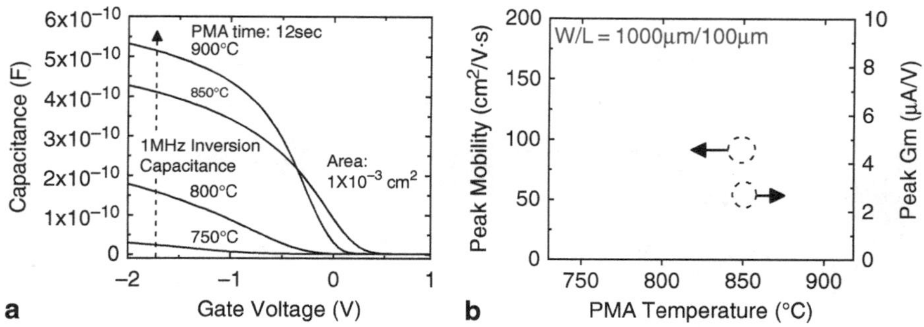

Fig. 11.35 **a** Inversion capacitance improvement **b** Mobility and peak G_m improvement with different PMA 750–900 °C 12 s

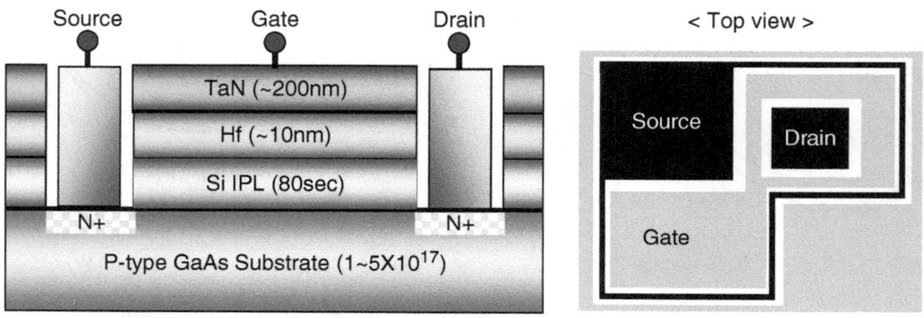

Fig. 11.36 Schematic cross section of n-MOSFET with Si passivation 80 s on p-type GaAs substrate

thickness 80 s was deposited for transistor fabrication to prevent Fermi level pinning after high temperature PMA at the p-type GaAs–HfO_2 interface [3, 4].

Figure 11.37a shows I_d and G_{m_max} versus different PMA condition. For S/D activation, Zn was more easily activated than Si (p-MOSFET) and Si was more easily activated in undoped GaAs than p-type GaAs (n-MOSFET). Using 80 s Si IPL, Excellent C–V characteristics in both accumulation and inversion region after PMA 950 °C 1 min were observed in Fig. 11.37b.

Accumulation C–V frequency dispersion was reasonable, however, inversion C–V curves for frequency between 1 MHz and 1 KHz had large difference because source and drain region were not fully activated (Fig. 11.35a). I_d–V_g and I_d–V_d characteristics are shown in Fig. 11.38. Drain current along with G_{m_max} for n-MOSFET on undoped GaAs and PMA (900 °C 60 s) thermal budget shows better characteristics than n-MOSFET on p-type GaAs with PMA of 950 °C 1 min. n-MOSFET with p-type GaAs shows V_{th} of 0.24 V, off current of ~10^7, transconductance G_{m_max} of 516.7 μA/V and swing S of 121.81 mV/decade with width 800 μm and length 10 μm at drain voltage (V_d = 50 mV); along with excellent I_d–V_d characteristics that I_{d_max} = 12.2 mA at $V_g - V_{th}$ = 2.0 V. n-MOSFET with

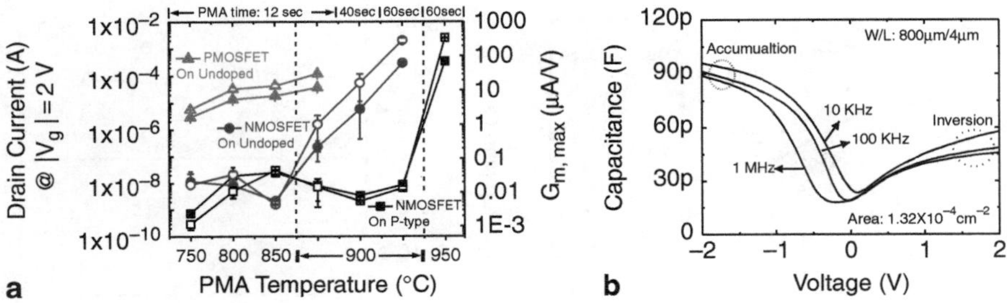

Fig. 11.37 **a** I_d at $|V_g| = 2$ [V] (*filled symbol*) and G_{m_max} (*empty symbol*) versus PMA temperature and time (750 °C 12 s, 800 °C 12 s, 850 °C 12 s, 900 °C 12 s, 900 °C 30 s, 900 °C 60 s, and 950 °C 60 s, respectively) with Si IPL (80 s deposition time) for p-MOSFET and n-MOSFET **b** frequency dispersion for accumulation and inversion C–V of n-MOSFET with 80 s Si IPL on p-type (PMA 950 °C 1 min) GaAs

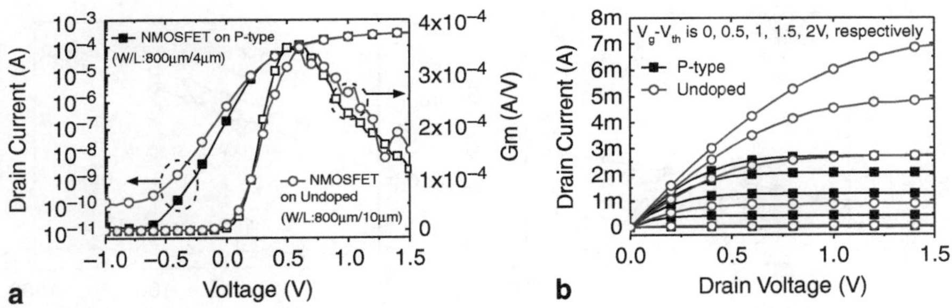

Fig. 11.38 **a** I_d–V_g and G_m **b** I_d–V_d of n-MOSFET with Si passivation with undoped GaAs (W 800 μm × L 10 μm) with p-type GaAs substrate (W 800 μm × L 4 μm)

undoped GaAs shows V_{th} of 0.24 V, off current of ~10^7, transconductance G_{m_max} of 516.7 μA/V and swing S of 121.81 mV/decade with width 800 μm and length 10 μm at drain voltage (V_d = 50 mV); along with excellent I_d–V_d characteristics that I_{d_max} = 12.2 mA at $V_g - V_{th}$ = 2.0 V.

After 950 °C PMA, leakage current remains low (Fig. 11.39a). Using charge-pumping technique with n-MOSFET on p-type GaAs, D_{it} value of ~1.2×10^{12}cm^{-2}eV^{-1} was obtained (Fig. 11.39b), which is similar to what was obtained from conductance method (Fig. 11.28a).

Electrical stress has been applied to examine reliability characteristics of these high-k GaAs MOSFETs (e.g., stress-induced V_{th} shift in Fig. 11.40). The threshold voltage shift magnitude was similar to stress induced V_{fb} on MOSCAP on n-type GaAs (Fig. 11.40b). Voltage shift was more depended on Hf thickness than on Si IPL thickness. After PMA V_{th} shift was slightly reduced due to increasing of leakage current.

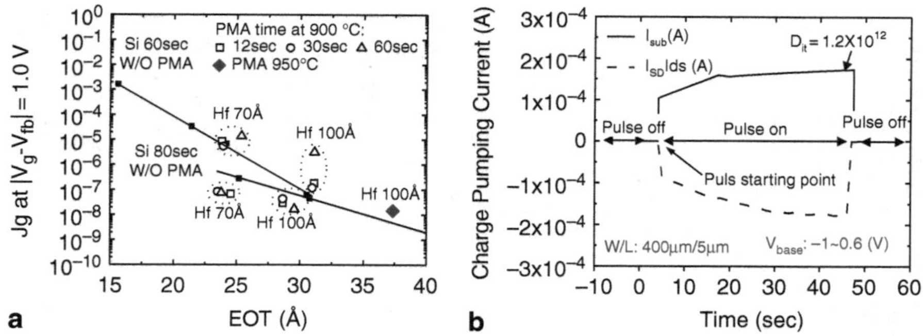

Fig. 11.39 **a** Leakage current density versus EOT of n-MOSFET with 80 s Si IPL after 950 °C PMA (PDA and 900 °C PMA condition were shown for comparison). **b** Charge pumping currents measurement for evaluation of the interface density (D_{it}) with n-MOSFET (W 400 μm × L 5 μm) on p-type GaAs substrate with PMA of 950 °C 60 s

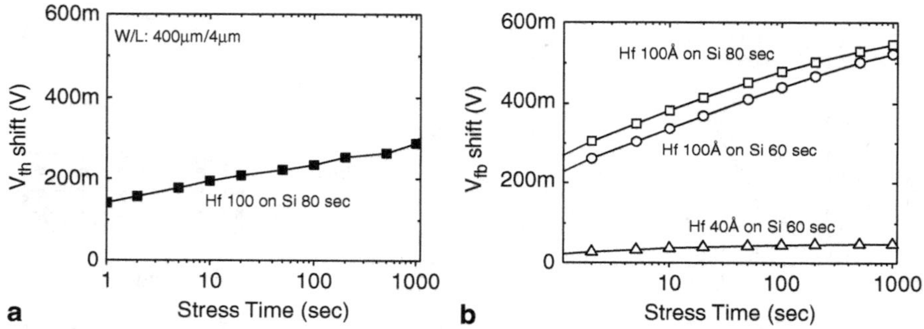

Fig. 11.40 **a** Stress induced V_{th} shift of n-MOSFET with Si passivation 80 s (W 400 μm × L 4 μm) on p-type GaAs substrate with 950 °C PMA. **b** Stress induced V_{fb} shift of MOSCAP with Si 60 s and 80 s thickness passivation and Hf 40 Å and 100 Å on n-type GaAs wafer without PMA

11.2.4 Temperature Effects of Si IPL Deposition on GaAs MOS Characteristics

In this chapter, we present the electrical characteristics of TaN/HfO_2/GaAs MOS capacitors with Si IPL under various PDA (post-deposition anneal) condition and various Si deposition temperature/time. Using optimal Si IPL under reasonable PDA conditions and various Si deposition temperatures, electrical characteristics will be presented. It was found that higher temperature of Si IPL deposition (at 300–400 °C) and longer PDA time at 600 °C improved EOT and leakage current.

Figure 11.41 shows C–V characteristics on the MOSCAP with 10 nm thickness of HfO_2 with Si 80 s deposition time (1.2 nm) and PDA 600 °C 15 min condition varying deposition temperature for Si IPL. Figure 11.41b depicts EOT versus Si IPL deposition temperature with increasing deposition temperature, accumulation capacitance increased (i.e., EOT decreased), possibly due to densification of Si IPL

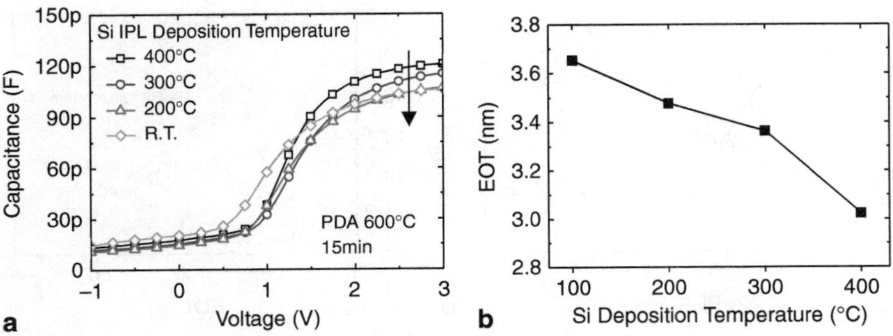

Fig. 11.41 **a** 1 MHz C–V characteristics of MOSCAPs with 1.2 nm thick of amorphous Si IPL as function of Si IPL deposition temperature. **b** EOT versus Si deposition temperature

layer at higher temperature. It has also been reported that the Si IPL with lower deposition temperature could be easier to be oxidized, which would result in thicker EOT. Figure 11.42 summarizes the frequency dispersion characteristics versus PDA time as a function of Si deposition temperature. With longer PDA time, low frequency dispersion (<5% and <50 mV) can be obtained.

This is due to more complete oxidation of the Si IPL layer and consequently reduced charge trapping. It has been found that the thinner the remaining unoxidized Si IPL layer is the higher interface quality (Fig. 11.11b). This can be explained by the quantum control well model or simply reduced charge trapping due to the reduced thickness of the highly trapped unoxidized Si layer. PDA time has more impact on frequency dispersion than Si IPL deposition temperature.

In general, longer PDA time lead to lower hysteresis for n-MOSCAP possibly due to the reduction of charge trapping in the high-k layer (Fig. 11.43a). Figure 11.43b shows the leakage current density at $V_g - V_{fb} = 1$ versus equivalent oxide thickness (EOT) with different Si deposition temperature. In general, higher Si deposition temperature led to lower J_g, possibly due to densified Si IPL that prevents As out-diffusion.

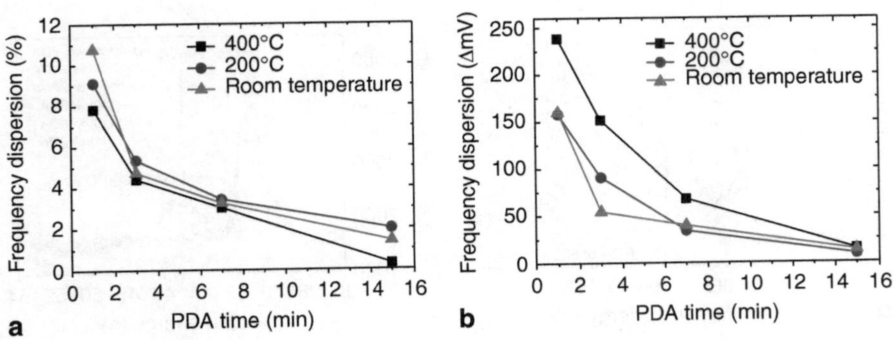

Fig. 11.42 Frequency dispersion versus PDA time as a function of Si IPL deposition temperature: **a** capacitance difference (%) and **b** voltage difference (mV) at flat band voltage between 1 MHz and 1 KHz

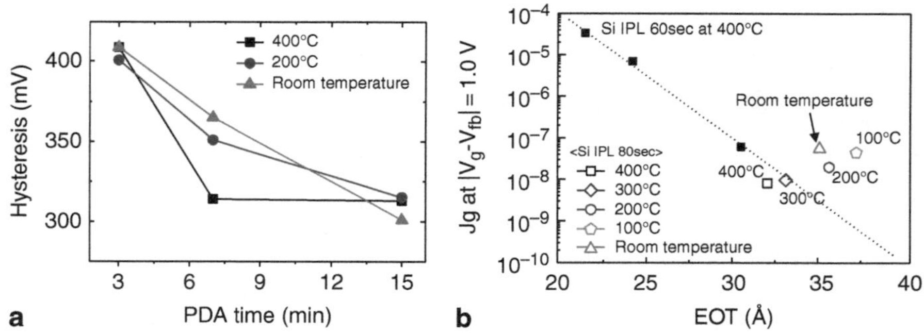

Fig. 11.43 **a** Hysteresis for 1 MHz CVs (sweep rate: 500 mV/s) versus PDA time. **b** Leakage current density versus EOT as a function of Si IPL deposition temperature

11.2.5 Influence of the Substrate Orientation on the Electrical and Material Properties of GaAs MOSCAPs and Self-Aligned Transistors Using HfO_2 and Silicon IPL

In this section, we will present the influence of the substrate orientation on the electrical and material characteristics of TaN/HfO_2/GaAs GaAs n-type Metal-Oxide-Semiconductor Capacitors (NMOSCAP) and PMOSCAP and self-aligned n-MOSFETs. There have been only a few reports on the orientation dependence of GaAs substrates [5–8]. The results suggested that orientation dependence could be explained by using GaAs/Si coupling energy [6, 7].

MOS capacitors were fabricated on n-type and p-type GaAs doped with Si and Zn for different orientation (100), (110), and (311). Figure 11.44 shows XPS results for different substrate on (100), (110), and (311) with and without PDA. As shown in Fig. 11.44, As_2O_3 and SiO_2 were obtained due to the PDA. The formation of SiO_2 and As_2O_3 was similar for different substrate after PDA of 600 °C 7 min (Fig. 11.44b).

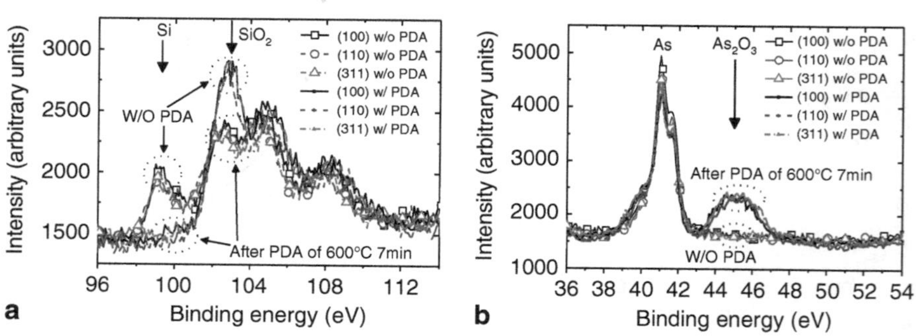

Fig. 11.44 XPS spectra of **a** Si and SiO_2 **b** As and As_2O_3 with (*line*) and without (*empty*) PDA of 600 °C 7 min on GaAs with Si IPL of 60 s deposition time

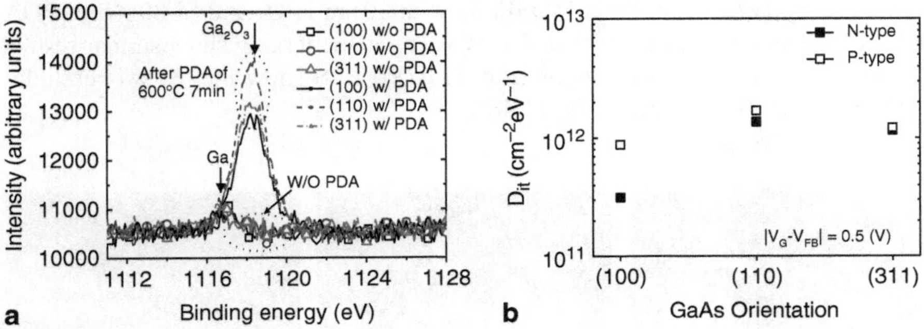

Fig. 11.45 XPS spectra of **a** Ga and Ga_2O_3 with (*line*) and without (*empty*) PDA of 600 °C 7 min on GaAs with Si IPL of 60 s deposition time. **b** D_{it} measured by conductance method at $|V_g - V_{fb}| = 0.5$ (V) using parameter adjustment on n-type and p-type GaAs substrate

In particular, the (110)-orientation shows the highest intensity and the (100) shows the lowest intensity of Ga_2O_3 spectra (Fig. 11.45a). Interface state density measurement (D_{it}) shows that the (100) oriented substrate exhibits slightly better interface quality on both n-type and p-type in Fig. 11.45b.

In terms of frequency dispersion of MOSCAPs, the (100) oriented substrate also exhibits slightly lower ΔmV values (Fig. 11.46b). The result suggests that (100) orientation exhibits the lowest interface defect density and PDA can be used to reduce these defects. However, the frequency dispersion (%) of MOSCAPs on each different orientation substrates was found to be very similar with different PDA condition (Fig. 11.46a).

GaAs n-MOSFET was fabricated employing the same gate stack using ring-FET and Si IPL (80 s) and 900 °C PMA on undoped GaAs substrate with different orientation [(100), (110) and (311)]. Si IPL layer was deposited for transistor fabrication to prevent Fermi level pinning at the GaAs-HfO_2 interface. A cross-sectional high-resolution transmission electron microscopy (HRTEM) image and energy loss

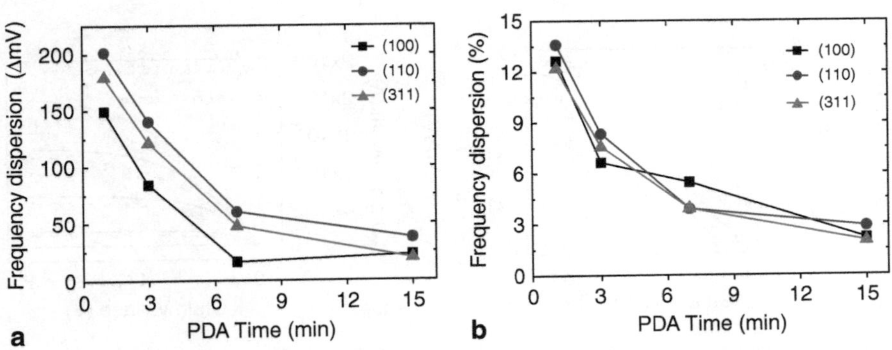

Fig. 11.46 **a** Frequency dispersion of ΔmV **b** Frequency dispersion of % with different PDA time at 600 °C and Si 60 s deposition thickness for different orientation

spectroscopy (EELS) on the n-MOSFETs with 10 nm HfO_2 and Si 80 s with PDA of 600 °C 3 min after PMA of 900 °C 60 s after gate patterning show similar results for different substrate orientations (Fig. 11.47). EELS show that Si was partially oxidized after PMA process (Fig. 11.48).

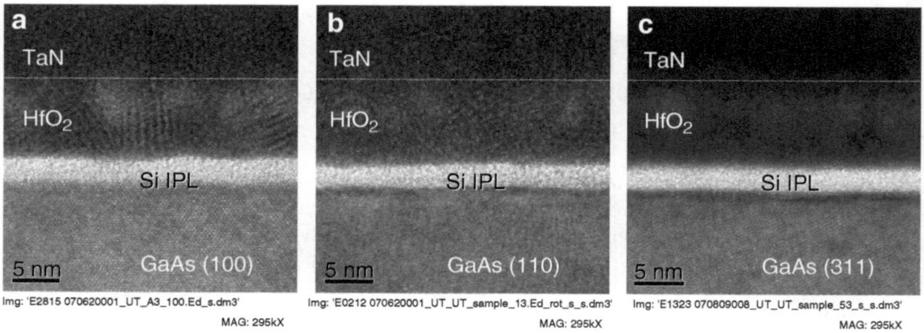

Fig. 11.47 A cross-sectional high resolution transmission electron microscopy (HRTEM) image for **a** (100) **b** (110) **c** (311) on the n-MOSFETs with 10 nm HfO_2 and Si 1.5 nm with PDA of 600 °C 3 min after PMA of 900 °C 60 s

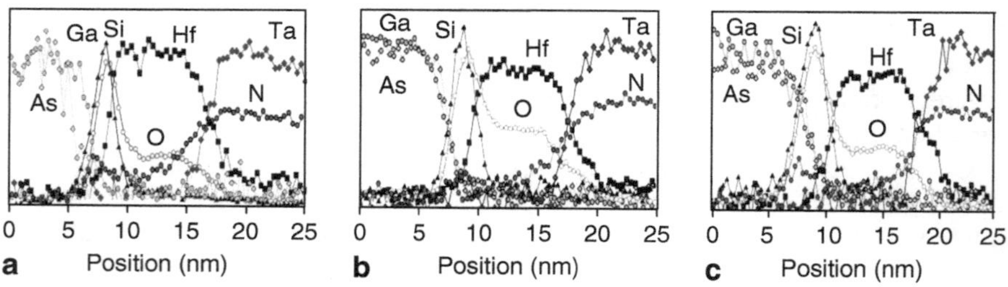

Fig. 11.48 Electron energy loss spectroscopy (EELS) for **a** (100) **b** (110) **c** (311) on the n-MOSFETs with 10 nm HfO_2 and Si 1.5 nm with PDA of 600 °C 3 min after PMA of 800 °C 30 s

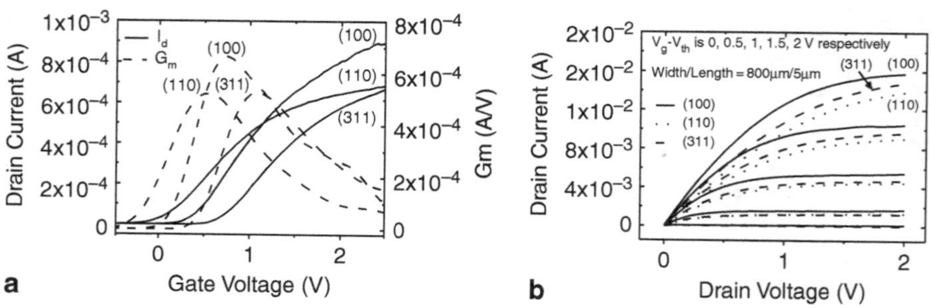

Fig. 11.49 **a** Drain current versus gate voltage with G_m **b** Drain current versus drain voltage with different substrate orientation with (100), (110), and (311) for n-MOSFET with Si Passivation (80 s) on undoped GaAs substrate

Orientation of (100) substrate shows better transistor characteristics in terms of drain current versus gate voltage, G_m, and drain current versus drain voltage (Fig. 11.49). The improved MOSFET characteristics of (100) devices can be attributed to better interface quality.

11.2.6 Metal Gate—HfO_2 MOS Structures on GaAs Substrate with SiGe IPL for Scaling Down

Using different Si and Ge deposition rate, we adjust the Si:Ge ratio in the SiGe IPL. MOS structures were fabricated on n-type GaAs wafer. SiGe IPL was deposited by sputtering of Si and Ge. Figure 11.50 shows C–V characteristics on the MOSCAP with 7 nm thickness of HfO_2 with optimum deposition time and PDA condition at 600 °C varying Si:Ge ratio.

As Ge content increases, V_{fb} shifts in negative direction while inversion capacitance increases. EOT reduction with increasing Ge content is shown more clearly in Fig. 11.50b. For Ge:Si ≥ 18:1, EOT is ~1 nm. Frequency dispersion characteristics (% and ΔmV) versus different IPL deposition condition are summarized in Fig. 11.51.

Generally, Ge IPL MOSCAP exhibited larger hysteresis than Si IPL counterpart. Hysteresis also can be adjusted with different Ge:Si ratio (Fig. 11.52a). Pure Si IPL devices exhibited the least hysteresis, as expected. Capacitors with Ge-rich IPL have very small process window compare to Si IPL. The variation of frequency dispersion for Ge-rich IPL is also larger possibly because the deposition rate is higher than Si deposition. With appropriate thickness of SiGe IPL with optimum PDA condition, low frequency dispersion (<5%) can be obtained. Leakage current also can be changed with different Ge:Si ratio (Fig. 11.52b). Ge-rich IPL shows slightly higher leakage current (~10^{-4} A/cm^2) than Si-rich IPL (~10^{-5} A/cm^2).

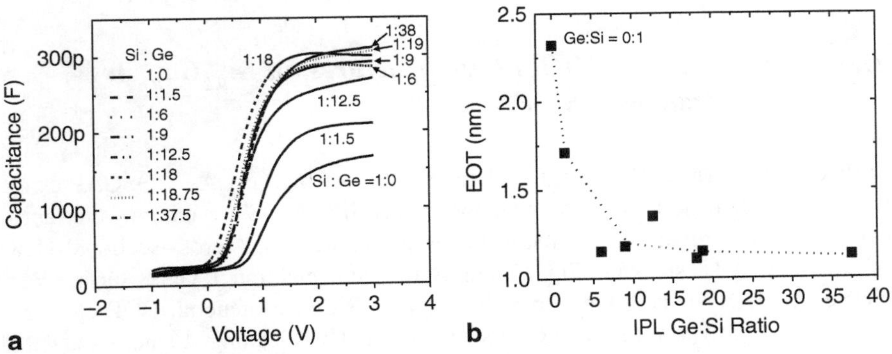

Fig. 11.50 **a** C–V with various Si:Ge ratio IPL for area 1.23 × 10^{-4}. **b** EOT (nm) versus Ge:Si ratio in IPL

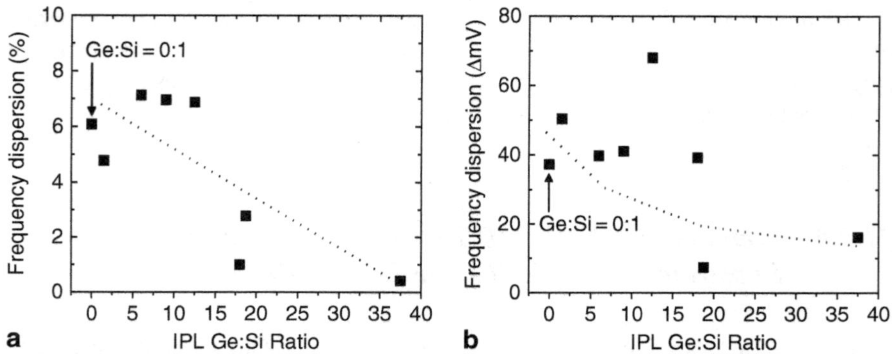

Fig. 11.51 **a** Frequency dispersion (%) between 1 KHz and 1 MHz at 2.5 V on n-type GaAs versus Ge:Si ratio in IPL. **b** Frequency dispersion (ΔmV) between 1 KHz and 1 MHz at flat band voltage on n-type GaAs versus Ge:Si ratio in IPL

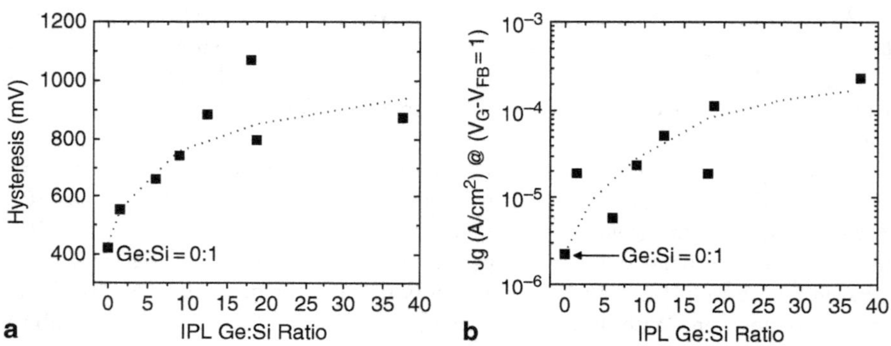

Fig. 11.52 **a** Hysteresis versus Ge:Si ratio in IPL. **b** Leakage current density J_g (A/cm^2) at $|V_g - V_{fb}| = 1$ versus Ge:Si ratio in IPL

11.3 MOSCAPs and MOSFETs on InGaAs with Si IPL

11.3.1 Metal Gate—HfO_2 MOS Structures on $In_{0.2}Ga_{0.8}As$ Substrate with Si IPL

In this section, we will present the characteristics of $In_{0.2}Ga_{0.8}As$ MOSCAP using the same oxide of HfO_2 as gate insulator with Si IPL. MOS capacitors were fabricated on MBE grown n-type InGaAs wafer doped with Si. A cross-sectional view of the MOSCAP is shown in Fig. 11.53a. Before Si deposition, InGaAs surface was cleaned by BOE or HCl with or without $(NH_4)_2S$ dip. In general, BOE cleaning resulted in slightly lower frequency-dispersion than HCl cleaning. As increased PDA temperature, $(NH_4)_2S$ dip resulted in a more stable interface between Si and InGaAs surface due to sulfide passivation (Fig. 11.53b). Thus, the frequency-dispersion

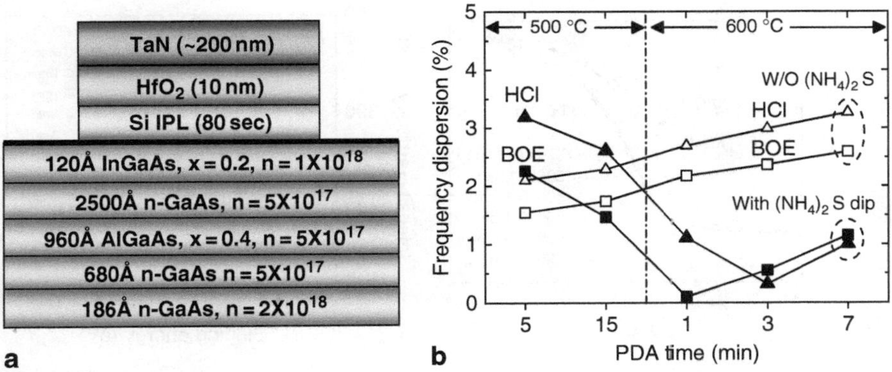

Fig. 11.53 **a** A cross-sectional view of the n-$In_{0.2}Ga_{0.8}As$ MOSCAP structure **b** frequency-dispersion (%) with different chemical cleaning methods as a function of PDA time (500 °C 5 min, 500 °C 15 min, 600 °C 1 min, 600 °C 3 min, and 600 °C 7 min)

remained low even with higher PDA temperature and time. After Hf deposition following by PDA, X-ray photoelectron spectroscopy (XPS) was measured. As shown in Fig. 11.54a, Ga_2O_3 was obtained through the PDA and the PDA above 600 °C was shown that Ga_2O_3 intensity was much higher than Ga_2O_3 intensity with 500 °C PDA as well as As_2O_3 (Fig. 11.54).

While the As–O and Ga–O bonds were observed after anneal, no other distinguishable bonds were observed and In *3d* remained unchanged with different annealing condition (data are not shown). XPS showed that with 500 °C and 600 °C 1 min PDA, Si was partially oxidized, and after 600 °C 3 min PDA, the Si IPL was oxidized to SiO_2 (Fig. 11.55b). Figure 11.55a shows C–V characteristics with 80 s Si IPL and different PDA condition. C–V characteristics of TaN/HfO_2/Si IPL/InGaAs/GaAs MOSCAP show similar behavior as those on GaAs-only substrate even though the depletion region might extend throughout the InGaAs layer and into GaAs substrate. As increase the PDA temperature and time, flat band voltage shift negative direction (Fig. 11.55a) with SiO_2 formation (Fig. 11.55b) [9].

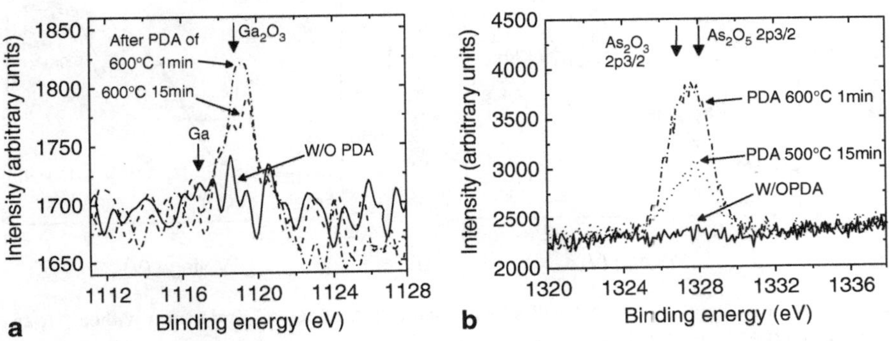

Fig. 11.54 XPS spectra of the Si *2p* (**a**) and As *2p* (**b**) for 80 s Si IPL thickness with different PDA time

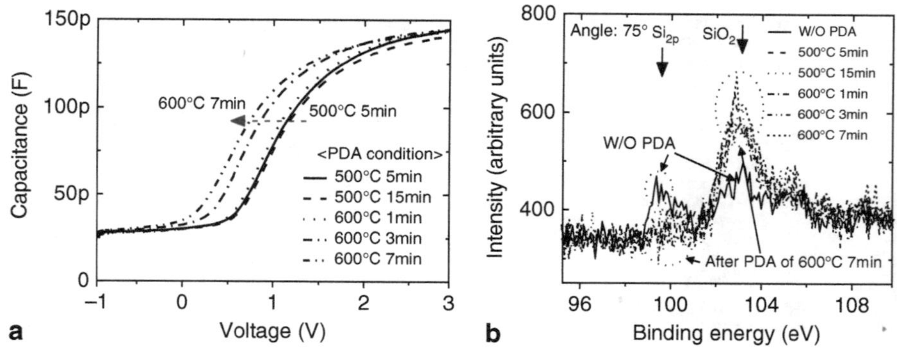

Fig. 11.55 **a** 1 MHz CVs for InGaAs and **b** XPS: Si spectra as a function of different PDA condition

Hydrogen incorporation effects was also investigated for $In_{0.2}Ga_{0.8}As$ MOSCAP using the same oxide of HfO_2 as gate insulator with Si IPL and H_2 annealing at 500 °C for 30 min. Electrical characterization was performed on MOS capacitors before and after H_2 annealing. Typical C–V characteristics of TaN/HfO_2/Si/$In_{0.2}Ga_{0.8}As$ as function of frequency with PDA condition of 600 °C 7 min before and after H_2 annealing for 80 s Si IPL are shown in Fig. 11.56. The slopes of the flat band voltage shift for with Si IPL and without Si IPL structure are similar to each other. The results suggest that flat band voltage shift was more affected by interfacial InGaAs oxide than Si oxidation and one of the clear contributions was As_2O_3, Ga_2O_3 (Fig. 11.54) and hydrogen incorporation compensated the As oxide contributions (Fig. 11.57). Hydrogen incorporation resulted in similar hysteresis (mV) (Fig. 11.57b). Thin amorphous Si IPL between HfO_2 and $In_{0.2}Ga_{0.8}As$ substrate with hydrogen incorporation did not significantly contribute to hysteresis characteristics which are more significantly affected by bulk HfO_2 than interface.

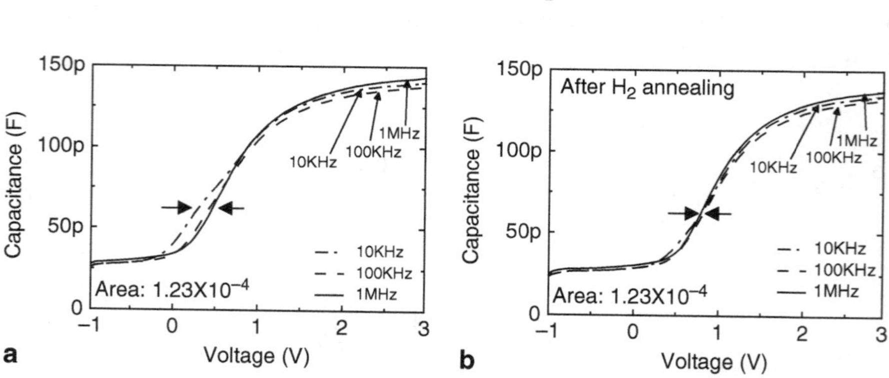

Fig. 11.56 Frequency dispersion for 80 s Si IPL with PDA of 600 °C 7 min for **a** without H_2 annealing and **b** with H_2 annealing

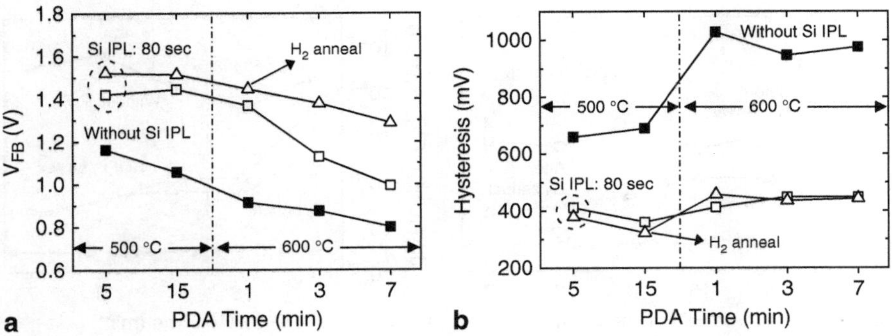

Fig. 11.57 **a** V_{fb} shift and **b** Hysteresis for 1 MHz CVs (sweep rate: 500 mV/s) as a function of PDA time (500 °C 5 min, 500 °C 15 min, 600 °C 1 min, 600 °C 3 min, and 600 °C 7 min)

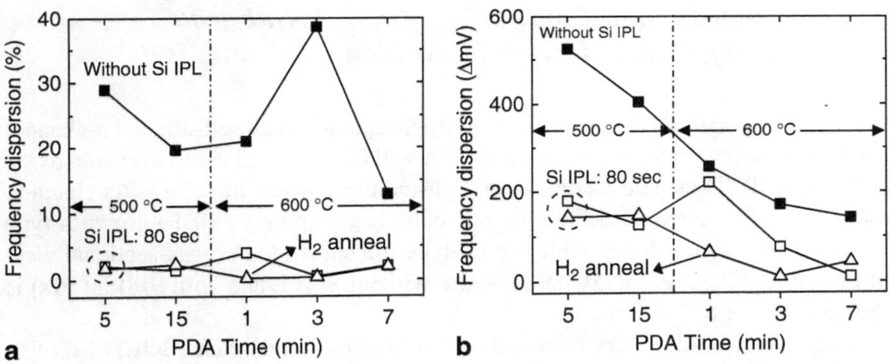

Fig. 11.58 **a** Frequency dispersion (%) **b** Frequency dispersion (ΔmV) as a function of PDA time with different Si IPL 80 s

With appropriate thickness of Si IPL and PDA condition, reduced frequency-dispersion (<5%) can be obtained. In contrast, without Si IPL, frequency-dispersion was around 15–40%. Si IPL was more effective to improve interface quality than different surface cleaning (Figs. 11.58 and 11.53b). In general, hydrogen incorporation resulted in slightly lower frequency dispersion (% and ΔmV).

Si deposition time of 80 s with 7 min PDA time resulted in ~10^{13} D_{it} value which was similar to GaAs based structure using conductance method (Fig. 11.12). Hydrogen incorporation samples resulted in lower D_{it} value ~2.8×10^{12} (~four times less than W/O H_2 annealing) (Fig. 11.59a). Si IPL led to lower leakage current and Si IPL with hydrogen incorporation led to similar leakage current (Fig. 11.59b) which was slightly less than the same structure with GaAs substrate.

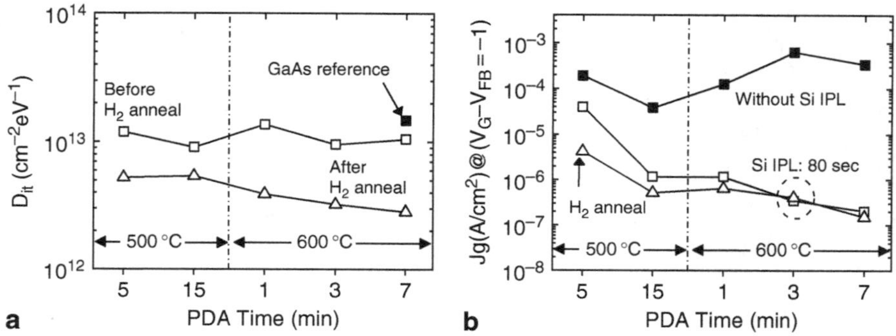

Fig. 11.59 **a** D_{it} and reference D_{it} of the same structure on GaAs substrate **b** Leakage current density as a function of PDA time with Si IPL thickness 80 s

11.3.2 Metal Gate—HfO$_2$ MOS Structures on High-Indium-Content In$_{0.53}$Ga$_{0.47}$As Substrate Using Physical Vapor Deposition

Electrical and material characteristics of MOS capacitors and self-aligned n-channel MOS transistor with high temperature PMA will be presented using the same oxide of HfO_2 by PVD as gate insulator on high-indium-content $In_{0.53}Ga_{0.47}As$ channel. MOS capacitors were fabricated on molecular beam epitaxy (MBE) grown n-type $In_{0.53}Ga_{0.47}As$ wafers doped with Si on n-type InP substrate. A cross-sectional view of the n-$In_{0.53}Ga_{0.47}As$ MOSCAP structure without Si IPL and with Si IPL (80 s) is shown in Fig. 3.8.

The surface oxides were removed with a buffered-oxide-etch (BOE) clean followed by $(NH_4)_2S$ dip, resulting in a clean S-passivated $In_{0.53}Ga_{0.47}As$ surface. For the purpose of improving the interface quality as in the case of GaAs, Si IPL was used for capacitors and transistor structures (Fig. 11.60b). However, contrary to the case of GaAs, the interface between HfO_2 and $In_{0.53}Ga_{0.47}As$ surface show no distinct interface layer (IL) or any degradation even without IPL, as shown in a cross-sectional high resolution transmission electron microscopy (HRTEM) image of the gate stack with PDA 3 min at 600 °C (Fig. 11.61a). Electron energy loss spec-

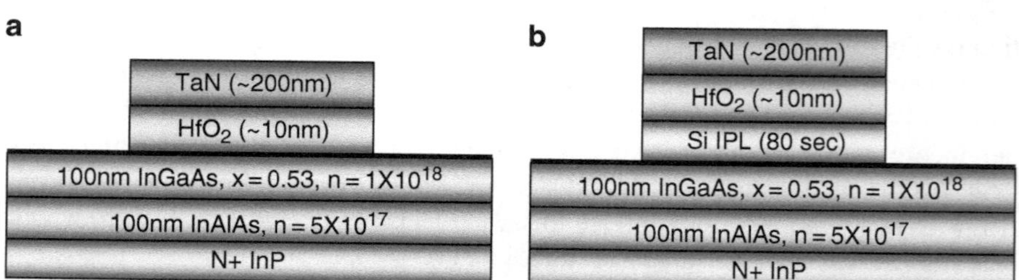

Fig. 11.60 A cross-sectional view of the n-$In_{0.53}Ga_{0.47}As$ MOSCAP structure **a** without Si IPL and **b** with Si IPL (80 s)

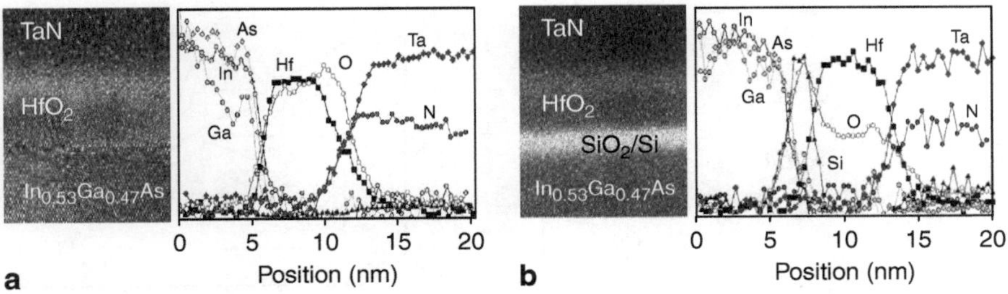

Fig. 11.61 A cross-sectional high resolution transmission electron microscopy (HRTEM) image of the gate stack on $In_{0.53}Ga_{0.48}As$ with PDA of 600 °C 3 min **a** without Si IPL **b** with Si IPL (80 s) with EELS analysis on both (**a**) and (**b**)

troscopy (EELS) shows that n-$In_{0.53}Ga_{0.47}As$ surface was oxidized due to oxidation process after Hf deposition (Fig. 11.61).

XPS showed that as increased PDA time at 600 °C, Ga_2O_3 and As_2O_3 peak from reference (minimum line of each XPS intensity) on GaAs were much higher than on $In_{0.53}Ga_{0.47}As$ (Fig. 11.62a). The variation for Ga_2O_3 peak between before and after PDA was small for both GaAs and $In_{0.53}Ga_{0.47}As$ samples. However, for high-indium-content $In_{0.53}Ga_{0.47}As$, Ga_2O_3 and As_2O_3 peaks were suppressed possibly due to indium or indium oxide (Fig. 11.62b). EOT of samples is shown as a function of oxide physical thickness in Fig. 11.63a in which PDA of 500 °C 5 min samples exhibit slightly smaller EOTs as compared to PDA of 600 °C 1 min samples. EOT of sample with Hf 3 nm and PDA 500 °C 5 min shows ~1.0 nm EOT value due to less SiO_2 formation during PDA. The linear dependence of EOT on the physical thickness of HfO_2 give dielectric constant of HfO_2 around 23–25 with $K_{SiO_2} = 3.9$.

Figure 11.63b illustrates the leakage current density (J_g) at $V_g = V_{fb} + 1\,V$ versus EOT. Leakage current was reduced (to ~$7.5 \times 10^{-3}\,A/cm^2$) for 3.0 nm thickness of high-k dielectric on n-type $In_{0.53}Ga_{0.47}As$ wafer. High-indium-content $In_{0.53}Ga_{0.47}As$ led to lower leakage current compare to low indium content $In_{0.2}Ga_{0.8}As$ and GaAs.

Figure 11.64a summarizes the frequency dispersion characteristics versus PDA time with different indium content substrate. The frequency dispersion could be as

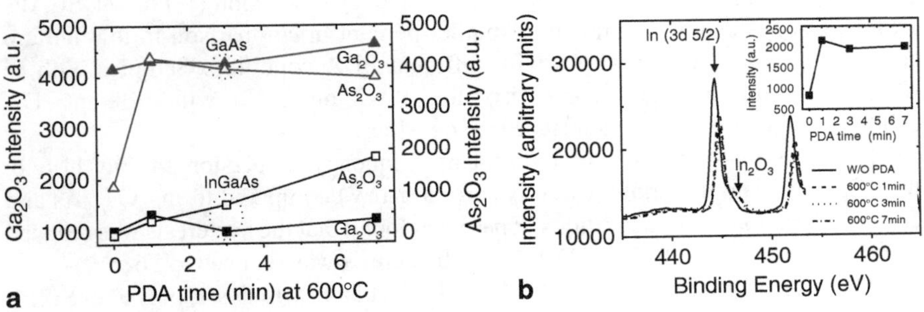

Fig. 11.62 **a** XPS spectra: peak value of the Ga_2O_3 (*2p* 3/2) and As_2O_3 (*2p* 3/2) from reference to peak with different PDA time (min) on GaAs and $In_{0.53}Ga_{0.47}As$ **b** XPS spectra of the In (*3d* 5/2) and In_2O_3 with different PDA time on $In_{0.53}Ga_{0.47}As$. *Inset* shows In_2O_3 (*3d* 5/2) peak value from the reference to peak with different PDA time

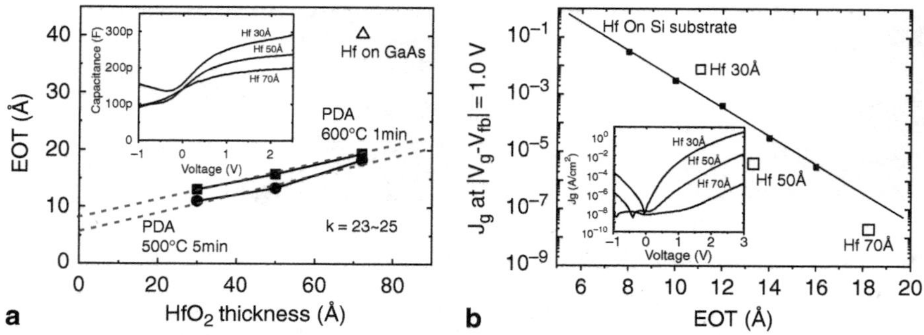

Fig. 11.63 **a** Equivalent oxide thickness (EOT) versus oxide (HfO_2) physical thickness with different PDA condition **b** leakage current density (J_g) versus EOT (Å) at V_{fb} + 1 [V] with $In_{0.53}Ga_{0.47}As$

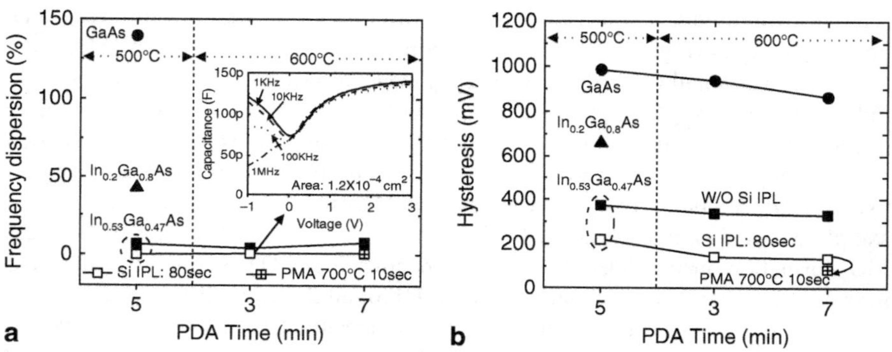

Fig. 11.64 **a** Comparison of frequency dispersion **b** Hysteresis for 1 MHz CVs (sweep rate: 500 mV/s) versus PDA time on substrates with different indium content without and with Si IPL 80 s and HfO_2 10 nm and PMA (700 °C 10 s) for the sample with PDA of 600 °C 3 min

large as 140% for GaAs MOSCAPs in the frequency ranging from 1 KHz to 1 MHz. This implies high interface trap densities. Frequency dispersion (%) of MOSCAPs on $In_{0.53}Ga_{0.47}As$ show significantly smaller percent in comparison to low indium content $In_{0.53}Ga_{0.47}As$ and GaAs with different PDA condition. Samples with Si IPL show smaller percent in comparison to that without Si IPL with different PDA condition. This implies low interface trap densities.

Hysteresis also shows similar trend as frequency dispersion in Fig. 11.64b. $In_{0.53}Ga_{0.47}As$ samples show lower hysteresis (ΔmV) compared to $In_{0.2}Ga_{0.8}As$ and GaAs with various PDA conditions. It has been found that the hysteresis was reduced significantly using the Si IPL. With PMA, hysteresis was reduced to 86 mV.

EOT is shown as a function of PDA in Fig. 11.65a in which samples with Si IPL exhibit larger (~2.9 nm) EOTs as compared to samples without Si IPL (~1.6 nm) due to SiO_2 formation during PDA. Unoxidized thin amorphous Si IPL between HfO_2/SiO_2 and $In_{0.53}Ga_{0.47}As$ substrate did not significantly contribute EOT change

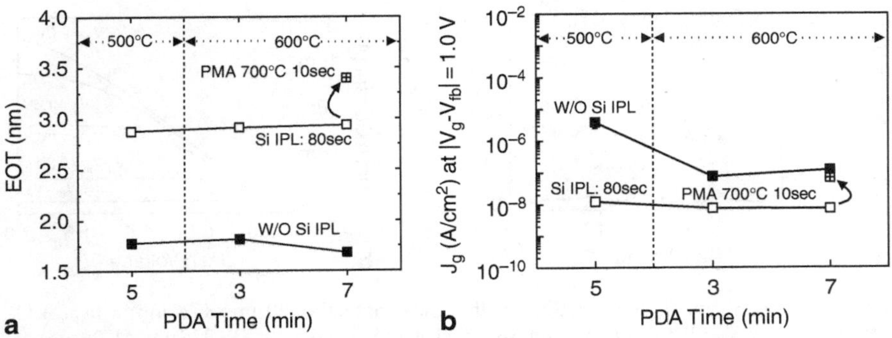

Fig. 11.65 **a** EOT and **b** leakage current at 1.5 V with different PDA time (500 °C 5 min, 600 °C 3 min, and 600 °C 7 min) without and with Si IPL 80 s and HfO_2 10 nm and PMA (700 °C 10 s) for the sample with PDA of 600 °C 3 min

after PDA. EOT was mostly related to HfO_2 thickness and interface oxide including SiO_2. The EOT values were found to be fairly independent of PDA condition.

This suggests that the formation of SiO_2 and native oxide such as In_2O_3, Ga_2O_3 and As_2O_3 was suppressed. The EOT was increased by 0.45 nm after PMA of 700 °C 10 s for the sample with Si IPL of 80 s, HfO_2 10 nm and PDA of 600 °C 7 min.

Figure 11.65b illustrates the leakage current density (J_g) as a function of PDA at V_g = 1.5 V. Leakage current was reduced (to ~7.4×10^{-9} A/cm²) for 10.0 nm thickness of high-k dielectric with 80 s of Si IPL on n-type $In_{0.53}Ga_{0.47}As$ wafer. Using the Si IPL reduced leakage current. In general, longer PDA time, thicker Si IPL led to lower leakage current. With PMA, the leakage current was increased to 6.8×10^{-8} A/cm² from 7.4×10^{-9} A/cm².

We will present the results of $In_{0.53}Ga_{0.47}As$ n-MOSFETs employing the same gate stack using a ring-FET. The schematic cross section and top view of n-MOSFET with Si passivation is shown in Fig. 11.66. Si IPL layer with thickness of 80 s deposition time was deposited for transistor fabrication.

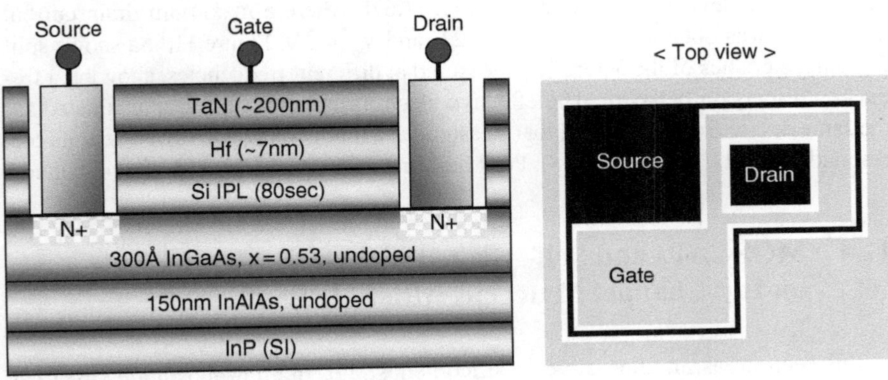

Fig. 11.66 Schematic cross section and top view of n-MOSFET with Si passivation 80 s

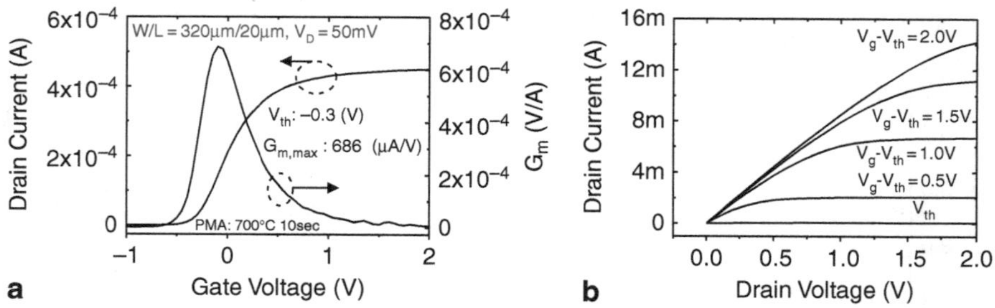

Fig. 11.67 **a** I_d–V_g and G_m of n-MOSFET with Si passivation (W 320 μm × L 20 μm) with peak G_m 686 (μA/V) and V_{th} −0.3 V. **b** I_d–V_d of n-MOSFET with Si passivation (W 320 μm × L 20 μm)

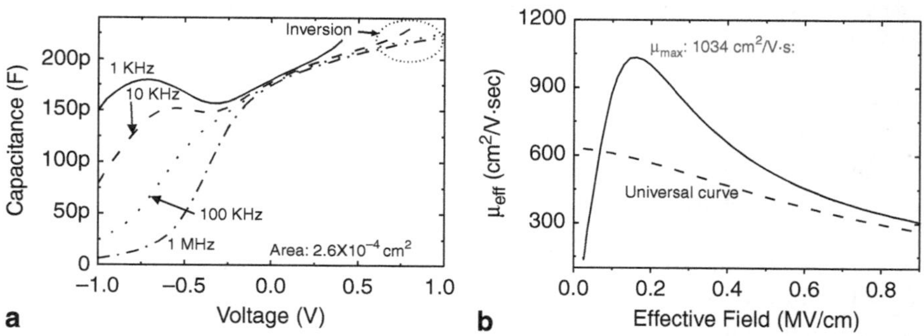

Fig. 11.68 **a** Frequency dispersion of n-MOSFET with Si 1.5 nm and PMA 700 °C 7 s. **b** Mobility of n-MOSFET with Si passivation (W 320 μm × L 20 μm) from split CV and I_d–V_g curve (*dashed line*: mobility of universal curve on Si substrate)

Figure 11.67a shows the transfer characteristics of an n-MOSFET with a gate width of 320 μm and length of 20 μm. A threshold voltage of −0.3 V was extracted by linear extrapolation technique. A maximum transconductance (G_m) of 686 (μA/V) was obtained for the device at a drain voltage of 0.05 V. The output characteristics of the MOSFET device are illustrated in Fig. 11.67b where a maximum drain current of 14.2 mA was obtained at a V_g–V_{th} of 2 V and V_d of 2 V. Figure 11.68a shows split CV characteristics of the MOSFET, measured at different frequencies, showing a frequency dispersion behavior. The effective electron mobility (μ_{eff}) was evaluated for the same device from the I_d–V_g plot (measured at a drain voltage of 50 mV) and its corresponding split C–V (Fig. 11.68). Peak mobility of 1034 cm^2/Vs has been obtained.

11.4 MOSCAPs and Self-Aligned n-channel MOSFETs on InP Channel Materials with Si IPL

In this section, as an alternative of silicon substrate, Indium-phosphide (InP) has been studied. MOS capacitors were fabricated on n-type (1~5 × 10^{17} cm^{-3}) InP (100) doped with Si. A cross-sectional view of the MOSCAP structure is shown in inset of

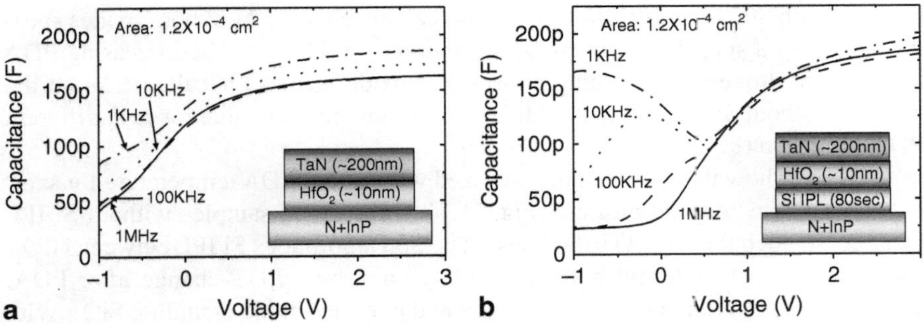

Fig. 11.69 A cross-sectional view of the n-InP MOSCAP structure **a** without Si IPL and **b** with Si IPL after PDA 1 min at 600 °C

Fig. 11.69. The surface oxides were etched with a buffered oxide etch (BOE) clean followed by $(NH_4)_2S$ dip, resulting in a clean S-passivated InP surface. Figure 11.69 illustrates CV characteristics of the MOS capacitors, fabricated on n-type substrates with different frequencies for sample without and with 80 s of Si IPL and PDA of 600 °C for 1 min. The capacitance increase in the inversion region is due to minority carrier generation in the substrate.

Figure 11.70 illustrates transmission electron microscopy (TEM) on the MOSCAP with 10 nm HfO_2 with PDA of 600 °C 3 min for the sample with and without Si IPL. Approximately 0.8 nm of the interface layer (IL) was formed between HfO_2 and InP surface during 600 °C 3 min PDA for sample without Si IPL (Fig. 11.70a). To improve the interface quality, Si IPL was used for capacitors and transistor structures (Fig. 4.2b).

Electron energy loss spectroscopy (EELS) shows that InP surface was oxidized due to oxidation process after Hf deposition. Most of oxygen signal near the substrate interface overlapped with indium signal while very small area of phosphorus signal overlapped with oxygen signal (Fig. 11.69a). After Hf deposition following by PDA, X-ray photoelectron spectroscopy (XPS) was measured. XPS showed that with increasing PDA temperature, In_2O_3 spectra increased. On the other hand,

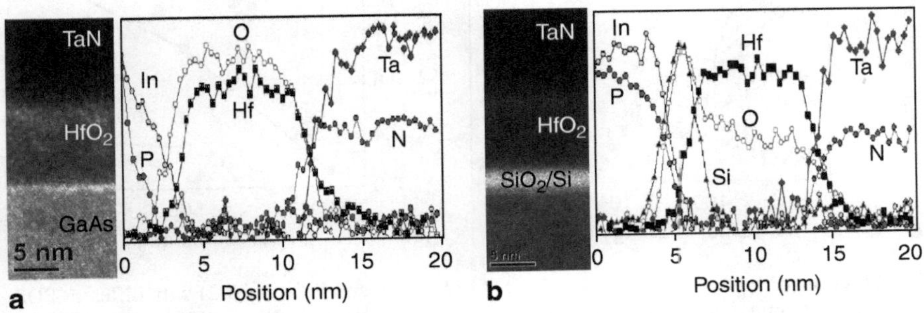

Fig. 11.70 **a** HRTEM image of the gate stack on InP with ~0.8 nm of native oxide after PDA of 600 °C 3 min without Si IPL **b** on InP with 80 s of Si IPL after PDA of 600 °C 3 min with EELS analysis on both (**a**) and (**b**)

samples with 80 s of Si IPL showed no appreciable changes in indium oxides spectra indicating a suppression of In_2O_3 formation (Fig. 11.70a). The increasing PDA temperature drove the remaining phosphorus oxide increase but the peaks on the samples without Si IPL and with Si IPL show almost similar value for each different PDA condition (Fig. 11.70b).

XPS also show that Si was more oxidized with higher PDA temperature for samples with 60 s Si deposition time (Fig. 11.71a). However, samples without Si IPL resulted in much thinner EOT thickness. The thin amorphous Si IPL between HfO_2/SiO_2 and InP substrate did not significantly contribute EOT change after PDA. EOT was mostly related to HfO_2 thickness and interface oxide including SiO_2. With increasing PDA time, Si layer was changed to SiO_2 and increased EOT with formation of In_2O_3 and P_2O_5 (Fig. 11.71b). So In_2O_3, P_2O_5 and SiO_2 caused the change to EOT and the differences between samples without Si IPL and with Si IPL were mainly come from SiO_2 formation.

The frequency dispersion on accumulation capacitance is another important issue for high-k dielectrics on III-V and the trend of frequency dispersion has been cor-

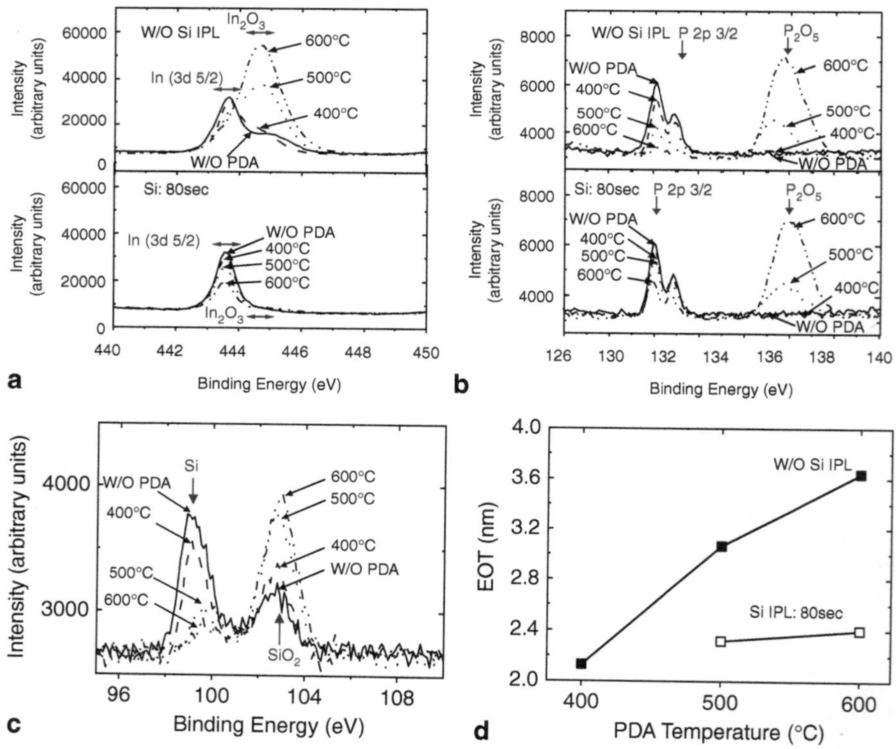

Fig. 11.71 **a** XPS spectra of the In and In_2O_3 (*3d* 5/2) **b** P and P_2O_5 (*2p* 3/2) with different PDA temperature (3 min at each temperature) on InP with and without Si IPL. **c** XPS spectra of the Si (*2p*) with different PDA temperature on InP with 1.2 nm of Si IPL. **d** Equivalent oxide thickness (EOT) versus different PDA temperature for 3 min with 1.2 nm and 1.5 nm of Si IPL

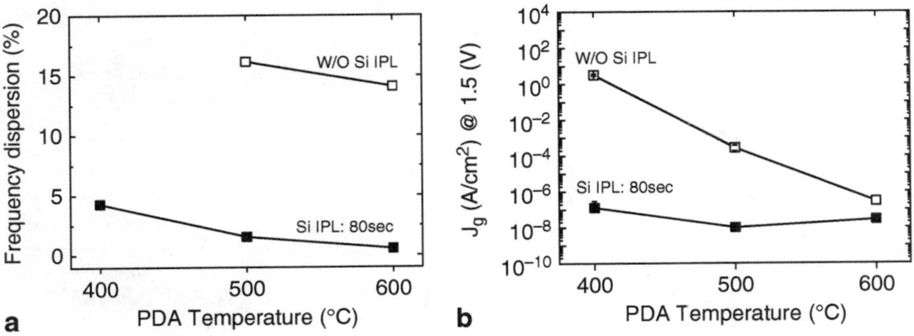

Fig. 11.72 **a** Frequency dispersion (%) **b** Leakage current density (J_g [A/cm²]) at 1.5 [V] versus different PDA temperature for 3 min with 80 s of Si IPL and without Si IPL

related with the interface quality. Figure 11.72a summarizes the frequency dispersion characteristics versus PDA temperature. With appropriate thickness of Si IPL and PDA condition, low frequency dispersion (<5%) can be obtained. In contrast, without Si IPL, frequency dispersion was around 15%. This implies that ~0.8 nm of native oxide (Fig. 11.69a) might consist of high trap densities. Figure 11.72b illustrates the leakage current density (J_g) at $V_g = V_{fb} + 1$ [V]. Si IPL and higher temperature PDA led to lower leakage current (Fig. 11.72b). Leakage current was reduced (to ~9.7×10^{-9} A/cm²) for 10 nm thickness of high-k dielectric with 80 s of Si IPL and PDA of 500 °C 3 min on n-type InP wafer.

InP n-MOSFETs were fabricated employing the same gate stack as GaAs using a ring-FET. The schematic cross section and top view of n-MOSFET with Si passivation is shown in inset of Fig. 11.73a.

Si IPL layer with thickness ~1.5 nm (PVD deposition time 80 s) was deposited for transistor fabrication to prevent Fermi level pinning at the InP-HfO_2 interface with PDA 600 °C 3 min. Figure 11.73a shows the transfer characteristics of an n-MOSFET with a gate width of 1000 μm and length of 100 μm. A subthreshold slope (SS) of 178 mV/decade were determined from the I_d–V_g plot. A threshold voltage of

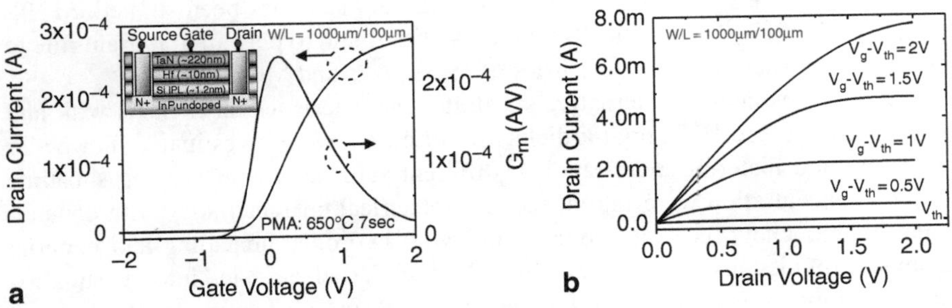

Fig. 11.73 **a** I_d–V_g and G_m of n-MOSFET with Si passivation (W 1000 μm × L 100 μm) with peak G_m 232.5 (μA/V) (*Inset:* Schematic cross section and top view of n-MOSFET with Si passivation 1.5 nm). **b** I_d–V_d of n-MOSFET with Si passivation (W 1000 μm × L 100 μm)

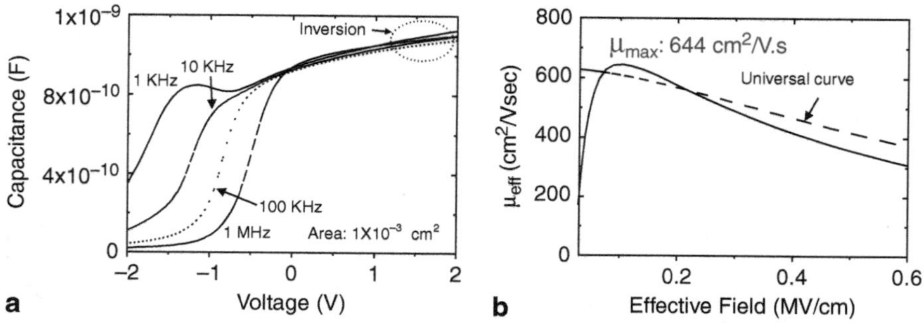

Fig. 11.74 **a** Frequency dispersion of n-MOSFET with Si 1.5 nm and PMA 650 °C 7 s. **b** Mobility of n-MOSFET with Si passivation (W 1000 μm × L 100 μm) from split CV and I_d–V_g curve

−0.24V was extracted by linear extrapolation technique. A maximum transconductance (G_m) of 232.5 (μA/V) was obtained for the device at a drain voltage of 0.05 V. The output characteristics of the MOSFET device are illustrated in Fig. 11.73b where a maximum drain current of 6.87 mA was obtained at a V_g–V_{th} of 2 V and V_d of 2 V. Figure 11.74a shows split CV characteristics of the MOSFET, measured at different frequencies, showing a frequency dispersion behavior. Peak mobility of 644 cm²/V s has been obtained. The effective electron mobility (μ_{eff}) was evaluated for the same device from the I_d–V_g plot (measured at a drain voltage of 50 mV) and its corresponding split C–V (Fig. 11.74b).

11.5 Conclusions

In this chapter, our discussions were devoted to three major investigations based on understanding of electrical and material characteristics with some reliability, and process development for GaAs, InGaAs and InP. The first one is that the understanding of HfO_2 high-k dielectric material and its relationship with Si IPL (interfacial passivation layer) in terms of evaluation in MOS device characteristics. Here, the inserted Si IPL with HfO_2 high-k dielectric in GaAs has been studied. Si IPL with partially SiO_2 or Ge with HfO_2 on GaAs has found to play an important role to determine the electrical characteristics of MOSCAPs and MOSFETs.

Secondly, as another alternative substrate, characteristics on InGaAs with and without Si IPL and HfO_2 high-k dielectric MOSFET were investigated. For better electrical and material understanding, different substrates with varying substrate doping concentration were introduced. The electrical characteristics with dependency of frequency, flat band voltage shift with hydrogen annealing and material analysis provided insight of trapping in HfO_2 high-k dielectric and interface quality. Here the improvements of MOSFET characteristics such as output current, transconductance, channel mobility, etc., were especially emphasized with ultra-thin EOT regime (~10 Å) in order to overcome the drawback of HfO_2 on GaAs. Demonstration of n-MOSFET on InGaAs with high-k dielectric can provide optimistic result near the future and possibility as alternative substrate to substitute conventional SiO_2 on Si substrate.

The third major discussion is on the process development of another alternative substrate with high-k gate dielectric material on InP for future scaled-down CMOS technology and applications in high-frequency digital circuits, microwave power amplifiers, and monolithic optoelectronics circuits. MOSFET characteristics with output current, transconductance, channel mobility, etc., were also emphasized. Using optimum thickness of the Si IPL preventing Fermi level pinning at InP-HfO_2 interface, ~1.2 nm (PVD deposition time 60 s) for MOSCAP and 1.5 nm (PVD deposition time 80 s) for MOSFET, material and electrical characteristics of TaN/HfO_2/InP self-aligned InP n-MOSFET with PVD Si IPL under various post deposition anneal (PDA) conditions was studied with transistor behavior.

References

1. Byoung Hun Lee, Laegu Kang, Wen-Jie Qi, Renee Nieh, Yongjoo Jeon, Katsunori Onishi, and Jack C. Lee, "Ultrathin hafnium oxide with low leakage and excellent reliability for alternative gate dielectric application," IEDM Tech. Cig., pp. 133–136, 1999.
2. S. Koveshnikov et al., Appl. Phys. Lett., 88, 022106, 2006.
3. InJo Ok, H. Kim, M. Zhang, T. Lee, F. Zhu, G. Thareja, L. Yu, S. Koveshnikov, W. Tsai, V. Tokranov, M. Yakimov, S. Oktyabrsky, and Jack C. Lee, IEEE Device Res. Conf., pp. 45–46, 2006.
4. InJo Ok, H. Kim, M. Zhang, T. Lee, F. Zhu, L. Yu, S. Koveshnikov, W. Tsai, V. Tokranov, M. Yakimov, S. Oktyabrsky, and Jack C. Lee, IEEE Int. Electron Devices Meet., pp. 829–832, 2006.
5. L. Paves, F. Piazza, M. Henini, and I. Harrison, Semicond. Sci. Technol., 8, No. 2, pp. 167–171, February 1993.
6. A. Zdyb, J. M. Olchowik, D. Szymczuk, J. Mucha, K. Zabielski, M. Mucha, and W. Sadowski, Crystal Res. Technol., 37, No. 8, 2002.
7. A. Zdyb, J. M. Olchowik, and M. Mucha, Mater. Sci. Pol., 24, No. 4, 2006.
8. Y. Wang, T. P. Ma, and R. C. Barker, IEEE Trans. Nucl. Sci., 36, No. 6, December 1989.
9. T. Hori, Gate Dielectrics and MOS ULSIs, p. 29, 1997.

Chapter 12
p-type Channel Field-Effect Transistors

Serge Oktyabrsky

Abstract Development of p-type MOSFETs using new materials is an important goal to provide a further scaling of CMOS circuits. Although bulk transport properties of Ge make it the main candidate for p-channel, strained III-As and, in particular, III-Sb are good competitors in particular for deeply scaled devices due to lower hole effective mass in strained InSb heterostructures, simplicity of band engineering with variety of material choices to create high barriers, reduce leakage, improve ohmic contacts, etc. The chapter begins with physics of scattering in bulk semiconductors, describes how the MOSFETs figures-of-merit change with scaling, reviews physics of strain effects in quantum wells with the emphasis on effective mass, density of states and mobility. Technologies and results on p-channel heterojunction field-effect transistors and MOSFETs are reviewed.

12.1 Introduction

Very attractive electron transport properties of group III-V semiconductors instigate the research of n-channel MOSFETs and are responsible for use of these materials in microwave and ultra-fast digital communication circuits. The relative simplicity and associated high density and low power of CMOS circuitry drive the development of p-channel III-V FETs, although the speed capability of p-type channels has fallen well short of that obtained in n-type channels. Nevertheless, the complimentary FET technologies based on high-electron mobility transistors (HEMTs) were developed and demonstrated [1, 2].

Table 12.1 summarizes transport properties of bulk group IV and group III-V semiconductors. It is clear that the bulk hole mobility in Ge makes it the primary candidate for p-channel devices. In the IC industry, the number of different materials integrated on the chip is continually increasing and it is worth expecting

S. Oktyabrsky (✉)
College of Nanoscale Science and Engineering, University at Albany—SUNY, Albany, NY 12203, USA
e-mail: soktyabrsky@uamail.albany.edu

S. Oktyabrsky, P. D. Ye (eds.), *Fundamentals of III-V Semiconductor MOSFETs*, DOI 10.1007/978-1-4419-1547-4_12,

Table 12.1 Transport properties of bulk semiconductors: E_g – bandgap; m_e – electron effective mass (isotropic for direct bandgap); m_t and m_l – transverse and longitudinal mass of electrons (for indirect bands); m_{hh} and m_{lh} – heavy and light hole masses; μ_e and μ_h – electron and hole mobility; $V_{e,sat}$ and $V_{h,sat}$ – electron and hole saturation velocities

	Electrons					Holes			
	E_g eV	m_e or m_t	m_l	μ_e cm²/V s	$V_{e,sat}$ 10^7 cm/s	m_{hh}	m_{lh}	μ_h cm²/V s	$V_{h,sat}$ 10^7 cm/s
Si	1.12	0.19	0.98	1350	1.0	0.54	0.15	460	0.72
Ge	0.66	0.082	1.64	3900	0.7	0.34	0.043	1900	0.63
GaAs	1.42	0.067	–	8500	1.2	0.53	0.074	400	0.9
$In_{0.53}Ga_{0.47}As$	0.8	0.05	–	14000	3	0.36	0.041	400	0.4
GaSb	0.73	0.041	–	3750	–	0.4	0.05	680	–
InAs	0.36	0.027	–	33000	3.5	0.4	0.026	460	–
InSb	0.17	0.013	–	77000	5.0	0.42	0.016	850	–

new channel materials integrated with Si wafer. Within this scenario the III-V's are competing with Ge for the p-channel. An important target for CMOS technology is a high-mobility p-type channel matching within a factor of 3–5 the mobility of the n-channel. Although it is hard to expect that the hole mobility of any III-V channel can be made higher than that of Ge, there are many other properties that can make III-V's more attractive for deeply scaled p-channels: low hole effective mass in strained heterostructures, simplicity of band engineering with variety of material choices to create high barriers, reduce leakage, improve ohmic contacts, etc.

12.2 Low-Field Hole Mobility in Bulk Semiconductors

The valence band structures of all the group IV and group III-V semiconductors (excluding III-nitrides) are very similar due to similarity of their diamond or zinc-blende crystal lattices. Figure 12.1 shows a schematic of the bandstructure of direct bandgap III-V's, and also the shift and deformation of the energy bands due to biaxial strain. The valence bandedge states consists of heavy hole (HH) and light hole (LH) bands which are degenerate in the center of the Brillion zone in bulk material with no strain. The spin-orbit split-off band are located quite deep in the semiconductors constructed from atoms with large atomic numbers, (0.3–0.7 eV below the valence band maximum for $Z > Z_{Ga} > 31$), and do not contribute to transport.

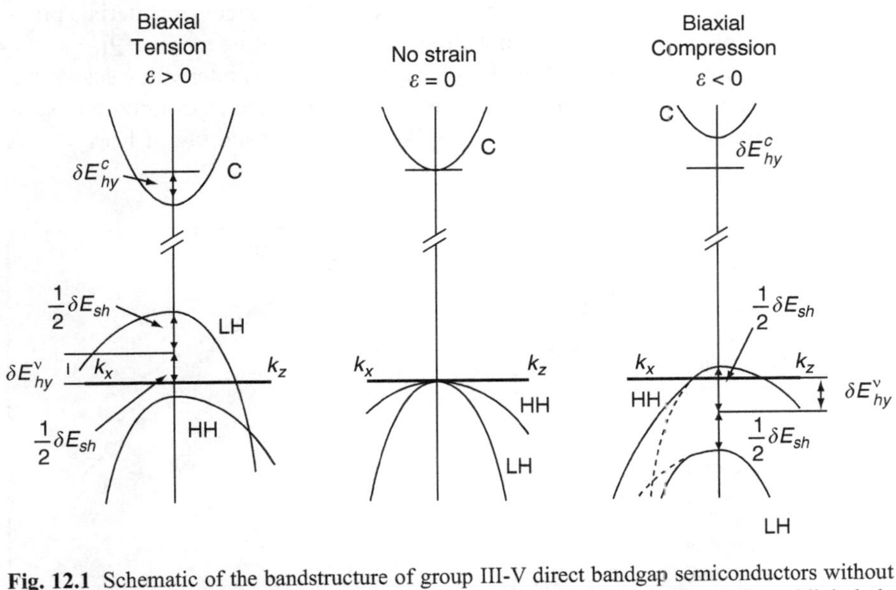

Fig. 12.1 Schematic of the bandstructure of group III-V direct bandgap semiconductors without and with applied biaxial stress. C, HH and LH denote conduction band, heavy hole and light hole valence bands, respectively. Stress is in the *x-y* plane. *Dashed lines* show parabolic non-interacting bands, HH-LH coupling removes their intersection. δE_{hy} and δE_{sh} are hydrostatic and shear strain-related components of energy shifts

Ge and GaAs are particularly very alike semiconductors that includes similar spin-orbit splitting (0.30 and 0.34 eV, respectively), similar phonon dispersion characteristics, because of close similarity of Ga, As and Ge atoms. Hole effective mass is just 1.6 times lower in Ge than in GaAs. A five times higher hole mobility in Ge over GaAs (Table 12.1) is quite surprising because the room temperature mobility is mainly determined by phonon scattering with the strongest contribution of non-polar (deformation potential) optical and acoustical phonon scattering [3].

Because of the difference of group III and group V atoms, the bonds are slightly ionic, and the compound semiconductors are polar materials, this results in two additional scattering mechanisms: polar optical phonon scattering and piezoelectric acoustical phonon scattering. However, these scattering mechanism are relatively weak mostly because of p-type symmetry of valence states and reduction of the electron-phonon interaction with large change of the carrier wavevector. It was estimated that polar optical phonon scattering mechanism contributes as low as 10–35% to the resultant mobility at room temperature in p-GaAs [4–6] and p-InSb [7, 8]. The piezo-acoustical phonon scattering is even weaker. Temperature dependence of mobility (Fig. 12.2) is determined by the relative contribution of various scattering processes. Most of the group III-V and group IV semiconductors show a similar temperature slopes of mobility at room temperature, $\mu \propto T^{-\beta}$, $\beta = 2.2{-}2.4$. The slope was shown to depend on relative contributions of non-polar optical and acoustical phonon scattering mechanisms [3]. It is equal to $\beta = 1.5$ for acoustical phonon scattering, and increasing with optical phonon scattering contribution.

The significantly higher mobility of holes in Ge is, therefore, not the result of difference in scattering mechanisms, but rather small differences in materials properties, such as phonon deformation potential and hole effective masses [9].

The complex valence band structure gives rise to phonon intervalley scattering. In bulk materials, due to degeneracy of the valence band, the most important scattering rates limiting the hole mobility are the intraband scattering of heavy holes

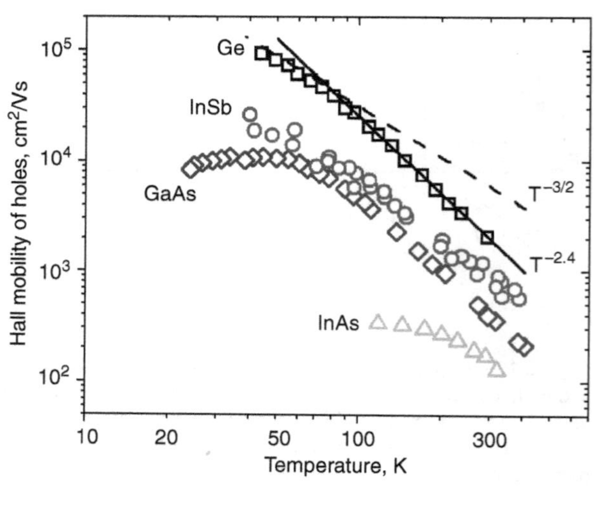

Fig. 12.2 Temperature dependence of Hall mobility in bulk p-type semiconductors. Data for Ge and GaAs are from Ref. [9], and for InSb and InAs from Ref. [10]. At room temperature the slope $\mu{\sim}T^{-\beta}$, $\beta = 2.2{-}2.4$ characteristic of non-polar (deformation potential) optical phonon scattering [9]. The *dashed line* with $\beta = 1.5$ slope corresponds to acoustical phonon scattering

and interband scattering of light holes [3]. The contribution of the light holes to the overall mobility is, therefore, quite low: 10–30% depending on temperature and material [3].

As can be observed from the Table 12.1, the hole mobility is more sensitive to the group V elements (As or Sb, and also P) than to the group III cations. The physical reason for that is that the valence band Bloch functions are constructed mostly from p-orbitals of the anions, and therefore, the scattering matrix elements are very similar in semiconductors with the same group V elements. The basic differences in hole effective masses and momentum scattering rates in different semiconductors remain similar in strained layers and quantum wells, however, one should expect changes in performance figures of merit with scaling of the channel length.

12.3 p-channel: Figures of Merit with Scaling of Channel Length

High mobility is still the primary parameter that drives the research of new channel materials for FETs. However, with further scaling of the channel length, other figures of merit become more important and challenging to meet. Table 12.2 gives some guidelines for channel materials parameters.

For many previous generations of MOSFETs, the most important metric was energy-delay product, EDP (the lower–the better):

$$EDP = \frac{CV_d^2}{W}\frac{CV_d}{I_d}$$

with V_d – the power supply voltage, I_d – drain saturation current, and W – channel width. In the intrinsic EDP_i, the first term is the energy needed to charge the

Table 12.2 Primary figures-of-merit for channel materials resulting from gate length: Q – channel charge, C_g, C_{ext} – gate and extrinsic (parasitic) capacitance, I_d – drain saturation current, V_d – power supply voltage, m^* – effective mass, n – carrier density, μ– mobility, $\langle\tau\rangle$ – average momentum scattering time, E_g – bandgap energy, F – electric field in the channel, J_g – gate leakage current

Critical transport mechanism	Approximate gate length	Main figure-of-merit	Channel material-related parameters
Drift	>50 nm	Energy-delay product (EDP) (intrinsic)	$EDP_i \propto \frac{Q^3}{C_g I_d} \propto \frac{n^2}{\mu} \propto \frac{m^{*3}}{\langle\tau\rangle}$
Ballistic	10–50 nm	EDP (extrinsic)	$EDP_{ext} = \frac{C_{ext}V^2}{W}\frac{C_{ext}V}{I} \propto \frac{V_d{}^3}{nv_{inj}} \propto \frac{V_d{}^3}{\sqrt{m^*}}$, $V_d - V_t \propto m^*$
Tunnel	<5–10 nm	Leakage current: BTBT and J_g	$J_{BTBT} \propto \exp\left(-\frac{4\sqrt{2m^* E_g^3}}{3\hbar e F}\right)$

gate capacitor $C = C_g$, and the second term is the intrinsic delay time needed to charge the C_g with the drain current. The charge in the channel or carrier density, $Q \propto n \propto m^*$, is limited by the density of states (DOS) in the semiconductor, and for 2D channel is proportional to the effective mass, m^*. In this case, the advantages of the high-mobility, low-effective mass materials are obvious.

As the gate length is scaled down, the parasitic capacitance starts dominating, in 50 nm-long MOSFETs the extrinsic capacitance $C_{ext} >> C_g$. In addition, the transport becomes almost ballistic and the current is proportional to the source injection velocity and the concentration of carriers close to the source, $I_d \propto v_{inj}n$. The injection velocity is usually taken as the thermal velocity (although it is close to Fermi velocity in highly degenerate carrier gas [11]) and, therefore, $v_{inj} \propto 1/\sqrt{m^*}$. The resulting EDP_{ext} direct dependence on effective mass ($\sim m^{*\,-1/2}$) is weak as compared to indirect dependence on mass through the drain voltage. Interestingly, in the ballistic regime with large parasitics, the EDP benefits result from the ability to obtain high drain current at low voltage overdrive $V_d - V_t$, where V_t is the threshold voltage. The necessary voltage overdrive is proportional to the n/C_g, and therefore, scales roughly as 2D DOS or just effective mass, m^*. For example, the n-InSb channels were shown to operate at $V_d = 0.5$ V, and resulted in more than an order of magnitude improvement of EDP over strained Si transistors [12]. On the other hand, a significant reduction of V_d is difficult because V_t should be maintained high enough to keep the subthreshold leakage current low. Another important consequence of this rather simplistic discussion is that in the ballistic channels high mobility of a material is still important. However, the contribution from low effective mass becomes progressively more important than scattering time contribution as time-of-flight is approaching scattering time in short quasi-ballistic channels. This quite obvious argument makes Ge and III-V channels competitive for short channel devices.

The further scaling of the gate length results in increased OFF current due to both higher gate leakage current through the thinned down gate dielectric, and also due to band-to-band-tunneling (BTBT). The latter depends on parameters of the channel material, namely bandgap, E_g, and effective mass. The tunneling probability shown in the Table 12.1 for direct bandgap tunneling through triangular barrier is exponentially sensitive to these parameters [13, 14], but the tunneling current also depends on channel field F, density of states of the bands, quantization, and is difficult to estimate with reasonable accuracy. Moreover, other effects, such as tunneling through defects and impact ionization in the high-field regions of a channel may increase the gate current. Detailed computations taking into account BTBT into indirect bands [15, 16, and Chap. 4 of this book] show that significant BTBT current of 0.1 μA/μm is expected in 15 nm-long n-type InSb and Ge channels but is significantly less in GaAs channel due to wider bandgap. However, thin body quantum confinement that increases the effective bandgap may reduce the BTBT leakage by as much as 3–6 orders of magnitude. Although, there are many variables affecting BTBT, it is quite clear that the leakage current will become the main show stopper at some point of further FET channel scaling, and other materials with large m^*E_g product, such as GaAs, diamond or GaN, might be considered.

12.4 Strained Quantum Wells

12.4.1 Valence Band Under Strain

Stress applied to diamond or zinc-blende semiconductors is known to affect their bandstructure (Fig. 12.1). In general, the modification of the band structure under stress can be calculated using either tight-binding model or k–p approximation, as described in many papers and textbooks [17–20] and will not be repeated here. The actual valence band splitting and the curvature at $k = 0$ depends on the crystallographic orientation and direction and symmetry of the stress. In the unstressed bulk semiconductor, the dispersion of the valence band in the simplest 2-band approximation can be described by Kohn-Luttinger equation with the coordinates aligned with {100} crystallographic axes:

$$E(k) = -\frac{\hbar^2}{2m_0}\left[\gamma_1 k^2 \pm \sqrt{4\gamma_2^2 k^4 + 12\left(\gamma_3^2 - \gamma_2^2\right)\left(k_x^2 k_y^2 + k_y^2 k_z^2 + k_z^2 k_x^2\right)}\right], \quad (12.1)$$

with Luttinger parameters γ in Table 12.3, and wavevector k, and plus and minus signs corresponding to light and heavy mass bands, respectively. The corresponding light and heavy hole masses along <100> directions are:

$$\frac{m_{100}^{lh}}{m_0} = \frac{1}{\gamma_1 + 2\gamma_2} \quad \text{and} \quad \frac{m_{100}^{hh}}{m_0} = \frac{1}{\gamma_1 - 2\gamma_2}. \quad (12.2)$$

Due to valence band anisotropy (warping) the masses along <110> directions are different:

$$\frac{m_{110}^{lh}}{m_0} = \frac{1}{\gamma_1 + \sqrt{\gamma_2^2 + 3\gamma_3^2}} \quad \text{and} \quad \frac{m_{110}^{hh}}{m_0} = \frac{1}{\gamma_1 - \sqrt{\gamma_2^2 + 3\gamma_3^2}}. \quad (12.3)$$

Table 12.3 Lattice constant a_0, Elastic moduli C, Luttinger parameters γ, and deformation potentials a and b for room temperature

	a_0 Å	C_{11} GPa	C_{12} GPa	γ_1	γ_2	γ_3	a_c eV	a_v eV	b eV
Si	5.431	165.8	63.93	4.285	0.339	1.446	+4.18(X)	−2.46	−2.1
Ge	5.658	124.0	41.3	13.38	4.28	5.68	−1.54(L)	−1.24	−2.9
GaAs	5.6533	119	53.8	6.98	2.06	2.93	−7.17	−1.16	−2.0
$In_{0.53}Ga_{0.47}As$	5.869	100	49.3	13.6	5.42	6.27	−6.06	−1.07	−1.9
GaSb	6.096	88.42	40.26	13.4	4.7	6.0	−7.5	−0.8	−2.0
InAs	6.0584	83.29	45.26	20.4	8.3	9.1	−5.08	−1.00	−1.8
InSb	6.479	68.47	37.35	34.8	15.5	16.5	−6.94	−0.36	−2.0

Parameters for InGaAs are extrapolated between GaAs and InAs.

One of the technologically important stress field symmetries is biaxial stress which is easily obtained in epitaxially grown films with slightly different lattice parameters. The biaxial compressive stress was proposed [21] and proven to be useful in splitting the valence band and reducing the effective mass of holes. It splits the valence sub-bands in heavy hole (HH) band and light hole (LH) band that the HH band (which is assigned so due to 3/2 projection of the total angular momentum of the Bloch states) becomes the highest in energy (Fig. 12.1). This splitting results in a low in-plane effective mass of holes (large curvature) in the HH band and large component of the effective mass tensor normal to the plane.

In the case of (100) orientation of the substrate plane, the biaxial strain is given by

$$\varepsilon = \varepsilon_{xx} = \varepsilon_{yy} = \frac{a_{sub} - a_{film}}{a_{sub}} \quad \text{and} \quad \varepsilon_{zz} = -2\frac{C_{12}}{C_{11}}\varepsilon, \tag{12.4}$$

and all the non-diagonal strain components equal zero. The isotropic (hydrostatic) and anisotropic (shear) strain components result in a change of the bandap and splitting of the valence band, respectively, as shown in the Fig. 12.1:

$$\begin{aligned} \delta E_{hy}^{v} &= a_v(\varepsilon_{xx} + \varepsilon_{yy} + \varepsilon_{zz}), \\ \delta E_{sh} &= -\frac{b}{2}(\varepsilon_{xx} + \varepsilon_{yy} - 2\varepsilon_{zz}). \end{aligned} \tag{12.5}$$

Under biaxial strain in (100) plane, the split bands keep their curvature in the normal direction (and the normal mass tensor component) the same as in the bulk. But the in-plane dispersion becomes non-parabolic with significant dependence of effective masses on the in-plane k-vector [22, 23]. The in-plane (transversal) effective masses for biaxial compressive strain are

$$\frac{m_t^{hh}}{m_0} = \frac{1}{\gamma_1 + \gamma_2} \ \text{(topmost band)}; \quad \frac{m_t^{lh}}{m_0} = \frac{1}{\gamma_1 - \gamma_2}. \tag{12.6}$$

And for biaxial tensile strain

$$\frac{m_t^{lh}}{m_0} = \frac{1}{\gamma_1 - \gamma_2} \ \text{(topmost band)}; \quad \frac{m_t^{hh}}{m_0} = \frac{1}{\gamma_1 + \gamma_2}. \tag{12.7}$$

Note, that in-plane masses are different from the bulk effective masses of either valence band.

Since the strain changes the semiconductor bandgap, the barrier height at the heterostructure interface is also affected. Figure 12.3 shows the valence band energy barrier variation due to splitting of HH and LH bands in biaxially strained bulk $In_xGa_{1-x}As$ grown on InP substrate as the indium content is varied. The valence band offsets for two barriers: InP and $In_{0.52}Al_{0.48}As$ are taken as 0.6 and 0.3 of the bandgap differences, respectively [24].

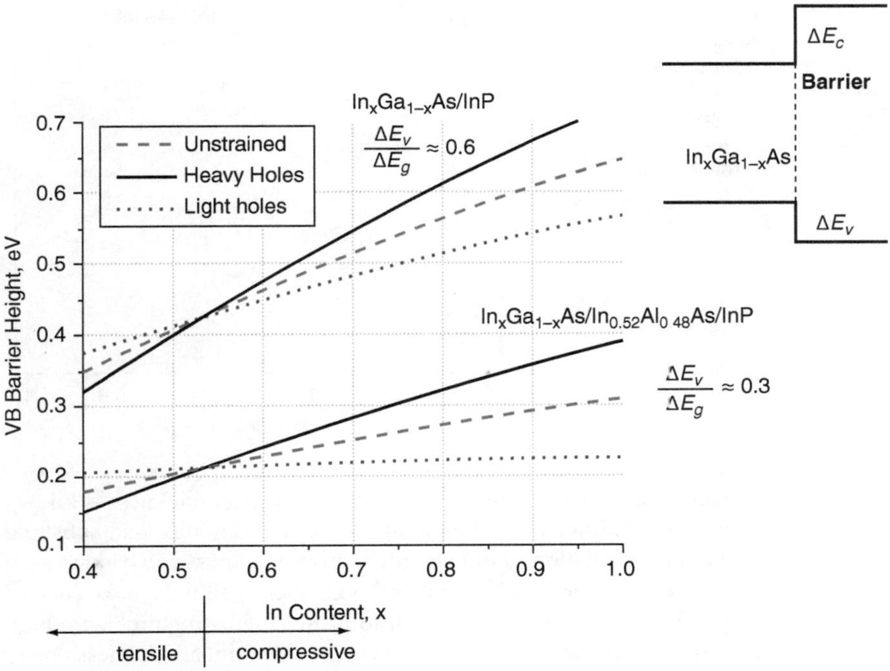

Fig. 12.3 Valence band (VB) barrier height ΔE_V variation with indium content in the bulk heterostructure. Graphs show splitting of bands and VB offset with two barriers: InP and isomorphic InAlAs

For indium composition higher than 53%, the induced strain is compressive and the uppermost valence band is the HH band. The heterojunction barrier height is increasing with In content because of the compositional bandgap reduction in InGaAs and additionally due to the biaxial strain. The valence band barrier is higher in InGaAs/InP than in InGaAs/$In_{0.52}Al_{0.48}As$ heterostructure due to larger valence band offset with InP. In general, the barrier height and HH-LH band separation have to be as large as possible in order to achieve high hole concentration with low effective mass in the uppermost subband; this requires increase of the strain in the channel.

12.4.2 Strain and Quantum Confinement

To increase the strain in heterostructure, one should reduce the thickness of the conduction channel below the critical thickness for dislocation nucleation. Figure 12.4 shows this critical thickness dependence on the alloy content, x, for $In_xGa_{1-x}As$/InP and $In_xGa_{1-x}As$/GaAs heterostructures obtained using the model by People and Bean [25] based on energy balance, and Matthews and Blakeslee [26] who considered mechanical equilibrium. The PB model usually gives better results for lower strain

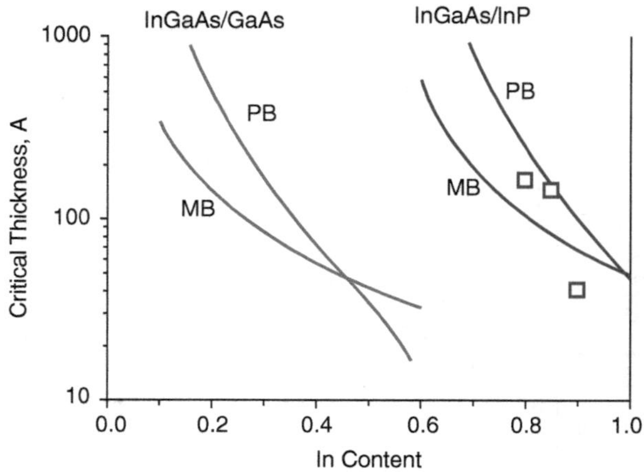

Fig. 12.4 Critical layer thickness as a function of In content in the InGaAs grown on GaAs and InP substrates. *Squares* [27] show the critical thickness for the onset of surface roughening at 450 °C growth temperature

and the MB model for higher strain, however, the actual critical thickness is known to depend on growth conditions (such as temperature and doping), and substrate properties (such as misorientation). For example, critical thickness can be increased by reduction of growth temperature [27] but still high stress will induce the surface roughening (Fig. 12.4). Typically it is difficult to grow heterostructures with high strain ($\varepsilon > 2\%$) even when the layer thickness is below the critical thickness due to In surface segregation [28, 29] and strain-induced formation of 3D islands [30].

Small thickness of the layers and/or normal field usually present in the FET structures further modifies the valence band structure due to quantum confinement. This problem was theoretically solved in many publications [31–37], and usually requires self-consistent calculations of Schrödinger equation with Kohn-Luttinger Hamiltonian and boundary conditions at the interfaces. The problem can be solved analytically if infinite barrier is considered [31, 38].

An example of the dependence of quantization as a function of $In_{0.83}Ga_{0.17}As/In_{0.52}Al_{0.48}As$ quantum well (QW) thickness is shown in Fig. 12.5a. Besides the HH-LH splitting due to the biaxial strain, the quantum confinement introduces splitting into subbands. Thinner strained epitaxial layers lead to a larger splitting between the low in-plane mass HH1 subband and the next subband (HH2 or LH1) with heavy in-plane mass. The large normal component of the effective mass of the top-most HH1 subband improves confinement of holes, which can be observed as low quantum confinement energy of the HH1 subband even in very thin QWs (Fig. 12.5a). This allows for keeping the barrier high in narrow QWs.

The reduced in-plane effective mass in strained QWs also leads to lower density of states (DOS) for holes, which is proportional to the mass: $DOS = m^*/\pi\hbar^2$ for isotropic parabolic dispersion. The dispersion becomes more complicated in the QW valence band under strain. Figure 12.5b shows the in-plane dispersion relation for a 5 nm thick $In_{0.83}Ga_{0.17}As/In_{0.52}Al_{0.48}As/InP$ QW. In narrow QW structures, HH1 non-parabolicity at low in-plane k_t-values is reduced leading also to a reduction in the transverse effective mass of holes in the HH1 subband when it is filled up to

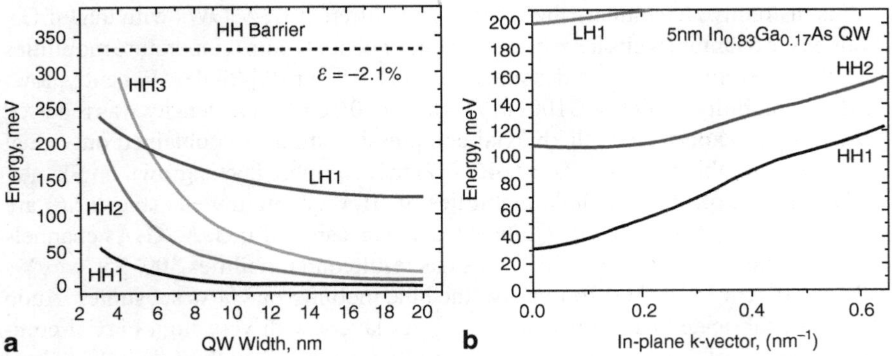

Fig. 12.5 **a** Energies at $k_t = 0$ of the hole sub-bands in $In_{0.83}Ga_{0.17}As/In_{0.52}Al_{0.48}As$ QW as a function of QW thickness. Energy is plotted to be positive for holes. **b** In-plane hole dispersion in a 5 nm–wide $In_{0.83}Ga_{0.17}As/In_{0.52}Al_{0.48}As$ QW. Energy is plotted to be positive for holes. The HH1 effective mass is ~0.07 m_0

k_t~0.5 nm^{-1} (hole concentration ~4 × 10^{12} cm^{-2}) as shown in Fig. 12.5b. For longer in-plane wavevectors, holes will occupy HH2 subband with larger mass and also the HH1 dispersion curve becomes non-parabolic and the effective mass increases.

12.4.3 Mobility in Quantum Wells

Carrier scattering in quantum wells is somewhat modified from bulk scattering [39]. First of all, the wavevectors of quantum confined carriers are confined in a plane, and therefore, the Boltzman scattering integral is modified. This, however, just change (typically increase at room temperature) the scattering rates quantitatively. There are yet qualitative changes and new mechanisms of scattering that are characteristic to a 2D hole gas. The quantum confinement causes splitting of the HH and LH valence band just because the confinement energy depends on effective mass of carriers normal to the QW plane. This splitting is qualitatively similar to the one induced by biaxial compression with the topmost HH band. The valence band splitting results in reduction of the interband scattering which makes an essential contribution to the bulk hole mobility. The 2D gas (both electron and hole) also experience scattering from various 2D scattering sources such as interface roughness [40], 2D phonons localized in the heterojunction plane [41] or sources associated with the presence of surfaces, interfaces, gate dielectric [42], etc. High-quality epitaxial QW heterostructures show temperature behavior of hole mobility similar to bulk materials ($\mu \propto T^{-\beta}$, $\beta \approx 2.4$) in 100–300 K temperature range [43, 44]. That strongly confirms that the major scattering mechanisms at room temperature are the same as in the bulk, namely non-polar (deformation potential) optical and acoustical phonon scattering. Reduction of β usually indicates the presence of other sources of scattering, such as surface roughness.

It is instructive to compare the mobilities in different III-V QWs with that of Ge. Strained Ge quantum wells have demonstrated excellent room temperature mobilities of 2700–3000 $cm^2/V\,s$ at hole densities $(0.5–1) \times 10^{12}\,cm^{-2}$ [45–48]. Recently, low-field drift mobility as high as 3100 $cm^2/V\,s$ at $4 \times 10^{12}\,cm^{-2}$ hole density was reported [49]. In these experiments, the biaxial compressive strain was obtained on virtual substrates with thick $Si_{1-x}Ge_x$ (x = 0.45–0.7) metamorphic layers grown on Si.

As can be expected, the hole mobilities of III-V quantum wells (Fig. 12.6) are well below that of strained Ge. Quite extensive research of InGaAs/GaAs channels for complimentary MODFETs in 1990's has resulted in mobilities 300–350 $cm^2/V\,s$ at $1.2 \times 10^{12}\,cm^{-2}$ [50, 51]. To improve the hole mobility on GaAs substrates, Kudo et al. [50] fabricated 6 nm-thick $In_{0.35}Ga_{0.65}As$ layers with very high biaxial compressive strain of 2.5%. Low temperature (420 °C) MBE growth of the QW helped to suppress the In surface segregation. Almost symmetrical double Be-modulation-doped structure was used to improve mobility, likely due to the more symmetrical QW potential with a higher hole density in the QW center. The resultant RT hole mobility of 354 $cm^2/V\,s$ at $1.2 \times 10^{12}\,cm^{-2}$ hole density was reported. A MODFET with 0.4 μm gate with transconductance of 114 mS/mm was demonstrated using this channel structure.

Similarly to the bulk mobilies, hole mobilities in antimonide QWs are higher than in III-arsenides. Recently, room temperature mobilities as high as 1350 $cm^2/V\,s$ at $1.1 \times 10^{12}\,cm^{-2}$ in GaSb [60] and 1500 $cm^2/V\,s$ at $7 \times 10^{11}\,cm^{-2}$ hole density in InGaSb [67] was reported by NRL team. The collaboration between INTEL and QinetiQ has recently resulted in demonstration of the best InSb QW hole mobility, 1230 $cm^2/V\,s$ at $1.1 \times 10^{12}\,cm^{-2}$. Interestingly, both teams used (100)GaAs substrates and metamorphic buffer layers (AlGaSb for GaSb and InGaSb channels and AlInSb for InSb channel) to create a "virtual substrate" with the required lattice parameters. The maximum mobilities were observed in InSb and InGaSb QWs with about 2% biaxial compressive strain, and with about 1% for GaSb QWs.

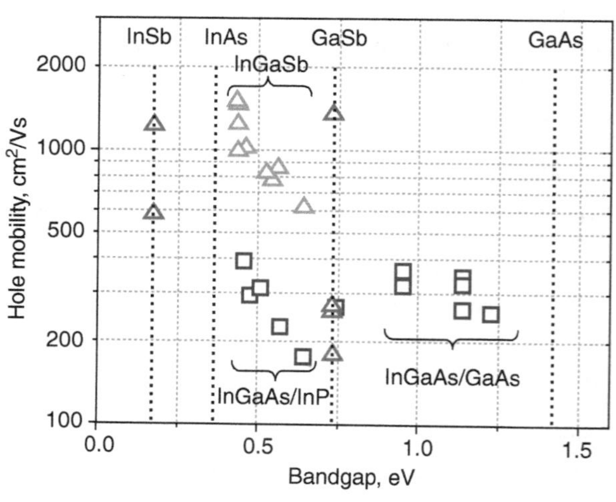

Fig. 12.6 Room-temperature hole mobility in quantum wells vs. bandgap of the bulk semiconductor material used in the QW. *Squares* are for III-As and *triangles* are for III-Sb QWs: InGaAs/GaAs [50–57]; GaSb [58–60]; InGaSb [61–64]; InSb [65, 66]. Note, that strain and quantum confinement typically increase the effective bandgap

The calculated dependence of hole mobility on biaxial strain is shown in Fig. 12.7 [68]. As discussed in the previous section, the biaxial compressive strain causes valence band splitting with topmost HH band having lower in-plane effective mass. The HH band effective mass does not depend on the strain directly to the first approximation, however, it slightly changes with the composition of the layer when the strain is induced by the lattice mismatch. Most of the mobility enhancement with strain is related to hole re-population towards the topmost subband with low in-plane mass and reduction of the interband and intersubband scattering due to valence band splitting and quantization (Fig. 12.5). Thus, to make this improvement in mobility significant, the second topmost valence subband should be separated from the HH1 significantly further than optical phonon energy (30–40 meV) and Fermi energy of holes (~70 meV at $2 \times 10^{12}\,\text{cm}^{-2}$ and $m^* = 0.07$).

Figure 12.8 shows a typical behavior of mobility in an $In_xGa_{1-x}As$/InAlAs/InP QW when the composition and hence the strain is varied. As expected, the mobility increases with increasing of In content from the lattice-matched x = 0.53. For a QW thickness of 10 nm, the mobility shows the peak at x = 0.77 and drops with further increasing of the strain due to plastic relaxation. The mobility can be further increased to the peak value of $390\,\text{cm}^2/\text{V s}$ at a concentration of $1.9 \times 10^{12}\,\text{cm}^{-2}$ at x = 83% if a thinner QW (6 nm) is used. Further increasing of the strain does not lead to higher mobility even if very thin QWs are grown most likely because of increased stress-induced In segregation and associated scattering. Similar behavior of mobility showing maximum at 2% biaxial compressive strain but naturally with higher mobility values was found in InGaSb QWs [67].

Because of non-parabolicity of in-plane effective mass and occupation of upper subbands, hole mobility drops with the increase of the hole density in a quantum

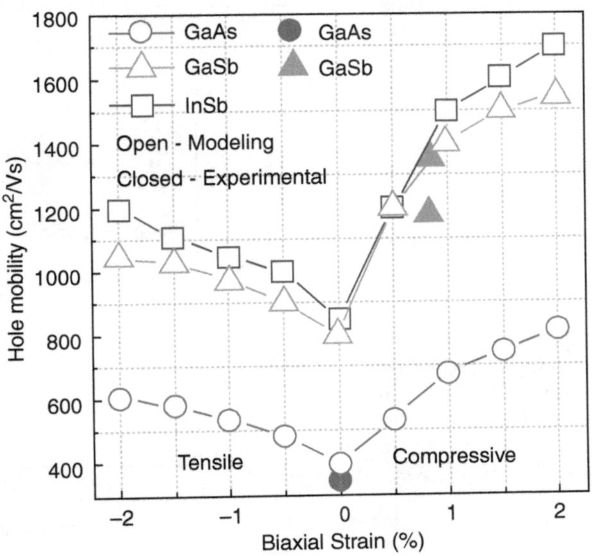

Fig. 12.7 Effect of biaxial [100] strain on hole mobility with strained band structure calculated using 8 band kp approach and mobility with Monte-Carlo simulations. The experimental points: GaAs from Ref. [72] and GaSb from Ref. [60] (Reprinted from [68] with permission. Copyright 2009 IEEE)

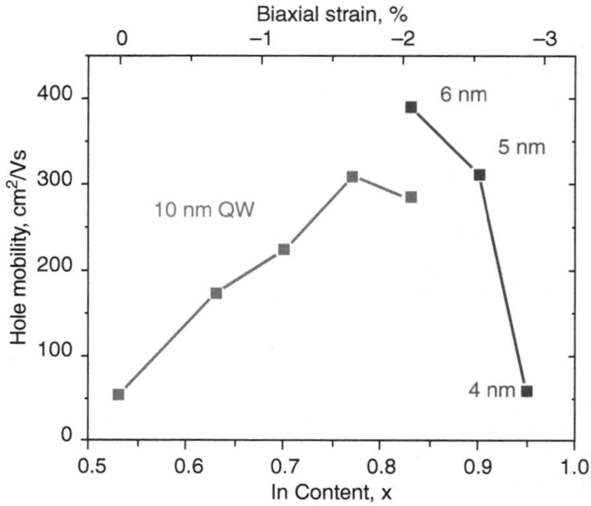

Fig. 12.8 Hall mobility vs. alloy composition in InGaAs/InAlAs/InP at room temperature. The carrier concentration in all the samples is about 2×10^{12} cm^{-2} [57]

well. Figure 12.9 shows the dependence of hole mobility on sheet carrier density in the strained QWs. The biaxial compressive strain is close to 2% in all the QWs shown.

It should be noted that the density of carriers in the inversion layers of 45 nm Si MOSFETs is as high as 2×10^{13} cm^{-2} in order to keep the drain current high to compensate for relatively large effective mass, low mobility and low carrier velocity. At these concentrations, the advantages of III-V's would be significantly reduced. However, the low effective mass allows for trading velocity for concentration with the aim to reduce the switching energy while keeping the drain current high, which is particularly important when the circuit speed is limited by charging of parasitic capacitance.

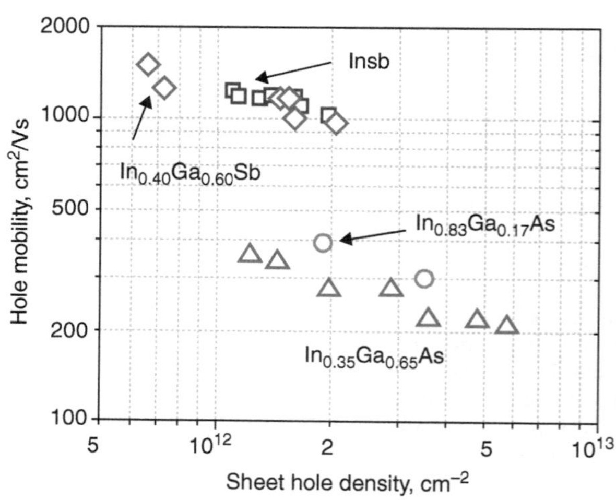

Fig. 12.9 Room temperature mobility vs. sheet carrier density in biaxially compressed quantum wells: InSb, $\varepsilon = -1.9\%$ [65], $In_{0.4}Ga_{0.6}Sb$, $\varepsilon = -2\%$ [67] $In_{0.35}Ga_{0.65}As$, $\varepsilon = -2.5\%$ [50], and $In_{0.83}Ga_{0.17}As$, $\varepsilon = -2\%$ [57]

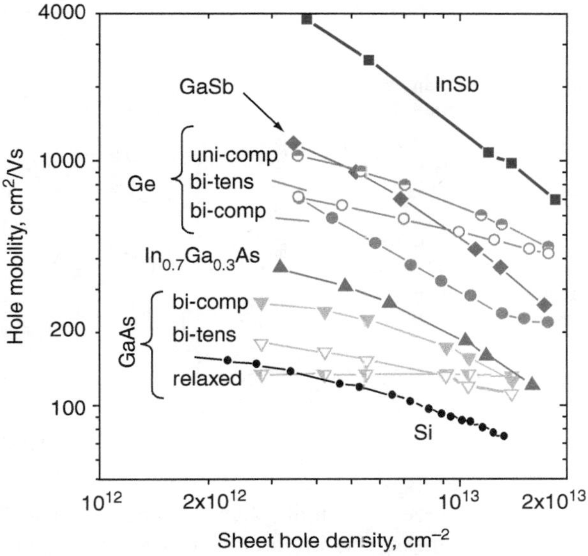

Fig. 12.10 Calculated hole mobility in 2% biaxially (on the (001) surface) compressively strained semiconductors. For comparison, mobility of Ge under 2% uniaxial (along [110] direction) compressive strain and 2% biaxial tensile strain, GaAs under 2% biaxial tensile strain, relaxed GaAs and universal Si curve are shown [69] (Reprinted with permission. Copyright 2009 IEEE)

Recently Zhang and Fischetti [69] has computed hole mobilities in inversion layers of strained semiconductors with SiO_2 gate oxide (Fig. 12.10). Phonon scattering processes, alloy scattering (for InGaAs) and surface roughness scattering were included in the calculation. Interestingly, GaSb is expected to have mobility similar to that of Ge, and InSb mobility exceeds it significantly even at high hole densities.

Although biaxial strain is still the easiest method to improve hole mobility, recently uniaxial compressive strain was also considered as a more efficient way to improve hole transport. In fact, uniaxial compressively strained Si is the dominant technology for high performance p-MOSFETs [70]. Figure 12.10 shows that uniaxial compressive strain provides about 2× improvement over relaxed Ge at 10^{13} cm^{-2} hole density. A similar trend is expected in III-V's due to similarity of the valence bands. Interestingly, the Ge inversion layer mobility is shown to be higher for biaxial tensile strain than for biaxial compressive strain for carrier densities of interest (Fig. 12.10) contrary to the III-V's as shown in the Figure for GaAs.

12.4.4 *Effective Mass in Strained QWs*

As discussed earlier, low field mobility is not a good metric for scaled devices with ballistic channel. As the mobility is determined by average scattering time and effective mass, $\mu \propto \langle \tau \rangle / m^*$, the scattering contribution into mobility becomes less important with the reduction of the channel length. Keeping m^* low is needed for increasing source injection velocity and reduction of the drain voltage overdrive. On the other hand, large drain current requires high density of holes with low

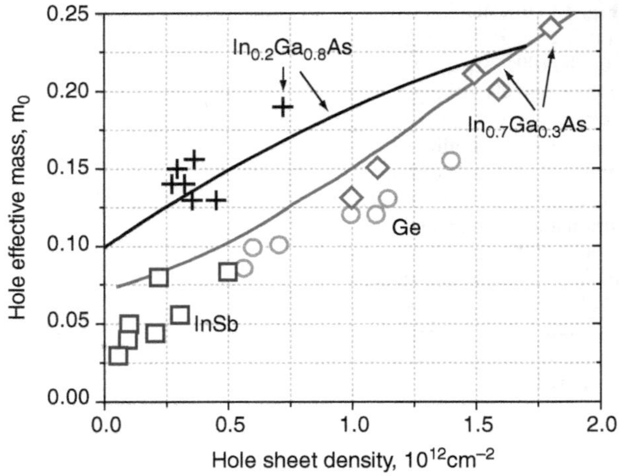

Fig. 12.11 Effective mass vs. sheet carrier density in biaxially compressed quantum wells from Shubnikov-de Haas or cyclotron resonance measurements at low temperature: Ge [71], $In_{0.2}Ga_{0.8}As$ [72–75], $In_{0.7}Ga_{0.3}As$ [76], InSb [77, 78]. *Solid curves* are theoretical predictions for strained InGaAs QWs

in-plane effective mass in the channel. Since the density of states is proportional to the effective mass, it is important to increase the splitting between the HH1 subband and other subbands with higher in-plane mass. There are three parameters to increase these splittings: employ a barrier material with large valence band offset, increase HH-LH splitting using high strain (Fig. 12.3) and HH1–HH2 splitting by reduction of the QW width (Fig. 12.5a).

However, even if the carriers are occupying just the HH1 band, their effective mass is increasing with density due to non-parabolicity of the in-plane band dispersion. The effect of nonparabolicity of in-plane effective mass is shown in Fig. 12.11 for different materials.

The hole effective mass in strained QW are typically measured at cryogenic temperatures using Shubnikov-de Haas or cyclotron resonance experiments. At low temperature, the thermal spreading of carriers is low and the transport is due to the holes at the Fermi energy level in a QW. While the hole density increases, the Fermi level moves further up in the band, with a decreasing curvature of the energy vs. k_t and, consequently, higher carrier mass. Typically, semiconductors with lower effective mass show higher non-parabolicity, and the masses of different semiconductors tend to equalize at higher carrier density. A significant message from the Fig. 12.11 is that contrary to the low-field mobility, the values of the in-plane HH1 effective masses are very similar in most of the III-V semiconductors and Ge. In other words, the mobility benefit of Ge is significantly reduced when deeply scaled channels are considered.

12.5 p-channel HFETs

Similar to the n-type FETs, most of the fabricated III-V p-FETs employed Schottky gates with heterostructures or quantum wells to localize holes in the channel, and are referred to as heterostructure FET (HFET) or QWFET, respectively. A strong driver for p-channel HFET development in 1990's was the objective to build fast and low-power complimentary circuits.

The primary approach has relied on strained InGaAs/AlGaAs QW grown on GaAs for both n- and p-channels [52, 79–81] Alternatively, heterostructures with InAlAs/InGaAs QWs grown on InP substrate was studied with a promise of improved both electron and hole transport [82, 83]. However, the improvements were demonstrated for n-channel but not for p-channel, likely due to low strain introduced into the structures. A comparison of a DC transonductance, g_m, and saturation current of p-HFETs fabricated with different materials systems is shown in Fig. 12.12. A transistor delay time $\tau = C/g_m$ (where C is the total input capacitance, gate and parasitics) is inversely proportional to the transconductance that makes the latter a reasonably good metric for channel comparison. On the other hand, the maximum drain current characterizes the carrier transport of the channel, and is good metric for channel and source-drain resistances. However, both the transconductance and drain current are inversely proportional to the gate length in channels with drift transport and low source-drain resistance and should be compared accordingly. It should be also noted, that the maximum drain current and trunsconductance values are obtained at lower supply voltages for shorter channels.

The InGaAs/AlGaAs p-channel HFETs with transconductance as high as 113 mS/mm and saturation current of 94 mA/mm for 0.8 µm-long channel was reported by Ruden et al. and consisted of 20 nm-thick undoped $In_{0.2}Ga_{0.8}As$ channel, 25 nm-thick $Al_{0.47}Ga_{0.53}As$ top barrier layer with 5 nm-thick GaAs capping layer [52]. A WSi gate metal was used as a self-aligned mask for source-drain ion-implantation.

Low barrier heights in the valence band of InGaAs/AlGaAs required quite thick top barrier layers to reduce the gate leakage current. On the other hand, the transconductance (and therefore speed) of the transistor is inversely proportional to the top barrier thickness (equivalent to $g_m \propto C_{ox}$ in MOSFETs). To enhance the barrier

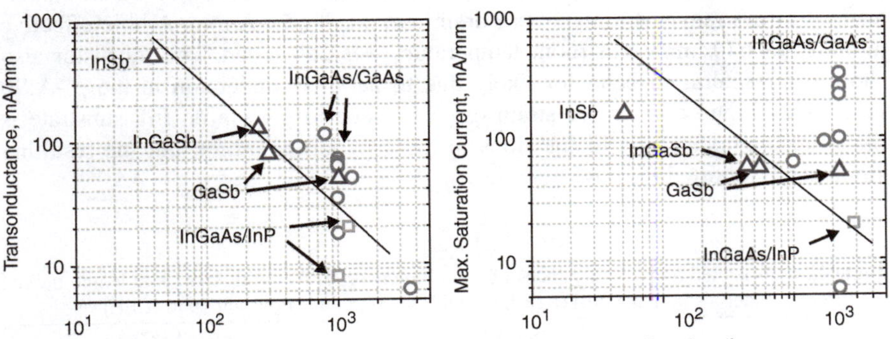

Fig. 12.12 Comparison of maximum transconductance data (*left*) and maximum saturation drain current (*right*) vs. channel length for p-channel HFETs built with different materials systems: InGaAs channel on GaAs substrate [51, 52, 84–90]; InGaAs pseudomorphic channel on InP substrate [82, 83]; GaSb-channel [64, 91, 92]; InGaSb channel [63]; and InSb channel [65]. *Solid lines* show linear scaling trend typical for long channels

height the n^+-doping of the top layers under the gate (anisotype-gate) was employed [84]. The top barrier consisting of (graded n^+-InGaAs)/GaAs/$Al_{0.3}Ga_{0.7}As$/(optional AlAs)/(InGaAs-QW) layers resulted in 1000 times reduction of leakage current, when compared to $Al_{0.75}Ga_{0.25}As$/InGaAs HFET. The fabricated device had a compact design with self-aligned gate and ion-implanted source and drain contacts. The maximum transonductance for 1.3 µm–long gate was 50 mS/mm, and the cut-off frequency $f_t = 5$ GHz was measured for 1 µm device.

More recently, $p^+/n^+/p$ camel-gate HFETs with $In_{0.15}Ga_{0.85}As/In_{0.49}Ga_{0.51}P$ channel were proposed to increase potential barrier heights and gate voltage swings for improved linearity of analogue circuits and higher noise margins [90, 93–95]. This structure resulted in a depletion mode HFET with extremely high gate turn-on voltage (+2 V). It is of great interest, that the p-type HFET with 1 µm showed maximum transconductance of 85 mS/mm and very high saturation current of 345 mA/mm at $V_d = -5$ V [90, 94].

An n-InAs and p-GaSb complementary HFET pair was proposed [96] to improve transistor characteristics over the InGaAs n- and p-pair (Table 12.4). The large valence-band offset of the AlSb/GaSb interface allows for utilization of a relatively thin barrier layer of 10 nm (additional 5 nm in the table is coming from the carrier distribution in the QW) keeping the tunneling gate current low, and therefore, significantly improving the value of transconductance.

An HFET with GaSb p-channel was implemented by Longenbach et al. in 1990 (Fig. 12.13a) [91]. An AlSbAs/AlSb/GaSb p-channel FET structure was grown by MBE on InP substrate at 480 °C. The growth was initiated with a GaSb/AlSb superlattice followed by a 1µm AlGaSb buffer layer. Following these layers, a 10 nm GaSb undoped channel is grown followed by a 4 nm AlSb spacer, a 10 nm Be-doped p^+-AlSb layer, a 10 nm p^+-$AlSb_{0.9}As_{0.1}$ layer and finally a 5 nm GaSb cap layer. The structure had 29 nm separation between the channel QW and a gold Schottky gate, had no gate recess and alloyed AuZn source and drain contacts. The FET with source to drain spacing of 3 µm and 1 µm gate length demonstrated transconductance of 50 mS/mm and a maximum drain current of 55 mA/mm. A hole effective mobility of 640 $cm^2/V\,s$ at room temperature was calculated. The transistor also showed nonsaturating behavior likely due to parallel conduction in the p^+-AlSb layer. Very high compressive strain of ~3.8% due to growth on InP substrate is likely accommodated by dislocations in the superlattice buffer, but the resulting strain is not reported for this structure.

Table 12.4 Projected transconductances of complimentary HFETs with 1 µm gate length based on different materials systems [96] (Reprinted with permission. Copyright 1990 IEEE)

	AlSb/GaSb	AlSb/InAs	AlGaAs/GaAs	
	p-channel	n-channel	p-channel	n-channel
Drift velocity 10^7 cm/s	0.5	3.6	0.24	1.2
Dielectric constant	14.4	14.4	11	11
Gate-channel separation, nm	15	35	36	36
Transconductance, mS/mm	425	1310	68	324

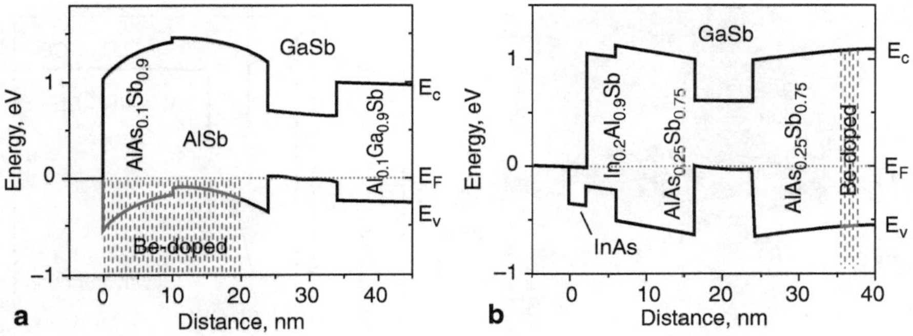

Fig. 12.13 Energy band diagrams of QW GaSb p-channel HFETs: **a** with AlSb/AlGaSb barrier on InP substrate [91] (Reprinted with permission. Copyright 1991 IEEE); **b** with AlAsSb barrier on GaAs [60] (Reprinted with permission. Copyright 2008 Elsevier). Note that the band diagram is shown for the area under the source/drain contact; the p^+-InAs is recessed under the Schottky gate

Recently the NRL group reported several novel Sb-based transistor structures and demonstrated excellent performance of p-channel quantum well HFETs [97]. The HFET with GaSb p-channel was fabricated on a GaAs(001) substrate with 1 μm–thick $AlAs_{0.24}Sb_{0.76}$ buffer layer consisting of AlSb/AlAs short period superlattice. The buffer layer was shown to accommodate the 7% lattice mismatch almost entirely, and provided a virtual substrate for the 1.2% compressively strained GaSb channel. The AlAsSb barrier were chosen because of large band offset (~0.6 eV) with GaSb to provide good hole confinement (Fig. 12.13b). A GaSb QW HFET with 0.3 μm gate demonstrated a maximum transconductance of 80 mS/mm and subthreshold slope of 104 mV/decade at the drain voltage of −1 V. The maximum cut-off frequency, f_t, and maximum oscillation frequency, f_{max}, were 6 and 18 GHz, respectively.

Another material system investigated by the NRL group is InGaSb/AlGaSb heterostructure [63, 67]. Increase of the In content results in reduction of the hole mass and improvement of the transport properties. The heterostructure was also metamorphic and contained plastically relaxed (lattice mismatch is 8%) 1.5 μm–thick $Al_{0.7}Ga_{0.3}Sb$ buffer layer on GaAs(001) substrate. A 7.5 nm-thick $In_{0.41}Ga_{0.59}Sb$ QW had 0.4 V valence band offset with the barrier and was under 2% compressive biaxial strain. A starting heterostructure with this design exhibited the room-temperature Hall mobility of 1020 $cm^2/V\,s$ at sheet hole concentration of $1.6 \times 10^{12}\,cm^{-2}$. The transistor had a typical HEMT design with T-gate and source/drain contacts on a narrow-bandgap InAs capping layer. It was a depletion mode device with threshold voltage of +0.14 V. The HFET with 0.25 μm recessed T-gate length has demonstrated maximum transconductance133 mS/mm and maximum saturated current of about 60 mA/mm at −2.5 V drain voltage. The extrapolated f_t and f_{max} were found to be 25 and 27 GHz, respectively [63].

The most advanced p-channel FET was recently produced jointly by the researches from Intel and QinetiQ (Fig. 12.14) [65]. The device utilized InSb channel with the lowest within the group III-V semiconductors hole mass and the highest

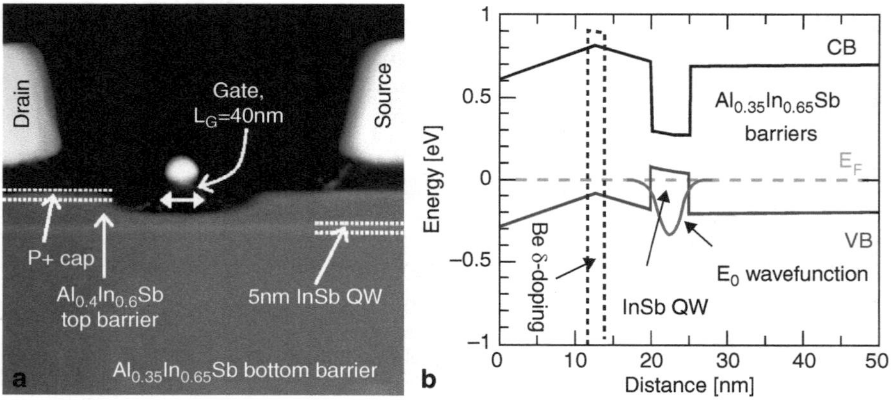

Fig. 12.14 **a** TEM image of 40 nm gate length InSb p-channel strained QWFET, compressively strained QWFET structure. **b** Band diagram of InSb p-channel 1.9% compressively strained QWFET [65] (Reprinted with permission. Copyright 2008 IEEE)

bulk hole mobility. In fact, the mobility in 1.9% compressively strained QW was as high as 1230 $cm^2/V\,s$ at $1.1 \times 10^{12}\,cm^2$ hole concentration or a factor of five higher than that in strained Si. The structure contained a 3 μm-thick $Al_{0.35}In_{0.65}Sb$ metamorphic buffer grown on a GaAs substrate, 5 nm InSb QW, and $Al_{0.4}In_{0.6}Sb$ top barrier with Be δ-doping layer separated by 7 nm from the channel. A Ti/Au recessed T-gate was separated by about 10 nm from the channel.

The HFET (or QWFET as it was named in the paper) with 40 nm gate length has shown record high characteristics for p-type channels: transconductance of 510 mS/mm, maximum f_t = 140 GHz, both obtained at 0.5 V drain voltage. The transistor characteristics are shown in Fig. 12.15 [65]) and illustrate important challenges.

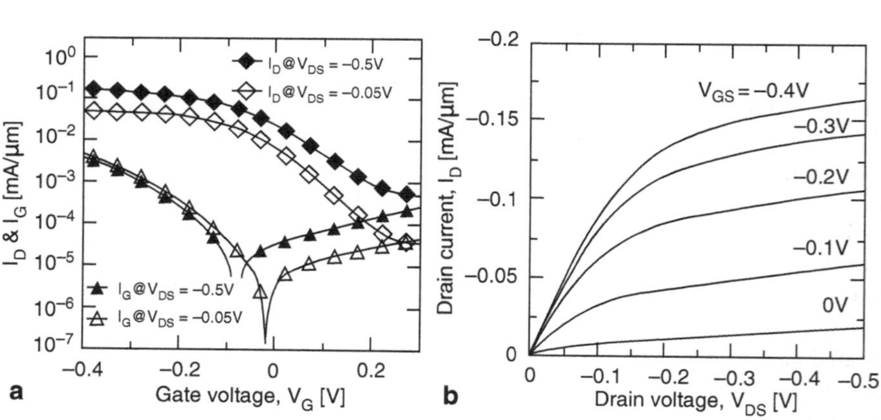

Fig. 12.15 **a** Drain (I_D) and gate (I_G) currents vs. gate voltage (V_G), and **b** drain current vs. drain voltage (V_{DS}) of InSb p-channel compressively strained QWFET with 40 nm gate length [65] (Reprinted with permission. Copyright 2008 IEEE)

The device drain current is lower than may be expected when comparing the maximum currents in Fig. 12.12b. The authors do no not describe the possible reasons for that, but it is likely due to source and drain resistances of the order of 0.5 Ω mm as can be estimated from Fig. 12.15b. The FETs with longer gates of 125 nm have demonstrated impressive short channel performance: subthreshold slope (SS) of 90 mV/decade. and drain-induced barrier lowering (DIBL) of 80 mV/V. The 40 nm gate devices have shown SS = 160 mV/decade. and DIBL = 220 mV/V indicating that 10 nm-thick top barrier layer is too thick to keep electrostatic integrity for this aggressively scaled channel. A simple reduction of the top barrier thickness is hardly possible, because it will inevitably increase the gate current, which is already limiting the ON/OFF ratio of the 40 nm FET to about 200.

This highly successful QWFET provides a good benchmark for p-channel FETs for logic applications. The energy-delay product comparison of HFETs built from different material systems is shown in Fig. 12.16. Compared to strained Si, the InSb channel demonstrates 10× lower power at the same speed, or 2× higher speed at matched power [65]. The authors also determined the 2× improvement of effective hole velocity over that in the strained Si at matched DIBL.

As discussed earlier in this book (Chap. 3), the scaling capabilities of HFETs with Schottky gate are quite limited, mainly due to fast increase of gate leakage current when the top barrier layer is thinned down to keep electrostatic integrity of the transistor. This requires the replacement of part or the entire top semiconductor barrier with a material with wider bandgap and high dielectric constant, and, therefore, the high-k oxides come to play.

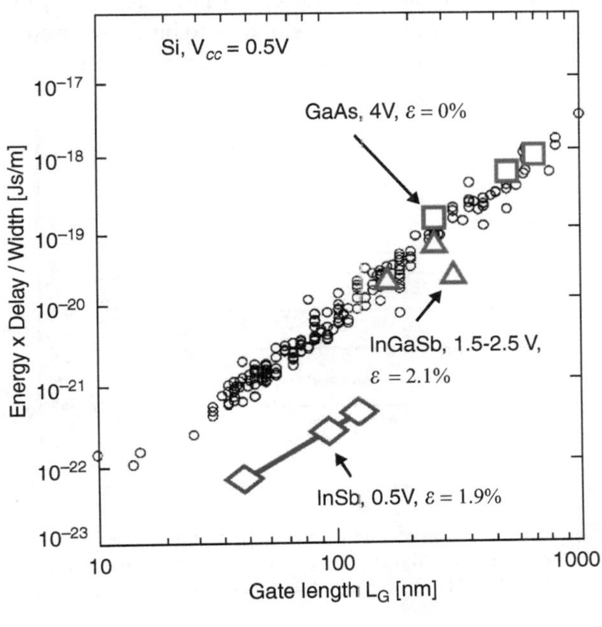

Fig. 12.16 Energy-delay product vs. gate length comparing InSb p-channel QWFET [65] with Si p-MOSFETs. Other group III-V channel p-QWFETs [63, 98] are included for reference [65] (Reprinted with permission. Copyright 2008 IEEE)

12.6 p-type MOSFETs

The results on p-type MOSFET are very limited. Most of the earlier work on p-channels focused on group III-Sb MOSFETs due to difficulties to make low-leakage Schottky barrier HFETs in these narrow bandgap semiconductors. The feasibility of MOSFETs with InSb and GaSb p-channels were first demonstrated in 1975 [99] and 1977 [100], respectively. The transistors employed 100 nm-thick low-temperature-grown SiO_2 gate oxide deposited on n-InSb and n-GaSb substrates. The InSb MOSFET with recessed gate showed FET characteristics at 77 K, and the GaSb MOSFET with Be-implanted source and drain has demonstrated transistor characteristics at room temperature with hole effective mobility of 166 $cm^2/V\,s$.

Later, various technologies for gate stack on InSb were tested, and p-type inversion channel MOSFETs were demonstrated. The main driving force for these studies was the need of switches for InSb monolithic infrared focal plane arrays. Since the arrays were typically cooled down, the transistor characteristics were presented at 77 K. The gate oxides on n-InSb were fabricated by thermal oxidation [101], anodic oxidation and SiO evaporation [102], photo-enhanced chemical vapor deposition of SiO_2 [103]. At 77 K the switches showed ON/OFF ratio over 10^4, but had very large subthreshold swing of about 1.5 V/decade [104].

Okamura et al. [105] demonstrated an InP p-channel MOSFET as early as 1980 using in-situ HCl vapor etching (~150 nm in depth) of the substrate prior to chemical vapor deposition of 120 nm-thick Al_2O_3, and Be-implanted source-drain contacts. Using Terman method, the authors estimated the interface trap density of $1 \times 10^{12}\,cm^{-2}\,eV^{-1}$ in both the upper and the lower parts of the bandgap. The inversion channel was demonstrated with maximum transonductance of about 0.7 mS/mm and maximum drain current of 1 mA/mm for 3 μm channel. The value of channel effective mobility was estimated to be about 16 $cm^2/V\,s$. Interestingly the in-situ HCl etching prior to oxide deposition was essential to reduce the interface state

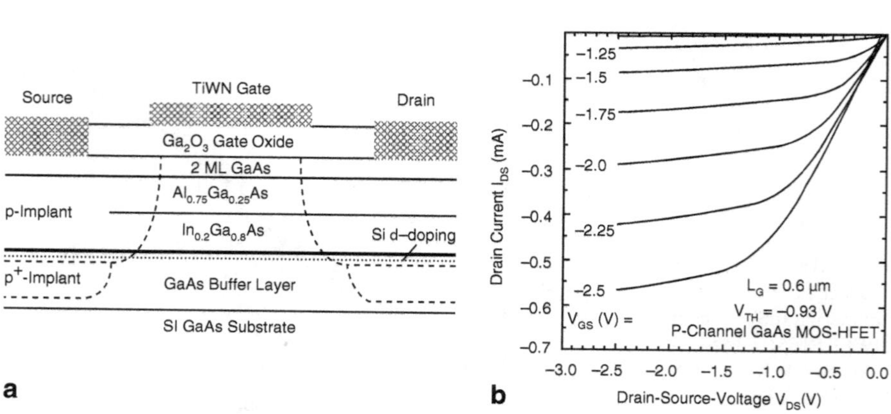

Fig. 12.17 **a** Cross-section of self-aligned enhancement mode InGaAs p-MOSFET. **b** DC output characteristics of a 0.6 μm device showing enhancement mode operation with the threshold voltage of −0.93 V [107] (Reprinted with permission. Copyright 2002 IEEE)

density close to the valence band. Without etching, the device did not show inversion channel behavior.

Just a few more recent papers reported p-type MOSFETs on GaAs and InGaAs [106–110]. Tametou et al. [108] reported a p-channel enhancement/inversion mode GaAs-MISFET with oxy-nitrided gate insulation film formed by nitriding after oxidation of the GaAs surface. The device with 0.7 μm gate had transconductance of 9 mS/mm. p-channel MOSFETs on semi-insulating GaAs with HfO_2 gate oxide and Si [101] or Ge [109] interface passivation layers were reported. Devices with Si or Ge passivation demonstrated DC transconductance of about 0.1 mS/mm for 100 μm gate and subthreshold swing of about 170 mV/decade.

Ren et al. [106] demonstrated p-MOSFET on GaAs with mixed Ga_2O_3–Gd_2O_3 gate oxide. The MOSFET with 40 μm gate showed DC transconductance of 0.3 mS/mm.

The highest transconductance in III-V p-type MOSFETs was demonstrated by Passlack et al. [107] The device structure is shown in Fig. 12.17 and includes an HFET-type modulation-doped heterostructure with 15 nm-thick strained $In_{0.2}Ga_{0.8}As$ QW channel and 15 nm-thick $Al_{0.75}Ga_{0.25}As$ top barrier layer. A 9 nm-thick Ga_2O_3 gate oxide was deposited by thermal evaporation from an effusion cell. The device had self-aligned ion-implanted source-drain contacts which provided

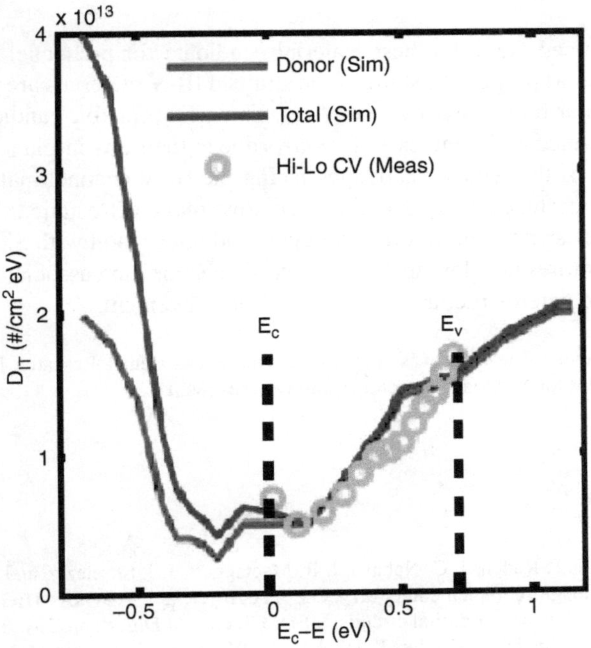

Fig. 12.18 Measured using Hi-Lo CV technique and simulated interface trap density profile at $In_{0.65}Ga_{0.35}As/Al_2O_3$ gate interface. A good match is obtained by using donor-like traps below the InGaAs conduction band [113] (Reprinted with permission. Copyright 2008 IEEE)

the contact resistance of 1.05 Ωmm after 700 °C annealing. The transonductance was measured as high as 51 mS/mm for 0.6 μm gate length. The authors do not report subthreshold characteristics of this device, therefore, it is not possible to assess interface trap density close to the valence band.

It should be noted, that to the best of the author's knowledge there are no reports on successful operation of a p-MOSFET with a strained high-In content InGaAs channel grown on InP. Two potential problems may be respopnsible for this difficulty. Firstly, it is well-documented that in high-In content InGaAs the surface position of the Fermi level moves towards the conduction band. This results in low (or even negative for InAs) Schottky barrier heights in these alloys [111] and formation of the conduction channel at zero bias in inversion mode n-MOSFET fabricated on p-type InAs substrate [112]. Secondly, the InGaAs alloys possibly have a relatively higher interface trap density in the lower part of the bandgap as compared to the upper part close to the conduction band. Some evidence for this was recently reported by Varghese et al. [113] as shown in Fig. 12.18. These two challenges, as well as really high interface trap density of $\sim 10^{13}\,cm^{-2}\,eV^{-1}$, should be addressed by interface engineering before the p-InGaAs channel may be considered as feasible.

12.7 Conclusions

Although strained Ge is the best material candidate for p-channel because of its intrinsic transport properties, some of the strained III-V materials are good competitors in particular for deeply scaled devices. The most probable candidates are compressively strained antimonides and InGaAs due to their low in-plane hole effective mass. On top of the typical challenges facing the III-V channel materials, such as high-quality interface with a gate dielectric, low total electrostatic thickness of dielectric, low-resistance source-drain contacts, and integration with Si, the p-channel in addition requires development of high-strain channel and associated with it band engineering of heterostructures and materials improvement.

Acknowledgement The work was supported by the Focus Center Research Program (FCRP) through the center for Materials, Structures, and Devices (MSD).

References

1. D. E. Grider, P. P. Ruden, J. C. Nohava, I. R. Mactaggart, J. J. Stronczer, and R. H. Tran, "0.7 micron gate length complementary $Al_{0.75}Ga_{0.25}$ As/$In_{0.25}Ga_{0.75}$ As/GaAs HIGFET technology for high speed/low power digital circuits," IEDM Technical Digest, pp. 331–334 (1992).
2. A. Leuther, A. Thiede, K. Kohler, T. Jakobus, G. Weimann, K. H. Ploog, G. Trankle, and G. Weimann, "Complementary HFETs on GaAs with 0.2 μm gate length," Compound Semiconductors 1999. Proc. 26th International Symposium on Compound Semiconductors, Bristol, UK, pp. 313–316 (2000).

3. J. D. Wiley, "Mobility of holes in III-V semiconductors," in *Semiconductors and Semimetals, vol.10 Transport Phenomena* (Academic Press, New York, NY, 1975), pp. 91–174.
4. J. D. Wiley, "Polar mobility of holes in III-V compounds," *Phys. Rev. B*, 4(8), 2485–2493 (1971).
5. M. Costato, C. Jacoboni, and L. Reggiani, "Hole transport in polar semiconductors," *Phys. Stat. Sol. B*, 52(2), 461–473 (1972).
6. H. J. Lee and D. C. Look, "Hole transport in pure and doped GaAs," *J. Appl. Phys.*, 54(8), 4446–4452 (1983).
7. A. S. Filipchenko and L. P. Bolshakov, "Mobility of holes in p-InSb crystals," *Phys. Stat. Sol. B*, 77(1), 53–58 (1976).
8. M. Zimpel, M. Oszwaldowski, and J. Goc, "Mobility of holes in InSb," *Acta Physica Polonica A*, A75(2), 297–300 (1989).
9. J. D. Wiley and M. DiDomenico, Jr., "Lattice mobility of holes in III-V compounds," *Phys. Rev. B*, 2(2), 427–433 (1970).
10. O. Madelung, *Data in science and technology* (Springer-Verlag, Berlin, 1991).
11. S. Takagi, T. Mizuno, T. Tezuka, N. Sugiyama, S. Nakaharai, T. Numata, J. Koga, and K. Uchida, "Sub-band structure engineering for advanced CMOS channels," *Solid State Electron.*, 49(5), 684–694 (2005).
12. R. Chau, S. Datta, and A. Majumdar, "Opportunities and challenges of III-V nanoelectronics for future high-speed, low-power logic applications," IEEE Compound Semiconductor Integrated Circuit Symposium (IEEE, Piscataway, NJ, 2005).
13. E. O. Kane, "Theory of tunneling," *J. Appl. Phys.*, 32(1), 83–91 (1961).
14. Y. Sun, S. E. Thompson, and T. Nishida, "Physics of strain effects in semiconductors and metal-oxide-semiconductor field-effect transistors," *J. Appl. Phys.*, 101(10), 104503-1-22 (2007).
15. T. Krishnamohan, D. Kim, C. D. Nguyen, C. Jungemann, Y. Nishi, and K. C. Saraswat, "High-mobility low band-to-band-tunneling strained-Germanium double-gate heterostructure FETs: Simulations," *IEEE Trans. Electron Device*, 53(5), 1000–1009 (2006).
16. D. Kim, T. Krishnamohan, L. Smith, H.-S. P. Wong, and K. C. Saraswat, "Band to band tunneling study in high mobility materials: III-V, Si, Ge and strained SiGe," 65th DRC Technical Digest, pp. 57–58 (2007).
17. P. Y. Yu, and M. Cardona, *Fundamentals of Semiconductors* (Springer, New York, NY, 2005).
18. J. Singh, *Electronic and Optoelectronic Properties of Semiconductor Structures* (Cambidge University Press, Cambridge, UK, 2003).
19. J. Piprek, *Semiconductor Optoelectronic Devices: Introduction to Physics and Simulation* (Academic Press, London, UK, 2003).
20. S. L. Chuang, *Physics of Optoelectronic Devices* (John Wiley & Sons, New York, NY, 1995).
21. G. C. Osbourn, "$In_xGa_{1-x}As$-$In_yGa_{1-y}As$ strained-layer superlattices: A proposal for useful, new electronic materials," *Phys. Rev. B*, 27(8), 5126–5128 (1983).
22. B. K. Ridley, "The in-plane effective mass in strained-layer quantum wells," *J. Appl. Phys.*, 68(9), 4667–4673 (1990).
23. I. Suemune, "Band-edge hole mass in strained-quantum-well structures," *Phys. l Rev. B*, 43(17), 14099–14106 (1991).
24. J. Bohrer, A. Krost, T. Wolf, and D. Bimberg, "Band offsets and transitivity of $In_{1-x}Ga_xAs$/$In_{1-y}Al_yAs$/InP heterostructures," *Phys. l Rev. B*, 47(11), 6439–6443 (1993).
25. R. People and J. C. Bean, "Calculation of critical layer thickness versus lattice mismatch for $GexSi_{1-x}$/Si strained-layer heterostructures," *Appl. Phys. Lett.*, 47(3), 322–324 (1985).
26. J. W. Matthews and A. E. Blakeslee, "Defects in epitaxial multilayers. I. Misfit dislocations," *J. Cryst. Growth*, 27(1), 118–125 (1974).
27. X. Wallart, B. Pinsard, and F. Mollot, "High In content pseudomorphic InGaAs layers for high-mobility heterostructures on InP (001)," *J. Cryst. Growth*, 278(1), 516–520 (2005).
28. H. Toyoshima, T. Niwa, J. Yamazaki, and A. Okamoto, "In surface segregation and growth-mode transition during InGaAs growth by molecular-beam epitaxy," *Appl. Phys. Lett.*, 63(6), 821–823 (1993).

29. H. Toyoshima, T. Niwa, J. Yamazaki, and A. Okamoto, "Suppression of In surface segregation and growth of modulation-doped N-AlGaAs/InGaAs/GaAs structures with a high In composition by molecular-beam epitaxy," *J. Appl. Phys.*, 75(8), 3908–3913 (1994).
30. M. Yakimov, V. Tokranov, G. Agnello, J. van Eisden, and S. Oktyabrsky, "In situ monitoring of formation of InAs quantum dots and overgrowth by GaAs or AlAs," *J. Vac. Sci. Tech. B*, 23(3), 1221–1225 (2005).
31. S. S. Nedorezov, "Size quantization in semiconducting films," *Fizika Tverdogo Tela*, 12(8), 2269–2276 (1970).
32. M. Alttarelli, "Heterojunctions and semiconductor superlattices," in *Heterojunctions and Semiconductor Superlattices*, Eds.: G. Allan, G. Bastard, N. Boccora, and M. Voos (Springer, Berlin, 1986).
33. G. Bastard and J. A. Brum, "Electronic states in semiconductor heterostructures," *IEEE J. Quantum Electron.*, QE-22(9), 1625–1644 (1986).
34. L. C. Andreani, A. Pasquarello, and F. Bassani, "Hole subbands in strained GaAs-$Ga_{1-x}Al_xAs$ quantum wells: exact solution of the effective-mass equation," *Phys. Rev. B*, 36(11), 5887–5894 (1987).
35. S. L. Chuang, "Efficient band-structure calculations of strained quantum wells," *Phys. Rev. B*, 43(12), 9649–9661 (1991).
36. D. A. Broido and L. J. Sham, "Effective masses of holes at GaAs-AlGaAs heterojunctions," *Phys. Rev. B*, 31(2), 888–892 (1985).
37. M. Jaffe, Y. Sekiguchi, and J. Singh, "Theoretical formalism to understand the role of strain in the tailoring of hole masses in p-type $In_xGa_{1-x}As$ (on GaAs substrates) and $In_{0.53+x}Ga_{0.47-x}As$ (on InP substrates) modulation-doped field-effect transistors," *Appl. Phys. Lett.*, 51(23), 1943–1945 (1987).
38. B. Laikhtman, R. A. Kiehl, and D. J. Frank, "Strained quantum well valence-band structure and optimal parameters for AlGaAs-InGaAs-AlGaAs p-channel field-effect transistors," *J. Appl. Phys.*, 70(3), 1531–1538 (1991).
39. T. Ando, A. B. Fowler, and F. Stern, "Electronic properties of two-dimensional systems," *Rev. Mod. Phys.*, 54(2), 437–672 (1982).
40. M. V. Fischetti, Z. Ren, P. M. Solomon, M. Yang, and K. Rim, "Six-band k·p calculation of the hole mobility in silicon inversion layers: dependence on surface orientation, strain, and silicon thickness," *J. Appl. Phys.*, 94(2), 1079–1095 (2003).
41. B. K. Ridley, "Electron scattering by confined LO polar phonons in a quantum well," *Phys. Rev. B*, 39(8), 5282–5286 (1989).
42. M. V. Fischetti, D. A. Neumayer, and E. A. Cartier, "Effective electron mobility in Si inversion layers in metal-oxide-semiconductor systems with a high-k insulator: The role of remote phonon scattering," *J. Appl. Phys.*, 90(9), 4587–4608 (2001).
43. I. J. Fritz, T. J. Drummond, G. C. Osbourn, J. E. Schirber, and E. D. Jones, "Electrical transport of holes in GaAs/InGaAs/GaAs single strained quantum wells," *Appl. Phys. Lett.*, 48(24), 1678–1680 (1986).
44. D. Lancefield, A. R. Adams, A. T. Meney, W. Knap, E. Litwin-Staszewska, C. Skierbiszewski, and J. L. Robert, "The light-hole mass in a strained InGaAs/GaAs single quantum well and its pressure dependence," *J. Phys. Chem. Solids*, 56(3), 469–473 (1995).
45. M. Myronov, T. Irisawa, S. Koh, O. A. Mironov, T. E. Whall, E. H. C. Parker, and Y. Shiraki, "Temperature dependence of transport properties of high mobility holes in Ge quantum wells," *J. Appl. Phys.*, 97(8), 83701-1-6 (2005).
46. H. von Kanel, M. Kummer, G. Isella, E. Muller, and T. Hackbarth, "Very high hole mobilities in modulation-doped Ge quantum wells grown by low-energy plasma enhanced chemical vapor deposition," *Appl. Phys. Lett.*, 80(16), 2922–2924 (2002).
47. R. J. H. Morris, T. J. Grasby, R. Hammond, M. Myronov, O. A. Mironov, D. R. Leadley, T. E. Whall, E. H. C. Parker, M. T. Currie, C. W. Leitz, and E. A. Fitzgerald, "High conductance Ge p-channel heterostructures realized by hybrid epitaxial growth," *Semicond. Sci. Tech.*, 19(10) (2004).

48. M. Myronov, T. Irisawa, O. A. Mironov, S. Koh, Y. Shiraki, T. E. Whall, and E. H. C. Parker, "Extremely high room-temperature two-dimensional hole gas mobility in Ge/$Si_{0.33}Ge_{0.67}$/Si(001) p-type modulation-doped heterostructures," *Appl. Phys. Lett.*, 80(17), 3117–3119 (2002).
49. M. Myronov, K. Sawano, Y. Shiraki, T. Mouri, and K. M. Itoh, "Observation of two-dimensional hole gas with mobility and carrier density exceeding those of two-dimensional electron gas at room temperature in the SiGe heterostructures," *Appl. Phys. Lett.*, 91(8), 082108-1-3 (2007).
50. M. Kudo, H. Matsumoto, T. Tanimoto, T. Mishima, and I. Ohbu, "Improved hole transport properties of highly strained $In_{0.35}Ga_{0.65}As$ channel double-modulation-doped structures grown by MBE on GaAs," *J. Cryst. Growth*, 175–176, 910–914 (1997).
51. T. J. Drummond, T. E. Zipperian, I. J. Fritz, J. E. Schirber, and T. A. Plut, "p-channel strained quantum well, field-effect transistor," *Appl. Phys. Lett.*, 49(8), 461–463 (1986).
52. P. P. Ruden, M. Shur, D. K. Arch, R. R. Daniels, D. E. Grider, and T. E. Nohava, "Quantum-well p-channel AlGaAs/InGaAs/GaAs heterostructure insulated-gate field-effect transistors," *IEEE Trans. Electron Device*, 36(11), 2371–2379 (1989).
53. N. Kudo, T. Mishima, H. Matsumoto, I. Ohbu, and T. Tanimoto, "Highly strained $In_{0.35}Ga_{0.65}As$/GaAs layers grown by molecular beam epitaxy for high hole mobility transistors," *J. Electron. Mater.*, 25(6), 944–947 (1996).
54. H. J. Kim, D. M. Kim, D. H. Woo, S.-I. Kim, S. H. Kim, J. I. Lee, K. N. Kang, and K. Cho, "High photoresponsivity of a p-channel InGaP/GaAs/InGaAs double heterojunction pseudomorphic modulation-doped field effect transistor," *Appl. Phys. Lett.*, 72(5), 584–586 (1998).
55. R. T. Hsu, W. C. Hsu, M. J. Kao, and J. S. Wang, "Characteristics of a Î′-doped GaAs/InGaAs p-channel heterostructure field-effect transistor," *Appl. Phys. Lett.*, 66(21), 2864–2866 (1995).
56. R. T. Hsu, Y. S. Lin, J. S. Su, W. C. Hsu, Y. H. Wu, and M. J. Kao, "Study of two-dimensional hole gas concentration and hole mobility in zinc delta-doped GaAs and pseudomorphic GaAs/$In_{0.2}Ga_{0.8}As$ heterostructures," *Superlattices and Microstructures*, 24(2), 175–180 (1998).
57. P. Nagaiah, V. Tokranov, M. Yakimov, and S. Oktyabrsky, "Strained quantum wells for p-channel InGaAs CMOS," *MRS Proc.*, 1108, A12–01 (2009).
58. K. Yoh, K. Kiyomi, H. Taniguchi, M. Yano, M. Inoue, and G. B. Stringfellow, "A p-channel GaSb heterojunction field-effect transistor based on a vertically integrated complementary circuit structure," Gallium Arsenide and Related Compounds 1991. Proc. 18th International Symposium (Bristol, UK) pp. 173–178 (1992).
59. L. F. Luo, K. F. Longenbach, and W. I. Wang, "p-channel modulation-doped field-effect transistors based on $AlSb_{0.9}As_{0.1}$/GaSb," *IEEE Electron Device. Lett.*, 11(12), 567–569 (1990).
60. B. R. Bennett, M. G. Ancona, J. B. Boos, C. B. Canedy, and S. A. Khan, "Strained GaSb/AlAsSb quantum wells for p-channel field-effect transistors," *J. Cryst. Growth*, 311(1), 47–53 (2008).
61. J. F. Klem, J. A. Lott, J. E. Schirber, S. R. Kurtz, and S. Y. Lin, "Strained quantum well modulation-doped InGaSb/AlGaSb structures grown by molecular beam epitaxy," *J. Electron. Mater.*, 22(3), 315–318 (1993).
62. B. R. Bennett, M. G. Ancona, J. B. Boos, and B. V. Shanabrook, "Mobility enhancement in strained p-InGaSb quantum wells," *Appl. Phys. Lett.*, 91(4), 042104-1-3 (2007).
63. J. B. Boos, B. R. Bennett, N. A. Papanicolaou, M. G. Ancona, J. G. Champlain, R. Bass, and B. V. Shanabrook, "High mobility p-channel HFETs using strained Sb-based materials," *Electron. Lett.*, 43(15), 834–835 (2007).
64. B. R. Bennett, M. G. Ancona, and J. B. Boos, "Compound semiconductors for low-power p-channel field-effect transistors," *MRS Bulletin*, 34, 530–536 (2009).
65. M. Radosavljevic, T. Ashley, A. Andreev, S. D. Coomber, G. Dewey, M. T. Emeny, M. Fearn, D. G. Hayes, K. P. Hilton, M. K. Hudait, R. Jefferies, T. Martin, R. Pillarisetty, W. Rachmady, T. Rakshit, S. J. Smith, M. J. Uren, D. J. Wallis, P. J. Wilding, and R. Chau, "High-performance 40 nm gate length InSb p-channel compressively strained quantum well field effect transistors for low-power (VCC = 0.5 V) logic applications," IEDM Technical Digest, 2008.

66. M. Edirisooriya, T. D. Mishima, C. K. Gaspe, K. Bottoms, R. J. Hauenstein, and M. B. Santos, "InSb quantum-well structures for electronic device applications," *J. Cryst. Growth*, 311(7), 1972–1975 (2009).
67. B. R. Bennett, M. G. Ancona, J. B. Boos, and B. V. Shanabrook, "Mobility enhancement in strained p-InGaSb quantum wells," *Appl. Phys. Lett.*, 91(4), 042104-421 (2007).
68. A. Nainani, T. Krishnamohan, D. Kim, and K. Saraswat, "Hole mobility and its enhancement with strain for technologically relevant III-V semiconductors," 2009 SISPAD, September 9–11, 2009, San Diego, CA, 2006.
69. Y. Zhang and M. V. Fischetti, "Calculation of hole mobility in Ge and III-V p-channels," 13th International Workshop on Computational Electronics (IWCE 2009), Beijing, China, 2009.
70. S. Takagi, T. Tezuka, T. Irisawa, S. Nakaharai, T. Numata, K. Usuda, N. Sugiyama, M. Shichijo, R. Nakane, and S. Sugahara, "Device structures and carrier transport properties of advanced CMOS using high mobility channels," *Solid State Electron.*, 51(4), 526–536 (2007).
71. T. Irisawa, M. Myronov, O. A. Mironov, E. H. C. Parker, K. Nakagawa, M. Murata, S. Koh, and Y. Shiraki, "Hole density dependence of effective mass, mobility and transport time in strained Ge channel modulation-doped heterostructures," *Appl. Phys. Lett.*, 82(9), 1425–1427 (2003).
72. M. Jaffe, J. E. Oh, J. Pamulapati, J. Singh, and P. Bhattacharya, "In-plane hole effective masses in $In_xGa_{1-x}As/A_{l0.15}Ga_{0.85}As$ modulation-doped heterostructures," *Appl. Phys. Lett.*, 54(23), 2345–2346 (1989).
73. J. E. Schirber, I. J. Fritz, and L. R. Dawson, "Light-hole conduction in InGaAs/GaAs strained-layer superlattices," *Appl. Phys. Lett.*, 46(2), 187–189 (1985).
74. D. Lancefield, W. Batty, C. G. Crookes, E. P. O'Reilly, A. R. Adams, K. P. Homewood, G. Sundaram, R. J. Nicholas, M. Emeny, and C. R. Whitehouse, "Pressure dependence of light-hole transport in strained InGaAs/GaAs," *Surf. Sci.*, 229(1), 122–125 (1990).
75. E. D. Jones, R. M. Biefeld, J. F. Klem, S. K. Lyo, T. Ikoma, and H. Watanabe, "Strain and density dependent valence-band masses in InGaAs and GaAs/GaAsP strained-layer structures," Gallium Arsenide and Related Compounds 1989. Proc. 16th International Symposium, Karuizawa, Japan, pp. 435–440 (1990).
76. S. Rapp, V. Harle, H. Bolay, A. Hangleiter, F. Scholz, W. Limmer, E. Vasiliadou, P. Grambow, and D. Weiss, "Evaluation of the effective hole masses in pseudomorphic compressively strained $Ga_xIn_{1-x}As/InP$ quantum wells," *Appl. Phys. Lett.*, 67(1), 67–69 (1995).
77. J. E. Schirber, C. P. Tigges, L. R. Dawson, H. P. Hjalmarson, and I. J. Fritz, "Effective masses and g factors for 2D light holes in InSb," *Semicond. Sci. Technol.*, 5(3), pp. 192–194 (1990).
78. C. K. Gaspe, M. Edirisooriya, T. D. Mishima, A. R. Dedigama, S. Q. Murphy, R. E. Doezema, M. B. Santos, L. C. Tung, and Y.-J. Wang, "InSb and $In_xGa_{1-x}As$ quantum wells remotely doped with Be," *J. Vac. Sci. Technol. B*, (May 2010, in press).
79. P. P. Ruden, A. I. Akinwande, D. Narum, D. E. Grider, and J. Nohava, "High performance complementary logic based on GaAs/InGaAs/AlGaAs HIGFETs," IEDM Technical Digest, pp. 117–120 (1989).
80. D. E. Grider, I. R. Mactaggart, J. C. Nohava, J. J. Stronczer, P. P. Ruden, T. E. Nohava, D. Fulkerson, and D. E. Tetzlaff, "A 4 kbit synchronous static random access memory based upon delta-doped complementary heterostructure insulated gate field effect transistor technology," 13th Annual GaAs IC Symp. Technical Digest, pp. 71–74 (1991).
81. D. E. Grider, P. P. Ruden, J. C. Nohava, I. R. Mactaggart, J. J. Stronczer, and R. H. Tran, "0.7 micron gate length complementary $Al_{0.75}Ga_{0.25}As/In_{0.25}Ga_{0.75}As/GaAs$ HIGFET technology for high speed/low power digital circuits," IEDM Technical Digest, pp. 331–334 (1992).
82. M. D. Feuer, Y. He, D. M. Tennant, S. C. Shunk, K. F. Brown-Goebeler, R. E. Behringer, and T. Y. Chang, "InP-based HIGFETs for complementary circuits," *IEEE Trans. Electron Device.*, 36(11), 2616 (1989).
83. S. Swirhun, T. Nohava, J. Huber, S. Bounnak, and P. Joslyn, "P- and n-channel InAlAs/InGaAs heterojunction insulated gate FETs (HIGFETs) on InP," Indium Phosphide and Related Materials. 3rd International Conference (IEEE, Cardiff, UK) pp. 235–238 (1991).

84. J. K. Abrokwah, J. H. Huang, W. Ooms, and J. A. Hallmark, "Anisotype-gate self-aligned p-channel heterostructure field-effect transistors," IEDM Technical Digest, pp. 315–318 (1992).
85. D. E. Grider, R. D. Horning, D. K. Arch, P. P. Ruden, T. E. Nohava, D. N. Narum, and R. R. Daniels, "Study of strain in pseudomorphic InGaAs heterostructures related to the enhanced performance of p-channel heterostructure field effect transistor devices," *J. Vac. Sci. Technol. B*, 7(2), 371–375 (1989).
86. R. R. Daniels, P. P. Ruden, M. Shur, D. Grider, T. E. Nohava, and D. K. Arch, "Quantum-well p-channel AlGaAs/InGaAs/GaAs heterostructure insulated-gate field-effect transistors with very high transconductance," *IEEE Electron Device Lett.*, 9(7), 355–357 (1988).
87. C.-P. Lee, H. T. Wang, G. J. Sullivan, N. H. Sheng, and D. L. Miller, "High-transconductance p-channel InGaAs/AlGaAs modulation-doped field effect transistors," *IEEE Electron Device Lett.*, EDL-8(3), 85–87 (1987).
88. J. K. Abrokwah, R. Lucero, J. A. Hallmark, and B. Bernhardt, "Submicron p-channel (Al,Ga)As/(In,Ga)As HIGFETs," *IEEE Trans. Electron Device*, 44(7), 1040–1045 (1997).
89. J.-H. Tsai, C.-M. Li, W.-C. Liu, D.-F. Guo, S.-Y. Chiu, and W.-S. Lour, "Integration of n- and p-channel InGaP/InGaAs doped-channel pseudomorphic HFETs," *Electron. Lett.*, 43(13), 732–734 (2007).
90. J.-H. Tsai, T.-Y. Weng, and K.-P. Zhu, "Investigation of InGaP/InGaAs n- and p-channel pseudomorphic modulation-doped field effect transistors with high gate turn-on voltages," *Superlattices and Microstructures*, 43(2), 73–80 (2008).
91. K. F. Longenbach, X. Li, Y. Wang, and W. I. Wang, "Polytype InAs/AlSb/GaSb for field-effect transistor applications," Proc. IEEE/Cornell Conference on Advanced Concepts in High Speed Semiconductor Devices and Circuits, Ithaca, NY, pp. 270–279 (1991).
92. L. F. Luo, K. F. Longenbach, and W. I. Wang, "p-channel modulation-doped GaSb field-effect transistors," *Electron. Lett.*, 27(5), 472–474 (1991).
93. J.-H. Tsai, K.-P. Zhu, Y.-C. Chu, and S.-Y. Chiu, "InGaP/InGaAs/GaAs camel-gate p-channel pseudomorphic modulation-doped field effect transistor," *Electron. Lett.*, 39(22), 1611–1612 (2003).
94. J.-H. Tsai, K.-P. Zhu, Y.-C. Chu, and S.-Y. Chiu, "High gate turn-on voltages of InGaP/InGaAs camel-gate n- and p-channel pseudomorphic modulation-doped field effect transistors prepared by low-pressure MOCVD," IVESC 2004. The 5th International Vacuum Electron Sources Conference Proceedings, Beijing, China. pp. 368–370 (2004).
95. J.-H. Tsai, W.-S. Lour, and W.-C. Liu, "InGaP/GaAs/InGaAs Î′-doped p-channel field-effect transistor with p+/n+/p camel-like gate structure," *Electron. Lett.*, 45(11), 572–573 (2009).
96. K. F. Longenbach, R. Beresford, and W. I. Wang, "A complementary heterostructure field effect transistor technology based on InAs/AlSb/GaSb," *IEEE Trans. Electron Device*, 37(10), 2265–2267 (1990).
97. J. B. Boos, B. R. Bennett, N. A. Papanicolaou, M. G. Ancona, J. G. Champlain, D. Park, W. Kruppa, B. D. Weaver, R. Bass, and B. V. Shanabrook, "Sb-based n- and p-channel HFETs for high-speed, low-power applications," DRC Technical Digest, PA, 2009.
98. H. Park, P. Mandeville, R. Saito, P. J. Tasker, W. J. Schaff, and L. F. Eastman," RF and DC characterization of P-channel $Al_{0.5}Ga_{0.5}As$/GaAs MODFETs with gate lengths as small as 0.25 μm," Proc. IEEE/Cornell Conference on Advanced Concepts in High Speed Semiconductor Devices and Circuits Ithaca, NY, pp. 101–110 (1989).
99. J. Shappir, S. Margalit, and I. Kidron, "p-channel MOS transistor in indium antimonide," *IEEE Trans. Electron Device*, ED-22(10), 960–961 (1975).
100. E. E. Barrowcliff, L. O. Bubulac, D. T. Cheung, W. E. Tennant, and A. M. Andrews, "GaSb metal-insulator-semiconductor field-effect-transistors," 1977 IEDM Technical Digest, pp. 559–562 (1977).
101. T. Takahashi, O. Sugiura, I. Watanabe, and M. Matsumura, "SiO_2/native-oxide double-gate InSb MOSFETs," *Electron. Lett.*, 21(12), 545–547 (1985).
102. S. L. Tu, W. H. Lan, T. S. Chiou, S. J. Yang, and K. F. Huang, "High breakdown p-channel InSb MOSFET," *Jpn. J. Appl. Phys.*, 29(3), L398–L400 (1990).

103. B.-D. Liu, S.-C. Lee, K.-C. Liu, T.-P. Sun, and S.-J. Yang, "High breakdown voltage InSb p-channel metal-oxide-semiconductor field effect transistor prepared by photo-enhanced chemical vapor deposition," *Proc. SPIE*, 2225, 215–226 (1994).
104. B.-D. Liu, S.-C. Lee, T.-P. Sun, and S.-J. Yang, "Detailed investigation of InSb p-channel metal-oxide-semiconductor field effect transistor prepared by photo-enhanced chemical vapor deposition," *IEEE Trans. Electron Device.*, 42(5), 795–803 (1995).
105. M. Okamura and T. Kobayashi, "Reduction of interface states and fabrication of p-channel inversion-type InP-MISFET," *Jpn. J. Appl. Phys.*, 19(10), L599–L602 (1980).
106. F. Ren, M. W. Hong, W. S. Hobson, J. M. Kuo, J. R. Lothian, J. P. Mannaerts, J. Kwo, Y. K. Chen, and A. Y. Cho, "Enhancement-mode p-channel GaAs MOSFETs on semi-insulating substrates," IEDM Technical Digest, pp. 943–945 (1996).
107. M. Passlack, J. K. Abrokwah, R. Droopad, Z. Yu, C. Overgaard, S. I. Yi, M. Hale, J. Sexton, and A. C. Kummel, "Self-aligned GaAs p-channel enhancement mode MOS heterostructure field-effect transistor," *IEEE Electron. Device Lett.*, 23(9), 508–510 (2002).
108. M. Tametou, M. Takebe, N. C. Paul, K. Iiyama, and S. Takamiya, "N-channel p-channel enhancement/inversion mode GaAs-MISFETs with gate insulating layers formed by dry oxinitridation," *Electronics and Communications in Japan, Part 2 (Electronics)*, 89(10), 18–26 (2006).
109. H.-S. Kim, I. Ok, F. Zhu, M. Zhang, S. Park, J. Yum, H. Zhao, and J. C. Lee, "n- and p-channel TaN/HfO_2 MOSFETs on GaAs substrate using a germanium interfacial passivation layer," 65th DRC Technical Digest, pp. 99–100 (2007).
110. I. Ok, H. Kim, M. Zhang, T. Lee, F. Zhu, L. Yu, S. Koveshnikov, W. Tsai, V. Tokranov, M. Yakimov, S. Oktyabrsky, and J. C. Lee, "Self-aligned n- and p-channel GaAs MOSFETs on undoped and p-type substrates using HfO_2 and silicon interface passivation layer," IEDM Technical Digest (2006).
111. K. Kajiyama, Y. Mizushima, and S. Sakata, "Schottky barrier height of n-$In_xGa_{1-x}As$ diodes," *Appl. Phys. Lett.*, 23(8), 458–459 (1973).
112. N. Li, E. S. Harmon, J. Hyland, D. B. Salzman, T. P. Ma, and P. D. Ye, "Properties of InAs metal-oxide-semiconductor structures with atomic-layer-deposited Al_2O_3 dielectric," *Appl. Phys. Lett.*, 92(14), 143507-1-3(2008).
113. D. Varghese, Y. Xuan, Y. Q. Wu, T. Shen, P. D. Ye, and M. A. Alam, "Multi-probe interface characterization of $In_{0.65}Ga_{0.35}As/Al_2O_3$ MOSFET," IEDM Technical Digest, pp. 379–382 (2008).

Chapter 13
Insulated Gate Nitride-Based Field Effect Transistors

M. Shur, G. Simin, S. Rumyantsev, R. Jain and R. Gaska

Abstract "Polarization doping" related to the piezoelectric and spontaneous polarization induced electric fields in nitride-based (III-N) semiconductors and large conduction and valence band discontinuities at the heterointerfaces in these materials enable extremely high sheet carrier densities in device channels. As a consequence, insulated gate III-N field effect transistors are quite tolerant of the interface states at semiconductor-dielectric interfaces. High breakdown fields of III-N materials allow achieving high power operation, and superior transport properties of nitride semiconductors make them suitable for high frequency operation. We describe materials growth, deposition and fabrication technology, device characteristics, reliability, and applications of insulated gate III-N field effect transistors and discuss future trends in this technology development.

13.1 Introduction

AlN/GaN/InN materials system has unique properties that enable the development of superior electronic and optoelectronic devices, including Metal Oxide Semiconductor Field Effect Transistors (MOSFETs), Metal Oxide Semiconductor Heterostructure Field Effect Transistors (MOSHFETs), Metal Insulator Semiconductor Field Effect Transistors (MISFETs) and Metal Insulator Heterostructure Field Effect Transistors (MISHFETs). For semiconductors in this materials system, the energy gap varies from 6.2 eV for AlN to 3.4 eV for GaN and 0.65 eV for InN, allowing for great flexibility in the energy band engineered structures [1]. Polarization doping [2, 3] allows for achieving extremely high carrier concentrations in the device channel without introducing dopants and related defects. The sheet carrier concentration in the GaN-based device channel can easily exceed $10^{13}\,\mathrm{cm}^{-2}$ and could be

M. Shur (✉)
ECSE Department and Broadband Center, Rensselaer Polytechnic Institute, Troy, NY 12180, USA
e-mail: shurm@rpi.edu

S. Oktyabrsky, P. D. Ye (eds.), *Fundamentals of III-V Semiconductor MOSFETs*, DOI 10.1007/978-1-4419-1547-4_13, as high as $5 \times 10^{13}\,cm^{-2}$ [2]. The electron mobility in the 2D electron gas (2DEG) at the GaN/AlGaN interface exceeded $2000\,cm^2/V\,s$ [4] at room temperature (with the record value estimated at 2,650 $cm^2/V\,s$ according to Frayssinet et al. [5]) The mobility-sheet carrier concentration product for these 2D systems exceeds those for GaAs/AlGaAs heterostructures and can be further enhanced by doping the conducting channels and by using "polarization" doping [2], which takes advantage of high piezoelectric constants of GaN and related materials and their large spontaneous polarization. High field characteristics predicted by detailed Monte Carlo simulations show record breaking values of the electron peak and transient velocities [6], especially for InN and InGaN devices [7]. High breakdown field exceeding 2.5 MV/cm [8], decent thermal conductivity (2.25 W/cm K compared 1.3 W/cm K silicon) [9], a relatively good lattice match between AlGaN and GaN, AlInN and GaN, and InGaN and AlInN, and ability to use quaternary materials system AlGaInN [10–12] for optimizing the materials properties and band mismatches make this materials systems to be a dream system for a FET designer.

The key problem with III-V MOS and insulated gate FETs is a large concentration of surface states at the dielectric-semiconductor interface. In nitride MOS devices, the carrier concentration in the device channels is much higher than for MOS devices implemented in other material systems, such as GaAs or InGaAs. Therefore, a relative impact of the interface state density at the semiconductor-dielectric interface might be not as severe as for GaAs or InGaAs MOSFETs. The first evidence of the existence of the 2DEG at the GaN/AlGaN heterointerface was provided by a large mobility enhancement at the heterointerface. In 1995, Khan et al. [13] observed a large mobility enhancement in the 2D-electron gas at the AlGaN/GaN interface. They measured the 2DEG Hall mobility around $5{,}000\,cm^2/V\,s$ at 80 K, compared to the maximum electron mobility of approximately $1{,}200\,cm^2/V\,s$ in their bulk doped GaN samples. Gaska et al. [4] reported on the electron mobility in the 2DEG at the GaN/AlGaN interface exceeding $10{,}000\,cm^2/V\,s$ at cryogenic temperatures and exceeding $2000\,cm^2/V\,s$ at room temperature. These values were observed in the samples with very high sheet carrier concentrations (on the order of $10^{13}\,cm^{-2}$).

Binari et al. [14] were the first to report on an insulated gate GaN FET, which was a GaN-based MISFETs with Si_3N_4 insulator. Khan et al. [15] reported on the first enhancement mode AlGaN/GaN HFET. Later, Hu et al. [16] reported on an enhancement mode AlGaN/GaN HFET using a p-n junction gate, and Gaska et al. [17] reported on Doped Channel GaN MOSFETs (DC-MOSFET) and MESFETs. The threshold voltage for MESFETs and DC-MOSFETs ranged from −1.5 to −10 V, and from −4 to −20 V, respectively, with the maximum drain currents up to 300 mA/mm and transconductances up to 60 mS/mm for 1 micron gate devices. The gate leakage current in DC-MOSFETs was more than three orders of magnitude lower than in MESFETs. Frayssinet et al. [5] reported on the first AlGaN/GaN heterostructures grown on bulk GaN substrates.

For many applications, a new device—AlGaN/GaN MOSHFET—[12, 18–25] has several advantages compared to a MOSFET. In MOSHFETs, the dielectric/semiconductor interface is separated from the device channel by a wide band gap barrier. This allows for achieving a much higher mobility compared to a MOSFET (~1,200 $cm^2/V\,s$ or more compared to 100–200 $cm^2/V\,s$ in the best MOSFETs). However, it is more difficult to implement an enhancement mode device in a MOSHFET structure. Huang et al. [26] compared characteristics of MOS capacitors on n- and p-type GaN, which was important for design of enhancement mode devices. Matocha et al. [27] reported on the MOS capacitor flatband voltage shift versus temperature and used the results to determine a pyroelectric voltage coefficient of GaN to be 7.0 $10^4\,V\,m\,K$. Since 2004, several groups reported on enhancement mode GaN MOSFETs [28–35]. Such devices are needed for power switches. However, for applications, such as power amplifiers or microwave switches, MOSHFETs have far superior characteristics and perform (or expected to perform) much better than conventional HFETs. A very promising direction is using HfO_2 as a part of the dielectric stack that might not only dramatically improve the device characteristics but also improve the reliability [36].

The chapter is organized as follows. Section 13.2 describes key materials growth and deposition technologies for GaN-based MOSFETs and MOSHFETs. Section 13.3 deals with transport properties followed by Sect. 13.4 on device design and fabrication and Sect. 13.5 on device characteristics including noise properties. Section 13.6 reviews a huge body of work on non-ideal effects, which still hinder many emerging applications of this technology, and reliability issues. Section 13.7 deals with the device performance and applications. Section 13.8 discusses future trends in this technology development.

13.2 Materials Growth and Deposition Technologies

13.2.1 Material Growth Techniques

Molecular beam epitaxy (MBE) and metalorganic chemical vapor deposition (MOCVD) are the two common methods used for growth of III-nitride based devices layers. MBE can produce high-quality layers with very abrupt interfaces and good control of thickness, doping, and composition. It involves evaporation of the source materials and layer-by-layer growth on a hot substrate. Typically, atoms are delivered as a beam of gas onto the substrate under extremely high vacuum. Growth temperatures are usually much lower than for MOCVD to avoid evaporation of the group III material. Since N_2 cannot be dissociated by using conventional effusion cells, alternate nitrogen sources are usually employed. Use of ammonia as a nitrogen source results in very low growth rates as ammonia is very stable at lower temperatures. Use of plasma sources (radio frequency or electron cyclotron resonance generated nitrogen plasma) can be used to grow high quality GaN at growth rates comparable to MOCVD. However, due to the need for ultra-high vacuum in MBE, MOCVD still remains the most common method for growth of III-nitrides. Morkoc reviewed III-nitride semiconductor growth by MBE [37]. Pei et al. [38] reported on the power performance of deep submicron AlGaN/GaN high electron mobility transistors grown by ammonia MBE. At 10 GHz, 70% power-added-efficiency (PAE) and 3 W/mm power-density were demonstrated at a drain bias of 20 V.

MOCVD has developed over the past two decades into the premier technique for epitaxial growth of the group III-nitrides. Growth by MOCVD involves gas phase transport of metalorganics, hydrides and carrier gases to a heated substrate. Higher growth temperatures allow the volatile precursors to pyrolize at the substrate and deposit a nonvolatile solid film. The group III sources are usually Trimethylgallium (TMGa), Trimethylaluminium (TMAl) and Trimethylindium (TMIn) whereas high-purity ammonia (NH_3) is used as the hydride source. Silicon (Si) is the most common n-type dopant and is delivered in hydride form, such as silane (SiH_4) and disilane (Si_2H_6).

Sensor Electronic Technology, Inc. (USA) developed a new growth technique called Migration Enhanced Metalorganic Chemical Vapor Deposition (MEMOCVD®) [39]. MEMOCVD® is an improved version of Pulsed Atomic Layer Epitaxy (PALE) [40], which deposits ternary $Al_xGa_{1-x}N$ or quaternary $Al_xIn_yGa_{1-x-y}N$ layers by repeats of a unit cell grown using sequential metalorganic precursor pulses of Al-, In-, Ga- and NH_3. In MEMOCVD®, the durations and waveforms of precursor pulses can be overlapped, providing a continuum of growth techniques ranging from PALE to conventional MOCVD. This technique enhances the mobility of precursor species on the surface and thus allows better atomic incorporation and improved surface coverage. MEMOCVD® grown layers exhibit much longer lifetimes and narrower photoluminescence (PL) lines proving the superiority of this epitaxial technique [41].

13.2.2 Substrate Issues

III-nitrides have faced a very unique challenge that is not seen in epitaxy of other III-V semiconductors i.e., the lack of a native substrate. Although native substrates (GaN and AlN) have now become available, the high costs and smaller sizes have delayed their commercial viability. HFET growth on both GaN [42] and AlN [43] substrates has been reported. However, alternate substrates, such as sapphire, 6H-SiC, 4H-SiC [44] and Si [45] are usually used for heteroepitaxial deposition of III-nitride films.

Sapphire has a large lattice and thermal mismatch with GaN (see Table 13.1), leading to high defect density (~10^{10} cm^{-2}) in the GaN film. Sapphire is electrically isolating but has a poor thermal conductivity, which limits the power handling capability of devices. Usually, GaN is grown on the c-plane of sapphire. Sapphire is a non-polar substrate. Films deposited by MOCVD on c-plane sapphire are normally Ga-face, however with MBE the polarity can be chosen. An AlN nucleation layer gives Ga-face polarity, whereas a GaN layer results in N-face polarity.

From the viewpoint of thermal conductivity and lattice mismatch, 6H or 4H polytypes of SiC are a good choice of substrate for heteroepitaxy. Although the lattice mismatch is only 3.51%, it is still large enough to cause high dislocation densities on the order of 10^9–10^{10} cm^{-2}, similar to GaN films grown on sapphire. AlN

Table 13.1 Substrate properties and lattice mismatch with GaN

Material	Crystal type	Lattice constant (nm) [46]	Lattice-mismatch with GaN	Thermal conductivity at 300 K (W/cmK) [46]
GaN	Wurtzite	a = 0.31891; c = 0.51855	0 %	1.3 2.25 [9]
AlN	Wurtzite	a = 0.3112; c = 0.4982	2.48 %	2
Al_2O_3	Rhombohedral	a = 0.4765; c = 1.2982	13.9 %	0.3
6H-SiC	Wurtzite	a = 0.3081; c = 1.5117	3.51 %	4.9
Si	Cubic	a = 0.5431	−16.96 %	1.3

nucleation layers are used to improve the quality of the epitaxial film. The much higher thermal conductivity of SiC allows for improved heat dissipation leading to better power performance. A very impressive power performance of a FET (38 W at 10 GHz) has been reported for AlGaN/GaN HFET on SiC substrate [47].

The use of Si as a substrate is an interesting alternative to SiC or sapphire. Si substrates are cheap, have a high degree in crystal perfection and are available in very large sizes. It also introduces the possibility of combining GaN and Si devices on the same wafer. Unfortunately, the lattice and thermal mismatch is quite large and special growth techniques are needed to overcome these problems. Nitronex, Inc. (USA) has been successful in developing proprietary growth techniques for improving GaN on Si material quality [48], and power performance of 12 W/mm at 2 GHz has been reported for AlGaN/GaN HFETs on Si substrates [49].

13.2.3 Growth of HFET Structures

The layer structure of a conventional HFET is shown in Fig. 13.1a. HFETs are usually grown using a two-step procedure: deposition of a thin initiation or nucleation

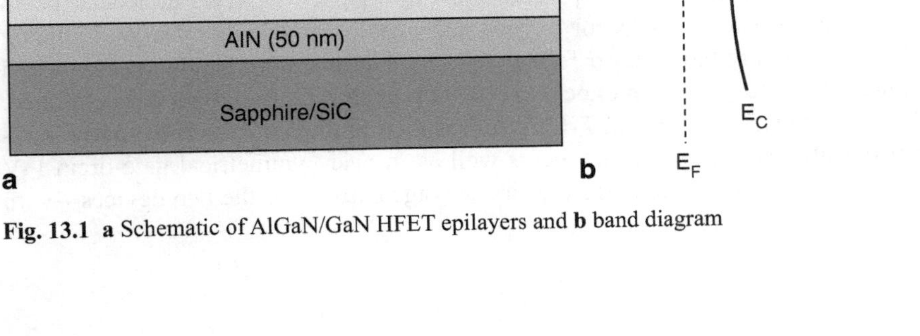

Fig. 13.1 **a** Schematic of AlGaN/GaN HFET epilayers and **b** band diagram

layer is followed by the growth of semi-insulating (SI) GaN and AlGaN barrier layers. Choice of substrate dictates the growth parameters of nucleation layer. Typically structures grown on sapphire use a thin (20–50 nm) GaN or AlN layer deposited at low temperatures (500–600 °C). The substrate is then heated up to about 1000 °C for SI-GaN and AlGaN deposition. The major difference between the AlGaN/GaN growth on sapphire and SiC substrates is the thickness of the insulating GaN layer.

The cross-sectional TEM analysis of GaN grown on sapphire and SiC reveals strong dependence of growth defect distribution along the growth direction on the substrate material. The significant reduction in the number of threading dislocations in GaN on sapphire is observed for layer thicknesses above 2 μm. The similar improvement in material quality for GaN grown on SiC was achieved at a thickness as low as 1 μm or even lower [50]. This is important for power devices because the active channel of HFETs is closer to SiC, which has a high thermal conductivity. Thus, the performance of the devices with higher levels of dissipated power can be improved by effective heat sinking through the SiC substrate. Semi-insulating (SI) or high resistivity GaN ensures proper drain-source current saturation, complete channel pinch-off, low loss at high frequencies, and low cross-talk between adjacent devices.

Heteroepitaxy of GaN at high temperatures generates vacancy defects and dislocations. Additionally, oxygen incorporated from sapphire and other unintentional impurities create high levels of defect states within the bandgap [51]. The majority of these states tend to be donor like, leading to a high level of unintentional n-type doping (N_D–N_A: 10^{16}–$10^{17}\,cm^{-3}$). This high level of background n-doping is detrimental to HFETs. High resistivity GaN is typically grown by optimizing the growth conditions and by tuning the parameters in order to self-compensate the material [52]. Another method involves intentional doping with carbon [53] or heavy metals such as Fe [54].

13.2.4 Gate Dielectrics

Deposition of gate dielectrics is a very important and critical step for fabricating MOSFETs, MISFETs, MOSHFETs and MISHFETs. Dielectric layers must be high quality insulators to decrease gate leakage current and to sustain large gate bias voltage. Unlike silicon technology, it is challenging to oxidize GaN into high quality native oxide (Ga_2O_3) because of the strong bond strength between Ga and N. A large variety of other gate dielectrics has been used for insulated gate FETs.

Gu et al. [55] reported on epitaxial growth of ZrO_2 by oxides molecular beam epitaxy using reactive H_2O_2 for oxygen and metalorganic source for Zr. Utilizing a low temperature buffer layer followed by high temperature in situ annealing and high-temperature growth, monoclinic (100)-oriented ZrO_2 thin films were obtained. The employment of epitaxial ZrO_2 layer resulted in the increase of saturation-current density and pinch-off voltage as well as in near symmetrical gate-drain I–V behavior. Figure 13.2 compares the gate leakage current for the two devices—with

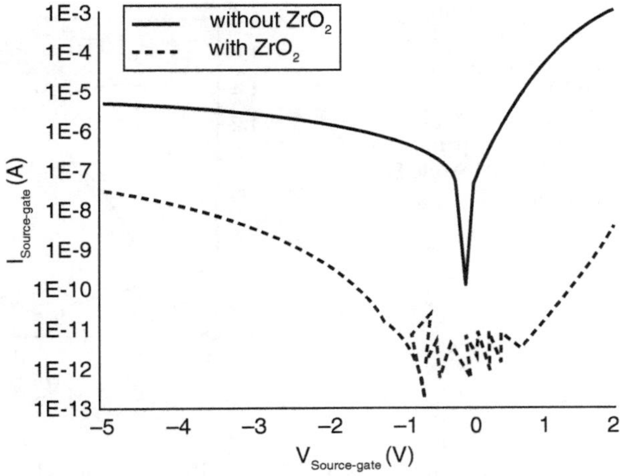

Fig. 13.2 Source-gate I–V characteristics of the HFET with and without ZrO_2 gate dielectric [55]

and without ZrO_2 gate dielectric. The forward/reverse gate-source I–V characteristics became nearly symmetrical for the devices fabricated with the ZrO_2 layer, which confirms the high resistivity nature of the ZrO_2 gate dielectric.

A similar dependence for Ga_2O_3 gate dielectric was observed, as shown in Fig. 13.3. Wu and Peng [56] reported the use of photo-enhanced chemical (PEC) technique to deposit Ga_2O_3. Gate leakage current density as low as 2×10^{-7} A/cm^2 at a bias field up to 2 MV/cm was observed in the GaN MOS devices formed by PEC wet etching. These devices had the top surface and mesa sidewall passivated by the photogrown Ga_2O_3. Ren et al. deposited Ga_2O_3/Gd_2O_3 stack as the MOSFET gate insulator, but the device performance was poor [57].

However, the best performance has been achieved for SiO_2 and Si_3N_4 gate dielectrics. Simin et al. [22] reported on MOS devices using a thin (~10 nm) SiO_2 layer, which was deposited on AlGaN/GaN heterostructure using plasma enhanced chemical vapor deposition (PECVD). Figure 13.4 shows the transfer characteristics

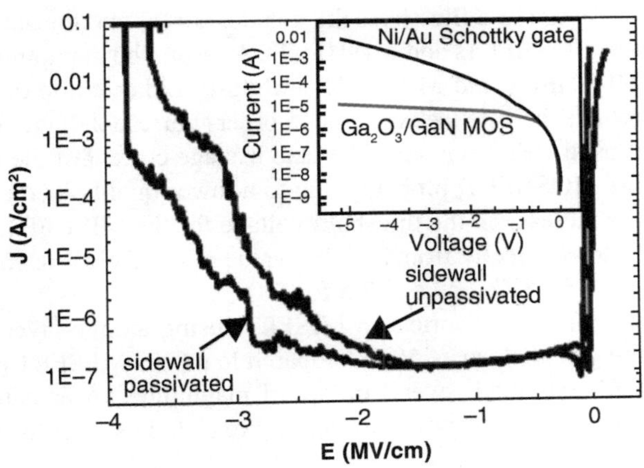

Fig. 13.3 Leakage current density in MOS devices with/without sidewall passivation by Ga_2O_3. *Inset:* The leakage current in Schottky-gate and MOS devices [56] (Reprinted with permission. Copyright 2006 Wiley-VCH Verlag GmbH & Co.)

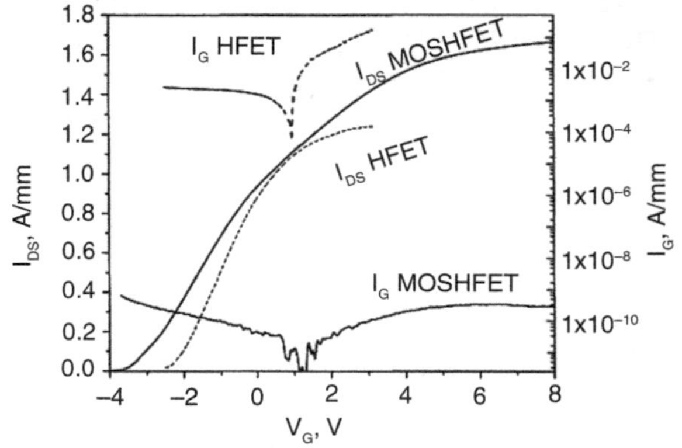

Fig. 13.4 Maximum saturation and gate leakage currents in 1.5 μm gate MOSHFET with SiO_2 and HFET devices [22] (Reprinted with permission. Copyright 2004 World Scientific Publishing Co.)

for the 1.5 μm gate MOSHFET and HFET measured at the drain voltage sufficient to shift the operating point into saturation regime. The figure also shows the gate bias dependence of the HFET and MOSHFET current in the saturation regime (for the MOSHFET the gate current remains in the low *nA* range). As seen, the gate voltage corresponding to the maximum of I_{DS} in the HFETs also corresponds to a sharp increase of the gate leakage current. This indicates that the mechanism responsible for the I_{DS} saturation at high gate bias is the gate leakage current. In the MOSHFETs, where the gate leakage is suppressed, the 2D electrons spillover into the AlGaN barrier becomes a limiting mechanism. A larger gate-channel separation in MOSHFET contributes to a higher value of gate voltage. Due to these factors, both the saturation gate voltage and the saturation current for the MOSHFETs are higher than those for the HFETs.

Simin et al. [22] also reported on insulated gate HFETs using Si_3N_4. Two sets of devices with identical geometry were fabricated on the same wafer. They consisted of MOSHFETs (10 nm SiO_2 under the gate and in the source-gate and drain-gate regions) and MISHFETs (10 nm Si_3N_4 insulator replacing SiO_2). Both the SiO_2 and the Si_3N_4 layers were deposited using PECVD.

Figure 13.5 illustrates the transfer curves and the gate-leakage current curves for the MOSHFETs and MISHFETs. As seen, the maximum saturation currents in both MOSHFET and MISHFET are close. Either oxide or the nitride insulator layers reduce the gate leakage by 6–5 orders correspondingly below that measured for the typical HFET devices. The gate leakage current of the MISHFET is higher than of the MOSHFET, probably due to a lower quality of the thin Si_3N_4 layer. However, the increase in the threshold voltage for the MISHFET device is not as large. This follows directly from a higher value of the dielectric constant of the Si_3N_4 layer (ε_r = 3.9 for SiO_2 and ε_r = 7.5 for Si_3N_4).

Wu et al. reported on MISFETs using atomic-layer-deposited (ALD) Al_2O_3 as the gate dielectric [58]. Compared to a GaN MESFET of similar design, the MOSFET exhibited several orders of magnitude lower gate leakage and nearly three times higher channel current. Figure 13.6 illustrates the saturated (V_{ds} = 16 V) drain

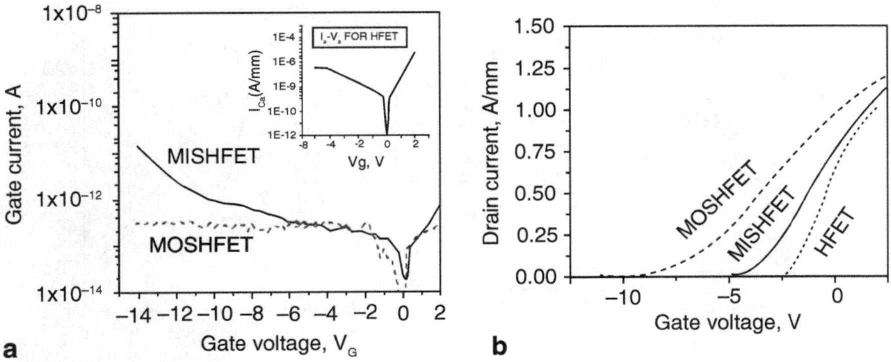

Fig. 13.5 Gate leakage current (**a**) and transfer characteristics (**b**) comparison for the MOSHFET and MISHFET fabricated on the same wafer. The *inset* shows the gate leakage current for a regular HFET [22] (Reprinted with permission. Copyright 2004 World Scientific Publishing Co.)

Fig. 13.6 The transfer and transconductance characteristics measured in the GaN MOSFET with Al_2O_3 and MESFET in saturation (V_{ds} = 16 V) [58] (Reprinted with permission. Copyright 2006 Elsevier)

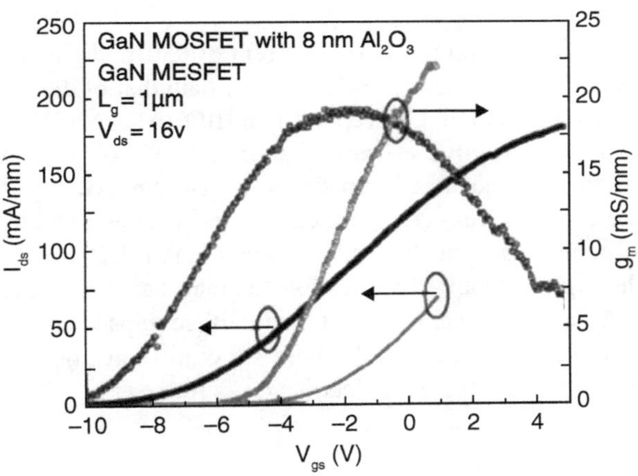

current density and extrinsic transconductance g_m as a function of gate bias for a GaN MOSFET and a MESFET of similar designs. The drain current density of the MESFET is limited to 70 mA/mm at V_{gs} = 1 V. The gate leakage current of the MESFET becomes unmanageable once the gate bias sweeps above 1 V due to low Ni/n-GaN Schottky barrier. By contrast, the drain current density of the MOSFET is 180 mA/mm at V_{gs} = 5 V and can be further increased under higher V_{gs}. The higher drain current achieved by employing Al_2O_3 gate oxide can be used to enhance the output power of the MOSFET compared to that of the MESFET.

More recently, very encouraging results were obtained for the structures employing HfO_2 as the gate dielectric. Figure 13.7 shows the transistor characteristic of AlGaN/GaN MOSHFET (referred to as MOS-HEMT by the authors) using reactive-sputtered HfO_2 as the gate dielectric from ref. [59]. The dielectric constant of HfO_2 found from the capacitance was estimated to be equal to 21. MOSHFETs

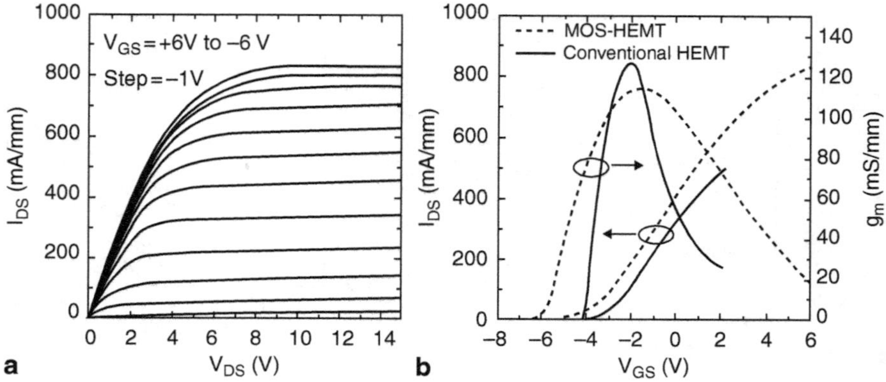

Fig. 13.7 Typical output characteristics of AlGaN/GaN MOS-HEMT with 23 nm reactive sputtered HfO_2 as the gate dielectric [59] (Reprinted with permission. Copyright 2006 American Institute of Physics)

exhibited a maximum drain current of 830 mA/mm and the gate leakage current at least five orders of magnitude lower than that of the reference HEMTs.

Tokranov et al. [36] reported on HfO_2/AlGaN/GaN structures using HfO_2 deposited by a reactive e-beam evaporation of Hf with oxygen. The structures were studied by means of impedance measurements. Figure 13.8 shows the results of the capacitance—voltage measurements (a) and DC leakage current of the studied structures (b). The dielectric constant of the HfO_2 $\varepsilon = 23, 24$ was found to be close to the highest reported values for this material. The conductance measurements indicated a low concentration of the interface traps in comparison with the electron concentration in the channel. These structures have been used to fabricate MISHFETs with HfO_2/SiO_2 gate dielectric stacks [60, 61].

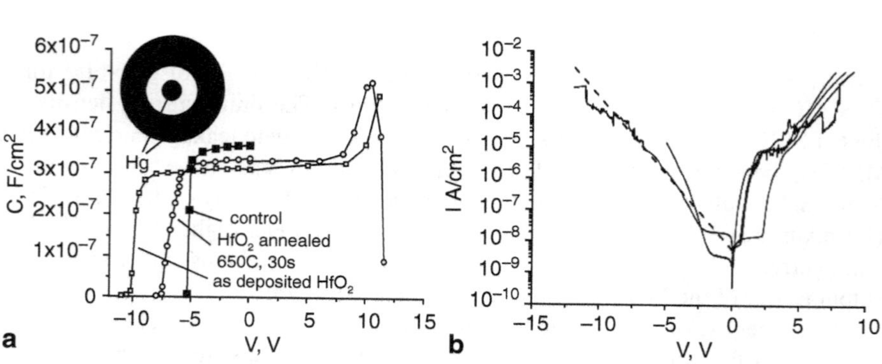

Fig. 13.8 **a** Capacitance as a function of the gate voltage for the control (no HfO_2), as-deposited, and annealed samples. The *inset* shows the contact configuration: $D_1 = 7.27 \times 10^{-2}$ cm, $D_2 = 0.153$ cm. **b** DC leakage current measured at different locations on the wafer with annealed HFO_2 [36] (Reprinted with permission. Copyright 2007 Wiley-VCH Verlag GmbH & Co.)

13.3 Transport Properties

First Monte Carlo calculations for GaN were done by Littlejohn et al. in 1975 [62]. Later, the Monte Carlo simulations have been used to simulate the electron transport within GaN [63–70], AlN [69, 71, 72] and InN [6, 7, 69, 73–76]. The Monte Carlo simulation approach has also been used to analyze the electron transport within the 2DEG at the AlGaN/GaN interface [6, 7, 77, 78]. Figure 13.9 compares computed velocity-field characteristics of nitride materials [69]. As seen, InN and GaN are faster materials than GaAs and, in transient regime they can be even faster (see Fig. 13.10).

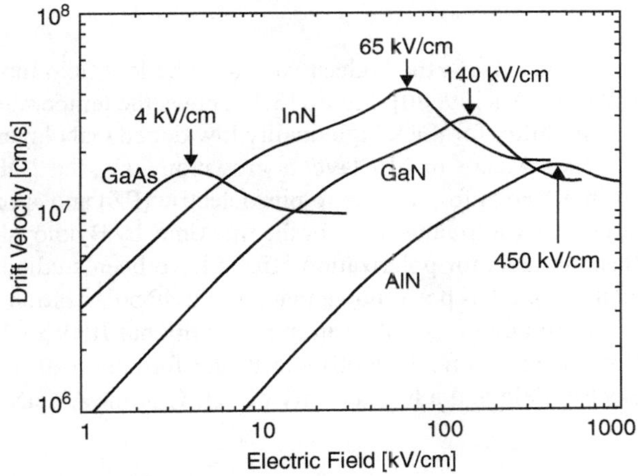

Fig. 13.9 Monte-Carlo simulation of the drift velocity versus electric field for several III-V compounds (InN is the fastest nitride material) [69] (Reprinted with permission. Copyright 1999 American Institute of Physics)

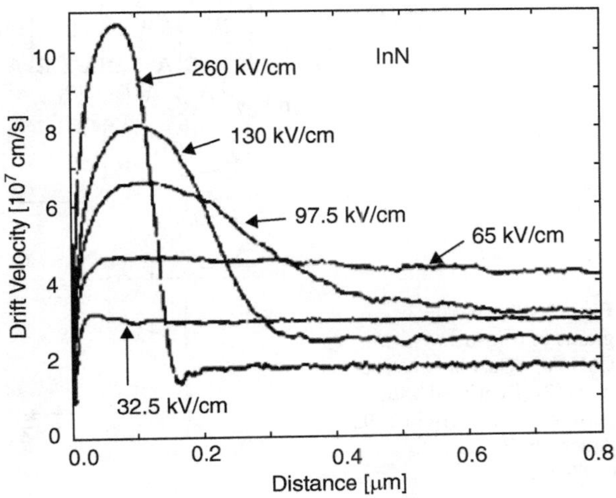

Fig. 13.10 Average electron velocity as a function of the displacement for different electric fields for InN [69] (Reprinted with permission. Copyright 1999 American Institute of Physics)

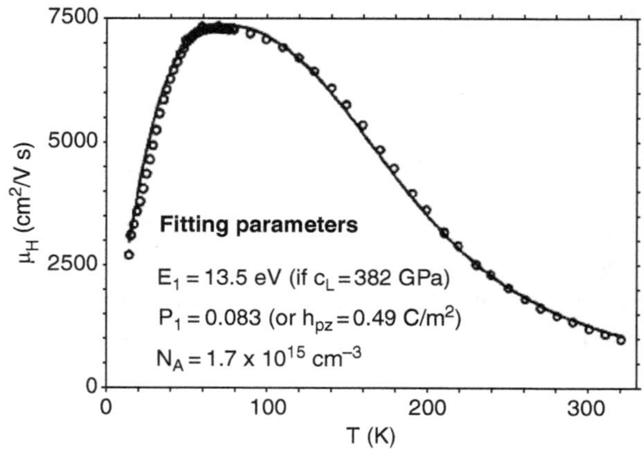

Fig. 13.11 Temperature dependence of the electron mobility in GaN, theory (*solid line*) and experiment (*circles*) [81] (Reprinted with permission. Copyright 2001 American Institute of Physics)

For bulk GaN, the highest values of the low field mobility at 300 K are close to 1,200 cm^2/V s [79, 80]. Figure 13.11 shows the temperature dependence of the electron mobility for thick high quality low doped GaN layer [81].

When AlGaN or AlN layer is grown on GaN, the 2DEG at the AlGaN/GaN heterointerface is formed due to piezoelectric (PZ) and spontaneous polarization (SP) effects, as was pointed out, for the first time, by Bykhovski et al. [2] (see Fig. 13.12). Basic models for polarization effects have been studied extensively by Ambacher et al. [82]. It has been shown that PZ and SP polarization constants are over an order of magnitude greater than in more traditional III-V or II-VI semiconductors [83]. Both of these effects contribute to the formation of a large polarization-induced electric field and a high-density of 2DEG at the AlGaN/GaN interface. In contrast

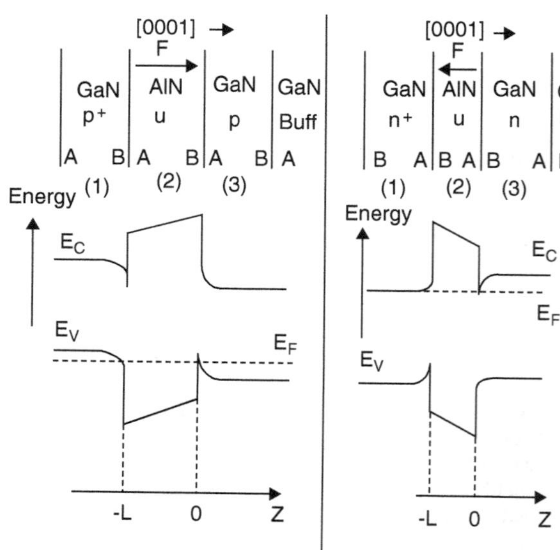

Fig. 13.12 Band diagrams of p^+-n-p and n^+-n-n GaN/AlN/GaN heterostructures [2] (Reprinted with permission. Copyright 1993 American Institute of Physics)

to traditional HFETs, barrier doping for III-nitrides is not necessary to obtain high sheet charge density [84].

Due to the existence of polarization in III-nitride heterostructures, the 2DEG electron concentration and subsequent device performance depend on a number of physical properties including polarity, strain, thickness, and barrier doping. The strain in the AlGaN layer is related to its thickness and composition. Aluminum composition in the range of 20–30% results in a large band offset and has a reasonably high critical thickness allowing for pseudomorphic growth up to 50–100 nm. 20–30% Al content gives a good compromise between mobility and sheet charge. Higher Al content provides better carrier confinement and increased sheet charge, but increased alloy scattering degrades mobility. At lower Al content, mobility remains roughly the same, but the reduced carrier confinement results in low sheet carrier concentration.

For pseudomorphic AlGaN on GaN, the strain is tensile and both the PZ and SP are directed opposite to the growth direction if the material is Ga-face and along the growth direction if the material is N-face [82]. MOCVD growth resulting in smooth surface morphology is always Ga-face polarity. Thus, the polarization will induce a positive charge in the AlGaN. A positive fixed polarization charge in the AlGaN layer is compensated by free electrons in the channel region. Free electrons are provided by the unintentional doping of the heterostructure or through surface donors states of AlGaN [84]. Due to the surface donor states, the 2DEG density increases and then saturates with increasing AlGaN thickness.

As the electrons are confined in a Two-Dimensional (2D) quantum well, bulk scattering effects such as ionized impurity scattering are eliminated, resulting in much higher mobility than for bulk GaN. As the quantum well is formed at the AlGaN/GaN interface, the main factors influencing 2DEG mobility are interface, alloy and dislocation scattering. Modified AlGaN/AlN/GaN structures, which employ a thin AlN interfacial layer between AlGaN and GaN layers, show higher 2DEG properties than those of conventional AlGaN/GaN structures [85]. This high performance is achieved due to the increased ΔE_C, which effectively suppresses the electron penetration from the GaN channel into the AlGaN layer, and results in the reduction of alloy disorder scattering.

Figure 13.13 shows the polarization directions in AlGaN/GaN layers grown on Ga and N substrates [82] and Fig. 13.14 presents calculated 2DEG densities [86]. Figure 13.14 shows the computed values of the 2DEG concentration at the AlGaN/GaN heterointerface as a function of the Al molar fraction for different thicknesses of the AlGaN wide band gap barrier layer. The dashed lines do not account for the stress relaxation at the critical thickness, when strain leads to the development of the dislocation arrays. As seen, the 2DEG densities on the order of $3 \times 10^{12}\,\mathrm{cm^{-2}}$ can be reached. Even higher sheet carrier densities can be obtained using nearly lattice-matched AlInN/GaN heterostructures [87]. Figure 13.15 shows the computed 2DEG sheet density induced in AlInN/AlN/GaN heterostructures (without any additional doping).

Low field mobility in 2D channel on the GaN/AlGaN interface is quite high, exceeding for 2D electrons 2,000 $\mathrm{cm^2/V\,s}$ at 300 K [4, 88], see Fig. 13.16.

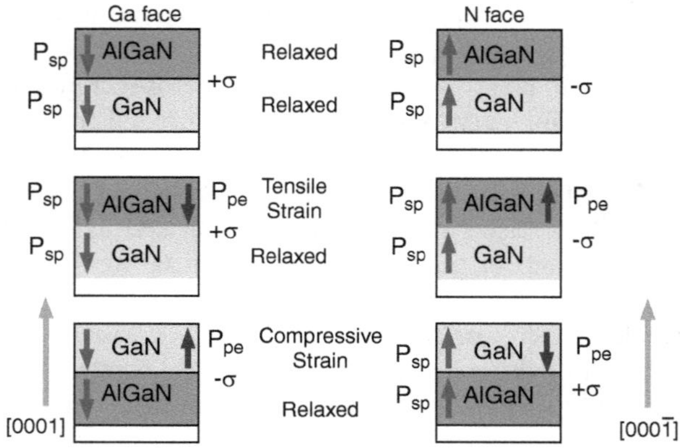

Fig. 13.13 Polarization directions in AlGaN/GaN layers grown on Ga and N substrates [82] (Reprinted with permission. Copyright 1999 American Institute of Physics)

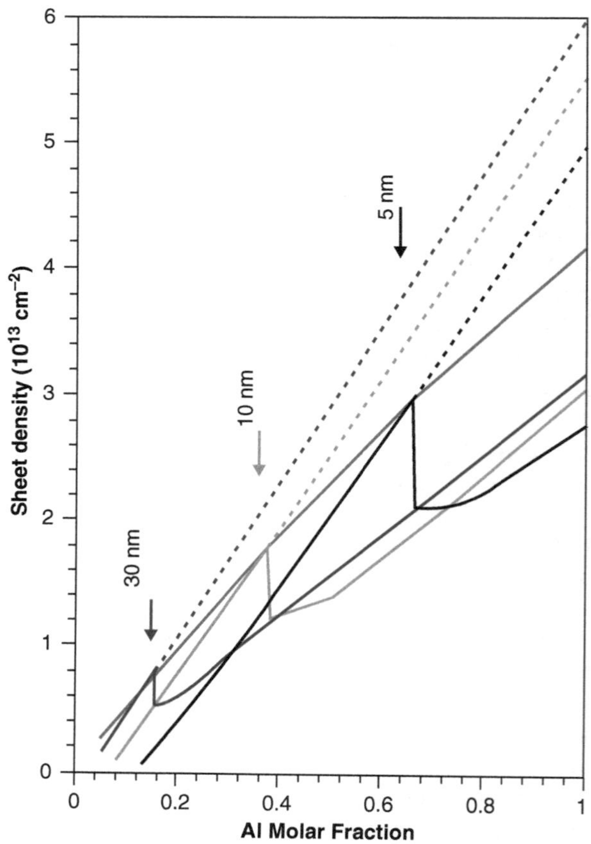

Fig. 13.14 Computed values of the 2DEG concentration at the AlGaN/GaN heterointerface as a function of Al molar fraction for different thicknesses of AlGaN barrier layer [86] (Reprinted with permission. Copyright 2000 Elsevier)

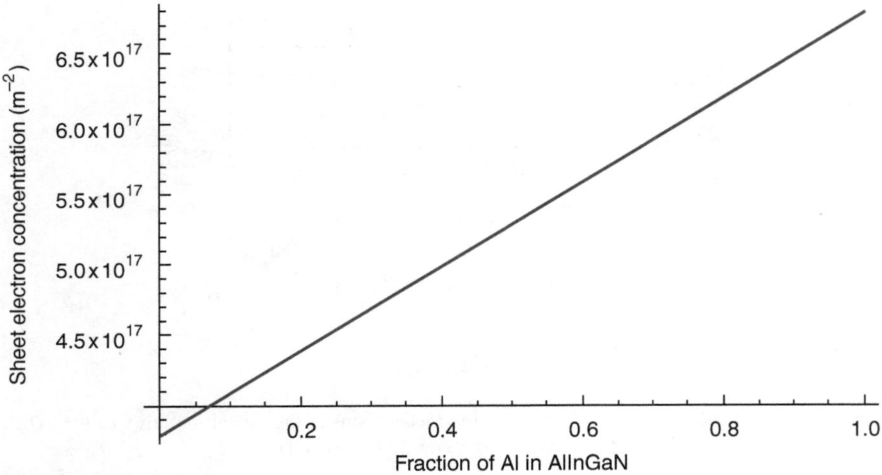

Fig. 13.15 Computed 2DEG sheet density in AlInN/AlN/GaN heterostructures

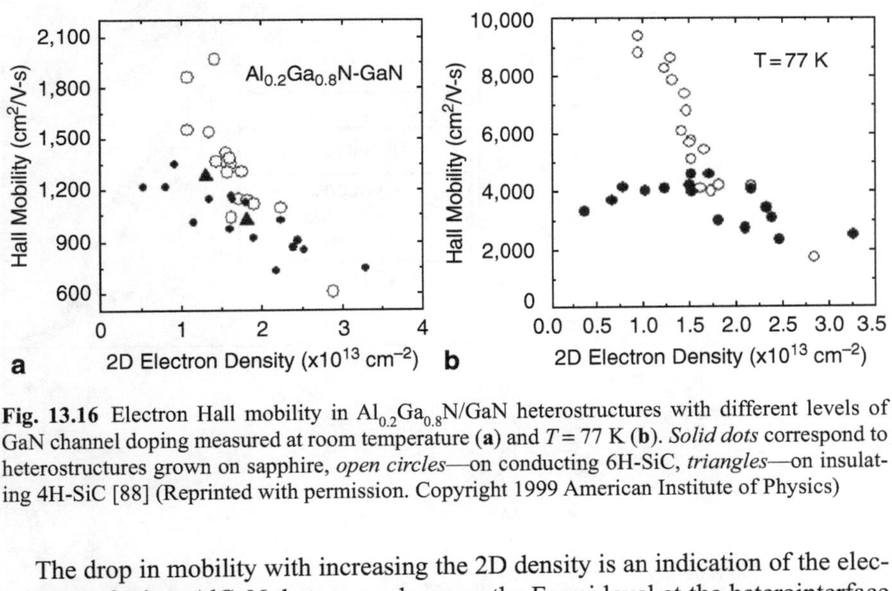

Fig. 13.16 Electron Hall mobility in $Al_{0.2}Ga_{0.8}N$/GaN heterostructures with different levels of GaN channel doping measured at room temperature (**a**) and $T = 77$ K (**b**). *Solid dots* correspond to heterostructures grown on sapphire, *open circles*—on conducting 6H-SiC, *triangles*—on insulating 4H-SiC [88] (Reprinted with permission. Copyright 1999 American Institute of Physics)

The drop in mobility with increasing the 2D density is an indication of the electron transfer into AlGaN that occurs because the Fermi level at the heterointerface is pushed too high and, at high densities, electrons cannot be fully contained in the quantum well formed at the heterointerface. Figure 13.17 shows the relative contributions of different scattering mechanisms to the overall mobility of the 2DEG in GaN [89].

Figure 13.18 compares temperature dependences of the 2DEG mobility in GaN grown on different substrates [5]. In field effect transistors, the electron mobility depends on the gate bias, which controls the 2DEG density in the channel.

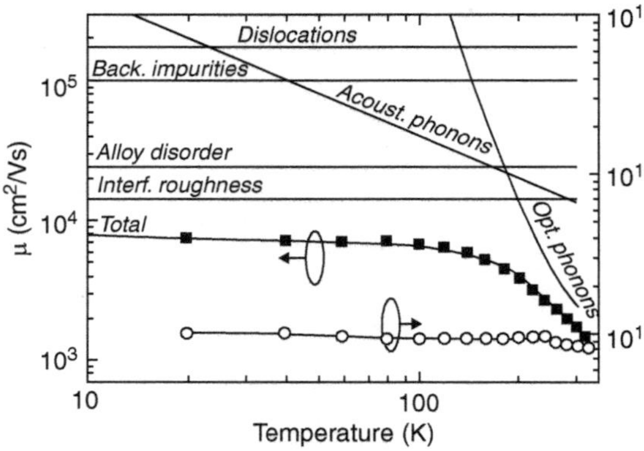

Fig. 13.17 Contributions of different scattering mechanisms to the overall mobility of the 2DEG in GaN [89] (Reprinted with permission. Copyright 2006 Elsevier)

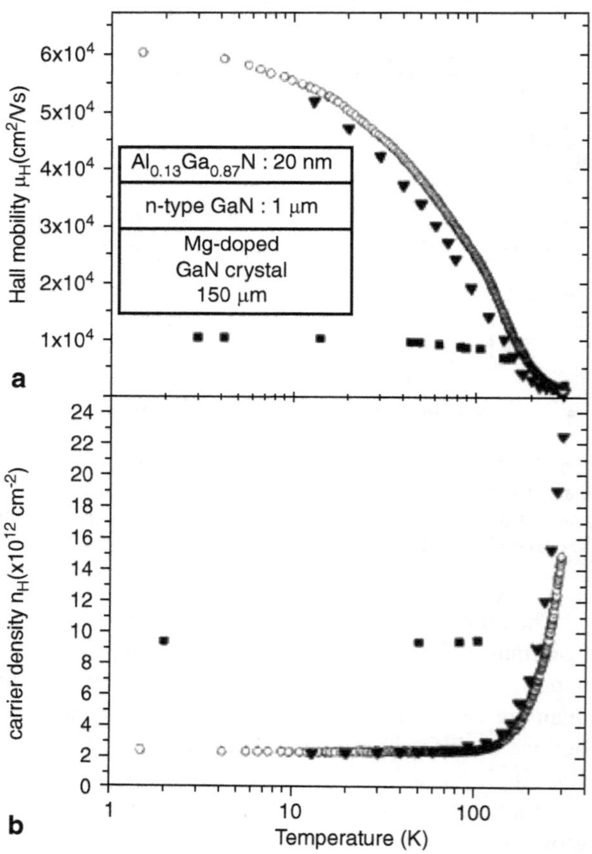

Fig. 13.18 **a** Hall mobility and **b** electron sheet density versus temperature for heterostructures deposited on GaN—*circles*; 6H-SiC—*squares*; sapphire—*triangles* [5] (Reprinted with permission. Copyright 2000 American Institute of Physics)

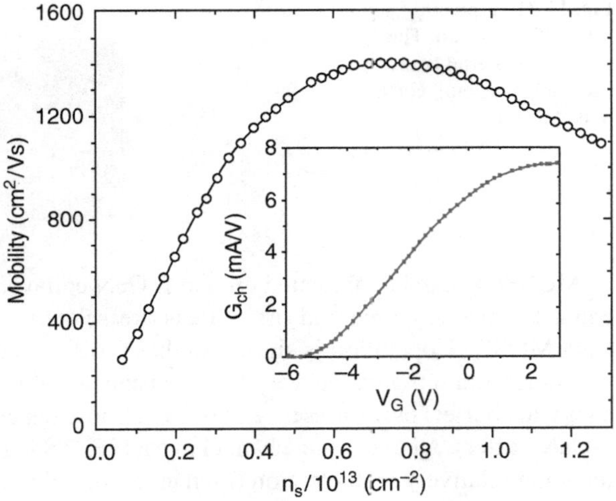

Fig. 13.19 The electron mobility dependence on sheet concentration of 2DEG in the channel of MOSHFET [90]

Ivanov et al. extracted effective mobility in the channel of GaN/AlGaN MOSHFET with 10 nm thick SiO_2 gate dielectric from the transfer current voltage characteristics [90]. Figure 13.19 shows the electron mobility as a function of the 2D concentration (changed by the gate voltage). As seen, the mobility is of the same order of magnitude as for regular HFETs, indicating that gate dielectric does not degrade the mobility. The inset shows the dependence of the channel conductivity G_{ch} versus gate voltage V_G.

13.4 Device Design and Fabrication

Typical GaN MOSFET and MOSHFET designs are illustrated in Figs. 13.20 and 13.21.

The gate oxide for MOSFETs and MOSHFETs is formed by deposition of SiO_2 [18, 32, 91], Ga_2O_3/Gd_2O_3 stack [57], Si_3N_4 [92, 93], Al_2O_3 [94], Sc_2O_3 [95], ZrO_2 [96], AlN [97, 98] and others.

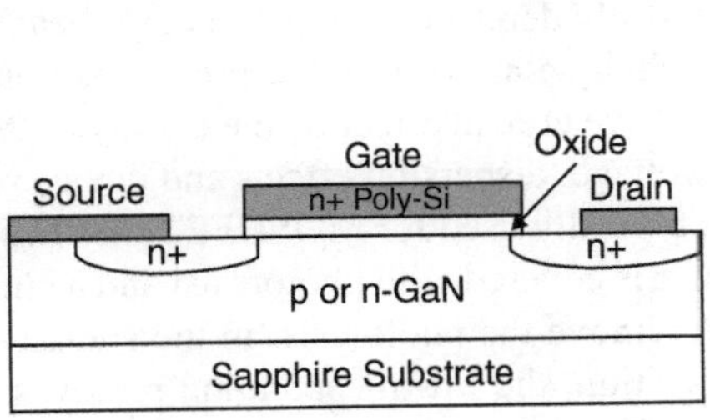

Fig. 13.20 Schematic cross section of a lateral n-channel GaN MOSFET [32] (Reprinted with permission. Copyright 2006 IEEE)

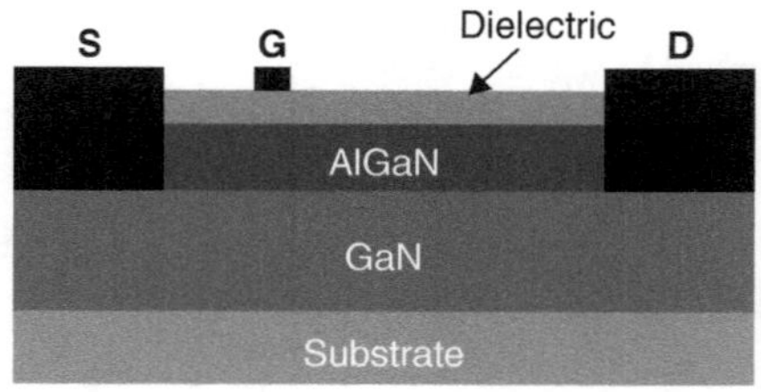

Fig. 13.21 Typical MOSHFET design. The substrate material could be SiC, sapphire, bulk GaN, bulk AlN, or Si

MOSFETs can be fabricated on p or n-GaN epilayers. To form the highly doped contact regions, source and drain areas are selectively implanted with Si atoms. GaN MOSFET operation is similar to that for Si-based devices. At zero gate bias, the source-drain region has very high resistance due to high resistivity of GaN (for an n-GaN buffer) or the presence of two back-to-back connected p-n junctions (for a p-GaN buffer). A conducting channel in MOSFETS is formed by applying gate bias inducing relatively high electron (n-channel) or hole (p-channel) concentration.

In MOSHFETs, as in regular AlGaN/GaN HFETs, the built-in channel is formed by the high-density 2D electron gas at the AlGaN/GaN interface. However, in contrast to a regular HFET, the gate metal is isolated from AlGaN barrier layer by a thin dielectric film (see Fig. 13.21). Thus, the MOSHFET gate behaves more like a MOS gate structure rather than a Schottky barrier gate used in regular HFETs. Since the properly designed AlGaN barrier layer is fully depleted, the gate insulator in the MOSHFET consists of two sequential layers: the dielectric film and the AlGaN epilayer. This double layer ensures an extremely low gate leakage current and allows for a large negative to positive gate voltage swing.

Typical MOSHFET fabrication process is close to that of regular Schottky gate HFETs. The ease of MOSHFET fabrication and compatibility with the HFET processing is another big advantage of the MOSHFET technology. Device fabrication normally starts with mesa isolation done using Reactive Ion Etching (RIE) or ion implantation to define the active area. Ohmic contacts are then being formed, most often using Ti/Al/Ti/Au stacks, although a number of more advance contact schemes have been reported to achieve lower contact resistance [99]. The next step is the gate dielectric formation. Typically, plasma-enhanced chemical vapor deposition (PECVD) is used to deposit films like SiO_2 or Si_3N_4. Other materials, like HFO_2 have been successfully deposited using ALD or e-beam techniques [36]. The dielectric deposition technique as well as the pre-deposition surface preparation, temperature regime etc. have crucial effect on the quality of the deposited films, the MOSHFET threshold voltage, dispersion effects and device reliability.

A lift-off process is typically used for MOSHFET fabrication. Prior to dielectric film deposition the wafer is covered with photoresist and a photolithography is used to pattern the wafer to remove the photoresist in the source—drain spacing. After the dielectric film deposition, the lift-off operation removes the dielectric outside the source-drain region. The remaining processing steps are no different from the HFET fabrication: metal gate deposition, normally Ni/Au, followed by optional passivation, field-plating, contact pad formation and electroplating.

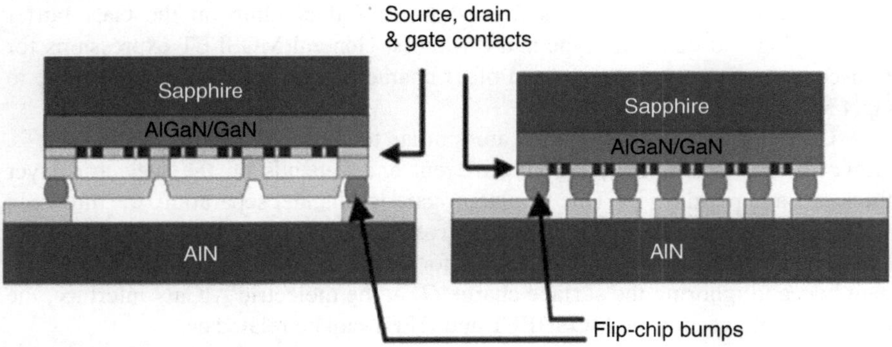

Fig. 13.22 Two flip-chip designs: First design has metal bumps placed on the source, drain and gate contact pad. The second design has additional bumps placed directly on the source and drain ohmic contacts to provide direct thermal and electrical contact between the ohmic contacts and the metal pads on the AlN carrier [100] (Reprinted with permission. Copyright 2006 IEEE)

Packaging for AlGaN/GaN power MOS and MOSHEFT devices is often similar to that used for silicon or III-V power devices. However, packaging for GaN-based devices is more challenging because of their higher power. Flip-chip mounting of the die is a preferred solution (see Fig. 13.22).

13.5 Device Characteristics

13.5.1 Current-Voltage Characteristics and Threshold Voltage

GaN MOSFETs are mostly normally-off (enhancement mode) devices [101]. Typical set of MOSFET drain current-voltage characteristics is shown in Fig. 13.23.

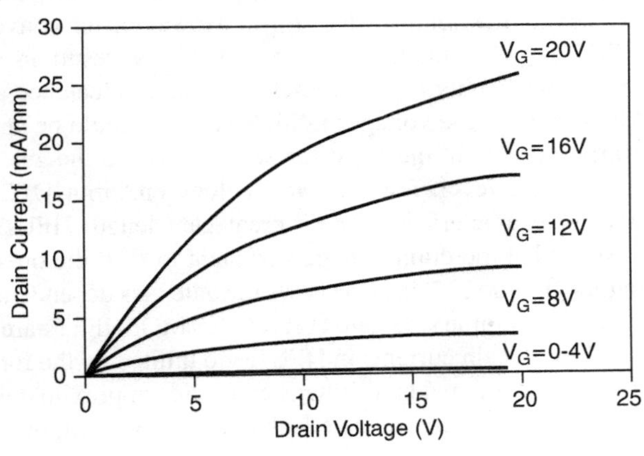

Fig. 13.23 Drain I–V characteristics of MOSFET on n-GaN [32] (Reprinted with permission. Copyright 2006 IEEE)

The threshold voltage varies from 0 to +5 V depending on the GaN buffer layer doping and dielectric type and thickness. General MOSFET expressions for transconductance, capacitances and other characteristics are as well applicable to GaN MOSFETs.

MOSHFET I–V characteristics are similar to those of Schottky-gate HFETs. However, the threshold voltage is different and depends on the dielectric layer thickness and permittivity. Due to a larger gate-to-channel separation, the threshold voltage of the MOSHFET is more negative than that of an HFET. Assuming the same sheet charge density in the channel for MOSHFET and HFET devices at zero gate bias and ignoring the surface charge Q_S at the dielectric/AlGaN interface, the threshold voltages for the MOSHFET and HFET can be related as:

$$Q_S = qN_S = C_{MOSH} \times V_{TMOS} = C_{MS} \times V_{TMS} \tag{13.1}$$

Or

$$V_{TMOS} = V_{TMS} C_{MS} / C_{MOSH} = V_{TMS} \times \left(1 + \frac{d_{OX}}{d_B} \cdot \frac{\varepsilon_B}{\varepsilon_{OX}}\right). \tag{13.2}$$

Here C_{MOSH} and C_{MS} are the capacitances of equal area pads on the oxide and non-oxide areas and ε_{OX} is the dielectric permittivity of the gate dielectric; d_{OX} and d_b are the thicknesses of dielectric and barrier layers correspondingly, V_{TMOS} and V_{TMS} are correspondingly the absolute values of the MOSHFET and HFET threshold voltages.

The oxide thickness d_{OX} can be extracted from the measured gate capacitances:

$$C_{MOS} = \frac{\varepsilon_0 \varepsilon_B}{d_B} \times \left(1 + \frac{d_{OX}}{d_B} \cdot \frac{\varepsilon_B}{\varepsilon_{OX}}\right)^{-1} = C_{MS} \times \left(1 + \frac{d_{OX}}{d_B} \cdot \frac{\varepsilon_B}{\varepsilon_{OX}}\right)^{-1} \tag{13.3}$$

The DC saturation drain current, I_{DS}, is a key parameter controlling the maximum output RF-power. This current I_{DS} increases with positive gate voltages until it reaches its maximum value, I_{DMAX}. However, for conventional AlGaN/GaN GaN HFET's, gate voltages in excess of +1.2 V result in an excessive leakage current, which limits I_{DS}, decreases the transconductance and increases the noise. In MOSHFETs, gate voltages as high as +10 V could be applied. This results in about 100% increase in the I_{DS} value with respect to the zero gate bias value. The gate leakage, however, remains well below 1 nA/mm [102]. Figure 13.24 shows the transfer characteristics for different gate length HFETs (a) and MOSHFETs (b) measured at the drain voltage sufficient to shift the operating point into saturation regime. Figure 13.24a also shows the gate bias dependence of the gate current in the saturation regime (the gate currents for MOSHFETs are very low and not shown). Maximum drain currents in HFETs are limited by the forward gate currents that are triggered at internal gate voltage exceeding approximately $V_{GM} \approx 1.7$ V [102]. In the MOSHFETs, where the gate leakage is suppressed, the 2D electrons spillover into

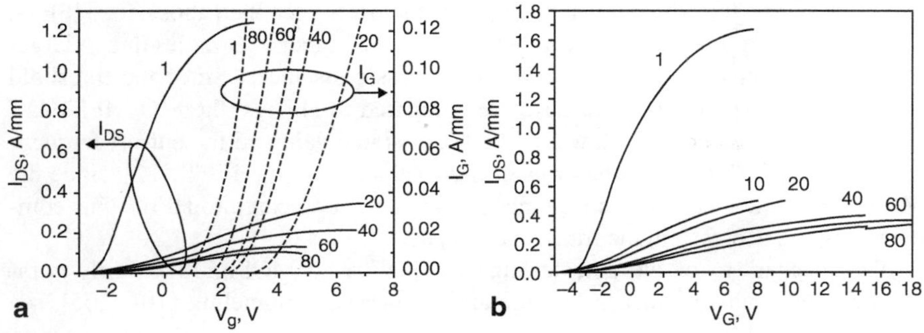

Fig. 13.24 Gate bias dependencies of the drain saturation current and gate leakage current for HFET (**a**) and MOSHFET (**b**) devices. For MOSHFETs gate leakage current is negligibly small and not shown. Drain bias corresponds to the saturation region of the device I–V characteristics. Gate length is given in micron next to the curves [102] (Reprinted with permission. Copyright 2002 American Institute of Physics)

the AlGaN barrier becomes a limiting mechanism. For SiO_2 based MOSHFETs, the corresponding maximum internal gate voltage was found to be around $V_{GM} \approx 5\,\mathrm{V}$ [102].

The MOSHFET saturation currents I_{DS0} for the zero gate bias are in a good agreement with the analytical model proposed in [103]:

$$I_{DS0} = \beta \cdot \frac{V_T^2}{1 + \beta \cdot R_s \cdot V_T + \sqrt{1 + 2\beta \cdot R_s \cdot V_T + \frac{V_T^2}{V_L^2}}} \tag{13.4}$$

Whereas maximum achievable drain currents I_{DM} can be found as:

$$I_{DM} = \beta \cdot \frac{V_{GMT}^2}{1 + \sqrt{1 + V_{GMT}/V_L^2}} \tag{13.5}$$

In these expressions, V_T is the threshold voltage, $V_{GMT} = V_{GM} - V_T$, R_s is the source-gate series resistance, $\beta = \frac{C_i \mu}{L_G}$, $V_L = \frac{v_s L_G}{\mu}$, where C_i is gate-channel capacitance per unit area, μ is the electron field effect mobility, v_s is the effective electron saturation velocity. Assuming that the maximum sheet carrier density in the 2DEG channel, n_s, is about $2 \times 10^{13}\,\mathrm{cm}^{-2}$ [86] and the effective electron drift velocity in the channel, $v = 5 \times 10^6\,\mathrm{cm/s}$, we estimate the maximum achievable channel current $I_{DM}/W = q \times n_s \times v \approx 1.6\,\mathrm{A/mm}$. The measured saturation current in MOSHFETs (Fig. 13.24b) is close to this maximum value.

The internal drain voltage for the drain current saturation (the knee voltage V_{KN}) for MOSHFETs is of the same order as that for HFETs for same drain currents [22].

However, since the MOSHFET drain currents are higher than those for HFETs, the MOSHFET current voltage characteristics have higher V_{KN} values due to larger voltage drop across the source and drain access resistances. Since the threshold voltage of MOSHFET is more negative compared to HFETs, the MOSHFET DC transconductance is lower. However the small-signal gain and the cut-off frequencies for MOSHFETs are same or even higher than those of HFET. [22], since the MOSHFET gate—channel capacitance is also lower compared to HFET thus compensating the decrease in the transconductance.

The mechanism of the gate leakage in GaN-based MOSFETs is fairly complicated and involves surface leakage and trap-assisted tunneling [104, 105], see Fig. 13.25.

Simin et al. [106] reported on the characteristics of AlGaN/GaN MOSHFETs were measured in the temperature range of 20–300 °C. At 300 °C, the leakage current of MOSHFET remained four orders of magnitude lower than that of regular HFET. The saturation current and transconductance for both types of transistors follow the temperature dependence of electron velocity in the channel.

The recovery of the current collapse (see Sect. 13.6) at elevated temperatures compensated the effect of the decrease of the steady-state saturation current with temperature. As a consequence, the saturation microwave power remained nearly constant in the temperature range 20–200 °C, varying only by about 20% or so. These results showed high potential of MOSHFETs for high-temperature microwave, digital and switching applications. This was further confirmed by Tarakji et al. [107], who studied DC and RF-characteristics of AlGaN/GaN MOSHFETs at elevated temperatures up to 300 °C, after a 36 h continuous operation at 200 °C and after a 1 min thermal stress at temperatures up to 850 °C. At 300 °C, the gate leakage current remained about 4 orders of magnitude lower than that for regular HFETs. At zero gate-bias, the saturation current decreased by only about 20% after 36 h of continuous operation at 200 °C. After a 700 °C, 1 min thermal stress, the gate leakage remained as low as 5 nA/mm, whereas the peak current and DC transconductance showed a 20% reduction. In spite of the decrease in the peak-current, the RF satura-

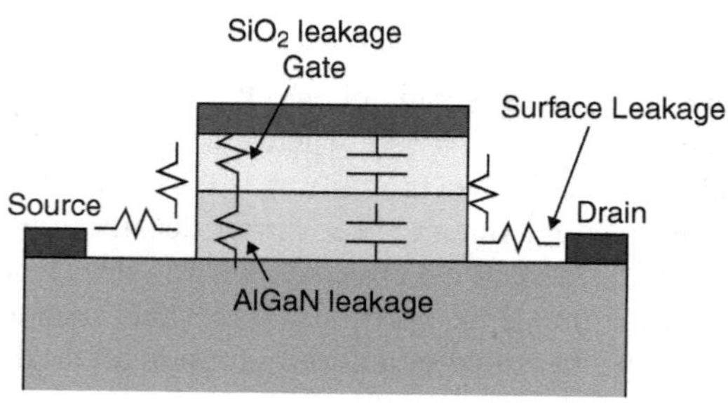

Fig. 13.25 Leakage current pass in AlGan/GaN MOSHFETs [104] (Reprinted with permission. Copyright 2002 Materials Research Society)

tion power remained nearly constant for operation at temperatures up to 200 °C that they also attributed to a reduction in the current collapse.

Simin et al. [108] used the oxide layer in MOSHFETs for bridging to increase the device periphery. They reported on AlGaN/GaN MOSHFETs over SiC substrates with peripheries from 0.15 to 6 mm. The devices featured a multigate (MG) design with source interconnections using a novel oxide-bridging approach. The saturation current scaled linearly with the gate width and reached 5.1 A for a 6 mm wide device with a 1.5 μm gate length in a 5 μm source—drain opening. The cutoff frequency of around 8 GHz was practically independent of periphery. Large-signal output RF-power as high as 2.7 W/mm was measured at 2 GHz. The RF-power also scaled linearly with device widths up to 2 mm.

13.5.2 Low Frequency Noise

The level of the low frequency (1/f and generation–recombination) noise is one of the most important parameters of semiconductor devices. In microwave and optical devices (generators, mixers, lasers) it up-converts to the phase noise and sets a lower limit on the signal level in broad-band circuits, Doppler locators, communication systems. The noise limits also the sensitivity of any detector and determines the signal to noise ratio. Therefore, noise is one of the crucial factors which determine the possibility of practical use of the device, especially in communication systems.

The low frequency noise is also a powerful tool to study deep levels, degradation, material structural perfection, mechanisms of current flow, recombination, light emission etc. The presence of generation–recombination noise indicates the well-defined local level. Measurements as a function of temperature of this noise allow us to find energy position, concentration and capture cross section of this level. In some cases, the temperature dependence of the capture cross section can also be determined. This is so-called Noise Spectroscopy [109, 110]. Refer to Refs. [111–113] regarding the generation–recombination noise in GaN-based devices.

The review of the noise properties of nitrides and GaN-based devices can be found in references [114–116]. There are several known noise sources in FETs, including the gate leakage current. As was shown by Rumyantsev et al. [117], the gate leakage current might significantly contribute to noise for low noise devices even for the relatively small gate currents (on the order of 0.01% of the drain current or so). Since FETs transistors with insulated gate have several orders of magnitude smaller gate leakage current, they are free of the noise source related to the gate current. On the other hand, traps in the gate dielectric and at the dielectric-semiconductor interface might cause additional noise. This noise mechanism known as McWhorter noise is the main noise source in Si MOSFETs [118, 119]. High-k dielectrics (used for the reduction of gate leakage current in submicron Si MOSFETs) cause elevated noise levels in those devices [120, 121].

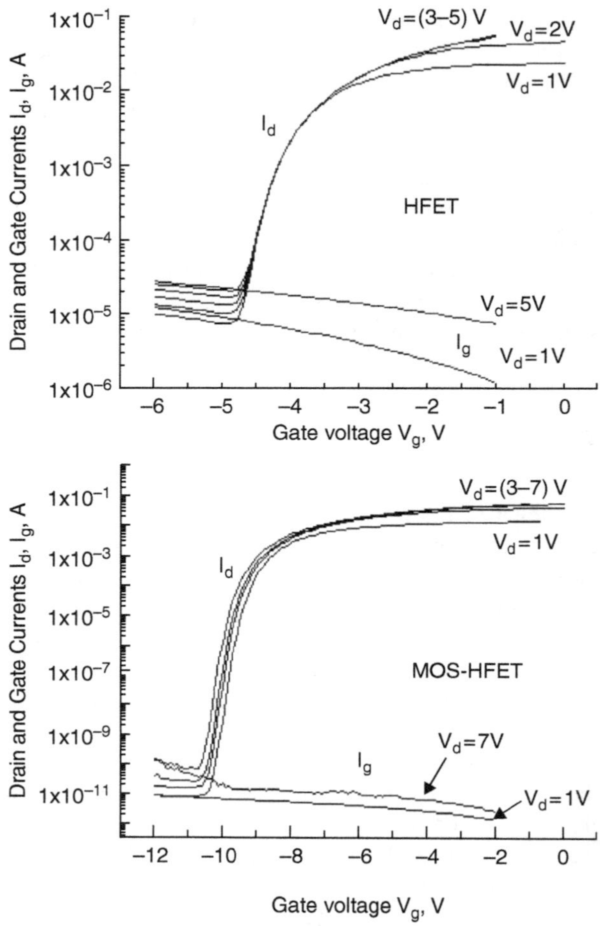

Fig. 13.26 Transfer current–voltage characteristics of HFETs and MOSHFETs [117] (Reprinted with permission. Copyright 2000 American Institute of Physics)

The noise properties of GaN-based HFETs with the SiO_2 insulated gate (MOSHFETs) were studied by Pala et al. [20] and Chiou et al. [122]. Pala et al. [20] fabricated the transistors with insulated and Schottky barrier gates on the same wafer. The SiO_2 layer was deposited on a part of the heterostructure using plasma enhanced chemical vapor deposition. Figure 13.26 shows the transfer current voltage characteristics of the HFETs and MOSHFETs. As seen, MOSHFETs are characterized by the extremely low gate current and high on-to-off ratio. Figure 13.27 shows the drain current noise S_I/I_d^2 as a function of the drain current at constant gate voltage for both type of transistors. As seen, MOSHFET are characterized by the same or smaller noise level as HFETs, i.e., silicon dioxide does not deteriorate the noise characteristics of MOSHFETs.

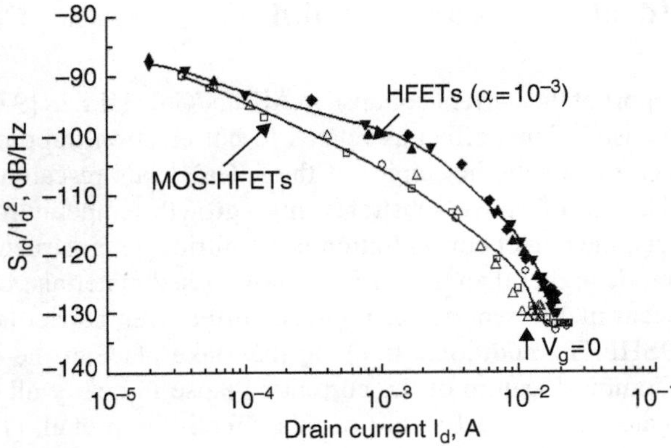

Fig. 13.27 Relative spectral noise density S_I/I_d^2 as a function of the drain current for HFETs and MOSHFETs. Frequency of analysis $f = 10$ Hz [117] (Reprinted with permission. Copyright 2000 American Institute of Physics)

Rumyantsev et al. [123] studied the noise properties of MESFETs and MOSFETs fabricated on the relatively high doped ($10^{18}\,cm^{-3}$) GaN layers. Figure 13.28 shows dependences of noise on the current at constant drain voltage for both types of the devices. As seen, in spite of the presence of oxide close to the channel noise properties of MESFETs and MOSFETs are identical.

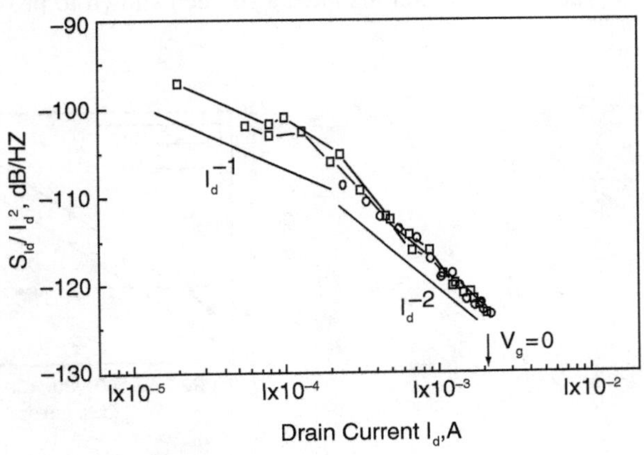

Fig. 13.28 Dependence of the of the relative spectral noise density of the drain current fluctuations on drain current. Drain voltage $V_d = 0.5$ V. Frequency of analysis $f = 200$ Hz. Different symbols show data for MESFETs and MOSFETs [123] (Reprinted with permission. Copyright 2001 American Institute of Physics)

13.6 Non-Ideal Effects and Reliability

Since the first report of the current collapse in AlGaN/GaN HFETs [97], which identified the mechanism of this effect as related to hot electron trapping, the current collapse has been one of the hot topics of the AlGaN/GaN research and development. Due to the lack of native substrates, high growth temperatures and significant piezo-effects, the defect concentration in III-nitride heterostructures is several orders of magnitude higher than that in Si or GaAs based materials. Due to this, the trapping may occur in different device regions: buffer layer, barrier layer and at the surface. In MOSHFETs, additional trapping may take place at the dielectric-barrier interface. Common feature of the current collapse in nearly all the III-nitride devices is the gate edge-related nature of this effect. Simin et al. [124] presented experimental evidence of the gate edges being responsible for the current collapse, as illustrated in Fig. 13.29.

As seen, during the transient, the portion of the resistance corresponding to the channel under the gate does not change (R(0) and R(τ) have the same slope); the difference comes from the gate length-independent shift which can be attributed to the time-dependent resistance at the gate edge. Similar behavior was found for GaN MESFETs and MOSFETs in [125]. Numerous factors may be responsible for the carrier trapping at the gate edges. These include strong electric fields at the drain edge of the gate, surface states [126], buffer trapping [127] and voltage induced strain in the layers forming the heterostructure [124].

Apart from the obvious way of reducing the dislocation density and defect concentration in the III-nitride heterostructures, an effective approach to reducing or eliminating the current collapse was found by surface passivation, typically using the Si_3N_4 layers [128]. Other materials have also been shown to provide the passi-

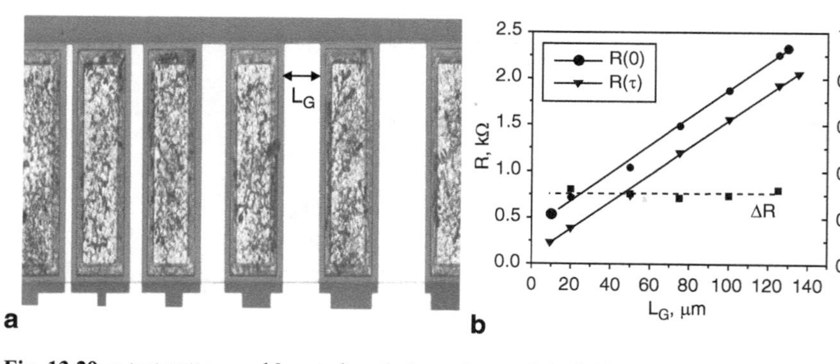

Fig. 13.29 **a** test pattern and **b** gate length dependence of the initial and steady state channel resistance. The test pattern consists of a set of transistors with constant source-gate and gate-drain spacing and variable gate length L_G. The channel resistance was measured in the beginning of the pulse applied to the drain: R(0) and after reaching a steady state condition: R(τ) [124] (Reprinted with permission. Copyright 2001 American Institute of Physics)

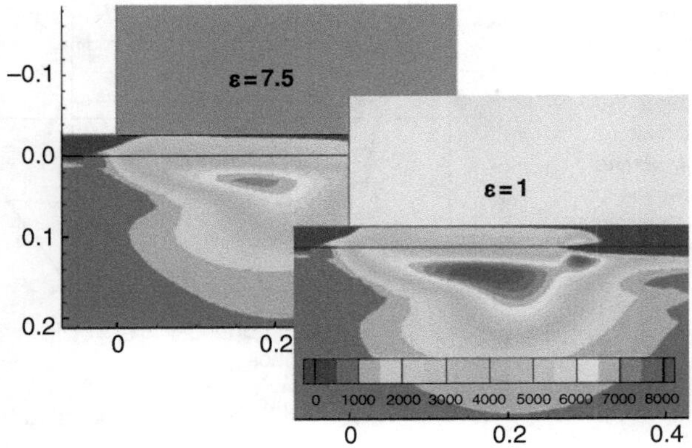

Fig. 13.30 Reduction in the surface peak field at the gate edge due to dielectric layer with high dielectric permittivity [130] (Reprinted with permission. Copyright 2006 IEEE)

vation effects and to eliminating or reducing the current collapse, such as SiO_2 and silicon-oxynitride layers [129], or HfO_2 layers [59]. The mechanism of surface passivation is related to several factors. One of them is the surface field reduction due to high dielectric constant of the deposited dielectric film (see Fig. 13.30).

The ability of the dielectric layer to mitigate the current collapse also depends on many other factors, such as polarization and built-in charges, lattice matching to the barrier layer material, conductivity, structural quality etc. Because of this complexity, surface passivation effects are still poorly reproducible. The above considerations are fully applied to the gate dielectric materials used in MOSFET/MOSHFET technology, which may or may not provide the passivation effects.

Another efficient approach to mitigating the current collapse is by implementing a so-called field-plating technology [131]. Field-modulating plate smoothens out strong electric field peak at the gate edge and thus increases the drain voltage needed to reach the threshold for hot electron trapping. This effect is illustrated in Fig. 13.31 [132].

The reliability and degradation processes in power GaN HFETs and MOSHFETs have been found to be strongly related to the carrier trapping and hence to the current collapse effects [132–134]. It has been shown that eliminating the current collapse using the above reviewed approaches allows for greatly increased lifetime of GaN based devices. The reliability of insulating gate devices is typically much better than those with Schottky based gates. The mechanism of enhanced MOSHFET reliability is related to low MOSHFET gate currents under dynamic forward gate biasing that is present in high power switches or microwave power amplifiers. In HFETs, such forward bias causes significant gate currents resulting in fast degradation. In MOSHFETs, the gate currents remain at low level allow for stable high-power operation [133, 135, 136]. Fujitsu Inc. has demonstrated and insulating gate AlGaN/GaN HFET with over 100 W output power and estimated lifetime exceeding one million hours [137].

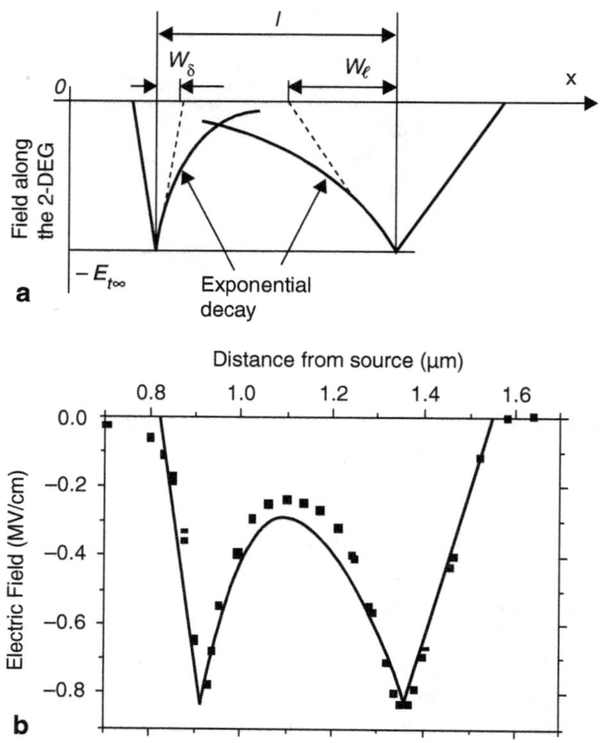

Fig. 13.31 The field distribution along the 2-DEG under optimum conditions for large plate extension *l*. **a** Model, **b** Comparison between model (*line*) and numerical calculations (*points*) of the *x*-component of the field [132] (Reprinted with permission. Copyright 2005 IEEE)

13.7 Applications and Performance

13.7.1 RF Amplifiers

RF power amplifier is the most important and widely used application of GaN based field-effect transistors [138]. Due to extremely high drain saturation currents and breakdown voltages, in combination with high operating temperatures, chemical stability and robustness, GaN power amplifiers have outperformed most other solid state amplifier types. Using GaN based amplifiers, for the first time the RF power densities as high as 30 W/mm (Watt per 1 mm of the device periphery) have been achieved [139, 140]. These power densities are one–two orders of magnitude higher than those obtained with Si, SiC or III-V semiconductor devices.

The use of insulated gate GaN devices in power amplifiers play crucial role in achieving simultaneously the highest power densities and high stability and reliability. HFET operation as an RF power amplifier is illustrated in Fig. 13.32. In a typical circuit configuration, the voltage applied at the gate is a superposition of the DC bias and the input RF signal (Fig. 13.32a). Changes in the HFET drain current

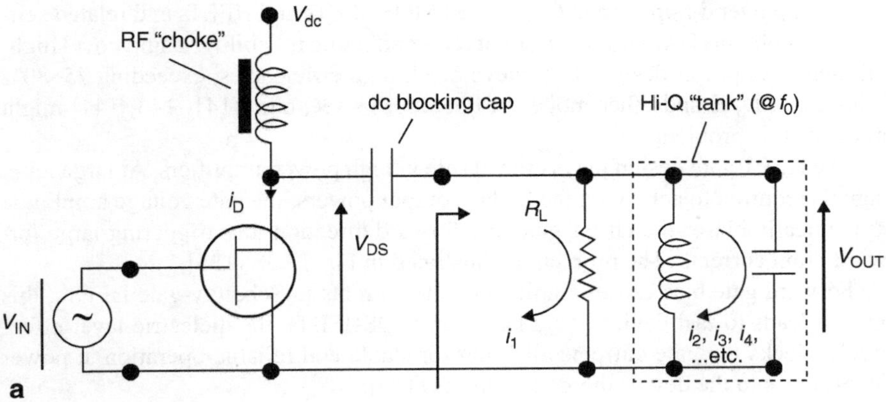

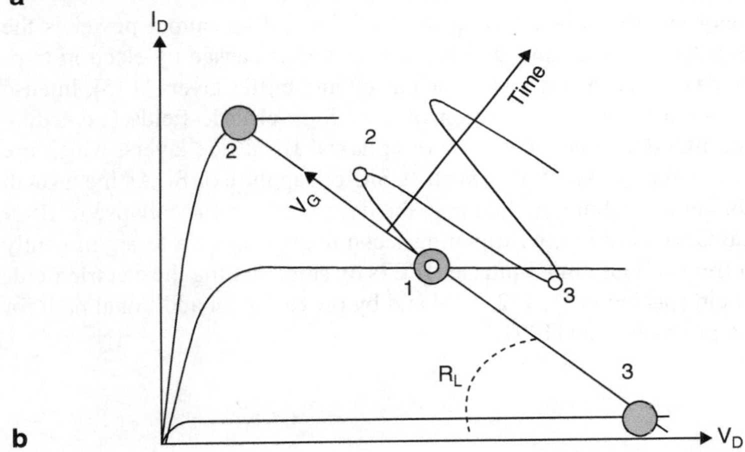

Fig. 13.32 **a** Typical power amplifier circuit (class AB), after [141] (Reprinted with permission. Copyright 2002 Artech House). **b** HFET load line in a power amplifier mode

caused by the gate voltage modulation, result in the modulated voltage across the load resistance R_L. The maximum linear output RF power can be estimated using the following simple approximate expressions. For linear mode amplification (so-called Class A mode), the DC bias corresponds to approximately half the maximum HFET drain current I_{DMAX}; as seen from the Fig. 13.32b, corresponding maximum amplitude of the RF current, $i_{MAX} = I_{DMAX}/2$. Peak drain voltage may not exceed the breakdown voltage V_{BD}; therefore, maximum amplitude of the output RF signal is $v_{MAX} = V_{BD}/2 - V_{KN}$, where V_{KN} is the drain voltage for the drain current saturation. Since typically, $V_{KN} << V_{BD}$, the maximum HFET output power, $P_{MAX} = i_{MAX} \times v_{MAX}/2 \approx I_{DMAX} \times V_{BD}/8$. In typical modern microwave HFETs, peak drain currents can be as high as $I_{DMAX} = 1.5$–$2\,\text{A/mm}$, with breakdown voltages $V_{BD} \approx 100$–$150\,\text{V}$. Hence maximum RF powers can be as high as 18–37 W/mm. The above example explains why III-nitride HFETs are capable of delivering record high output power.

Large power dissipation in GaN-based MOSHFETs and HFETs and related self-heating problems [142] is an important factor affecting reliability. Using novel high-efficiency amplifier designs to achieve much high efficiencies, exceeding 75–80% (classes C, D, E and other modes of operation)—see, e.g., [141, 143, 144] might alleviate this problem.

The HFET gate design plays crucial role in high power amplifiers. At large input signals required to achieving the highest output powers, the gate voltage amplitude dynamically biases the HFET gate into forward direction thus triggering large forward drain currents. The problem is illustrated in Fig. 13.33 [135].

Forward gate bias causes significant gate currents in Schottky-gate HFETs; this in turn leads to fast device degradation. In MOSHFETs, the dielectric layer effectively blocks the gate currents allowing for stable and reliable operation at power levels close to the device theoretical limits [139].

Another factor significantly affecting the HFET amplifier output power is the current collapse. As discussed above, this phenomenon is caused by electron trapping at the device surface, as well as in the barrier and buffer layers [145]. Intense trapping is triggered by the combination of very high electric fields (exceeding 3 MV/cm) with high defect concentration of epitaxial III-nitride layers, which are typically grown on foreign substrates, such as SiC or sapphire or Si. As the growth and device fabrication technology matures, the degree of current collapse in III-N devices decreases. In addition, field-plating technology is known to significantly mitigate the current collapse in III-nitride HFETs by smoothening the electric fields in the gate—drain spacing [131, 132, 146] and by providing an additional path for the trapped charge dissipation [139].

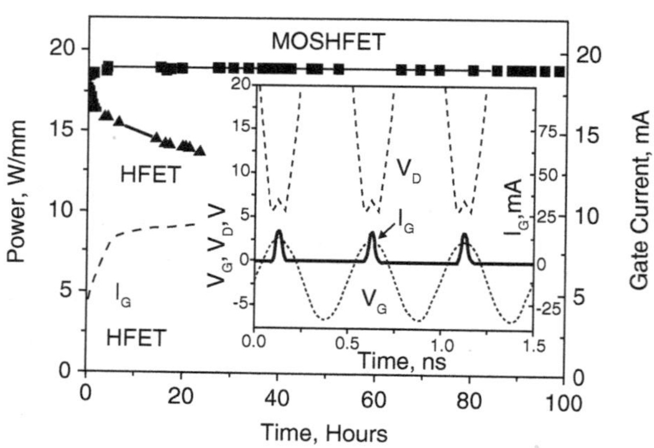

Fig. 13.33 Output powers at 2 GHz for identical geometry FP MOSHFET and HFET as a function of time. The drain bias for both FP device types was 55 V. Also shown is the time dependence of the HFET DC gate current. The *inset* shows Aim-Spice simulations for the HFET gate voltage, drain voltage and gate current at 40 V drain bias and −2 V gate bias [135] (Reprinted with permission. Copyright 2005 IEEE)

13.7.2 RF Switches

High-frequency switches are the key elements of many modern systems such as radars, phase arrays, and large variety of wireless communication systems ranging from mobile phones and PDAs to GPS receivers. RF switches are also essential elements of phase shifters, attenuators and other microwave components. Currently, pin-diodes, GaAs MESFETs or HEMTs and RF MEMS are commonly used as RF switching and control components. All of these devices have significant performance limitations arising from the fundamental materials properties and/or device design. Major limitations of pin-diode switches are related to (a) significant forward currents required to turn the device on; (b) minority carrier accumulation leading to slow turn-off times; and (c) vertical device structure complicating the use of pin-switches in MMICs. RF MEMS require high control voltages. Most RF MEMS cannot handle high RF powers due to self-actuation under large-signal stress and notorious reliability problems at high power levels. Direct-contact type RF MEMS suffer from these limitations and are not commonly used in power RF systems. Capacitive-type RF MEMS cannot be used in broad-band systems operating at DC or at low-frequencies. GaAs-based MESFETs and HEMTs have very limited operating voltage and current ranges. The breakdown voltages are typically below 30 V, maximum switching powers do not exceed 1 W (unless the devices are stacked, which significantly complicates the technology and increases insertion loss and reflection).

Very high densities of the 2D electron gas in AlInGaN based heterostructures (up to $2 \times 10^{13}\,cm^{-2}$ or even higher in MOSHFET structures enable a new paradigm in the RF switch design [147]. Using this technology, the RF switches with insertion below 0.2 dB at 2 GHz have been demonstrated [148].

A single FET device connected into a transmission line in series as a variable resistor is shown in Fig. 13.34a. Source and drain electrodes are connected to the line input and output correspondingly. The gate electrode is connected to a control voltage supply through a blocking resistor. Figure 13.34b shows more detailed equivalent circuit of the FET RF switch, including the variable, gate-voltage controlled channel resistance, parasitic device capacitances and inductors associated with the bonding wires.

In the "ON" state, the MOSHFET gate bias is zero or positive, the channel resistance R_{Ch} is low that ensures low-loss input-output transmission. The drain-source capacitance is shunted by the R_{Ch}, therefore the transmission is almost frequency independent. The value of the R_{Ch} can be estimated from the device parameters as $R_{Ch} = 2R_C + R_{GS} + R_{GD} + R_G(V_G)$, where the R_{GS} and R_{GD} are the resistances of the source-gate and gate- drain openings, R_C is the contact resistance and $R_G(V_G)$ is the voltage dependent resistance of the channel under the gate. At high positive gate bias, especially for the short gate devices, $R_G(V_G) << R_{Ch}$. Also, $R_{GS} = R_{SH} \times L_{GS}/W$ and $R_{GD} = R_{SH} \times L_{GD}/W$, where $R_{SH} = 1/(qN_S\mu_n)$ is the layer sheet resistance, L_{GS} and L_{GD} are the source-gate and gate-drain spacing. For a typical MOSHFET, $R_{SH} \approx 400\,\Omega$, $L_{GS} \approx L_{GD} \approx 1.5\,\mu m$ and the contact resistance, $R_C \approx 1\,\Omega/mm$, which results in $R_{Ch} \approx 3\,\Omega/mm$. The insertion loss of a series resistor R_{ON} connected into

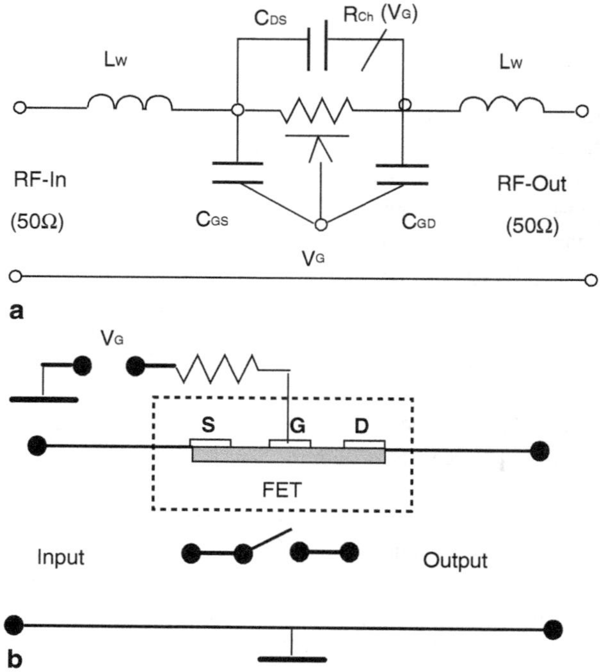

Fig. 13.34 **a** FET connected into transmission line as a voltage controlled variable resistor; **b** Equivalent circuit of a single-element FET switch including device parasitic capacitances and mounting wire inductances [147] (Reprinted with permission. Copyright 2006 World Scientific)

a transmission line with the characteristic impedance Z_0 ($Z_0 = 50\,\Omega$ in our experiment), assuming $R_{ON} << Z_0$ can be estimated as:

$$L_{Ins}(dB) = -20Log\frac{1}{1 + R_{ON}/2Z_0} \approx 0.087R_{ON} \tag{13.6}$$

At high frequencies, the isolation of a simple series switch degrades due to decreasing impedance of the device capacitance in the off-state. Referring to Fig. 13.35 and

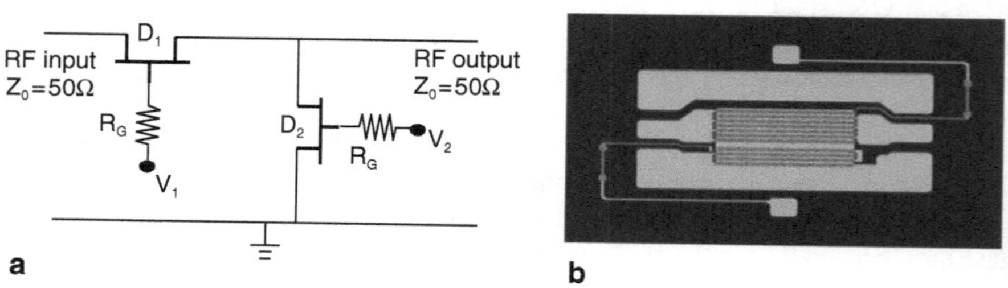

Fig. 13.35 **a** Equivalent circuit with series (D1) and shunting (D2) HFETs. **b** CCD image of low-loss SPST RF switch MMIC [148] (Reprinted with permission. Copyright 2008 IEEE)

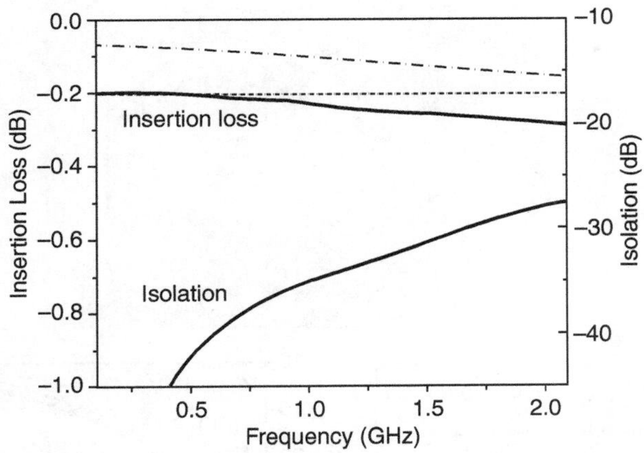

Fig. 13.36 Frequency dependencies of insertion loss and isolation of the single-pole-single throw (SPST) switch. *Solid lines*—experimental data. *Dotted line* shows the simulated insertion loss in the absence of shunt HFET capacitance. *Dashed line* shows the simulated insertion loss in the absence of current crowding in the switch metal electrodes [148] (Reprinted with permission. Copyright 2008 IEEE)

noting that $C_{GS} \approx C_{GD}$, $C_{OFF} \approx C_{DS} + C_{GS}/2$. The MOSHFET switch isolation can be significantly improved by using series-shunt configuration, see Fig. 13.35 [148].

Using series-shunt switch layout, one can simultaneously achieve low insertion loss and high isolation as illustrated in Fig. 13.36. Excessive insertion loss of around 0.2 dB, as compared to the simulated data, comes from the resistance of metal electrodes; this can be eliminated by increasing the metal thickness, e.g., using additional electroplated metal.

Recently novel type of RF switches employing capacitively-coupled contacts (C^3) has been demonstrated as an efficient alternative to conventional transistor-based switches for high frequencies, above 2 GHz [60, 149, 150]. In C^3 RF switches, the RF signal is injected via capacitive coupling between the metal electrode and the conducting channel, see Fig. 13.37; as a result no annealed ohmic contacts are required; hence the entire device or MMIC can be fabricated using self-aligned process. Effective contact impedance of C^3-s can be as low as 0.1–0.4 Ω mm depending on frequency and epilayer structure.

Flexible and robust C^3 technology allows for a number of RF switch performance improvements without significant technology complications. One example is multigate III-N RF switch demonstrated in [151]. Additional gates placed within a tight source-drain spacing of multigate C^3—switch using fully self-aligned technology, significantly improve the isolation and maximum switching power of RF switch (see Fig. 13.38).

An important feature of the C^3 electrodes making them fundamentally different from regular ohmic contacts is the impedance dependence on the DC bias between the electrode and the device channel. Unlike ohmic contact, whose impedance is

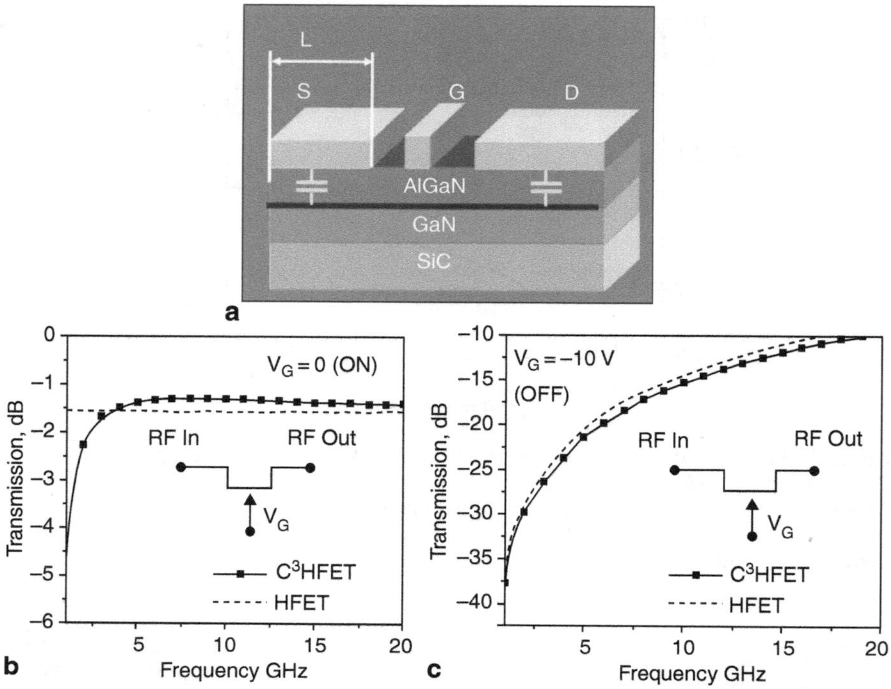

Fig. 13.37 C^3 HFET single-element RF switch: **a** layout; **b** insertion loss; **c** isolation. Insertion loss and isolation for a regular HFET with ohmic contacts is also shown [150] (Reprinted with permission. Copyright 2007 IEEE)

low and bias-independent, the impedance of the C^3 is low when its potential with respect to the channel is zero or positive. However, when negative bias exceeding the pinch-off voltage is applied to C^3-electrode, the channel under it is fully depleted and the contact impedance becomes very high. This feature allows for efficient control of the RF signal transmitted through C^3-device by changing the bias between the electrode and the channel. In other words, in C^3-devices, the control over the transmitted RF power can be achieved without placing the gate between the source and drain electrodes, which is the case for any type of regular FET-based switch with ohmic contacts. The first gateless microwave switch using III-nitride transistors with C^3 electrodes has been demonstrated in [152], see Fig. 13.39.

The absence of the gate in the source-drain spacing of RF switch tremendously simplifies the device layout and fabrication technology. Note that low-loss FET-based RF switches require large total device periphery W = 1–3 mm as seen from the above discussed example. Gate alignment in large periphery devices is a well-known technological challenge. In addition, the gate metal resistance along the large periphery device leads to gate potential non-uniformity and deteriorates the switch performance. Another problem is that in high-power switches source—gate and gate—drain

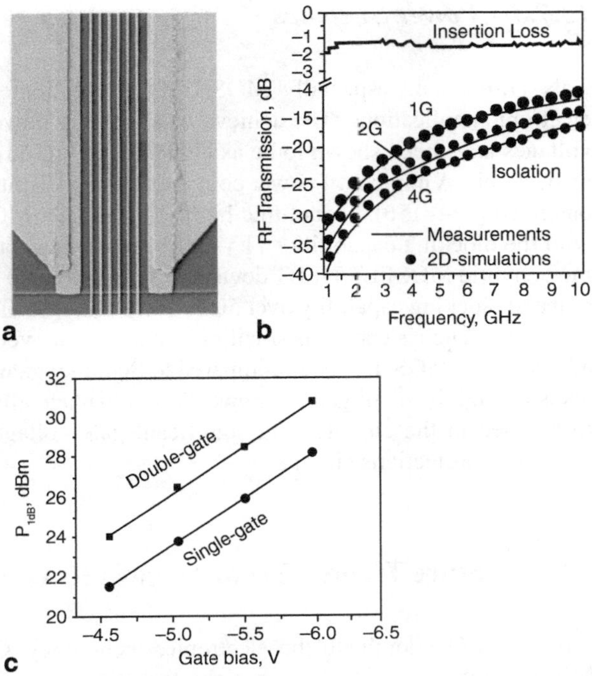

Fig. 13.38 **a** SEM-image of the source-drain region of a MG-C³-HFET; **b** Measured and simulated insertion loss and isolation for MG C³ RF switch connected in series in the 50 Ω transmission line. $V_G = 0$ for the insertion loss and $V_G = -8$ V for the isolation data; **c** Maximum RF power as a function of gate bias for single and double gate C³ RF switches. HFET threshold voltage $V_T \approx -4$ V. Signal frequency is 2 GHz [151] (Reprinted with permission. Copyright 2009 IEEE)

spacing needs to be sufficiently high to avoid gate premature breakdown; additional inter-electrode spacing leads to higher ON-resistance and increases the insertion loss. Finally, in the HFET pinched-off state, the capacitance between the 2D channel outside the gate and the gate electrode creates the path for RF signal, which limits the isolation. Gateless RF switches are free from most of these issues and therefore are very promising for microwave high-power switching applications.

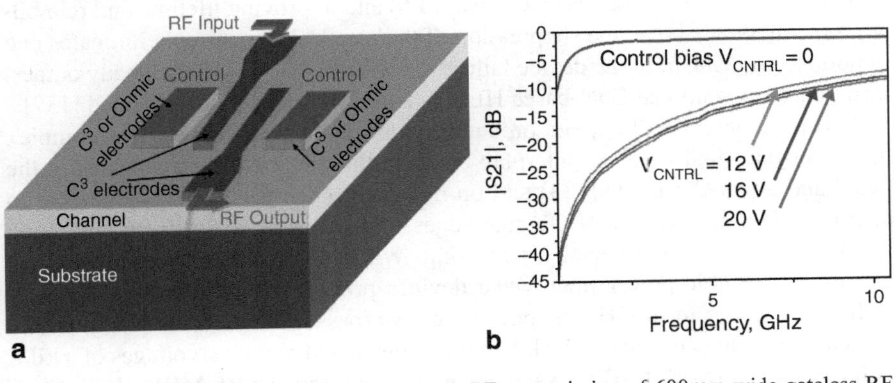

Fig. 13.39 **a** Gateless III-N RF switch layout; **b** RF transmission of 600 μm wide gateless RF switch at different control electrode voltages [152]

13.7.3 Power Switches

GaN HFETs and especially MOSHFETs have tremendous potential for power electronic applications. The achievable switching power densities and breakdown voltages have been shown to be as high as 2.5 kV [153] and several kW/mm^2 [18], respectively. A number of power converters using III-nitride HFETs have been demonstrated [154–156]. Monolithic high voltage GaN MOSFET/Schottky diode pair with the blocking capability ~1 kV has been demonstrated by [157]. Enhancement mode hybrid MOSFET/HFET devices demonstrated up to 70 A maximum drain current and blocking capability over 500 V [45]. The specific ON-resistance of this kind of hybrid switches can be as small as 20 mΩ cm^2 or even smaller [45, 158]. The key advantage of MOSHFETs as compared to Schottky gate HFETs for power electronics is negligibly small gate currents. This advantage allows III-nitride MOSHFETs to be used in the circuits with significant gate voltage overshoots and eliminate additional protections circuits.

13.8 Future Trends: From Megawatts to Terahertz

Silicon CMOS dominate the electronics technology. Compared to Si CMOS, Si MESFETs or Si-Ge HFETs are niche technologies commanding with tiny market share. However, III-V insulated gate transistors have difficult time competing with their MESFET and HFET counterparts. This is because of additional interface states at the semiconductor-insulator interface. In the nitride materials system, the situation is completely different. Huge densities of the 2D electrons in the device channels mask the effects the interface states that are further diminished in MOSHFET structures, where these interface states are separated from the active device channel by the wide band gap barrier layer. Moreover, the dielectric layer technology has proved to be necessary for the surface state passivation in the ungated regions for eliminating or reducing the current collapse (see Sect. 13.6) and improving lifetime and reliability. As mentioned above, the suppression of the gate leakage current eliminates one important mechanism of the device failure. GaN-based MOSHFETs already outperform more conventional GaN-based HEMTs and have been commercialized [137].

Further progress will depend on better understanding of the new and complex device physics that involves interplay between the dielectric surface states in the gated and ungated regions [134] and on more advanced gate edge engineering to alleviate high field effects at the gate edges. Using high-k dielectric stacks, such as including HfO_2, is especially promising. When this technology improves, we can expect reliable power microwave devices providing tens of watts of power at frequencies up to 60 GHz or more with unsurpassed efficiencies [144], power switches handling currents in the kilo ampere range and blocking voltages of a kilovolt or more. GaN-MOSHFET switches will mostly replace RF MEMs because of extremely low insertion loss, excellent linearity, and a very high isolation.

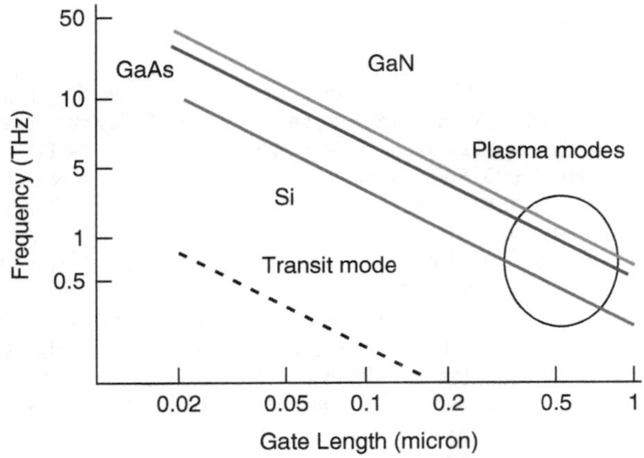

Fig. 13.40 Frequency of THz emission or detection versus gate length for Si, GaAs and GaN-based FETs [159] (Reprinted with permission. Copyright 2003 IEEE)

Even more challenging is to push this technology to extremely high frequencies (up to and including the THz range of frequencies). The THz emission from conventional GaN HEMTs [160, 161] and resonant and non-resonant THz detection [162] clearly show the potential for III-N semiconductors to penetrate the THz range. As seen from Fig. 13.40, GaN-materials system is expected to yield higher frequency devices for THz applications than II-V or Si technology.

We recently identified the detrimental effects of the ungated regions of the frequency performance [163] and proposed solutions involving re-engineering of these regions [164, 165] and even adding additional capacitively coupled contacts to boost the frequency response (see Fig. 13.41).

Great results already obtained for III-N insulated gate FETs stimulated research on other wide band gap materials systems, such as SiC HFETs [166] and ZnO [167]. There is no doubt that the most exciting results are still to come.

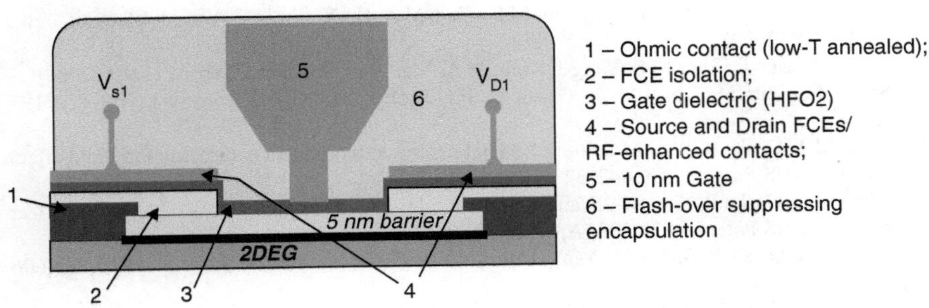

Fig. 13.41 Schematic structure of 5-terminal THz HFET with field-controlling electrodes (FCEs) [165] (Reprinted with permission. Copyright 2009 World Scientific)

References

1. Levinshtein M E, Rumyantsev S L, Shur M S, Properties of Advanced Semiconductor Materials: GaN, AlN, InN, BN, SiC, and SiGe, John Wiley and Sons, New York (2001)
2. Bykhovski A, Gelmont B, Shur M S, J. Appl. Phys., 74(11), 6734 (1993)
3. Ambacher O, Foutz B, Smart J, Shealy J R, Weimann N G, Chu K, Murphy M, Sierakowski A J, Schaff W J, Eastman L F, Dimitrov R, Mitchell A, Stutzmann M, J. Appl. Phys., 87, 334 (2000)
4. Gaska R, Yang J W, Osinsky A, Chen Q, Khan M A, Orlov A O, Snider G L, Shur M S, Appl. Phys. Lett., 72(6), 707 (1998)
5. Frayssinet E, Knap W, Lorenzini P, Grandjean N, Massies J, Skierbiszewski C, Suski T, Grzegory I, Porowski S, Simin G, Hu X, Khan M A, Shur M S, Gaska R, Maude D, Appl. Phys. Lett., 77(16), 2551 (2000)
6. O'Leary S K, Foutz B E, Shur M S, Eastman L F, J. Mater. Sci.: Mater. Electron., 17, 87 (2006)
7. O'Leary S K, Foutz B E, Shur M S, Eastman L F, Appl. Phys. Lett., 88, 152113 (2006)
8. Dyakonova N V, Dickens A, Shur M S, Gaska R, Yang J W, Appl. Phys. Lett., 72(20), 2562 (1998)
9. Liu W, Balandin A A, Lee C, Lee H Y, Phys. Stat. Sol. A, 202(12), R135 (2005)
10. McIntosh F G, Boutros K S, Roberts J C, Bedair S M, Piner E L, El-Masry N A, Appl. Phys. Lett., 68, 40 (1996)
11. Chitnis A, Kumar A, Shatalov M, Adiravahan V, Lunev A, Yang J W, Simin G, Khan M A, Gaska R, Shur M S, Appl. Phys. Lett., 77(23), 3800 (2000)
12. Khan M A, Gaska R, Shur M S, Yang J, Method of producing nitride-based heterostructure devices, Patent number: 6764888 (2004)
13. Khan M A, Chen Q, Sun C J, Shur M S, Gelmont B L, Appl. Phys. Lett., 67(10), 1429 (1995)
14. Binari S C, Rowland L B, Kelner G, Kruppa W, Dietrich H B, Doverspike K, Gaskill D K, Int. Symp. Compd. Semicond., San Diego (1994)
15. Khan M A, Chen Q, Sun C J, Yang J W, Blasingame M, Shur M S, Park H, Appl. Phys. Lett., 68(4), 514 (1996)
16. Hu X, Simin G, Yang J, Khan M A, Shur M S, Gaska R, Electron. Lett., 36(8), 753 (2000)
17. Gaska R, Khan M A, Hu X, Simin G, Yang J, Deng J, Rumyantsev S, Shur M S, Device research conference, 2000 Conference Digest, 43–44, Denver, CO, USA (2000)
18. Simin G, Hu X, Ilinskaya N, Kumar A, Koudymov A, Zhang J, Khan M A, Gaska R, Shur M S, Electron. Lett., 36, 2043 (2000)
19. Khan M A, Hu X, Tarakji A, Simin G, Yang J, Gaska R, Shur M S, Appl. Phys. Lett., 77(9), 1339 (2000)
20. Pala N, Gaska R, Rumyantsev S, Shur M S, Khan M A, Hu X, Simin G, Yang J, Electron. Lett., 36(3), 268 (2000)
21. Pala N, Gaska R, Shur M S, Yang J, Khan M A, Mat. Res. Soc. Proc, 595, W11.9.1 (2000)
22. Simin G, Shur M, Khan M A, Gaska R, Int. J. High Speed Electron Syst., 14(N1), 197 (2004)
23. Khan M A, Gaska R, Shur M S, Yang J, Metal oxide semiconductor heterostructure field effect transistor, US Patent 6,690,042 B2 (2004)
24. Khan M A, Gaska R, Shur M S, Yang J, Metal oxide semiconductor heterostructure field effect transistor, US Patent 6,878,593 (April 2005)
25. Khan M A, Gaska R, Shur M S, Yang J, Nitride based heterostructure devices, Patent 7,348,606 (2008)
26. Huang W, Khan T, Chow T P, J. Electron. Mat., 35(4), 726 (2006)
27. Matocha K, Chow T P, Gutmann R J, IEEE Electron Device Lett., 23(2), 79 (2002)
28. Irokawa Y, Nakano Y, Ishiko M, Kachi T, Kim J, Ren F, Gila B P, Onstine A H, Abernathy C R, Pearton S J, Pan C C, Chen G T, Chyi J I, Appl. Phys. Lett., 84(15), 2919 (2004)

29. Matocha K, Chow T P, Gutmann R J, IEEE Trans. Electron Devices, 52(1), 6 (2005)
30. Jang S, Ren F, Pearton S J, Gila B P, Hlad M, Abernathy C R, Yang H, Pan C J, Chyi J I, Bove P, Lahreche H, Thuret J, J. Electron. Mater., 35(4), 685 (2006)
31. Lee H B, Cho H I, An H S, Bae Y H, Lee M B, Lee J H, Hahm S H, IEEE Electron Device Lett., 27(2), 81 (2006)
32. Huang W, Khan T, Chow T P, IEEE Electron Device Lett., 27, 796 (2006)
33. Huang W, Chow T P, Khan T, Phys. Stat. Sol. A, 204(6), 2064 (2007)
34. Kambayashi H, Niiyama Y, Ootomo S, Nomura T, Iwami M, Satoh Y, Kato S, Yoshida S, IEEE Electron Device Lett., 28(12), 1077 (2007)
35. Chow T P, Naik H, Li Z, Phys. Stat. Sol. A, 1, doi: 10.1002 (2009)
36. Tokranov V, Rumyantsev S L, Shur M S, Gaska R, Oktyabrsky S, Jain R, Pala N, Phys. Stat. Sol. (RRL), 1(5), 199 (2007)
37. Morkoç H, J. Mater. Sci.: Mater. Elect., 12(12), 677 (2001)
38. Pei Y, Poblenz C, Cordon A L, Chu R, Shen L, Speck J S, Mishra U K, Electron Lett., 44(9), 598 (2008)
39. Fareed Q, Gaska R, Shur M S, Methods of growing Nitride-based film using varying pulses, US Patent 7192849 (2007)
40. Zhang J P, Kuokstis E, Fareed Q, Wang H M, Yang J W, Simin G, Khan M A, Tamulaitis G, Kurilcik G, Jursenas S, Zukauskas A, Gaska R, Shur M, Phys. Stat. Sol. A, 188, 95 (2001)
41. Kazlauskas K, Žukauskas A, Tamulaitis G, Mickevičius J, Shur M S, Fareed R S Q, Zhang J P, Gaska R, Appl. Phys. Lett., 87, 172102 (2005)
42. Khan M A, Yang J W, Knap W, Frayssinet E, Hu X, Simin G, Prystawko P, Leszczynski M, Grzegory I, Porowski S, Gaska R, Shur M S, Beaumont B, Teisseire M, Neu G, Appl. Phys. Lett., 76(25), 3807 (2000)
43. Hu X, Deng J, Pala N, Gaska R, Shur M S, Chen C Q, Yang J, Simin G, Khan M A, Rojo J C, Schowalter L J, Appl. Phys. Lett., 82, 1299 (2003)
44. Matocha K, Chow T P, Gutmann R J, 15th Int. Symp. Power Semicond. Devices ICs, ISPSD 2003, 54 (2003)
45. Kambayashi H, Satoh Y, Niiyama Y, Kokawa T, Iwami M, Nomura T, Kato S, Chow T P, 21st Int. Symp. Power Semicond. Devices ICs, ISPSD 2009, 21 (2009)
46. Liu L, Edgar J H, Mater. Sci. Eng. R, 37, 61 (2002)
47. Sheppard S T, Smith R P, Pribble W L, Ring Z, Smith T, Allen S T, Milligan J, Palmour J W, Dev. Res. Conf., Cat. No. 02TH8606, 175 (2002)
48. Rajagopal P, Roberts J C, Cook J W, Jr., Brown J D, Piner E L, Linthicum K J, Mater. Res. Soc. Symp. Proc., 798, Y7.2.1 (2003)
49. Johnson J W, Piner E L, Vescan A, Therrien R, Rajagopal P, Roberts J C, Brown J D, Singhal S, Linthicum K J, IEEE Electron Device Lett., 25, 459 (2004)
50. Khan M A, Simin Shur M S, Gaska R, Electronic device based on group III Nitrides. In: Buschow K H J, Cahn R W, Flemings M C, Ilschner B, Kramer E J, Mahajan S (eds) Encyclopedia of Materials: Science and Technology, 3, 2616 (2008)
51. Wetzel C, Suski T, Ager JW, III, Weber E R, Haller E E, Fisher S, Meyer B K, Molnar R J, Perlin P, Phys. Rev. Lett., 78, 3923 (1997)
52. Hubbard S M, Zhao G, Pavlidis D, Sutton W, Cho E, Mater. Res. Soc. Symp. Proc., 831, E11.11.1 (2004)
53. Haffouz S, Tang H, Bardwell J A, Hsu E M, Webb J B, Rolfe S, Solid-State Electron, 49(5), 802 (2005)
54. Iliopoulos E, Zervos M, Adikimenakis A, Tsagaraki K, Georgakilas A, Superlattices Microstruct., 40(4–6), 313 (2006)
55. Gu X, Izyumskaya N, Avrutin V, Xie J, Chevtchenko S, Xiao B, Morkoç H, Proc. SPIE, 6473, 64730S (2007)
56. Wu H M, Peng L H, Phys. Stat. Sol. C, 3(6), 2291 (2006)
57. Ren F, Hong M, Chu S N G, Marcus M A, Schurman M J, Baca A, Pearton S J, Abernathy C R, Appl. Phys. Lett., 73, 3893 (1998)
58. Wu Y Q, Ye P D, Wilk G D, Yang B, Mater. Sci. Eng. B, 135, 282 (2006)

59. Liu C, Chor E F, Tan L S, Appl. Phys. Lett., 88, 173504 (2006)
60. Koudymov A, Pala N, Tokranov V, Oktyabrsky S, Gaevski M, Jain R, Yang J, Hu X, Shur M, Gaska R, Simin G, IEEE Electron Device Lett., 30(5), 433 (2009)
61. Koudymov A, Pala N, Tokranov V, Oktyabrsky S, Gaevski M, Jain R, Yang J, Hu X, Shur M S, Gaska R, Simin G, IEEE Electron Device Lett., 30, 478 (2009)
62. Littlejohn M A, Hauser J R, Glisson T H, Appl. Phys. Lett., 26, 625 (1975)
63. Gelmont B, Kim K, Shur M, J. Appl. Phys., 74, 1818 (1993)
64. Mansour N S, Kim K W, Littlejohn M A, J. Appl. Phys., 77, 2834 (1995)
65. KolnıK J, Oguzman I H, Brennan K F, Wang R, Ruden P P, Wang Y, J. Appl. Phys., 78, 1033 (1995)
66. Foutz B E, Eastman L F, Bhapkar U V, Shur M S, Appl. Phys. Lett., 70, 2849 (1997)
67. Bhapkar U V, Shur M S, J. Appl. Phys., 82, 1649 (1997)
68. Albrecht J D, Wang R P, Ruden P P, Farahmand M, Brennan K F, J. Appl. Phys., 83, 4777 (1998)
69. Foutz B E, O'Leary S K, Shur M S, Eastman L F, J. Appl. Phys., 85, 7727 (1999)
70. O'Leary S K, Foutz B E, Shur M S, Eastman L F, J. Electron Mater., 32, 327 (2003)
71. O'Leary S K, Foutz B E, Shur M S, Bhapkar U V, Eastman L F, Solid State Commun., 105, 621 (1998)
72. Albrecht J D, Wang R P, Ruden P P, Farahmand M, Brennan K F, J. Appl. Phys., 83, 1446 (1998)
73. O'Leary S K, Foutz B E, Shur M S, Bhapkar U V, Eastman L F, J. Appl. Phys., 83, 826 (1998)
74. Bellotti E, Doshi B K, Brennan K F, Albrecht J D, Ruden P P., J. Appl. Phys., 85, 916 (1999)
75. Vasconcelos T F, Maia F F, Jr., Caetano E W S, Freire V N, da Costa J A P, da Silva E F, Jr., J. Cryst. Growth, 246, 320 (2002)
76. O'Leary S K, Foutz B E, Shur M S, Eastman L F, J. Mater. Sci. Mater. Electron, doi: 10.1007/s10854-009-9896-1 (2009)
77. Oberhuber R, Zandler G, Vogl P, Appl. Phys. Lett., 73, 818 (1998)
78. Krishnan M S, Goldsman N, Christou A, J. Appl. Phys., 83, 5896 (1998)
79. Heying B, Smorchkova I, Poblenz C, Elsass C, Fini P, DenBaars S, Mishra U, Speck J S, Appl. Phys. Lett., 77, 2885 (2000)
80. Saxler A, Look D C, Elhamri S, Sizelove J, Cull D, Mitchel W C, Callahan M, Bliss D, Bouthillette L, Wang S Q, Sung C M, Lee K Y, GaN and Related Alloys, Mater. Res. Soc. Symp. Proc., 639 (2000)
81. Look D C, Sizelove J R, Appl. Phys. Lett., 79, 1133 (2001)
82. Ambacher O, Smart J, Shealy J R, Weimann N G, Chu K, Murphy M, Schaff W J, Eastman L F, Dimitrov R, Wittmer L, Stutzmann M, Rieger W, Hilsenbeck J, J. Appl. Phys., 85, 3222 (1999)
83. Bernardini F, Fiorentini V, Vanderbilt D, Phys. Rev. B., 56, R10024 (1997)
84. Smorchkova I P, Elsass C R, Ibbetson J P, Vetury R, Heying B, Fini P, Haus E, DenBaars S P, Speck J S, Mishra U K, J. Appl. Phys., 86, 4520 (1999)
85. Smorchkova I P, Chen L, Mates T, Shen L, Heikman S, Moran B, J. Appl. Phys., 90, 5196 (2001)
86. Shur M S, Bykhovski A D, Gaska R, Solid-State Electron, 44, 205 (2000)
87. Liberis J, Matilioniene I, Matulionis A, Sermunksnis E, Xie J, Leach J H, Morkoc H, Phys. Stat. Sol. A, 206, 1385 (2009)
88. Gaska R, Shur M S, Bykhovski A D, Orlov A O, Snider G L, Appl. Phys. Lett., 74(2), 287 (1999)
89. Aggerstam T, Lourdudoss S, Radamson H H, Sjödin M, Lorenzini P, Look D C, Thin Solid Films, 515, 705 (2006)
90. Ivanov P A, Levinshtein M E, Simin G, Hu X, Yang J, Khan M A, Rumyantsev S L, Shur M S, Gaska R, Electron. Lett., 37(24), 1479 (2001)
91. Huang W, Chow T P, Niiyama Y, Nomura T, Yoshida S, 21st Int. Symp. Power Semicond. Devices IC's, ISPSD 2009. 29 (2009)

92. Binari S C, Dietrich H B, Kruppa W, Kelner G, Saks N S, Edwards A, Redwing J M, Wickenden A E, Koleske D D, Proc. 2nd Int. Conf. Nitride Semicond. (ICNS 97), 476 (1997)
93. Ochiai M, Akita M, Ohno Y, Kishimoto S, Maezawa K, Mizutani T, Jpn. J. Appl. Phys., 42, 2278 (2003)
94. Ye P D, Yang B, Ng K K, Bude J, Wilk G D, Halder S, Hwang J C M, Appl. Phys. Lett., 86, 063501 (2005)
95. Mehandru R, Luo B, Kim J, Ren F, Gila B P, Onstine A H, Abernathy C R, Pearton S J, Gotthold D, Birkhahn R, Peres B, Fitch R, Gillespie J, Jenkins T, Sewell J, Via D, Crespo A, Appl. Phys. Lett., 82, 2530 (2003)
96. Rai S, Adivarahan V, Huang P, Tipirneni N, Husna F, Simin G, Yang J W, Khan M A, Jpn. J. Appl. Phys., 45(6A), 4985 (2006)
97. Khan M A, Shur M S, Chen Q C, Kuznia J N, Electron. Lett., 30(25), 2175 (1994)
98. Cho D, Shimizu M, Ide T, Ookita H, Okumura H, Jpn. J. Appl. Phys. V., 41, 4481 (2002)
99. Mohammed F M, Wang L, Koo H J, Adesida I, J. Appl. Phys., 101, 033708 (2007)
100. Das J, Oprins H, Ji H, Sarua A, Ruythooren W, Derluyn J, Kuball M, Germain M, Borghs G, IEEE Trans. Electron Devices, 53(11), 2696 (2006)
101. Binari S C, Rowland L B, Kelner G, Kruppe W, Dietrich H B, Doverspike K, Gaskill D K, Int. Symp. Compound Semicond., Bristol, UK: IOP, 459 (1995)
102. Koudymov A, Fatima H, Simin G, Yang J, Khan M A, Tarakji A, Hu X, Shur M S, Gaska R, Appl. Phys. Lett., 80, 3216 (2002)
103. Shur M S, GaAs Devices and Circuits, Plenum Press, New York and London (1987)
104. Clarke F W, Ho F D, Khan M A, Simin G, Yang J, Gaska R, Shur M S, Deng J, Karmalkar S, Mat. Res. Symp. Proc., 743, L 9.10 (2002)
105. Karmalkar S, Sathaiya D M, Shur M S, Appl. Phys. Lett., 82, 3976–3978 (2003)
106. Simin G, Tarakji A, Hu X, Koudymov A, Yang J, Khan M A, Shur M S, Gaska R, Phys. Stat. Sol. A, 188(1), 219 (2001)
107. Tarakji A, Hu X, Koudymov A, Simin G, Yang J, Khan M A, Shur M S, Gaska R, Solid-State Electron, 46(8), 1211 (2002)
108. Simin G, Hu X, Ilinskaya N, Zhang J, Tarakji A, Kumar A, Khan M A, Gaska R, Shur M S, IEEE Electron Device Lett., 22(2), 53 (2001)
109. Copeland J A, IEEE Trans. Electron Devices, 18, 50 (1971)
110. Levinshtein M E, Rumyantsev S L, Semicond. Sci. Tech., 9(6), 1183 (1994)
111. Rumyantsev S L, Pala N, Shur M S, Borovitskaya E, Dmitriev A P, Levinshtein M E, Gaska R, Khan M A, Yang J, Hu X, Simin G, IEEE Trans. Electron Devices, 48(3), 530 (2001)
112. Pala N, Rumyantsev S, Shur M S, Gaska R, Hu X, Yang J, Simin G, Khan M A, Fluct. Noise Lett., 2(4), L349 (2002)
113. Rumyantsev S L, Pala N, Shur M S, Levinshtein M E, Gaska R, Khan M A, Simin G, Generation—Recombination noise in GaN-based devices. In: Shur M S, Davis R F (ed), GaN-Based Materials and Devices, Selected Topics in Electronics and Systems, World Scientific, 33 (2004)
114. Levinshtein M E, Balandin A A, Rumyantsev S L, Shur M S, Low-frequency noise in GaN-based field effect transistors. In: Balandin A A (ed), Noise and Fluctuations Control in Electronic Devices, American Scientific Publishers (2002)
115. Surya C, Fong W K, Leung B H, Material processing and low frequency noise in GaN thin films and devices. In: Balandin A A (ed), Noise and Fluctuations Control in Electronic Devices, American Scientific Publishers (2002)
116. Levinshtein M E, Rumyantsev S L, Shur M S, Gaska R, Khan M A, Spec. Issue IEE Proc. Circuits Devices Syst., 149(1), 32 (2002)
117. Rumyantsev S L, Pala N, Shur M S, Gaska R, Levinshtein M E, Khan M A, Simin G, Hu X, Yang J, J. Appl. Phys., 88(11), 6726 (2000)
118. McWhorter A L, Proc. Conf. Phys. Semicond. Surf., Philadelphia, 207 (1956)
119. Christensson S, Lundstrom I, Svensson C, Solid-State Electron., 11, 797 (1968)
120. Morshed T H, Devireddy S P, Celik-Butler Z, Shanware A, Green K, Chambers J J, Visokay M R, Colombo L, Solid-State Electron., 52, 711 (2008)

121. Rumyantsev S, Young C, Bersuker G, Shur M, 20th Int. Conf. Noise Fluct., ICNF 2009, Pisa, Italy, AIP Conf. Proc., 1129, 255 (2009)
122. Chiou Y Z, Su Y K, Gong J, Chang S J, Wang C K, Jpn. J. Appl. Phys. V., 45(4B), 3405 (2006)
123. Rumyantsev S L, Pala N, Shur M S, Gaska R, Levinshtein M E, Khan M A, Simin G, Hu X, Yang J, J. Appl. Phys., 90(1), 310 (2001)
124. Simin G, Koudymov A, Tarakji A, Hu X, Yang J, Khan M A, Shur M S, Gaska R, Appl. Phys. Lett., 79, 2651 (2001)
125. Pala N, Rumyantsev S L, Shur M S, Hu X, Tarakji A, Gaska R, Khan M A, Simin G, Yang J, Solid State Electron., 46, 711 (2002)
126. Vetury R, Zhang N Q, Keller S, Mishra U K, IEEE Trans. Electron Devices, 48, 560 (2001)
127. Binari S C, Klein P, Kazior T, Proc. IEEE, 90, 1048 (2002)
128. Green B M, Chu K K, Chumbes E M, Smart J A, Shealy J R, Eastman L F, IEEE Electron Device Lett., 21, 268 (2000)
129. Arulkumaran S, Egawa T, Ishikawa H, Jimbo T, Sano Y, Appl. Phys. Lett., 84, 613 (2004)
130. Braga N, Mickevicius R, Rao V, Fichtner W, 16th Biennial University Government Industry Microelectronics Symposium, San Jose, CA (2006)
131. Ando Y, Okamoto Y, Miyamoto H, Nakayama T, Inoue T, Kuzuhara M, IEEE Electron Device Lett., 24, 289 (2003)
132. Karmalkar S, Shur M S, Simin G, Khan M A, IEEE Trans. Electron Devices, 52, 2534 (2005)
133. Koudymov A, Wang C X, Adivarahan V, Yang J, Simin G, Khan M A, IEEE Electron Device Lett., 28, 5 (2007)
134. Koudymov A, Shur M, Int. J. High Speed Electron. Syst., 18(4), 935 (2008)
135. Adivarahan V, Yang J, Koudymov A, Simin G, Khan M A, IEEE Electron Device Lett., 26(8), 535 (2005)
136. Saygi S, Fatima H, He X, Rai S, Koudymov A, Adivarahan V, Yang J, Simin G, Khan M A, Phys. Stat. Sol. (c), 2(7), 2651 (2005)
137. Kanamura M, Kikkawa T, Iwai T, Imanishi K, Kubo T, Joshin K, Electron Devices Meeting, 2005, IEDM Tech. Dig., 5, 572 (2005)
138. Mishra U, Shen L, Kazior T E, Wu Y F, Proc. IEEE, 96, 287 (2008)
139. Simin G, Adivarahan V, Yang J, Koudymov A, Rai S, Khan M A, Electron Lett., 41(13), 774 (2005)
140. Wu Y F, Saxler A, Moore M, Smith R P, Sheppard S, Chavarkar P M, Wisleder T, Mishra U K, Parikh P, IEEE Electron Device Lett., 25, 117 (2004)
141. Cripps S C, Advanced Techniques in RF Power Amplifier Design, Artech House, Boston/London (2002)
142. Gaska R, Yang J, Osinsky A, Khan M A, Shur M S, IEDM-97 Tech. Dig., 565 (1997)
143. Saito W, Domon T, Omura I, Kuraguchi M, Takada Y, Tsuda K, Yamaguchi M, IEEE Electron Device Lett., 27, 326 (2006)
144. Koudymov A, Shur M, Two-stage amplification using intermediate non-linear square wave, US Patent 7,560,985 (2009)
145. Braga N, Mickevicius R, Fichtner W, Gaska R, Shur M S, 2005 IEEE CSICS Tech. Dig., Palm Springs, CA, 149 (2005)
146. Nakajima A, Itagaki K, Horio K, IEEE International Reliability Physics Symposium, 722 (2009)
147. Simin G, Khan M A, Shur M S, Gaska R, Int. J. High Speed Electron Syst., 16(N2), 455 (2006)
148. Yang Z, Wang J, Hu X, Yang J, Simin G, Shur M, Gaska R, IEEE Electron Device Lett., 29, 15 (2008)
149. Simin G, Yang Z J, Koudymov A, Adivarahan V, Yang J, Khan M A, Appl. Phys. Lett., 89, 033510 (2006)
150. Simin G, Yang G, Shur M, IEEE/MTT-S Int. Microwave Symp., 457 (2007)

151. Hu X, Gaevski M, Khan B, Wang J, Gaska R, Simin G, Koudymov A, Jain R, Yang J, Shur M, IEEE Electron Device Lett., 30(9), 895 (2009)
152. Simin G, Wang J, Koudymov A, Sattu A, Yang J, Shur M, Gaska R, 33rd Workshop on Compound Semiconductor Devices and Integrated Circuits, WOCSDICE 2009, Málaga (Spain), Extended Abstract Book, Sect. Monday 5, 10 (2009)
153. Huang W, Chow T P, Niiyama Y, Nomura T, Yoshida S, 20th Int. Symp. Power Semicond. Devices IC's, ISPSD 2008, 291 (2008)
154. Zhang N Q, Moran B, DenBaars S P, Mishra U K, Wang X W, Ma T P, Phys. Stat. Sol. A, 188(1), 213 (2001)
155. Pytel S G, Lentijo S, Rai S, Fatima H, Adivarahan V, Chitnis A, Yang J, Hudgins J L, Santi E, Monti A, Simin G, Khan M A, Power Electron. Spec. Conf. 2004 IEEE 35th Annual, 1, 579 (2004)
156. Saito W, Kuraguchi M, Takada Y, Tsuda K, Omura I, Ogura T, IEEE Trans. Electron Devices, 51, 1913 (2004)
157. Huang W, Chow T P, Proc. 19th Int. Symp. Power Semicond. Devices ICs, 265 (2007)
158. Huang W, Li Z, Chow T P, Niiyama Y, Nomura T, Yoshida S, 20th Int. Symp. Power Semicond. Devices IC's, ISPSD 2008, 295 (2008)
159. Ryzhii V, Shur M S, ISDRS Digest, Washington DC, WP7-07-10, 200 (2003)
160. Deng Y, Kersting R, Xu J, Ascazubi R, Zhang X C, Shur M S, Gaska R, Simin G S, Khan M A, Ryzhii V, Appl. Phys. Lett., 84(15), 70 (2004)
161. Dyakonova N, Fatimy A E, Lusakowski J, Knap W, Dyakonov M I, Poisson M A, Morvan E, Bollaert S, Shchepetov A, Roelens Y, Gaquiere Ch, Theron D, Cappy A, Appl. Phys. Lett., 88(14), 141906 (2006)
162. Fatimy A E, Tombet S B, Teppe F, Knap W, Veksler D B, Rumyantsev S, Shur M.S, Pala N, Gaska R, Fareed Q, Hu X, Seliuta D, Valusis G, Gaquiere C, Theron D, Cappy A, Electron. Lett., 42(23), 1342 (2006)
163. Turin V O, Shur M S, Veksler D B, Int. J. High Speed Electron. Syst., 17(1), 19 (2007)
164. Pala N, Yang J Z, Koudymov A, Hu X, Deng J, Gaska R, Simin G, Shur M S, Device Research Conference, 2007 65th Annual, 43 (2007)
165. Simin G, Shur M, Gaska R, Int. J High Speed Electron Syst., 19(1), 7 (2009)
166. Chandrashekhar M V S, Thomas C I, Lu J, Spencer M G, Appl. Phys. Lett., 91, 033503 (2007)
167. O'Leary S K, Foutz B E, Shur M S, Eastman L F, Appl. Phys. Lett., (2009, Submitted)

Chapter 14
Technology/Circuit Co-Design for III-V FETs

Jaydeep P. Kulkarni and Kaushik Roy

Abstract III-V semiconductor based Field Effect Transistors (FET) are conceived as a promising candidate for low voltage high performance applications. In this chapter, we show the complete technology-circuit assessment of III-V FETs. The co-design approach spans from the device/SPICE models, logic/memory circuit analysis to technology requirements. We show feasibility of the use of Si + III-V hybrid technology for future high speed low voltage applications. We prescribe the technology requirements as well as suggest the application space for III-V FETs.

14.1 Introduction

Aggressive scaling of transistor dimensions has resulted in increased integration density and improved device performance at the expense of increased power dissipation. Increased integration density along with increased leakage necessitates ultra low power operation in present power constrained design environment [1]. Power reduction methods include: voltage scaling, switching activity reduction, architectural techniques of pipelining and parallelism, and CAD issues of device sizing, interconnect and logic optimization [2]. Among these methods, power supply reduction has significant effect on power savings. Reducing the supply voltage reduces the dynamic power quadratically and leakage power linearly to the first order. Hence, supply voltage scaling has remained the major focus of low power design. However to improve the performance at a lower supply voltage, we need ultra fast transistors having better driving capability (ON current). Various materials and structures such as carbon nano-tubes, semiconducting nano-wires and III-V semiconductors are explored for high speed low voltage applications [3, 4]. These materials have higher carrier mobility (either electron or hole) compared to silicon, making them

K. Roy (✉)
School of Electrical and Computer Engineering, Purdue University, W. Lafayette, IN 47907, USA
e-mail: kaushik@ecn.purdue.edu

S. Oktyabrsky, P. D. Ye (eds.), *Fundamentals of III-V Semiconductor MOSFETs*,
DOI 10.1007/978-1-4419-1547-4_14,

Table 14.1 Semiconductor material properties at 300 K

	Si	GaAs	$In_{0.53}Ga_{0.47}As$	InAs	InSb
Electron mobility (cm^2/vs) $n_s = 10^{12}/cm^2$	600	4600	7800	20000	30000
Electron saturation velocity (10^7 cm/s)	1.0	1.2	0.8	3.5	5.0
Ballistic mean free path (nm)	28	80	106	194	226
Energy band-gap (eV)	1.12	1.42	0.72	0.36	0.18

potentially useful for high speed computations. Carbon nano-tubes and semiconducting nano-wires are fabricated using bottom-up chemical synthesis methods and are prone to the problem of parallel alignment and series contact resistance [3].

On the other hand III-V semiconductor based transistors are formed using conventional top-down patterning approach and hence can be easily fabricated using the present lithographical techniques. III-V semiconductors have superior electron mobility (50–100X) compared to Silicon (Si), giving higher ON current (I_{ON}) even at lower supply voltage. Table 14.1 shows the material properties of various III-V semiconductors along with silicon [5–8].

Among the available III-V semiconductors, Indium Antimonide (InSb) has the highest electron mobility, saturation velocity and ballistic length. Hence InSb based devices are good candidates for the ultra fast transistors. Recently researchers have successfully demonstrated InSb Quantum Well Field Effect Transistors (QWFET) with sub-100 nm channel length for high speed applications (Fig. 14.1) [9].

We choose InSb/AlInSb material system for subsequent logic/memory circuit analysis. Similar analysis can be done for other III-V semiconducting materials. In this chapter, we have analyzed the technology-circuit co-design aspects for III-V QWFETs [10]. The co-design approach spans Device/SPICE models → Logic and memory circuits → III-V technology requirements. We show the feasibility of the use of Si + III-V hybrid technology for high speed logic and memory applications. We suggest technology requirements for future deployment of III-V transistors. Further, we suggest the application space of III-V transistors.

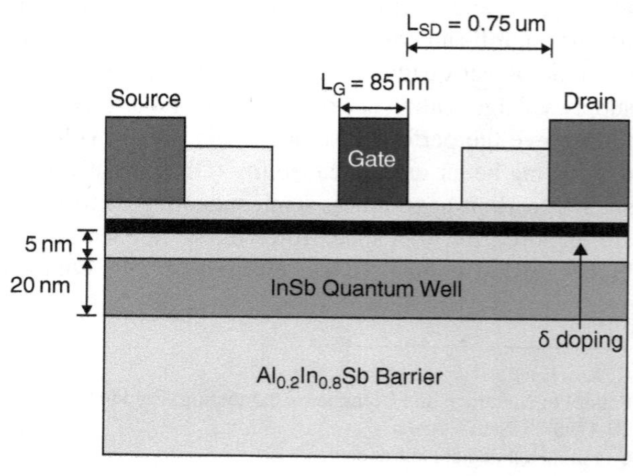

Fig. 14.1 85 nm L_G InSb/$Al_{0.2}In_{0.8}Sb$ Quantum Well Field Effect Transistor (QWFET): Layers from the *bottom* to *top* consist of an accommodation layer, $Al_yIn_{1-y}Sb$ buffer (3 μm), InSb quantum well (20 nm), $Al_{0.2}In_{0.8}Sb$ spacer (5 nm), a single Te-doped donor layer and $Al_{0.2}In_{0.8}Sb$ barrier layer (45 nm) [9]

14.2 Device/SPICE Models

The Hetero-structure Field Effect Transistor (HFET) structure consists of narrowband InSb quantum well sandwiched between wider bandgap $Al_xIn_{1-x}Sb$ barrier layers. Barrier layers achieve (1) carrier confinement in the quantum well (2) reduced junction/transistor off state leakage [11]. Band-to-band tunneling in these narrow bandgap semiconductors is reduced by lowering the drain side electric field in the quantum well by optimizing the barrier thickness [12]. Modulation doping in AlInSb barrier region is achieved with Te δ-doped donor layer [9]. SPICE compatible models are generated using Heterostructure FET physics (HFET) [13]. SPICE model includes Schottky diodes D1 and D2 to model Metal/AlInSb/InSb structure, shown in Fig. 14.2. Bias dependent G-S/D capacitance due to Schottky diodes is also included in the model. Various physical mechanisms related to small geometries and high electric fields are taken into account in the SPICE model [14]. Heterostructure interface at the S/D contacts are modeled as S/D series resistance. The source/drain parasitic resistance (R_{SD}) is significant due to Gate-S/D separation ($L_{SD} = 0.75$ mm). In addition we incorporate the effects of electron transfer into the barrier layer separating the InSb quantum well from the gate [14].

14.2.1 I-V Model

The drain current (I_{ds}) is given by:

$$I_{ds} = \frac{g_{ch} V_{ds}(1 + \lambda V_{ds})}{\left[1 + \left(V_{ds}/V_{sat}\right)^m\right]^{1/m}}$$

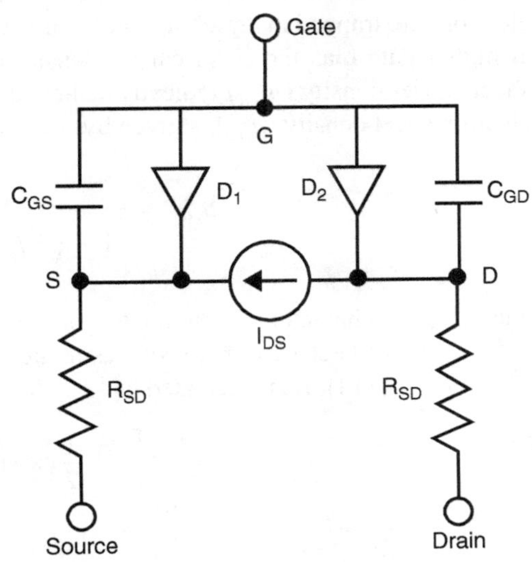

Fig 14.2 QWFET SPICE model based on Heterostructure FET physics: Diodes D1 and D2 model Schottky diodes present between Gate and Source/Drain. R_{SD} model the series resistance due to heterostructure interface as well as the separation between the Gate and Source/Drain. C_{GS} and C_{GD} model the capacitance between the Gate and Source/Drain

where g_{ch} = channel conductance including S/D resistance.
V_{ds} = Drain-to-source voltage
V_{sat} = Saturation drain voltage
λ = parameter describing the finite output conductance in the saturation region
m = parameter that determines the shape of the I_{ds}–V_{ds} characteristics in the knee region.

The channel conductance (g_{ch}) which includes source and drain resistance (R_S/R_D) is given by:

$$g_{ch} = \frac{g_{chi}}{1 + g_{chi}\,(R_S + R_D)}$$

where g_{chi} is the intrinsic channel conductance given by:

$$g_{chi} = \frac{q n_{stot} W \mu}{L}$$

where q = electronic charge
n_{stot} = total electron sheet density at the surface
W = width of the device
L = channel length
μ = low field mobility.

HFET transport characteristics are affected by the limitation of carrier sheet density in the quantum well due to the energy band discontinuity at heterostructure interface. At large V_{GS} the electron quasi Fermi level inside wide bandgap layer approaches the bottom of the conduction band. This causes significant transfer of electrons into the wide gap barrier layer. If barrier layer has defects present, these electrons are trapped in these traps and will not contribute to the current flow. Hence at higher gate bias the sheet carrier density (n_{stot}) gets saturated. The maximum sheet carrier density (n_{max}) typically achieved is $1\text{–}2 \times 10^{12}\,\text{cm}^{-2}$ [8, 14]. Thus total electron sheet density (n_{stot}) is given by:

$$n_{stot} = \frac{n'_s}{\left[1 + \left(n'_s/n_{\max}\right)^{\gamma}\right]^{1/\gamma}}$$

where γ = characteristic parameter for the transition to the saturation region.
n'_s = Sheet carrier density estimated by Unified Charge Control model [14]. It is calculated as

$$n'_s = 2n_0 \ln\left[1 + \frac{1}{2}\exp\left(\frac{V_{GS} - V_T}{\eta V_{th}}\right)\right]$$

where n_0 = sheet carrier density at threshold
V_T = Threshold voltage
η = bo dy effect parameter = $1 + C_d/C_i$

Gate current is modeled as diode current and given as:

$$I_{GS} = I_s \left[\exp\left({}^{V_{GS}}/_{mV_{th}} \right) - 1 \right]$$

I_S = diode saturation current
m = ideality factor
V_{th} = thermal voltage

14.2.2 C-V Model

The G/S and G/D capacitances are calculated using unified gate-channel capacitance with Meyer's capacitance model [15].

$$C_{GS} = \frac{2}{3} C_{ch} \left[1 - \left(\frac{V_{GS} - V_T - V_{DSe}}{2\,(V_{GS} - V_T) - V_{DSe}} \right)^2 \right]$$

$$C_{GD} = \frac{2}{3} C_{ch} \left[1 - \left(\frac{V_{GS} - V_T}{2\,(V_{GS} - V_T) - V_{DSe}} \right)^2 \right]$$

$$\begin{aligned} V_{DSe} &= V_{DS} \quad \text{for} \quad V_{DS} < V_{GS} - V_T \\ &= (V_{GS} - V_T) \quad \text{for} \quad V_{DS} > V_{GS} - V_T \end{aligned}$$

The channel charge C_{ch} is given by

$$C_{ch} = WLq \frac{dn_s}{dV_{GS}} \approx \frac{C'_{ch}}{\left[1 + (n'_s/n_{\max})^{\gamma}\right]^{1/\gamma}}$$

where C'_{ch} = channel charge without sheet carrier density saturation.

$$C'_{ch} = C_i \left[1 + 2\exp\left(-\frac{(V_{GS} - V_T)}{\eta V_{th}} \right) \right]^{-1}$$

where C_i = insulator capacitance given by

$$C_i = {}^{WL\varepsilon_i}/_{d_i}$$

ε_i and d_i are permittivity and the thickness of the wide bandgap barrier layer. The SPICE simulated I_{DS}–V_{DS} and I_{DS}–V_{GS} characteristics for an enhancement mode (e-mode) InSb NFET are matched with the experimental data (Figs. 14.3 and 14.4).

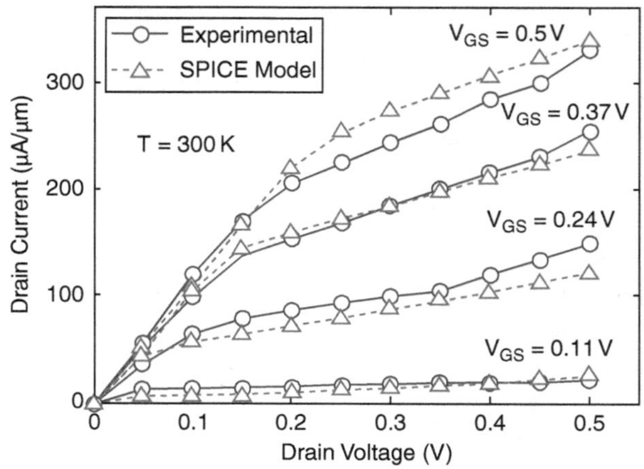

Fig. 14.3 Enhancement mode 85 nm InSb QWFET I_D–V_D characteristics: SPICE vs. Experimental data [9]

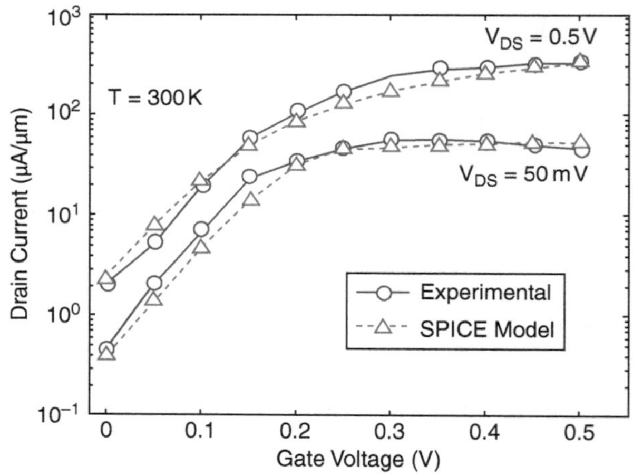

Fig. 14.4 Enhancement mode 85 nm InSb QWFET I_D–V_G characteristics: SPICE vs. Experimental data [9] for V_{DS} = 50 mV and V_{DS} = 0.5 V

The SPICE simulation gives 3% error in ON current compared to the experimental data. For the depletion mode devices (due to un-availability of published data) SPICE parameters are kept same as the enhancement device except the threshold voltage.

14.3 Logic Circuit Analysis

14.3.1 InSb Inverter Analysis

Hole mobility in InSb is very small compared to the electron mobility. Hence it is difficult to achieve high speed logic circuit using complementary InSb FETs. As a result, logic styles based on NFETs are implemented for high speed design using Direct Coupled FET Logic (DCFL) [16]. Figure 14.5 shows inverter schematics using DCFL. Depletion mode QWFET (d-mode) is used as resistive load. Gate and Source nodes of the d-mode QWFET are connected together. Enhancement mode (e-mode) QWFET is used as the driver transistor. Figure 14.5 also shows the voltage transfer characteristics for minimum sized inverter ($W = 2^*L_{min}$, L_{min} = 85 nm). When input gate voltage is logic '1' the output voltage is determined by the drive strengths of load device and driver device. The logic family is said to be '*ratio-ed*'. Hence output low voltage (V_{OL}) is not zero and the output voltage transition is not rail to rail. This results in degraded noise margins. As seen in Fig. 14.5b, low noise margin (NM_L) is lower than high noise margin (NM_H).

Parasitic S/D resistance plays a significant role in determining low output voltage (V_{OL}). Reducing the size of load device should result in lower V_{OL}. However it is found that $R_{S/D}$ limits output voltage and V_{OL} does not reduce even if the load device is weaker than the driver device (Fig. 14.6). SPICE simulations with $R_{S/D} = 0\,\Omega$ (ideal case) show that V_{OL} reduces significantly with reducing S/D resistance as shown in Fig. 14.6. Hence S/D resistance should be minimized for successful deployment of InSb QWFETs in logic circuits. The heterostructure interface at quantum well/SD contact and the separation between the Gate and S/D contacts results in higher $R_{S/D}$.

Due to absence of good quality gate dielectric, G/S and G/D diodes cause significant leakage current [17]. Figure 14.7 shows the effect of increasing fanout on output high voltage (V_{OH}). With increasing fanout, the input gate leakage current increases, which is supplied by the load device (d-mode QWFET) of the previous stage circuit. Increased fanout results in V_{OH} degradation and leads to functional

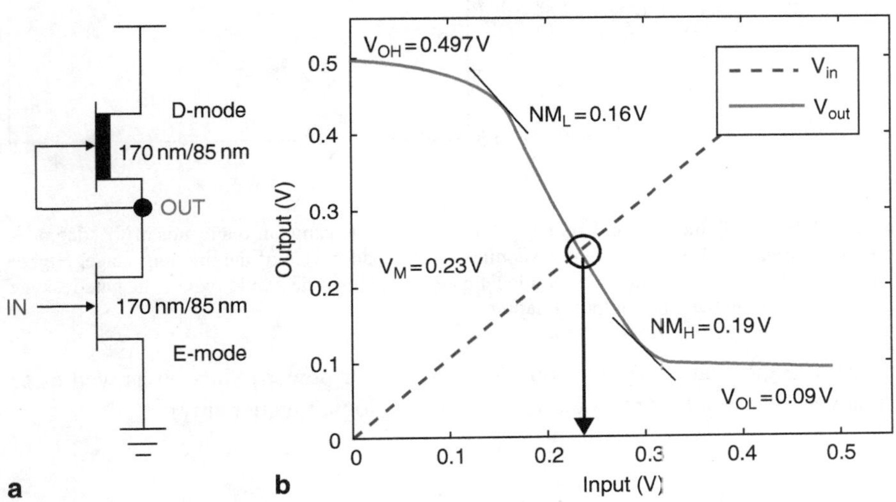

Fig. 14.5 Direct Coupled FET Logic (DCFL): **a** An inverter structure is formed using depletion mode (D-mode) transistor as load device while enhancement mode (E-mode) transistor acts as driver device. **b** Inverter input-output characteristics, Output high voltage ($V_{OH} = 0.497$ V), output low voltage ($V_{OL} = 0.09$ V), Switching threshold ($V_M = 0.23$ V), Low noise margin ($NM_L = 0.16$ V) and high noise margin ($NM_H = 0.19$ V) are indicated. Output voltage swing is reduced due to static current consumption resulting in degraded noise margin

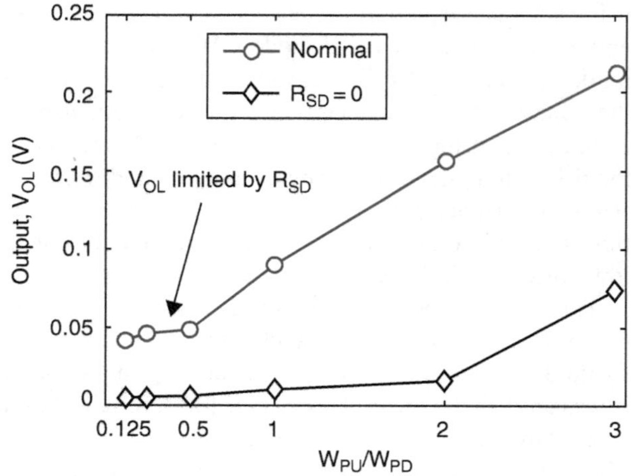

Fig. 14.6 Effect of series resistance (R_{SD}) on output low voltage (V_{OL}): Higher series resistance limits the low output voltage even if width of pull up device (W_{PU}) is smaller than the width of pull down device (W_{PD}). Nominal case refers to the experimental data [9]. In ideal case ($R_{SD} = 0\,\Omega$), V_{OL} reduces significantly with pull up size reduction

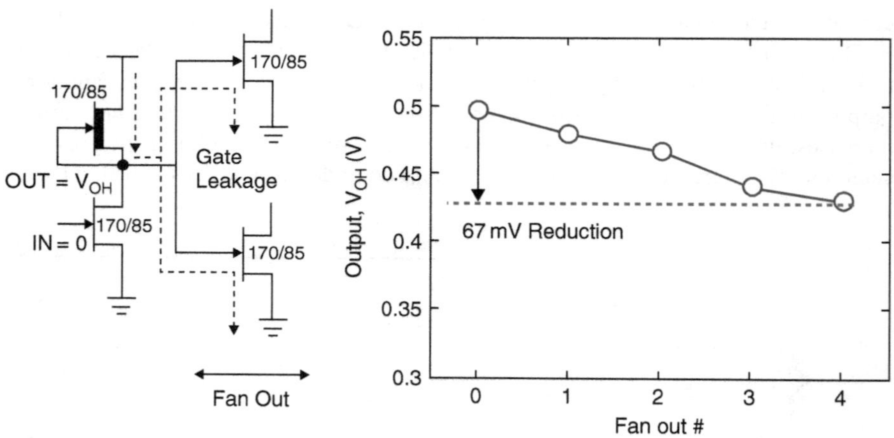

Fig. 14.7 Effect of fanout on high output voltage (V_{OH}): Increasing fan-out results in V_{OH} degradation. Increased gate leakage current is supplied by the load device of the previous stage. Higher fan-out may affect the functionality of a logic gate. Gate dielectric would reduce the gate leakage as well as improve the input-output isolation

failures. Hence gate leakage should be reduced for power reduction as well as to achieve better input-output isolation (for correct logic functionality).

14.3.2 Comparison with Silicon Technology

85 nm InSb QWFET technology is compared with 90 nm Si MOSFET technology for performance. Predictive technology models are used for 90 nm Silicon technology [18, 19]. Since InSb QWFET shows higher ON current as well as higher OFF current compared to the contemporary Si MOSFETs, it is worthwhile to compare

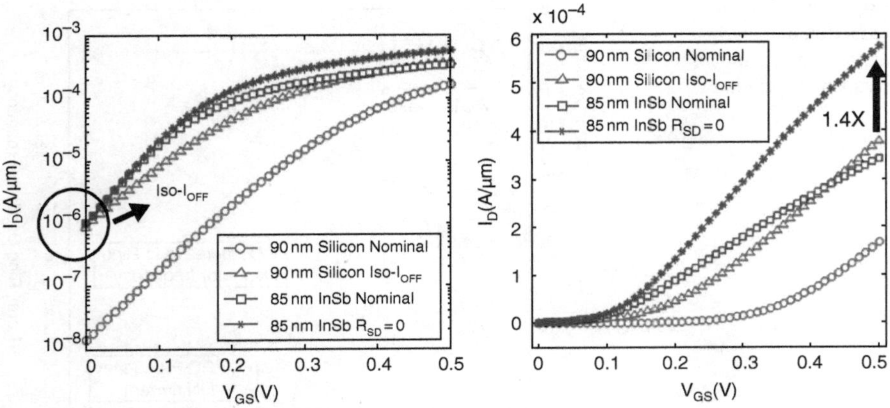

Fig. 14.8 90 nm Silicon MOSFET vs. 85 nm InSb QWFET I_D–V_G characteristics [9, 18]: log scale and linear scale. 90 nm nominal threshold voltage (V_T) Si MOSFET shows lower OFF state current compared to 85 nm InSb QWFET. Threshold voltage of Si-MOSFT is reduced, to match the off state current same as that of InSb QWFET. At Iso-OFF current, low V_T Si MOSFET gives marginally higher ON current compared to present InSb QWFET. Reduced series resistance in InSb QWFET gives much higher ON current compared to low V_T Si MOSFET

ON current improvement in InSb devices under identical off state current condition. OFF state current in 90 nm Si NMOS is matched with 90 nm InSb e-mode QWFET ($I_{OFF} \approx 1\,\mu A\,\mu m^{-1}$, $V_{DS} = 0.5$ V) by reducing the threshold voltage of Si NMOSFET. Figure 14.8 shows I_D–V_G characteristics of 90 nm InSb QWFET and 85 nm Si MOSFET under identical OFF state current condition. It is observed that lower V_T Si NMOS gives marginally higher ON current than the present 85 nm devices. Higher S/D resistance limits the ON current in InSb devices. If S/D resistance is reduced (ideal case being $R_{S/D} = 0$) then ON current in InSb QWFET is 40% higher than the corresponding Si MOSFET (At iso-I_{OFF} condition). This analysis indicates that InSb QWFETs with reduced S/D resistance can possibly result in higher speed computation units compared to low-V_T Si MOSFETs.

14.3.3 Effect of Source/Drain Resistance

As the QWFET fabrication process becomes mature, the S/D resistance would be reduced. In such case, it would be useful to evaluate the ON current improvement in InSb devices for different S/D resistance. Figure 14.9 shows variation of circuit parameters as a function of percentage reduction in S/D resistance. It is observed that ideal case ON current (with $R_{S/D} = 0\,\Omega$) is 77% higher than the present experimental device. The 9 stage ring oscillator circuit built using minimum sized DCFL inverters gives 67% improvement in the ring oscillator frequency compared the present experimental device. Further the power delay product (PDP) is reduced by ~10% and Energy delay product (EDP) is reduced by 46% (as seen in Fig. 14.10).

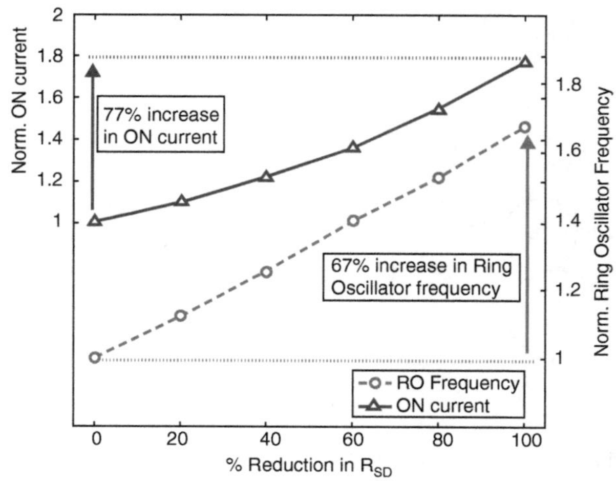

Fig. 14.9 Effect of R_{SD} reduction on current and ring oscillator frequency variation

Figure 14.11 shows normalized circuit metrics for 90 nm Si NMOS and 85 nm InSb QWFET under iso-I_{OFF} condition. It is observed that low V_T Si NMOS gives marginally higher ON current compared to present 85 nm InSb QWFET. However ring oscillator (9 stages) frequency in InSb devices is higher than Si NMOS due to lower gate capacitance present at the intermediate nodes of the ring oscillator. If S/D resistance in InSb devices is reduced by 50% (say); it results in 14% higher ON current than Si NMOS. This leads to 12.3X better ring oscillator frequency consequently 12X reduction in EDP. However due to higher static leakage current in DCFL, the power consumption is higher in InSb DCFL logic resulting in power-delay-product (PDP) comparable to Si NMOS. The Power-Delay-Product indicates the amount of energy consumed in logic computation. Thus 85 nm InSb DCFL (with 50% reduced $R_{S/D}$) shows ~12X faster computation compared to 90 nm Si CMOS

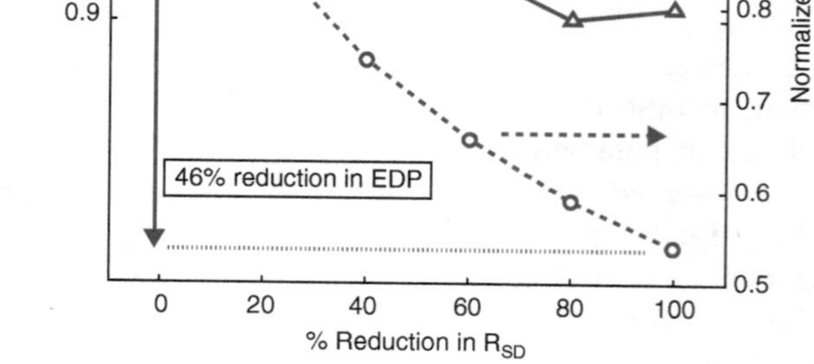

Fig. 14.10 Effect of R_{SD} reduction on Power Delay Product (PDP) and Energy Delay Product (EDP) variation

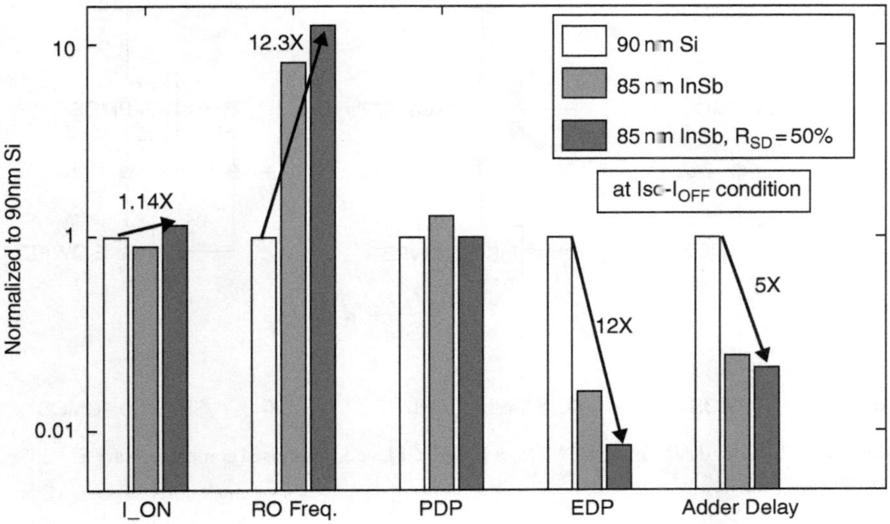

Fig. 14.11 Circuit metrics comparison of 90 nm silicon NMOS and 85 nm InSb e-mode QWFET: InSb QWFET with reduced R_{SD} results in 14% higher ON current, 12.3X higher ring oscillator frequency, ~12X lower energy delay product (EDP) compared to silicon NMOS at iso-OFF current condition. Carry propagation delay of a 4 bit ripple carry adder is reduced by 5X using InSb QWFETs

(operating at V_{DD} = 0.5 V) for same energy consumption. Moreover, for a 4 bit ripple carry adder, the carry propagation delay is reduced by 5X in InSb DCFL adder as shown in Fig. 14.11.

14.3.4 Si + InSb Hybrid Logic Technology

As DCFL incurs significant static leakage current and lower noise margins, we believe that hybrid technology consisting of InSb NFET + Si PMOS can be used along with the contemporary Si CMOS technology. Figure 14.12 shows the inverter schematics using Si CMOS, InSb DCFL and Si + InSb hybrid technology. Use of Si PMOS reduces the static leakage and results in complementary logic style.

Use of InSb devices would give better performance than "dual V_T" Si MOSFET approach [20]. In a Dual V_T design low V_T transistors are used in the critical paths to improve the performance while High V_T transistors are used in non-critical paths to reduce the leakage current [20]. Si + InSb CMOS can be used in *critical paths* while Si CMOS in non-critical paths (Fig. 14.13).

InSb QWFETs can be integrated with Si MOSFETs on a single Si substrate [21]. Figure 14.14 shows the cross section of Si + InSb devices on single Si wafer. Similar integration of InGaAs based devices on silicon substrate has been demonstrated [22].

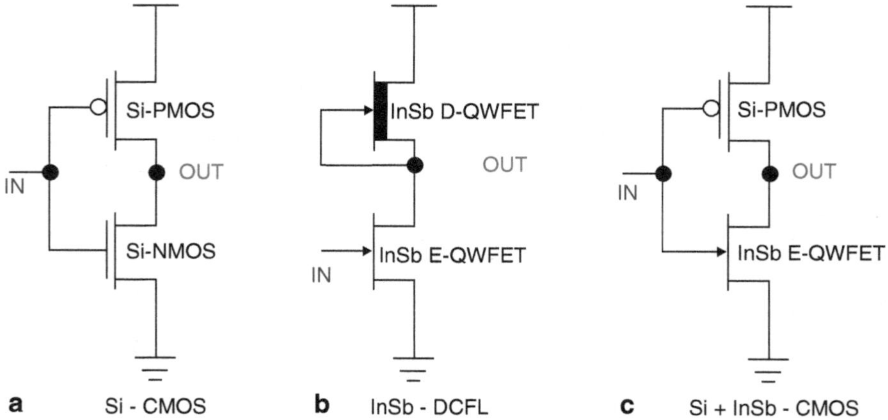

Fig. 14.12 Inverter design in **a** Si-CMOS, **b** InSb-DCFL and **c** proposed complementary Si + InSb hybrid technology

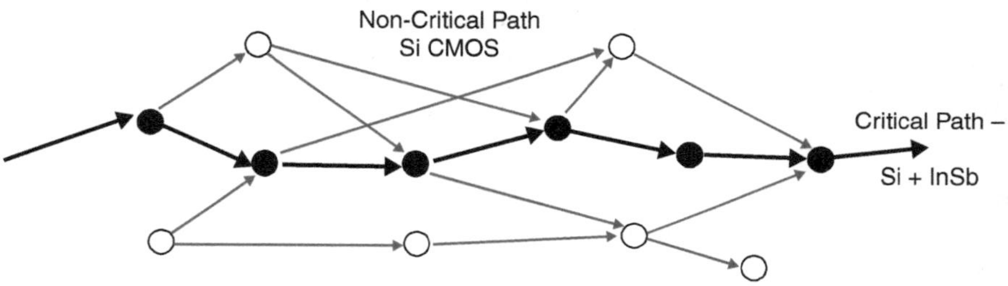

Fig. 14.13 InSb NFET + Si PMOS hybrid complementary logic can be used in critical paths for better performance while Si-CMOS with lower leakage current can be used in non-critical paths

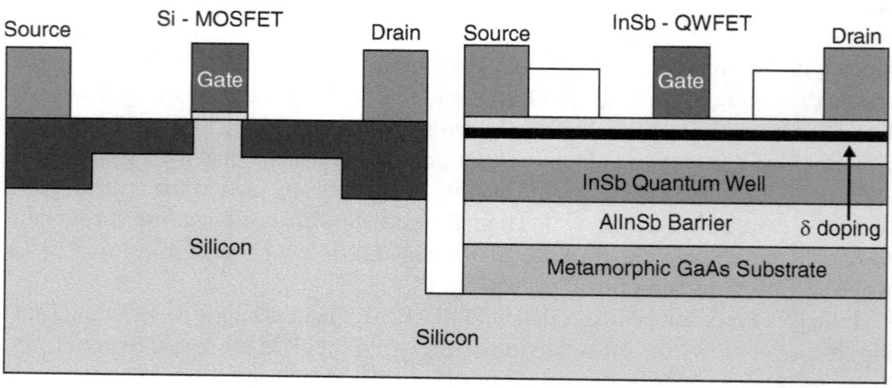

Fig. 14.14 Si + InSb hybrid technology integration on single silicon substrate [21]

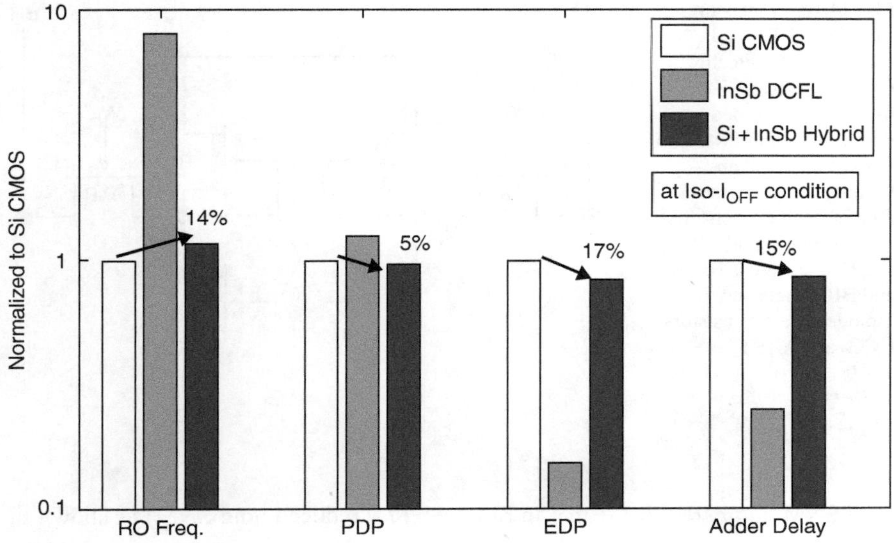

Fig. 14.15 Silicon CMOS and Si + InSb hybrid technology comparison

Figure 14.15 shows the comparison of circuit metrics for Si CMOS and Si + InSb hybrid technology (InSb $R_{S/D} = 50\%$). Si + InSb hybrid technology gives 14% higher ring oscillator frequency with 17% lower energy delay product. The carry propagation delay for Si + InSb 4 bit adder is 15% lower than Si CMOS based adder. At iso-OFF current; Si + InSb can give better performance compared to dual V_T Si CMOS.

14.4 Memory Circuit Analysis

14.4.1 A. 6T Bitcell Analysis

Embedded cache memories are expected to occupy 90% of the total die area of a system-on-a-chip [23]. We also evaluated InSb QWFETs for possible use in memory applications. 6T SRAM bitcells in Si and InSb technology are compared for stability and leakage. InSb 6T SRAM bitcell is formed using cross coupled DCFL inverter pair, shown in Fig. 14.16. This results in static current consumption in the branch storing '0' (i.e. node voltage is V_{OL}).

Static Noise Margin (SNM) is estimated graphically as the length of a side of the largest square that can be embedded inside the lobes of the butterfly curve [24]. SNM is estimated during read and hold case ($V_{DD} = 0.5\,V$). Due to higher R_{SD}, and ratio-ed logic family, InSb bitcell performs poorly in terms of SNM. At 0.5 V V_{DD}, InSb bitcell shows read SNM of 10 mV compared to 86 mv read SNM in 6T Si bitcell. Reducing $R_{S/D} = 0\,\Omega$ improves read SNM to 18 mV indicating marginal effect on InSb SRAM read stability. For the hold case, InSb bitcell shows 86 mv

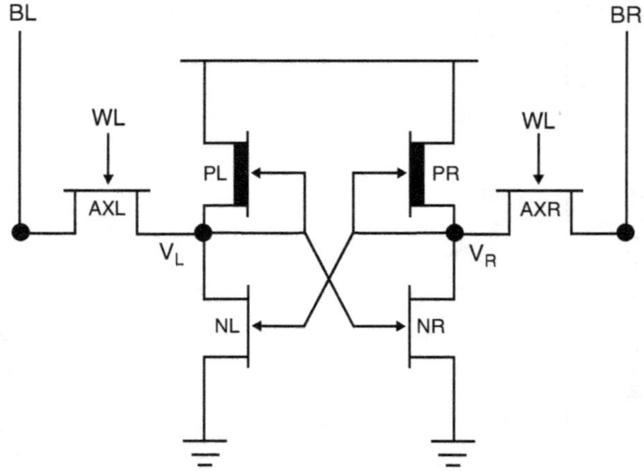

Fig. 14.16 Transistor InSb QWFET based DCFL SRAM bitcell: Cross coupled DCFL inverters are formed with PL-NL and PR-NR transistors. Read and write operation to the memory cell is controlled by wordline signal (WL) and bitline signals (BL and BR) connected to the e-mode access transistors (AXL and AXR)

hold SNM compared to 176 mV in Si bitcell (2X reduced hold SNM) as shown in Fig. 14.17.

Reducing S/D resistance helps in improving the hold SNM. Reducing $R_{S/D} = 0\,\Omega$ results in 114 mV hold SNM which is 1.3X better compared to the present experimental device (Fig. 14.18). Due to Schottky diode (G-S/D) leakage, ratio-ed logic family InSb bitcell incurs much higher leakage (~2 orders of higher magnitude) compared to Si bitcell. Sleep transistors are found effective in reducing InSb bitcell leakage by ~5X. Thus 6T Silicon bitcell gives better cell stability as well as reduced cell leakage compared to InSb based DCFL bitcell. For high density memory applications, bitcell area, cell stability and cell leakage are important criteria for robust low voltage operation. In such case, InSb QWFETs may find limited use in high density, low voltage SRAMs such as Level 2 (L2) and/or Level 3 (L3) caches.

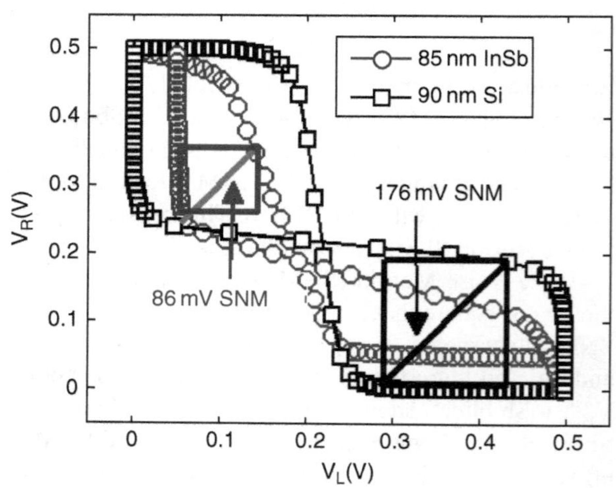

Fig. 14.17 Si vs. InSb technology Hold Static Noise Margin (SNM) comparison: SNM is estimated graphically as the length of a side of the largest *square* that can be embedded inside the lobes of the transfer characteristics [24]. V_L and V_R represent the storage node voltages. Due to higher series resistance and non-zero V_{OL} Hold SNM in InSb bitcell is 86 mV compared to 176 mV in Si bitcell

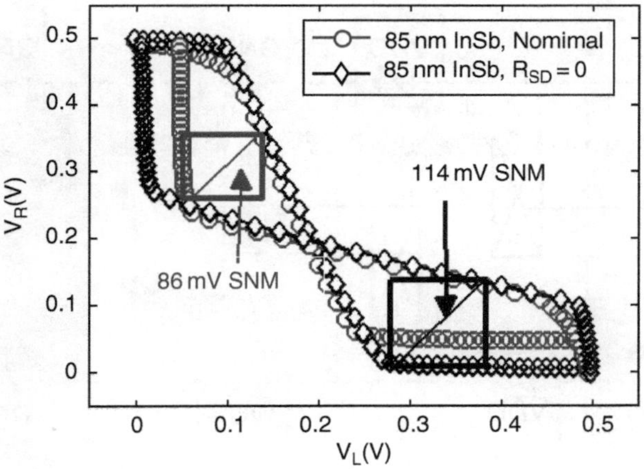

Fig. 14.18 Effect of series resistance on hold SNM in InSb SRAM cell: Reducing the series resistance improves hold SNM by 35% in InSb bitcell. Nominal case refers to 85 nm experimental data [9]

14.4.2 B. 8T Bitcell Analysis

For Level 1 (L1) data cache, access time is an important metric which has direct implications on the processor execution time and hence the clock speed. In such timing critical applications, higher ON current in InSb devices can be utilized for faster bitline discharge by employing an InSb read buffer in an 8T bitcell [25] (also called as Register File cell) (Fig. 14.19). The 8T bitcell stability is determined by the cross coupled inverter pair formed using contemporary Si MOSFETs. Figure 14.20 shows the "thin-cell" layout of the 8T bitcell employing InSb NFET based read buffer [26].

As InSb QWFET fabrication follows traditional "top-down" patterning approach which is compatible to present day technology. Figure 14.21 shows the L1 cache schematics utilizing the 8T Si + InSb hybrid bitcell. InSb QWFETs are used as

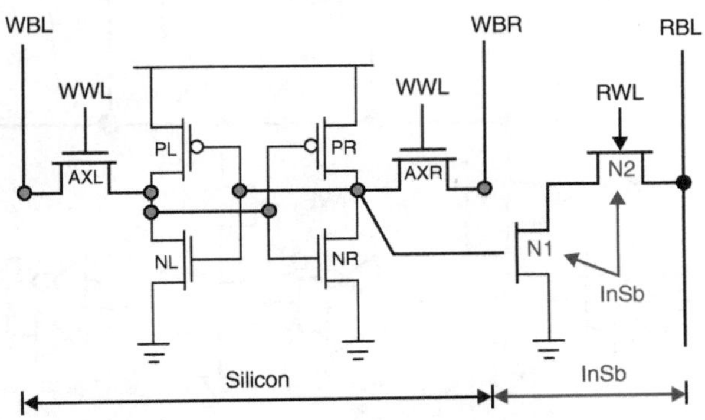

Fig. 14.19 Si + InSb hybrid 8T SRAM bitcell: bitcell stability is governed by the cross coupled inverter pair formed using contemporary silicon MOSFETs. The read port (N1 and N2) is formed using e-mode InSb QWFETs giving improved bitline sensing speed

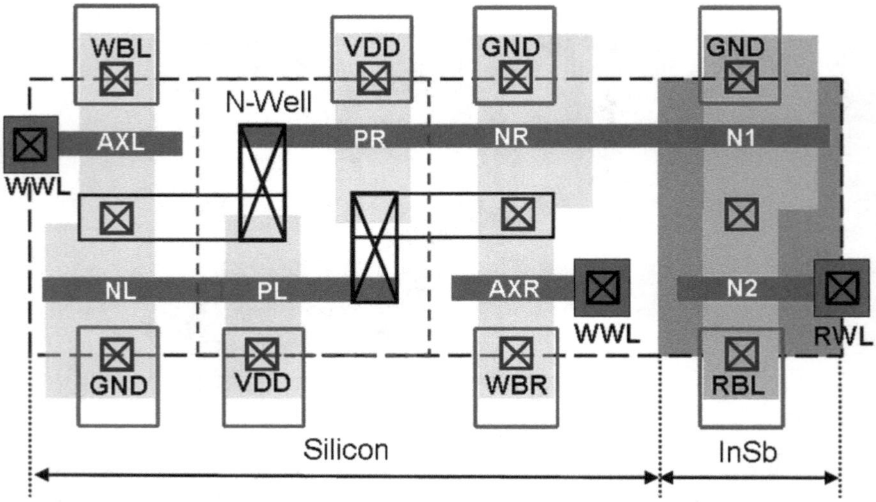

Fig. 14.20 8T hybrid Si + InSb thin-cell layout: InSb QWFETs follow traditional top-down patterning approach and hence can be integrated with the present silicon MOSFET technology

read port transistors of an 8T hybrid bitcell. The access transistors in cross coupled inverter pair are omitted for the clarity. The local bitline (LBL) is precharged to V_{DD} using a clock signal. During the evaluation phase, one of the read word line (RWL) turns ON and local bitline gets discharged if stored data (D) is '1'. Keeper PMOS is a weak device which tries to pull up the LBL node to prevent unintentional LBL collapse due to leakage and/or noise. Global bitline is then evaluated depending on the LBL evaluation.

Si + InSb hybrid 8T bitcell gives 18% lower LBL evaluation time (C_{BL} = 300fF, V_{DD} = 0.5 V). Further, with reduced $R_{S/D}$ LBL evaluation delay reduces by 60%.

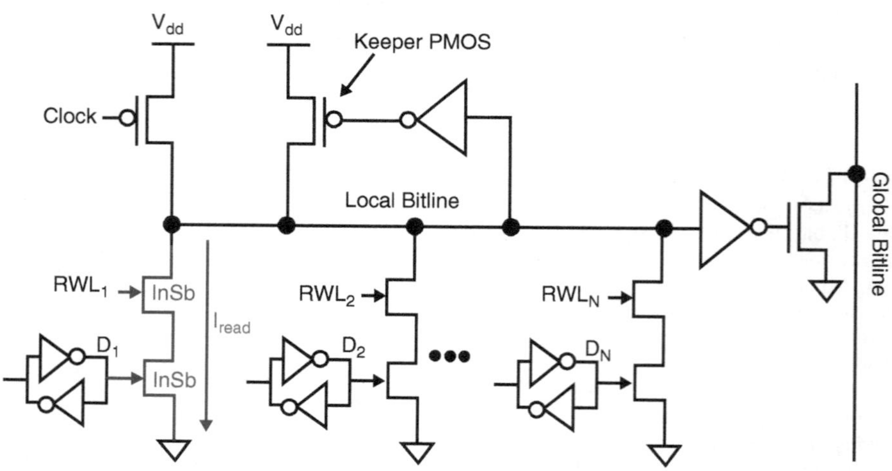

Fig. 14.21 L1 cache memory schematics employing 8T Si + InSb hybrid bitcell

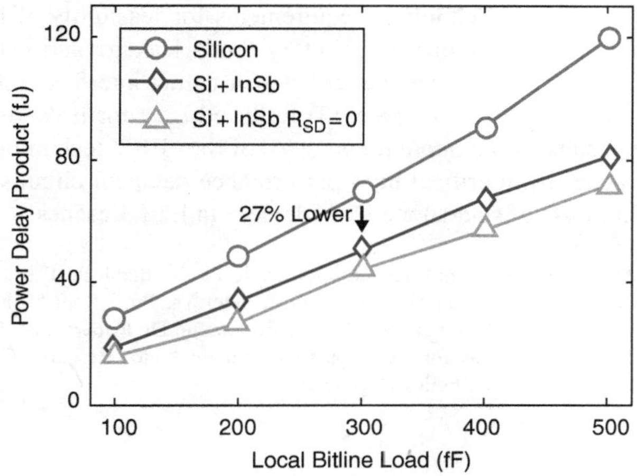

Fig. 14.22 Si 8T bitcell vs. Si + InSb hybrid 8T bitcell: Power Delay Product (PDP) comparison

Due to higher speed of operation, Si + InSb hybrid bitcell shows 27% lower Power-Delay-Product (PDP) compared to silicon 8T cell (Fig. 14.22). During the stand-by mode, the leakage current in the memory blocks can be reduced significantly using a sleep transistor. The sleep transistor technique can be applied to Si as well as hybrid Si + InSb bitcell. Thus III-V FETs can be useful in high speed L1 caches. This would result in faster memory accesses and would improve the processor clock speed.

14.5 Application Space of III-V QWFETs

With the addition of gate dielectric and minimization of S/D resistance would be beneficial for the scalability of ultra fast III-V QWFETs. Due absence of good P-type FET; high speed, rail-to-rail output CMOS logic family would be difficult to achieve using III-V QWFETs. Further higher leakage current and poor cell stability might limit the use of III-V devices in dense memory designs. High speed N-type III-V QWFETs would be used as a supplement to the present Si CMOS technology. III-V QWFETs can be useful in performance critical datapath circuits operating at ultra low voltage. A circuit designer would choose either a low V_T Si NMOS device or e-mode N-type III-V QWFET depending on the speed requirements. For high speed memory design, III-V QWFETs can be useful in the read port of an 8T SRAM cell for improved performance.

14.6 Conclusions

We have shown the use of Si + III-V hybrid technology for high speed logic and memory with a complete technology-circuit assessment. The co-design approach spans from the device/SPICE models to the logic and memory circuit analysis.

We provide technology requirements for feasibility of III-V circuits viz. (1) Gate insulator: for reducing Schottky diode leakage and better input-output isolation. (2) L_{SD} reduction and better heterostructure interface: for better R_{SD} to improve the speed and noise margins. (3) Reduced I_{OFF}: for leakage power reduction. Further, we suggest *the application space* of Si + III-V technology as: (a) in high speed L1 cache (b) in critical high performance datapath circuits (c) higher leakage would limit use of standalone III-V devices in L2/L3 caches.

Acknowledgements The authors would like to thank Prof. Muhammad. A. Alam, Prof. Mark Lundstrom, Prof. Supriyo Datta, Prof. Peter Ye, Prof. Ram Mohan, Prof. Trond Ytterdal, Prof. Suman Datta, Dr. Vivek De, Shekhar Borkar and Dr. Robert Chau for useful technical discussions. This research was funded in part by Semiconductor Research Corporation (SRC) and by Intel Foundation Ph.D. Fellowship Award.

References

1. De V. and Borkar. S.; *Technology and Design Challenges for Low Power and High Performance*, Proceedings of International Symposium on Low Power Electronics and design, 163–168 (1999)
2. Roy K. and Prasad S.; *Low Power VLSI Circuit Design*, Wiley (2000)
3. Bohr M.T.; *Nanotechnology Goals and Challenges for Electronic Applications*, IEEE Transactions on Nanotechnology, Vol. 1, No. 1, 56–62 (2002)
4. Chau R., Datta R. and Majumdar A.; *Opprtunities and Challenges of III-V Nanoelectronics for Future High Speed, Low Power Logic Applications*, Proceedings of Compound Semiconductor Integrated Circuit Symposium, 17–20 (2005)
5. Vurgaftman I., Meyer J. R., Ram-Mohan L. R.; *Band Parameters for III-V Compound Semiconductors and their Alloys*, Journal of Applied Physics, Vol. 89, 5815–5875 (2001)
6. Madelung O.; *Semiconductor Data Handbook*, Springer-Verlag, 3rd edition, New York (2003)
7. Levinshtein M., Rumyantsev S. and Shur M.; *Handbook Series on Semiconductor Parameters*, New Jersey: World Scientific, Singapore (1996)
8. Singh J.; *Physics of Semiconductors and Their Hetero-structures*, McGraw-Hill, Inc., New York (1993)
9. Datta S., Ashley T., Brask J., Buckle J., Doczy M., Emeny M., Hayes D., Hilton K., Jefferies R., Martin T., Phillips T. J., Wallis D., Wilding P. and Chau. R.; *85 nm Gate Length Enhancement and Depletion mode InSb Quantum Well Transistors for Ultra High Speed and Very Low Power Digital Logic Applications*, Proceedings of International Electron Device Meeting, 763–766 (2005)
10. Kulkarni J. P. and Roy K.; *Technology Circuit Co-Design for Ultra Fast InSb Quantum Well Transistors*, IEEE Transactions on Electron Devices, Vol. 55, No. 10, 2537–2545 (2008)
11. Krishnamohan T., Krivokapic Z., Uchida K., Nishi Y. and Saraswat K. C.; *Low Defect Ultrathin Fully Strained-Ge MOSFET on Relaxed Si with High Mobility and Low Band-to-Band-Tunneling (BTBT)*, Proceedings of VLSI Technology Symposium, 82–83 (2005)
12. Sze S. M., *High Speed Semiconductor Device*, Willey-Interscience, 1st Edition (1990)
13. Lee K., Shur M., Fjeldly T. A. and Ytterdal T.; *Semiconductor Device Modeling for VLSI*, Prentice Hall Inc, New Jersey (1993)
14. Byun Y., Lee K. and Shur M.; Unified Charge Control Model and Subthreshold Current in Heterostructure Field Effect Transistors, IEEE Electron Device Letters, Vol. EDL-11, 50–53 (1990)
15. Meyer J. E.; *MOS Models and Circuit Simulation*, RCA Review, Vol. 32, 42–63 (1971)

16. Long S. I. and Butner S. E.; *Gallium Arsenide Digital Integrated Circuit Design*, New York: McGraw-Hill (1990)
17. Chau R., Brask J., Datta S., Dewey G., Doczy M., Doyle B., Kavalieros J., Jin B., Metz M., Majumdar A. and Radosavljevic M.; *Application of High-*k *Gate Dielectrics and Metal Gate Electrodes to Enable Silicon and Non-silicon Logic Nanotechnology*, Microelectronics Engineering, Vol. 80, 1–6 (2005)
18. Zhao W. and Cao Y.; *New Generation of Predictive Technology Model for Sub-45nm Design Exploration*, Proceedings of International Symposium on Quality Electronics Design, 585–590 (2006)
19. Predictive Technology Models: Available online at http://www.eas.edu/ptm/
20. Karnik T., Ye Y., Tschanz J., Wei L., Burns S., Govindarajulu V., De V. and Borkar S.; *Total Power Optimization by Simultaneous Dual-V_T Allocation and Device Sizing in High Performance Microprocessors*, Proceedings of 39th Design Automation Conference, 486–491 (2002)
21. Ashley T., Buckle L., Datta S., Emeny M.T., Hayes D.G., Hilton K.P., Jefferies R., Martin T., Phillips T.J., Wallis D.J., Wilding P.J. and Chau R.; *Heterogeneous InSb Quantum Well Transistors on Silicon for Ultra-High Speed, Low Power Logic Applications*, Electronics Letters, Vol. 43, No.14 (2007)
22. Hudait M. K., Datta S., Dewey G., Fastenau J. M., Kavalieros J., Liu W. K., Lubyshev D., Pillarisetty, R., Radosavljevic M. and Chau R.; *Heterogeneous Integration of Enhancement Mode* $In_{0:7}Ga_{0:3}$*As Quantum Well Transistor on Silicon Substrate using Thin (<2 μm) Composite Buffer Architecture for High-Speed and Low-voltage (0.5 V) Logic Applications*, Proceedings of International Electron Device Meeting, 625–628 (2007)
23. Yoshinobu N., Masahi H., Takayuki K. and Itoh K.; *Review and Future Prospects of Low-Voltage RAM Circuits*, IBM Journal of Research and Development, Vol. 47, No. 5/6, 525–552 (2003)
24. Seevinck E., List F. and Lohstroh J., *Static Noise Margin Analysis of MOS SRAM Cells*, IEEE Journal Solid-State Circuits, Vol. SC-22, No. 5, pp. 748–754 (1987)
25. Chandrakasan A. and Bowhill W., Fox F., *Design of High Performance Microprocessor Circuits*, 1st edition, Wiley-IEEE press (2000)
26. Khare M., Ku S. H., Donaton R. A., Greco S., Brodsky C., Chen X., Chou A., DellaGuardia R., Deshpande S., Doris B., Fung S. K. H., Gabor A., Gribelyuk M., Holmes S., Jamin F. F., Lai W. L., Lee W. H., Li Y., McFarland P., Mo R., Mittl S., Narasimha S., Nielsen D., Purtell R., Rausch W., Sankaran S., Snare J., Tsou L., Vayshenker A., Wagner T., Wehella-Gamage D., Wu E., Wu S., Yan W., Barth E., Ferguson R., Gilbert P., Schepis D., Sekiguchi A., Goldblatt R.,Welser J., Muller K. P. and Agnello P., *A High Performance 90nm SOI Technology with 0.992* μm^2 *6T-SRAM Cell*, Proceedings of International Electron Device Meeting, 407–410 (2002)

Index